AN INTIMATE RELATION

BOSTON STUDIES IN THE PHILOSOPHY OF SCIENCE

VOLUME 116

AN INTIMATE RELATION

Studies in the History and Philosophy of Science
Presented to Robert E. Butts on his 60th Birthday

Edited by

JAMES ROBERT BROWN
University of Toronto, Canada

and

JÜRGEN MITTELSTRASS
Konstanz University, F.R.G.

KLUWER ACADEMIC PUBLISHERS
DORDRECHT / BOSTON / LONDON

Library of Congress Cataloging in Publication Data

An Intimate relation : studies in the history and philosophy of science : presented to Robert E. Butts on his 60th birthday / edited by James Robert Brown and Jürgen Mittelstrass.
p. cm. -- (Boston studies in the philosophy of science ; v. 116)
Includes bibliographical references.
ISBN 0-7923-0169-2
1. Science--Philosophy. 2. Science--History. 3. Butts, Robert E. I. Butts, Robert E. II. Brown, James Robert. III. Mittelstrass, Jürgen. IV. Series.
Q174.B67 vol. 116
[Q175.3]
001'.01 s--dc20
[509] 89-19859

ISBN 0-7923-0169-2

Published by Kluwer Academic Publishers,
P.O. Box 17, 3300 AA Dordrecht, The Netherlands.

Kluwer Academic Publishers incorporates
the publishing programmes of
D. Reidel, Martinus Nijhoff, Dr W. Junk and MTP Press.

Sold and distributed in the U.S.A. and Canada
by Kluwer Academic Publishers,
101 Philip Drive, Norwell, MA 02061, U.S.A.

In all other countries, sold and distributed
by Kluwer Academic Publishers Group,
P.O. Box 322, 3300 AH Dordrecht, The Netherlands.

printed on acid free paper

Printed in The Netherlands

TABLE OF CONTENTS

ROBERT E. BUTTS

(*Photo by Erik Singer, London, Canada, 1988*)

PREFACE

The best philosophy of science during the last generation has been highly historical; and the best history of science, highly philosophical. No one has better exemplified this intimate relationship between history and philosophy than has Robert E. Butts in his work. Throughout his numerous writings, science, its philosophy, and its history have been treated as a seamless web. The result has been a body of work that is sensitive in its conception, ambitious in its scope, and illuminating in its execution. Not only has his work opened new paths of inquiry, but his enthusiasm for the discipline, his encouragement of others (particularly students and younger colleagues), and his tireless efforts to build an international community of scholars, have stimulated the growth of HPS throughout Europe and North America. Many of the essays in this volume reflect that influence.

Our title, of course, is deliberately ambiguous. The essays herein are by colleagues and former students, all of us wishing to honour an intimate friend. Happy Birthday, Bob!

INTRODUCTION

The essays herein cover a variety of concerns: from Descartes to reduction, from Galileo to gambling, from Freud's psychoanalysis to Kant's thing-in-itself. But under this diversity there is an approach common to them all. Things are largely done with a concern for and a sensitivity to historical matters (including contemporary history, of course). Actual science as it was practiced and as it is presently being practiced plays a major role in the philosophical thinking of all the contributors. In this regard, the authors share with Robert Butts (see bibliography at the end of this volume) the belief that there is *an intimate relation* between the philosophy and the history of science.

The volume begins with Joe Pitt's "Apologia pro Simplicio: Galileo and the Limits of Knowledge" which attacks the common interpretation of Galileo as a platonist. Instead Pitt makes Galileo out to be an instrumentalist, rather than any sort of scientific realist. The focus of his attention is the notion in Galileo of nature being mathematical. Just what does this mean? Pitt takes Galileo as holding that humans can have god-like certain knowledge in some mathematical cases. This bit of mathematics which is known with certainty can be applied to analyzing various situations in the physical world, thus giving us certain knowledge in these domains. However, we are extremely limited in our capacities in this regard — at best, only a finite fragment of the infinite whole is accessible to us.

The theme of the mathematization of nature is continued in William Shea's "Cartesian Clarity and Cartesian Motion". Descartes, of course, held that matter is extension, or space, and hence that it can be completely understood geometrically. All the key concepts needed for grasping matter and motion are innate. Shea details the various connections between Descartes's matter-extension ontology and his epistemology of clear and distinct ideas.

Cartesian certainty has been a problem in philosophy for three hundred years. Descartes has long been the paradigm rationalist philosopher. The traditional picture has him holding that everything can be known with certainty. However, historically-oriented philosophers of science of the last generation have considerably revised this picture. Some Cartesian principles such as "matter is space" and "all phenomena are to be explained by matter in motion" are taken to be *a priori* certainties in Descartes's philosophy, but the details, such as the way that a magnet works, are held to be hypotheses, tested by their observable consequences. The authors of Descartes essays in this volume, Shea, Morrison, and Duncan would all, we think, subscribe to this newer picture of the Cartesian programme. Margarat Morrison, however, wants to make the revised account somewhat more subtle.

She argues for several different senses of certainty in Descartes, including the metaphysically certain (e.g., matter is space), the morally certain (e.g., magnets depend on screw-shaped particles) and finally, hypotheses of a rather general sort (e.g., light is an action, subtle matter exists) which have *a posteriori* certainty. Morrison argues forcefully that the difference between *a posteriori* and moral certainty is one of kind, not degree. Descartes is a willing fallibilist about the latter, but the whole Cartesian edifice would crumble if any of the former were false.

Howard Duncan is especially concerned with methodological issues in his "Descartes and the Method of Analysis and Synthesis". He agrees that Descartes is not a simple-minded a priorist who deduces all truths from self-evident first principles. There is a role for experience and for hypotheses, but the latter are not merely bold conjectures. They result from the method of analysis and synthesis. Analysis is the method starting with some desired result and working backward to first principles; from these the desired result follows. Synthesis starts with the first principles and on the basis of these establishes the desired result. Using nice illustrations, Duncan argues that Descartes both advocated and employed a version of this ancient methodological process.

Thomas Lennon surveys the fascinating debates on atomism that occurred in the latter 17th century in his essay, "Physical and Metaphysical Atomism: 1666—1682". The followers of Gassendi advocated an ontology of atoms and the void while Cartesians opted for a plenum view of matter. Lennon's rather bold thesis is that during the time period the debate became increasingly metaphysical and less focused on atomism as a physical theory of the world.

Newton's *Principia*, arguably the greatest scientific work of all time, just celebrated its 300th birthday. It is entirely fitting that it should be the subject of two essays in this volume. The first is Kathleen's Okruhlik's "The Foundations of All Philosophy: Newton's Third Rule".

The *Principia* contains four rules of philosophy, as Newton called them. Until recently these had been taken unproblematically as methodological rules. The third rule in particular — Newton called it the foundation of all philosophy — seemed to be a rule of inference for passing from the observed to the unobserved. Recent work by some Newton scholars has cast the third rule in quite a different light; it is now seen as deeply involving ontological concerns, especially pertaining to Newton's atomism. Okruhlik greatly clarifies what is at issue in prior debates and also what the role of the third rule is in Newton's *Principia*.

Though it is undoubtedly a work of genius, Newton's *magnum opus* remains something of an enigma. Its brilliance is conceded, but its meaning is often elusive. What role do the rules of philosophy play? What did Newton mean by '*Hypotheses non fingo*'? How do the phenomena support the law of universal gravitation? In spite of their importance, these and many other questions remain unsettled. In "Consiliance and Natural Kind Reasoning" William Harper addresses the last of these questions using the apparatus of "inductive consiliences" made central by Whewell in his account of science more than a century ago. What a consilience is is itself somewhat controversial, so Harper does much clarificatory work on the notion before applying it to Newton's reasoning.

Harper's result may be of importance for contemporary debates concerning scientific realism. Realists want the strength of a powerful, unifying theory while anti-realists have opted for the security of believing only what is observable. Harper argues that we need not be faced with making a choice between them. When a genuine consilience is achieved as in Newton's universal gravitation, strength and security come as a package. Realists can eat their cake and have it too.

We turn away from Newton now to his great continental rival, Leibniz. François Duchesneau, in "Leibniz's 'Hypothesis Physica Nova': A conjunction of Models for Explaining Phenomena", directs attention to the young Leibniz and attempts to make clear what is going on in the youthful essay "Hypothesis Physica Nova". Leibniz was trying to do justice to the same range of phenomena explained by the corpuscular philosophy; he was highly critical of the *ad hoc* approach of Descartes, Boyle, and Gassendi. Duchesneau argues forcefully and in considerable detail that Leibniz's own account found in the "Hypothesis" is an excellent example of theorizing which is dependent on the available phenomena conditioned by architectonic norms.

The second of the pair of Leibniz essays in this volume is Richard

Arthur's "Russell's Conundrum, or How are Leibniz's Monads Related to the Continuum?" In his classic book on Leibniz, Bertrand Russell posed the problem: if monads are not parts of continuous bodies, then why did Leibniz think they would solve the problem of the continuum; but if they are, then Leibniz has lapsed into inconsistency.

Arthur argues at length that Russell's conundrum is the result of not distinguishing different senses of "actual" in Leibniz's work and more importantly, of not understanding and appreciating Leibniz's doctrine of the continuum. As well as defending Leibniz from this misunderstanding made by Russell and other interpreters of Leibniz, Arthur goes on to attempt a resolution of the conundrum.

Nicholas Rescher introduces us to a number of philosophical views on chance in "The Philosophers of Gambling". In the early 17th century, just before the theory of probability was put on a firm mathematical footing, several philosophers addressed themselves to chance, especially in the form of gambling. Opinions of the day included that it was a way of peacefully settling disputes; that it mirrored real life; that it could be turned to theological advantage; that it did not really exist since there is a reason for everything that happens. Rescher ends his essay with interesting speculations on the social background (i.e., the English Civil War and the Thirty Years War) which may have given rise to the consciouness of chance.

Kant is the subject of the next pair of complementary papers by Gerd Buchdahl and by Jaakko Hintikka; not only are they about Kant, but the notorious "thing-in-itself" looms large in each.

We have the capacity to be affected by objects, but what is doing the affecting? Is it the transcendental object, the thing-in-itself, or is it the thing *qua* appearance? This is the problem Buchdahl poses in "Reductive Realism and the Problem of Affection in Kant". There are passages in Kant in support of each possible answer. Kant's somewhat confusing and contradictory views on affection lead Buchdahl to propose a novel interpretation, not only of "affection", but also of "appearance", and "thing-in-itself". Buchdahl proposes, among other things, that the transcendental object is not something which exists in addition to objects *qua* appearances, nor is it the external cause of those appearances; rather transcendental objects stand to appearances in a relation of correspondence, as he puts it. As Buchdahl interestingly remarks, his analysis leads to a view of the Kantian world as entirely enclosed within its transcendental framework.

In "The Paradox of Transcendental Knowledge" Hintikka expresses an attitude toward the thing-in-itself which is quite different from Buchdahl's. Whereas the latter is trying to consistently interpret various of Kant's remarks about the thing-in-itself and about appearances and affection, Hintikka wishes to reject the notion when understood as completely unknowable. Hintikka thinks it a great mistake on Kant's part to conceive the thing-in-itself this way. Though he is generally sympathetic to the Kantian enterprise, he wants to replace the unknowable with the inexhaustible. Hintikka concludes with the very optimistic remark that the prospects of transcendental knowledge and transcendental argumentation are thus virtually unlimited — the philosophy of logic and of language have, he remarks, provided many examples.

A long-standing aim of philosophy is to distinguish between good science and bad. Gereon Wolters's "Mesmer in a Mountain Bar: Anthropological Difference, Butts, and Mesmerism" is concerned with a fascinating special case of this problem. Kant himself worried about the difference between good and bad science; he attacked spiritualism as being, not merely wrong, but as being in principle misguided. One of the most interesting parts of Butt's *Kant and the Double Government Methodology* (Reidel, 1985) is devoted to this. Wolters takes up this thorny issue of science *vs* pseudo-science by looking at the case of Franz Anton Mesmer whose theory of animal magnitism was very controversial two centuries ago. Mesmer's theory, its theraputic applications, its reception by critics, its interaction with social forces, and its evaluation by a tribunal of Mesmer's contemporaries are all discussed by Wolters in fascinating detail.

We have arranged the essays in this volume in chronological order, but other groupings are possible and readers may wish to jump around. For example, Gereon Wolters and Adolf Grünbaum are both concerned with science vs pseudo-science; Tom Lennon and Lorenz Krüger are both interested in the theme of reduction; and the following paper by Andrew Lugg has much to say about Whewell, hence it has considerable overlap with the earlier paper by William Harper.

Whereas Harper's essay on Newton employs Whewell's consilience of inductions, Andrew Lugg ("History, Discovery, and Induction: Whewell on Kepler on the Orbit of Mars") investigates the more general Whewellian enterprise of acquiring philosophical ideas and methods from the study of history. William Whewell was one of the most historically sensitive philosophers of science and Robert Butts almost single-handedly in a series of essays and an anthology (see the Butts bibliography at the end of the volume), made the philosophy of science world aware of the importance of Whewell. Lugg gives a sympathetic reading to a number of Whewellian themes, especially those in which actual scientific practice as a datum loom large. *A priori* elements were not negligible in Whewell's thought; nevertheless, as Lugg notes, Whewell's views on the scientific process were largely informed by the detailed study of individual cases such as Kepler on the Orbit of Mars.

Though many philosophers turn to the history of science for inspiration and evidence, one stands out for the negative morals he draws. Paul Feyerabend has argued for epistemological anarchism, claiming that every methodological rule proposed has been broken somewhere in the history of science, and moreover, breaking those rules was essential to making progress. In "For Method, or, Against Feyerabend", Larry Laudan takes up the cudgels against Feyerabend's anti-methodology, "Anything Goes". Laudan examines Feyerabend's case in detail and forcefully argues that it is badly flawed, consisting largely of non-sequiturs. If we do not have the right methodological rules now, that only shows that methodology, like the rest of science, is very hard to do; it does not mean it is hopeless.

In "World Pictures: The World of the History and Philosophy of Science" Jürgen Mittelstrass takes aim at the common division of labour between the philosophy and the history of science, which makes the one (philosophy) an element in the progress of knowledge and the other (history) an element in the appreciation of knowledge. The relation is vastly more complicated, subtle, and intimate. Mittelstrass argues that, among other things, such discoveries as the non-cumulative character of science has lead philosophy to historicize science which in turn has made the history of science look somewhat like the philosophy of science. The consequences are a highly historical philosophy and a highly philosophical history. Looking at the history of science from the point of view of a historian of ideas reveals that each of the fundamental physical theories of the past gave birth to a general world view, a world-picture, and that a change in the former induced a change in the latter. In addition to this diachronic relativity of world pictures, there is also a synchronic one: A physical theory does not determine unambiguously its philosophical interpretation. Looking at the history of science from the point of view of a philosopher of science shows, however, that this double relativity of scientific world-pictures does not entail a relativism in science proper. Scientific theories can still be assessed unambiguously by invoking transparadigmatic methodological criteria. In this sense, reason retains the last word in deciding between competing scientific approaches.

"Learning From the Past" by James Robert Brown is also concerned with the relations between history of science and philosophy of science. Almost all of the essays in this volume have a historical orientation, but the four essays by Lugg, Laudan, Mittelstrass, and Brown are probably the most self-consciously concerned with the history-philosophy relation. Brown begins by noting that while many philosophers have conceded that we must "do justice to history", only a very few have actually offered a detailed account of what is involved in deriving methodological norms out of historical facts. Brown criticizes some existing accounts, including those of Lakatos, Laudan, and Butts, then offers his own version of how to learn from the past.

With Lorenz Krüger's "Reduction without Reductionism?" we turn our attention to a contemporary debate. The traditional account of reduction has it as being some sort of derivation relation between theories. Thermodynamics, for example, is reduced to mechanics (so the story goes) by being derived from it. Using this standard example, but examing it in considerable detail, Krüger evaluates several distinct approaches to reduction. He focuses on the character of ensembles in statistical mechanics and concludes that we must accept their distinctly physical character. The consequence is that rather than absorbing thermodynamics into classical mechanics, a kind of intermediate third theory, "interlevel" as he calls it, is created which embraces both. Krüger goes on to draw some interesting and important morals such as the rejection of the view in which nature is understood as a hierarchy having basic laws about the elements from which everything else follows.

In a wide ranging paper, Vadim Sadovsky, "Models of Scientific Knowledge" canvasses a number of issues. He sketches some of the chief characteristics of formal theories; he discusses some of the features and difficulties of such methodologies of science as the positivists' verificationism and Popper's falsificationism; and he also discusses a number of issues in reductionism. The main moral that Sadovsky draws is that rival philosophical conceptions of the nature of science (e.g., realism, instrumentalism) do not determine the methodology of science, which is something that surprisingly seems to develop with relative independence.

Martin Carrier examines the very thorny problem of theory-laden observation in his "Circles Without Circularity". This problem has been central in the philosophy of science for the past generation. If theory influences, conditions, or molds experience, then how can that very experiential data be used in any non-circular way to test the theory?

Carrier examines and rejects Sneed's notion of *theoretical* and employs instead a notion he call *Einstein-Feigl completeness.* A complete theory in this sense manages to analyze theoretically the measuring process connected to its own theoretical therms. In other words, it accounts for the means to establish its own observation basis. Examples from classical mechanics and special relativity are used to illustrate. Carrier argues for the use of slow-clock-transport in establishing non-conventional simultaneity relations in special relativity. He avoids the usual objections to this by invoking self-measured velocity. His analysis of theory-ladenness allows him to conclude that special relativity can be applied to its own observation basis in an unambiguous, non-circular way.

William Demopoulos, "On Applying Learnability Theory to the Rationalism-Empiricism Controversy", is interested in the contemporary rationalist-empiricist debate over language acquisition, within the framework of formal learning theory. He gives a brief synopsis of the important work of Gold which suggests that constraints on learning strategies are necessary in order to learn any sort of interesting language. Innate syntax-specific constraints are, of course, grist for the rationalist mill; Chomsky has long argued for their presence. Demopoulos argues that there is, however, a spectrum of views concerning the nature of these constraints — whether they are rightly classified "rationalist" or "empiricist", he says, remains to be seen.

While Demopoulos is concerned with language acquisition and the nature of the mental, Hubert Schleichert's "The Relationship Between Consciousness and Language" has a more general interest in language and its relation to consciouness. There is a "canonical phraseology", as he calls it, a common way of talking about consciouness. But is there anything which answers to the term "consciousness"? Schleichert's rather surprising answer is that there is such an object, language itself. All canonical phrases concerning consciousness, he argues, apply to language and to language alone.

John Nicholas turns our attention to the foundations of non-deductive inference in his paper "Realism for Shopkeepers: Behaviouralist Notes On Constructive Empiricism". In particular Nicholas advocates a behaviouralist account which he considers a brand of realism. His target is the anti-realist epistemology of Bas van Fraassen. He argues that the "leap" of inductive inference should not be seen as one from partial belief to full belief, but rather from partial belief to "act of assertion".

The status of Freud's psychoanalysis has been a perennial topic of debate. Is it a legitimate science, or not? No one has contributed more to clarifying these issues than Adolf Grünbaum. His essay "Why Thematic Kinships Between Events Do NOT Attest Their Causal Linkage" is a critical discussion of some of Freud's champions who try to side-step the science *vs* pseudo-science issue by claiming that psychoanalysis is a hermeneutical enterprise — it does not uncover "causal" connections, but rather "meaning" connections.

Grünbaum carefully considers inferences based on what he calls "thematic kinships"; he notes that with additional premisses thematic connections can be sometimes useful in establishing causal connections, though they are not without pitfalls. Grünbaum concludes, however, that "narratives replete with mere hermeneutic pablum are explanatorily bankrupt".

The final contribution to this volume is on the topic of biology, in particular the structure of Darwinian evolutionary theory. Michael Ruse and Paul Thompson begin their essay "Neo-Darwinism: Form and Content" with an account of the neo-Darwinian theory of evolution. This is Darwin's theory updated. We are taken carefully through a number of key changes which have improved the theory over the last century. This is intrinsically interesting, of course, but it also sets the stage for some key philosophical questions, such as, Is biology different than physics?, and What is the structure of biological theory? The authors discuss some of the pros and cons of various answers, not always agreeing between themselves. Ruse holds that Darwinian evolution is a system of postulates, to be understood hypothetico-deductively (the so called received view) while Thompson advocates the semantical approach which sees Darwin's theory as a (conjectured) model of biological reality. Readers may find this a helpful introduction to the present (syntax vs semantics) controversy over the nature of scientific theories in general.

As we mentioned in the preface, the best philosophy of science in recent years has been highly historical, that is, intimately connected to actual science; and the best history of science has been highly philosophical. This intimate relation has been beautifully exemplified by the work of Robert Butts, and we hope that the contributions to this volume are a worthy tribute to the man who contributed so much to making the pattern.

JAMES ROBERT BROWN
University of Toronto
Toronto, Ontario
Canada

JÜRGEN MITTELSTRASS
Konstanz University
Konstanz
West Germany

JOSEPH C. PITT

APOLOGIA PRO SIMPLICIO: GALILEO AND THE LIMITS OF KNOWLEDGE*

1. INTRODUCTION

The limits of knowledge are determined by the world, by the apparatus the investigator brings to bear on his problems, and the cognitive values and methods that govern the objectives of the process of inquiry. Kant was right, there is little we can say about the world *per se*. What is of epistomological interest is not the way the world is, but (a) the character of the inquiry as determined by the goals, values and methods of the inquirers, and (b) the way the world is *believed* to be.

It is not clear what Galileo believed about the constitution of the world. Instead of addressing this issue directly he chose metaphor, claiming that the book of nature was written in the language of geometry. In 1641, he presented his views on this topic in a long response to Fortunio Liceti, an Aristotelian professor of philosophy at the University of Padua, regarding Liceti's efforts to prove the earth the center of the universe. In this letter, written one year before his death, Galileo gives us his final statement on this issue.

> But I truly believe the book of philosophy to be that which stands perpetually open before our eyes, though since it is written in characters different from those of our alphabet it cannot be read by everyone, and the characters of such a book are triangles, squares, circles, spheres, cones, pyramids, and other mathematical figures, most apt for reading.[1]

The meaning of the passage is not clear since it leads to at least two questions regarding Galileo's position *vis a vis* the book of nature:

(1) Given that the characters of the book of nature are geometrical figures, does it follow that Galileo believed that the world is a geometrical structure?

Or, rather,

(2) Did Galileo believe the structure of the world could be revealed through the application of mathematical methods of analysis?

If we answer (1) positively, the it would be but a short step to conclude that Galileo was a Platonist. What we get with a positive answer to (2) is not so obvious.

J. R. Brown and J. Mittelstrass (eds.), An Intimate Relation, 1—22.

In what follows I try to answer (2) as I think Galileo would have and to explain the ramifications of that answer. To anticipate: yes, Galileo did think that some part of the structure of the world could be revealed through the appropriate use of mathematics. But his answer is carefully circumscribed by a number of caveats concerning the limits of knowledge, an account of how much can be known, and his firm commitment to the use of observation and mathematics. I argue (1) that he is not a platonist, (2) that he is an instrumentalist *of sorts*, (3) that mathematics alone reveals nothing about the world as such, and finally (4) that even using mathematics, we should not expect to find out very much. I will begin with an argument designed to shut down attempts to make Galileo a realist by first making him a platonist. In the remainder of the paper I attempt to explain why Galileo thought mathematics was useful and how it contributes to knowledge.

2. "PLATONISMS"

2.1. The question of Galileo's Platonism has been raised in different ways. Primarily it comes to us through the monumental work of Koyré. In many respects much modern scholarship on Galileo begins with Koyré. His *Etudes Galilean* is a sufficiently comprehensive treatment of Galileo's views, and his case for interpreting Galileo as a Platonist so strongly presented, that the dialectical nature of inquiry demands we address his views, if only to acknowledge them.[2] I, of course, want to do more, i.e. to reject some of them. To accomplish this I will look at some versions of Platonism and throughout the rest of the paper, concentrate on Galileo's conception of the nature and role of mathematics. Together, these arguments should put claims of platonism and its associated partner in crime, realism, in Galileo's thought to rest.[3]

2.2. There are at least two forms of Platonism that we need to indicate here. The first is the standard neo-Platonism of the 14th, 15th and 16th centuries, founded on a revival of the allegory of the cave from Plato's dialogue *The Republic* and transmuted into a general philosophical position. It was a view that took a certain degree of support from the Copernican system of the universe. The placement of the sun at the center of the system by Copernicus had more than symbolic value for the neo-platonist, who believed that the clear light of reason could by itself produce the answers man sought to the eternal questions. Given Galileo's emphasis on experience and experimentation, it is going to be very difficult indeed to make a case for him as a Platonist of this fashion.

There is a form of Platonism which is close to the view Galileo actually espoused. On this account, a "platonist" believes that the use of mathematics leads to truths about the real nature of the universe. This view can also be thought of as a form of neo-pythagoreanism since it has more to do with the priority of mathematics as an intellectual accomplishment than the mathematical nature of reality. Even so amended, there are two reasons why this view does not fully capture the spirit of Galileo's enterprise. First, for Galileo's, it was associated with a form of mysticism, any type of which he rejected. Thus, at the beginning of *Day 1*, Salviati rejects the "reasons" and "proofs" Simplicio offers in defense of Aristotle's proof that there are only three dimensions. He is then chastised by Simplicio because these are supposed to be the same sorts of reasons the pythagoreans use, and, Simplicio continues, he knows that as a mathematician Salviati holds many of the same opinions. It therefore seems strange to him that Salviati would "scorn their mysteries". In response, Salviati says,

> That the Pythagoreans held the science of number in high esteem, and that Plato himself admired the human understanding and believed it to partake of divinity simply because it understood the nature of numbers, I know very well; nor am I very far from being of the same opinion. But that these mysteries which caused Pythagoras and his sect to have such a veneration for the science of numbers are the follies that abound in the sayings and writings of the vulgar, I do not believe at all. (p. 11)[4]

Salviati then continues to explain that "he knows" that the pythagoreans spread rumors of these silly mysteries (now attributed as reasons to the Aristotelians) so as "to prevent the things they admired from being exposed to the slander and scorn of the common people."

Galileo's point here is not complicated: Yes, he is sympathetic to the pythagorean/platonic view of numbers and their relation to human knowledge, *but* he does not believe in mysteries which produce non-mathematical reasons such as the ones Simplicio proposed off on behalf of Aristotle.

The consequences of this position are several. First, it does no harm for Galileo to admit he is attracted to the power of numerical arguments. He is, after all, a mathematician. Second, it allows him a distinction between being a lover of mathematics and being a platonist. The "mysteries" which Simplicio associates with both Aristotle and the pythagoreans, Galileo has just dumped in the laps of the Aristotelians as so much nonsense. He is thereby left in the enviable position of having the mathematics and not the mysteries.

Furthermore, through this move, Galileo rejects the Platonic theory of reminiscence, since he appears to side with the pythagoreans and their concerns to keep the mysteries of number theory away from the common people. If Plato is right, even the slave boy can come up with this mysterious knowledge. So, it at least seems that here Galileo is rejecting this Platonic theory of mathematical knowledge. On the other hand, in the quote containing the claim that the secrets of the book of nature were not accessible to everyone, while Galileo seems to suggest that special tutoring is needed to read the book of nature, he also appears inclined to the idea that the tutoring is no more esoteric than learning a new language would require. It is not that one already knows what the book of nature says, it is, rather, that one can acquire the means to read it. Where that leaves us on the questions of the accessibility of nature for the common person is not clear.[5]

It was indicated above that there were two reasons why neo-pythagoreanism fails to do justice to Galileo's program. The first was it carried unwanted baggage, i.e. "mysteries", to which Galileo objected. The second reason is that stripped of its "mysterious" element, *according to the reasons Galileo himself provides*, the remaining doctrine is trivial; all it maintains is that the pythagoreans held the science of number in "high esteem". From that nothing follows. Our concern is to discover why Galileo believed mathematical reasoning provided privileged knowledge. Without the metaphysics of Platonism or the "mysteries" of Pythagoreanism, we are left with only the mathematics. Thus, while having endorsed mathematics very early on, Galileo also rejected two of the principle schools of thought which provided it with possible epistemological import.

3. REALISM, GEOMETRIC REALISM AND SCIENTIFIC REALISM

3.1. In the same sense that Everyman is a realist, so was Galileo. This is to say nothing more than he believed there was a world "out there" which, in some unspecified sense, is "responsible" for whatever experiential input our senses monitor. The big questions are two:

(3) What is the best method for analyzing experiential input so as to insure that the results qualify as knowledge?

(4) What can we conclude from the results of that analysis about the structure of the universe?

While Everyman is not concerned with these questions, Galileo was obsessed by (3) and used his answer to (4) to produce some of his more spectacular theories, among them his theory of the tides.

For Galileo, the answer to (3) seems simple: mathematics. But is not clear that matters can be resolved that easily, for if "mathematics" is the answer to (3) why did Galileo think mathematics was so special? Adopting some form of Platonism or Pythagoreanism would have answered that question. But, as we have seen, Galileo appears disinclined to buy into the metaphysics those positions entail.

Galileo's account of the special status of mathematics is based on his fundamental rejection of what he perceived as obscure metaphysical theories and his commitment to a form of common sense realism for which the observable world around us is the basis for all claims about the form and content of the universe. Elsewhere I have shown that Galileo believed his use of mathematics would be *vindicated* in terms of the explanations it produced.[6] This means he was committed to *two* things: mathematics *and* observation.

3.2. If we were to attempt to apply contemporary philosophical labels to Galileo, the closest we could come to accurately describing his position would be to classify him as an *instrumentalist.* In contemporary terms, an instrumentalist maintains that the real world is the world we have access to by means of our sensory apparatus. Scientific theories, on this view, are merely convenient *instruments* we construct in order to facilitate the process of making inferences from what we have observed in the past or are observing at present to what we can expect to observe in the future.

Instrumentalism is *as close as we can get* to classifying Galileo in contemporary terms with respect to his views on the reality of non-observable entities. Nevertheless, he doesn't even fit this category. For one thing, Galileo is not opposed to the existence of some kinds of theoretical entities. Take for example, his work on the rate of fall. "Rate of fall' is a theoretical concept. Galileo had no difficulty entertaining that concept. But he did have difficulty with such notions as a final cause. The basic difference for Galileo between the rate-of-fall case and the final-cause is that one could be measured and the other could not. Put straightforwardly: Galileo had no problems with theoretical notions which lent themselves to quantification, especially measurement. He expressed reservation when claims were made for the existence of entities, forces, and situations which could *not* be analyzed using his measurement techniques.

In this respect, Galileo reminds us of an early version of Hume. Hume, however, was the archtypical instrumentalist and would have taken Galileo one step further than he probably would have liked. If, for example, Hume had been told that Galileo had measured the rate of fall and, therefore, there was a rate of fall, his response would have been: "so, you have a measurement, now, where is the rate of fall?" Galileo, however, had no interest in pursuing uniquely philosphical hobby-horses for their own sake. His concern was to provide the underpinning for a theory of knowledge, not to undermine the project from the start. Thus, his instrumentalism was tempered by his prior commitment to the role of quantification in the acquisition of knowledge, i.e., there seems to be a pragmatic dimension to Galileo's methodology. It is too early to assert this conclusively, but it is a point worth keeping an eye on, since if we must label Galileo in some way or other, new labels may bear larger fruit than the old.

Earlier it was claimed that Galileo was committed to mathematics *and* observation. In particular, as we see later, he believed that the reliability of observational claims was to be guaranteed by providing a mathematical demonstration of those claims, suggesting some sort of self-adjusting process as yet unexplicated. This position leads us to two more questions:

(5) How much did Galileo insist on mathematical reasoning?

(6) In addition to the infinity of things to be known, did Galileo admit to any other limits to human knowledge than its quantification?

Let us attempt ot answer these in order.

4. WHY MATHEMATICS?

4.1. The answer to our question "why mathematics" can be found by a close analysis of parts of Galileo's *Dialogue on the Two Chief World Systems*. In that work his fundamental philosophical objective is the articulation and defense of his geometric realism. This is not to say that there aren't other significant philosophical issues — such as the consequences of his demonstration that the soundness of arguments for or against the earth's motion cannot be determined by appeal to earth-bound empirical evidence. Nevertheless, for our purposes, we need to separate what is germane to his discussion of the relative merits of the Copernican and Ptolemaic systems from the philosophical views revealed in his methodological assumptions and procedures. That is, we need to distinguish his proofs of specific claims from the methodology he employs in those proofs. It is in his arguments for the methodology that we find his philosophical position on mathematics.

Along with a series of other comments he makes on the limits of human knowledge, Galileo's initial maneuver in the *Dialogue*, and its follow-up, exhibit as clearly as possible some of his basic assumptions. Let us begin with a brief examination of his first use of geometry and then continue by considering some of Galileo's commentary on the possibilities of knowledge.

4.2. *Day 1* of the *Dialogue* begins straightforwardly enough. Galileo's three participants, Salviati, Sagredo, and Simplicio meet for the purpose of discussing

> the character and the efficacy of those laws of nature which up to the present have been put forth by the partisans of the Aristotelian and Ptolemaic position on the one hand, and by the followers of the Copernican system on the other. [p. 9]

Salviati suggests they begin by examining the Aristotelian arguments against the view that the earth moves. He then immediately introduces the Aristotelian distinction between celestial and terrestrial substances, "the former being invariant and eternal; the latter, temporary and destructible." (p. 9) Quickly summarizing Aristotle's proof for the perfection of the world, he then introduces his objection. The objection is somewhat muted insofar as it is phrased primarily as a complaint about the form of the argument and not its content. Here is Galileo's version of Aristotle's arguments:

> it is not a mere line, nor a bare surface, but a body having length, breadth, and depth. Since there are only these three dimensions, the world having these, has them all, and, having the Whole, is perfect. (p. 9)

Galileo's response is first show that the conclusion Aristotle wants (that there are only three dimensions and having these the world is perfect) can be obtained by other techniques, techniques which do not beg the questions. As Salviati puts it,

> I much wish Aristotle had proved to me by rigorous deductions that simple length constitutes the dimension which we call a line, which by the addition of breath becomes a surface; that by further adding altitude or depth to this there results a body, and that

after these three dimensions there is no passing farther — so that by these three alone, completeness, or, so to speak, wholeness is concluded. *Especially since he might have done so very plainly and speedily.* (pp. 9—10, italics added)

But, having established the base for introducing a simple geometric example, Galileo resists the *coup de grace*. He does not provide the quick demonstration that would seal his point. Instead he allows Simplicio to intercede on Aristotle's behalf. The point here is that Galileo wants to clarify what the war is about: this, and the remaining important battles in *The Dialogue*, is a confrontation over what is going to count as an adequate demonstration. For, at the heart of Galileo's epistemology is his conception of proof. If we recall, Salviati's objection was that he wished Aristotle would have provided a rigorous *deduction* of his conclusion. To this Simplicio responds with a catalogue of texts in which he maintains that Aristotle (a) provides "elegant demonstrations", (b) "first proved", (c) "confirmed", (d) provides "reasons", and (e) "clinch(es) the matter with another proof". All of which Salviati rejects.

But Galileo does more than simply dismiss the alleged proofs. He argues that Aristotle's beginning points are unintuitive assumptions.

I do not feel impelled by all these reasons to grant any more than this: that whatever has a beginning, middle, and end may and ought to be called perfect. I feel no compulsion to grant that the number three is a perfect number, nor that it has a faculty of conferring perfection upon its possessor. *I do not understand, let alone believe*, that with respect to legs, for example, that the number three is more perfect than four or two; neither do I conceive the number four to be an imperfection in the elements, nor that they would be more perfect if they were three. Therefore, it would have been better for him to leave these subtleties to the rhetoricians, and to prove his point by rigorous demonstrations such as are suitable to make in the demonstrative sciences. (p. 11, italics added)

Galileo feels "no compulsion" nor does he "understand" nor does he "believe" Aristotle's purported proofs about the perfection that follows from possessing three qualities. Thus we finally have a straightforward attack on the assumptions of Aristotle's arguments. It is a denial of the legitimacy of an appeal to the self-evident nature of the basic assumptions on which the remainder of the "proof" relies. In its place Galileo wants "rigorous demonstrations". To complete the attack, he proceeds to provide such a demonstration following Simplicio's challenge that "if a more cogent demonstration had existed, Aristotle would not have omitted it."

Salviati's elegant little proof that there are only three dimensions even convinces Simplicio — but Simplicio cannot admit it without indicating some reservation.

I shall not say this argument of yours cannot be conclusive. But I will say with Aristotle, that in physical (*naturali*) matters one need not always require a mathematical demonstration. (p. 14)

There are two ways to read this less than whole-hearted acceptance by Simplicio. First, one can take the traditional approach and view it in conjunction with other similar comments elsewhere in the *Dialogue* as a device by Galileo to make fun of the Aristotelian philosophers: i.e., even faced with the obvious they will blindly defer to Aristotle.

The second way to interpret Simplicio's hesitation is less orthodox but it fits the epistemological character of *Day 1* better. (We also should note that there is no reason why the same remark cannot perform both functions.) On the second reading, which admittedly is more to our own purpose, Simplicio's reluctance to quickly endorse Salviati's demonstration points to the fact that despite its simplicity, something is still missing: a justification for accepting a geometric demonstration *in place of* other forms of proof. In other words, Simplicio has a reasonable objection. He is not always the fool. Galileo does use him to set up the discussion, but from that it does not follow that he is continuously wrong-headed in his objections. Galileo also uses this ploy again in *Day 4*, as we shall see.

4.3. Galileo's justification for geometry is fairly long in coming. In fact, we don't really find anything beyond the constant reliance on a variety of proofs offered in support of a number of different arguments until the very end of *Day 1*. This final argument begins by acknowledging the limits of human knowledge. Next, Galileo notes that from these limits, it doesn't follow that genuine knowledge is not possible. The way in which real knowledge can be obtained by human beings is, Galileo argues, through arithmetic and geometry. This is the way God obtains certainty. Let us look at this is some detail.

The announced topic of *Day 1* is the Aristotelian distinction between celestial and terrestrial substances. With the conversation winding to a close, the discussants are found deep in the analysis of the differences between the earth and the moon, which subject was introduced following the attempt to resolve the problem of the illumination of the moon's surface. It only seems natural that once the topic of the *similarity* between the surfaces of the earth and moon is raised that the next question will concern other similarities such as atmospheric conditions and the possibility of other species. Following their speculations on this topic Galileo's spokesman, Salviati, concludes,

> Besides, if I were asked what my basic knowledge and natural reason told me regarding the production there [the moon] of things similar to or different from ours, I should always reply, "Very different and entirely unimaginable by us"; *for this seems to me to fit with the richness of nature and the omnipotence of the Creator and Ruler.* (p. 101, italics added.)

This confession by Salviati as to what he does not know is immediately followed by an elaboration of its basic message by Sagredo,

> It always seems to me extreme rashness on the part of some when they want to make human abilities the measure of what nature can do. On the contrary, there is not a single effect in nature, even the least that exists, such that the most ingenious theorists can arrive at a complete understanding of it. This vain presumption of understanding everything can have no other basis than never understanding anything. For anyone who had experienced just once the perfect understanding of one single thing and had truly tasted how knowledge is accomplished, would recognize that of the infinity of other truths he understands nothing. (p. 101)

Several things are going on here. First, there is Sagredo's denial of the protagorean maxim that man is the measure. This sort of denial, furthermore, is found over and over again in Galileo's work. In conjunction with his elaboration of the role of geometry, this recognition of the limits of human cognition is the basis of Galileo's epistemology. The second item to be noted here is the strong claim that there is no secret *of nature* that is totally knowable by any man or group of men. So, if Galileo is supposed to be an infallibilist (read "Platonist"), his infallibilism is going to have to be highly circumscribed. Finally, there is Sagredo's hint that he knows what real knowledge would be like. If that was all that Sagredo or Salviati had to say on the matter, then we would not be in very much of a position to say anything beyond acknowledging that Galileo seems to think man's knowledge is limited in some ways.

4.4. Fortunately, Galileo continues this line of development, but in typical fashion appears to back into his conclusion. Salviati picks up what first looked like a quasi-throwaway line about claims to omnipotence being based on ignorance and alludes to the Delphic Oracle's endorsement of Socrates as the wisest since he knew the least. Simplicio then jumps in with the observation that someone must be lying. Salviati disagrees and explains that the Oracle could be correct and that Socrates also need not be lying, provided the distinction between the limitations on what man can know and the infinity of things to be known was maintained. Examples follow but fail to convince Simplicio. Salviati's final example is of the difference in the amount of knowledge necessary for a sculptor to produce a statue of a man and nature to produce the living man. He concludes by asking, "May we not rightly say that the making of a statue yields by an infinite amount to the formation of a live man, even to the formation of the lowest worm?" The suggestion is that the amount of knowledge Nature has in order to create the things in the world simply cannot compare in quality and magnitude to the limited things *we* know about nature. The emphasis is once again on the contrast between the finite and the infinite.

Given Salviati's latest references to what it takes to create human understanding itself, Simplicio attempts to drive a wedge in Salviati's argument.

Among your greatest encomiums, if not indeed the greatest of all, is your praise for the understanding which you attribute to natural man. A little while ago you agreed with Socrates that his understanding was nil. Then you must say that not even Nature understood how to make an intellect that could understand. (pp. 102—103)

To answer Simplicio, Galileo has Salviati introduce a distinction through which he also reveals some essential aspects of his theory of the special status of geometry.

You put the point very sharply, and to answer the objection it is best to have recourse to a philosophical distinction and to say that the human understanding can be taken in two modes, the *intensive* or the *extensive*. *Extensively*, that is, with regard to the multitude of intelligibles, which are infinite, the human understanding is as nothing even if it understands a thousand propositions; for a thousand in relation to infinity is zero. But taking man's understanding *intensively*, in so far as this term denotes understanding some proposition perfectly, I say that the human intellect does understand some of them perfectly, and thus in these it has as much absolute certainty as Nature itself has. Of such are the mathematical sciences alone. That is, geometry and arithmetic, in which the Divine intellect indeed knows infinitely more propositions, since it knows all. But with regard to those few which the human intellect does understand, I believe that its knowledge equals the Divine in objective certainty, for here it succeeds in understanding necessity, beyond which there can be no greater sureness. (p. 103)

Galileo's point is that human knowledge *is* possible. We can achieve perfect understanding of some propositions, namely those of arithmetic and geometry. But, he does not go so far as to claim we can understand *all* of them. Galileo is quite cautious here. He admits there is a certain class of propositions of which we can understand a small subclass. Furthermore, we can understand these with the same degree of surety that the Divine has in objective knowledge. That means we can have objective knowledge with respect to a small subclass of propositions, i.e. a small subclass of the propositions of geometry and arithmetic.

The remaining question is how it comes to be that we can have such certainty. According to Salviati, it is because we understand necessity. The question now becomes "what kind of necessity?" While Galileo does not make this as explicit as we would like, it seems clear that the answer is logical necessity. Consider Galileo's distinction between how God comes to have objective certainty and how we do:

> as to the truth of the knowledge which is given by mathematical proofs, this is the same that Divine wisdom recognizes; but . . . the way in which God knows the infinite propositions of which we know some few is exceedingly more excellent than ours. Our method proceeds with reasoning by steps from one conclusion to another, while His is one of simple intuition. (p. 103)

Thus, according to Galileo, we have the same objective certainty as God once we arrive at a mathematical demonstration; hence, Galileo's emphasis on mathematical demonstration over discursive proof at the start of *Dialogue*. There are some differences, however, between God and us. First, we differ from the Divine in the manner by which we arrive at our conclusions. God knows all by simple intuition. We come to our knowledge by the process of *reasoning*, slowly establishing one point after another and moving through the process of producing proofs in a laborious fashion.

There is also a second difference which is a bit more difficult to unravel. Galileo's point here is muddled, but it is clear that he is pushing for something. One thing is fairly obvious. Since he already is in the theological waters, discussing man and God in this way, he must be careful to steer clear of pronouncements which will conflict with Dogma. Thus, he is careful to note that while we *can* have objective certainty, there are fundamental differences between the knowledge we have and that of God's. For, according to Galileo, the properties of any given object are infinite and God knows all of these by intuition. We, however, are extremely limited in terms of which properties we have knowledge. In other words, not only does our knowledge differ from

God's by way of the manner we acquire it, the *kind* of knowledge we possess also seems different from God's.

> We, for example, in order to win a knowledge of some properties of the circle (which has an infinity of them), begin with one of the simplest, and, taking this for the definition of circle, proceed by reasoning to another property, and from this to a third, and then to a fourth, and so on; but the Divine intellect, by a simple apprehension of the circle's essence, knows without time-consuming reasoning all the infinity of its properties. (pp. 103—4)

Thus, one difference in kind is through the manner by which God comes to know, i.e. without the trial of thinking, by sheer intuition. But there is also the second difference in kind. God not only knows instantaneously, but his knowledge of the item is total, encompassing the infinity of properties an object has. Our knowledge is necessarily limited to some finite subset of that infinite class. It follows that the "object" of which we have knowledge is different from the object God knows. Galileo does not come right out and say this; nevertheless, it is a clear consequence of the position he has been developing. The argument for this consequence of Galileo's distinction between intensive and extensive knowledge is quite simple. What we know of an object we know by virtue of our knowledge of its properties.[6]

The more properties of an object we know, the more we know about the object. Since with different properties, its behavior will vary and what we predict its reactions to various tests will be also will vary. Thus it would seem that using Galileo's own argument here we could conclude something like the following: we do not know the same objects as God does since we only know a finite number of the properties of the object, and even if we knew a thousand properties, a thousand relative to infinity is as nothing, thus the object God knows when he knows the infinity of its properties, must be different from the object we know. Or, a second possible conclusion is that given that ratio we know nothing.

Finally, as a limiting case with respect to our claim that, for Galileo, God's knowledge must be different in kind from ours, consider the following: one of the things we know is that there is more to know about the object in question, although we cannot explain what that is. God, on the other hand, knowing everything, does not know *that.* Thus, the world God knows is *necessarily* richer than ours. It has more objects with more properties, thus it admits of more relations, hence, different generalizations.

There is a danger here. Taken too far, this argument concerning the differences between the kinds of knowledge God and man possess may undermine Galileo's main point that knowledge is possible because what we know we know with the same objective certainty that God does. Whether or not Galileo is sensitive to this worry, at this point he appears to be more concerned about the kind of differences represented by the manner in which God and man achieve whatever knowledge they have. And *that* does bother him. For the more he pushes the point the more he seems forced to the conclusion that the *necessity* man understands when he achieves knowledge cannot be the same as God's since man's grasp of necessity is the product of his reasoning and God's is the product of simply having reality present before him. After considering the ramifications of this, Galileo concludes rather lamely with some remarks about the wonder of nature and God.

He begins this final point with the observation that the infinity of properties of objects is included in their definitions, ". . . and ultimately, through being infinite, are perhaps but one in their essence and in the Divine mind." Now, with respect to the position we have been arguing for, that surely represents a fundamental difference between man's knowledge and God's. And it is at that point in his presentation that I believe Galileo realizes he is stuck. In his argument he appeared headed toward the conclusion that there was something essentially the same in the kind of knowledge shared by man and God. But he seems to have driven himself into a corner, concluding this line of thought by repeating his desired conclusion, while admitting there is no argument for it.

> Nor is all the above entirely unknown to the human mind either, but is clouded with deep and thick mists, which become partly dispersed and clarified when we master some conclusions and get them so firmly established and so readily in our possession that we can run over them rapidly . . . .Now these advances, which our intellect makes laboriously and step by step, run through the Divine mind like light in an instant; which is the same as saying that everything is always present to it. (p. 104)

But clearly Galileo is fudging here. It is not the same to say, on the one hand, that everything is always present to the Divine mind and, on the other, that the Divine mind runs quickly through the steps leading to its conclusions. In fact, earlier Galileo denied that God reasons to his conclusions at all — God intuits them. The only explanation for this apparent shift is that it is a deliberate attempt by Galileo to muddy the waters so that his hand-waving is missed.

There are two conclusions that follow from Galileo's failure to establish some sort of correlation between the workings of God's mind and that of man. In the first place, paradoxically, it is possible to view this not so much as a failure but as a success. The position Galileo finally endorses here is the same one maintained by the Church. The discussion over the methods by which God and man arrive at their respective understandings of necessity only obscures the real point.

Prior to the discussion of these different methods of arriving at an understanding of necessity, Galileo's argument had been concerned to show that despite man's limitations, he is capable of acquiring knowledge. This is possible because knowledge consists in achieving certainty and irrespective of the difference in the methods man and God use, they can both achieve that state. Man achieves certainty by employing mathematical reasoning to produce knowledge of his object, God intuits it. So far so good, but we are still missing a premise or two. For it is still not clear why using mathematical techniques gives man the *same* kind of certainty God gets by intuition. This is also the same problem which Galileo faced at the beginning of *Day 1* when he urged the merits of geometrical demonstrations over Aristotelian methods. In both contexts, against the Aristotelians and when discussing the relation of man's knowledge to God's, the question remains the same: what is so special about mathematics?

In some respects this is merely a different way of noting that even if mathematics provides certainty, it doesn't follow that it provides *knowledge about the world.*

5. GEOMETRIC REALISM

Mathematical reasoning is special for Galileo because despite its failure to give apriori empirical knowledge, it remains a *means* of achieving knowledge of the world. As we have seen, he believed that man could acquire knowledge of the necessity of a conclusion by employing mathematical reasoning. But it is not through the nature of mathematics itself that man obtains knowledge of the world. The mathematics has to be interpreted. Galileo's method was to produce a geometrical demonstration and then in an informal manner correlate that demonstration with a discrete empirical situation. Thus, in his theory of tides, he gives a geometric account of the compound motion of the earth as it rotates around the sun and relates it back to an example of how water in the hold of a barge sloshes back and forth. In his proof of the existence of mountains on the moon, he demonstrates the effect of sunlight on large structures to produce shadows, correlates this with how mountains on the earth form shadows and then extends that analysis to the moon. It is his use of mathematics as a basis for the claims of certainty that he feels places him in a position to argue from some terrestrial example to the universality of the phenomenon in question. If the conclusion of a mathematical demonstration is truly certain, it must be universal — by definition. The way we get from the certainty to the world is to show some sort of parallel between features of the proof and features of the world. Admittedly, that is where problems arise, but not for Galileo. He not only has argued that we can know certain features of the observable world, but (mistakenly) argued that we can know them with as much surety as God can. Thus, from our certainty derived from mathematics and our similarity to God in respect of certain aspects of our acquisition of knowledge, it follows we can have certain knowledge of some features of the world. But from this it does not follow that the world is mathematical.

Galileo's argument for the possibility of knowledge about the world then looks like this:

(1) The number of propositions to be known is infinite.

(2) Human beings can know some subset of that total number of knowable propositions.

(3) Of those propositions human beings know, they have absolute certainty.

(4) Certainty is achievable only with understanding of necessity.

(5) Only mathematics yields understanding of necessity.

(6) The necessity of mathematical truths is self-evident.

(7) The structure of mathematical arguments can be mapped onto the world.

Therefore:

(8) When man correctly reasons about the world using mathematics he arrives at conclusions about which he is absolutely certain because he understands the necessity of the conclusion.

It may seem that there is also a sub-argument buried here to the effect that since mathematics yields necessity, what mathematical proofs are about must also be necessary. That lemma would have given Galileo the conclusion that the world has a necessary structure if he could have proved it, which he didn't. Furthermore, it is not possible to derive such a conclusion even if adding Galileo's belief that scientific truths "must be necessary and eternal". (p. 406) For from the conclusion that mathematical truths must have those properties, it does not follow that the world must, only that it can be understood as having that structure.

6. MATHEMATICS AND METHODOLOGY

6.1. Galileo's unwillingness to move from the certainty obtained by mathematical reasoning to the conclusion that the world is mathematical is further emphasized by a regular policy he employs. Galileo does not insist that *until* there is a mathematical demonstration available, the thing claimed to be known is invalid. Rather, he keeps emphasizing a weaker position: whenever possible, replace qualitative arguments with demonstrable deductive proofs. Not only is this a weaker claim, but it is vastly more interesting than a dogmatic assertion, for it amounts to a genuine *methodological principle.* It also deserves a name; *PQ*: *The Principle of Quantification.* Indeed, if we read Galileo's appeal to mathematics from a methodological point of view, rather than as embodying some metaphysical truth about the world, we can make considerable headway in understanding his epistemology. In this vein, it is even possible to construct a kind of proof for *PQ*. It is a proof using a generic *Principle of Practical Reasoning* (*PPR*):

PPR: given your goal, if you have shown that a given action or set of actions would bring that goal about, you ought to perform those actions, *ceteris paribus.*

Just earlier we produced an argument with the following conclusion:

(8) When man correctly reasons about the world using mathematics he arrives at conclusions about which he is absolutely certain because he understands the necessity of the conclusion.

Since it is clear that Galileo is concerned to obtain the highest degree of certainty possible where knowledge is concerned, then given *PPR* and the viability of 8 above, *PQ* follows.

6.2. With *PQ* in hand, it becomes possible to get a better sense of Galileo's own methodology. For example, he never rejects an argument *just because* it is not mathematical. If he rejects a proof, it is because it can be shown to be based on flawed assumptions or inconsistent principles. In other cases, such as in the first section of *Day 1*, he introduces geometry for the purpose of clarifying the structure of the point in contention; the rigor of the demonstration being the essential issue. Rigor again is the emphasis at the end of *Day 3* when he is discussing Gilbert's proof of the properties of the lodestone.

> I might have wished for in Gilbert would be a little more of the mathematician, and especially a thorough grounding in geometry, a discipline which could have rendered him less rash about accepting as rigorous proofs those reasons which he puts forward as *verae causae* for the correct conclusions he himself had observed. His reasons, candidly speaking, are not rigorous and lack that force which must unquestionably be present in those adduced as necessary and eternal scientific conclusions. [p. 406]

This has been a long way round the barn in order to answer Question 5 from Section 3. That questions was "How much did Galileo insist on mathematical reasoning?" The answer is "a lot" — but not completely. We still have one last question from that same Section 3 to answer: "In addition to the infinity of things to be known did Galileo admit to any other limits to human knowledge than its quantification. Again the answer is "yes."

6.3. Galileo believed that our ability to achieve knowledge, i.e. certainty about the world, was limited both by our capacity and by nature herself. How long Galileo held this view is not clear. But he is quite explicit as early as 1623, when he published *The Assayer*

> I might by many other examples make clear the bounty of nature in producing her effects by means which we would never think of if our senses and experiences did not teach us of them, though even those are sometimes insufficient to remedy our incapacity. Therefore, I should not be denied pardon if I cannot determine precisely the manner in which comets are produced, especially as I never boasted that I could, knowing that it may occur in some way beyond our power to imagine.

We have already examined the passage at the end of *Day 1* of *Dialogue* in which Sagredo says:

> It always seems to me extreme rashness on the part of some when they want to make human abilities the measure of what nature can do. On the contrary, there is not a single effect in nature, even the ideals that exist such that the most ingenious theorist can arrive at a complete understanding of it. [p. 100]

Finally, in *Discourses on Two New Sciences* we find the following:

> No firm science can be given of heaviness, speed, and shape which are variable in infinitely many ways. Hence, to deal with such matters scientifically, it is necessary to abstract from them.

This is not merely a recognition of the limits of human capacity. This observation also includes another methodological injunction, the Principle of Abstraction, PA. PA is the logical extension of the point Galileo made at the end of *Day 1* of *The Dialogue*, when he noted that man's way of proceeding toward knowledge was to isolate one property of an object and then reason slowly from it to the finite number of other properties he could deduce. As a methodological principle, PA takes the negative side of Galileo's acceptance of our finitude and turns it into a positive point. Most importantly, however, it is a methodological principle that follows from a recognition of the limits of human knowledge.

To secure this final point one only needs to turn to the final exchange between Simplicio and Salviati at the end of *Day 4* of the *Dialogue*. It is alleged that this set of comments so enraged Urban VIII that it precipitated Galileo's trial.[7] There is really no evidence to support that claim — and yet it persists as part of the mythology surrounding Galileo's history. Now, however, in the light of the argument that has been presented so far, the proper interpretation of this passage is possible. Having just concluded the discussion of the topic of *Day 4*, the tides, Simplicio is still not convinced. But he is willing to admit that

> if asked whether God in His infinite power and wisdom could have conferred upon the watery element its observed reciprocating motion using some other means than moving its containing vessels, both of you would reply that He could have, and that He would have known how to do this in many ways which are unthinkable to our minds. From this I conclude that, *this being so*, it would be excessive boldness for anyone to limit and restrict the Divine power and wisdom to some particular fancy of his own. (italics added, p. 464)

This is more than an admission by Simplicio that God could do whatever he wants. It includes the claim that neither Salviati nor Sagredo are committed to having some particular structure to the world necessarily revealed. And it is not a gratuitous claim, for this is what Galileo has been saying all along. Man's knowledge is limited and different in kind from God's. It should not come as a surprise that the real structure of the world is hidden from us.

Thus, it is appropriate that the closing word should echo that thought, for it has been the theme of the book. So we have Salviati extend Simplicio's comment, endorsing it while reminding us of the concluding comments of *Day 1*.

> An admirable and angelic doctrine, and well in accord with another one, also Divine, which, while it grants to us the right to argue about the constitution of the universe (perhaps in order that the working of the human mind shall not be curtailed or made

lazy) adds that we cannot discover the work of His hands. Let us, then, exercise these activities permitted to us and ordained by God, that we may recognize and thereby so much the more admire His greatness, however much less fit we find ourselves to penetrate the profound depths of His infinite wisdom. (p. 464).

Our conclusion then tells us, that in addition to revising the standard account of Galileo's epistemology, we ought to reconsider the bad press that Simplicio has received. While not a totally admirable intellect, he often reminds us that just because Galileo is opposed to the standard Aristotelian point of view, it doesn't follow that everything he says is justified or fully comprehensible. Finally, as the only representative of the Church's position, he also reminds us that the epistemological view Galileo has been developing throughout the discussion is consistent with the Church's own stance and may even augment it.

In conclusion we can observe that Galileo's understanding of knowledge was based on the conviction that certainty could only be achieved through the rigor of mathematical demonstration. We have seen how he emphasized proof and extolled the virtues of geometry, and how a proper appreciation of his understanding of the role of mathematics generated the methodological principle of quantification, *PQ*. We also saw what Galileo thought about the limits of human knowledge based on the extent and variety of nature and the possibility of the lack of sufficient imagination on the part of human investigators. This produced a second methodological principle, PA. These various points leave us with the following tentative conclusion: *The book of nature*, written *in the vocabulary of geometry*, is written by man and records the results of his efforts to use geometric techniques to examine the *behavior of physical phenomena*. Thus, to answer question (1), the world cannot be shown to be geometrically structured. In response to question (2) we can say that while it is still not clear if Galileo thought that *the* structure of the world could be completely revealed through the use of mathematics, the book of nature records at least *a* structure of the world.

Virginia Polytechnic Institute and State University
Blacksburg, Virginia
USA

NOTES

* An earlier version of this paper was presented to the Workshop on Early Modern

Science, Blacksburg, Virginia, 1985. My thanks to the participants for their suggestions and particularly to Roger Ariew for his rejection of my thesis and Robert Westman for the final connection to Simplicio.

[1] From S. Drake, *Galileo at Work*, Chicago: University of Chicago Press, 1978, p. 412. No honest reader is going to require Galileo to have carefully sorted through the variety of philosophical ideas that permeated both the literature he relied on and responded to and the intellectual atmosphere of his day in order to produce what *we* want by way of a philosophically consistent position, especially when Galileo was not intending to present his view as philosophy. We leave those sorts of demands to the neo-scholastics. The objective here is *not* to show how everything Galileo said of epistemological interest can be made consistent. It is, rather, to try to make sense of as much as possible in a way which provides insight into Galileo's methodological contribution to the scientific revolution. Likewise, I resist the view that the slightest mention of a prominent idea of a particular general philosophical theory automatically makes the speaker of that thought an advocate of that general philosophical view. If we applied those standards to everyone, we all would probably qualify at best as eclectics. Finally, if philosophy is a dialogue, then in a very important sense it matters not what views the positions under discussion borrow from. What is important is the end product and to what *that* commits us. Our interests in Galileo are many, but foremost among them is the extent to which his philosophical views provided an intelligible foundation for scientific methodology at the time when it was first coming to be in the modern manner. For such a time of intellectual turmoil it hardly seems reasonable to worry if the old philosophical labels fit. What is important is what Galileo did with the old ideas in order to make them work for the new science. If he transformed them, fine; if he amalgamated different sets of views, that too is fine; likewise if he merely adopted them without change. But let us see what he did. As it turns out, despite a lot of smoke, there is no fire. Platonism is not the issue, Galileo's approach to understanding the world is.

[2] A. Koyré, Etudes Galiléennes, Librarie Scientifique Hermann et Cie, 1939.

[3] All page references to the *Dialogue* are to *Dialogue on Two Chief World Systems*, translated by Stillman Drake, Berkeley: University of California Press, second revised edition, 1967.

[4] Galileo also gets a little heavy-handed here in his literary style. There is little doubt that his use of the dialogue format is influenced by the poopularity of Plato's dialogues. Nevertheless, Galileo cannot resist taking a bit of a dig at Plato's theory of "innate" knowledge (reminiscence) when he provides the proof that Simplicio claims would have been produced by Aristotle if it existed. In response to Sagredo's challenge to provide such a proof so that he can comprehend it, Salviati obliges, nothing that it can be understood "not only by you, but by Simplicio too; and not merely comprehended, but already known — though perhaps without your realizing it." (p. 12) It is no wonder that Galileo assumes Simplicio would not realize what he knew since Simplicio often even denies what he admits has been demonstrated before his very eyes! Unfortunately, what Galileo intended as wit has been misinterpreted as proof of his Platonistic leanings. This is a mistake. Just as Galileo's choice of the dialogue form for this popular work does not make him a Platonist, his jokes ought not to be taken as philosophical endorsements.

[5] Cf. J. C. Pitt, "The Untrodden Road: Rationality and Galileo's Theory of the Tides", in *Nature &System*, Vol. 4, 1982, pp. 87—99.

[6] This is not to beg the issue and say that an object is only the sum of its properties, but it is sympathetic to that point of view. It is also worth mentioning in passing that the "ideals" that cannot be known by man might be interpeted to be a reference to Platonic Ideals by Galileo, thereby being a further rejection of classic platonism, firming up the position discussed in Section 2.2.

[7] It appears that the story can be credited to Pierre Duhem. In his *To Save the Phenomena* (original published in French as *Zein Ta Philosophie Ainomena: Essai sur la motion de thérie physique de Platon à Galilée* in *Annales de Philosophie Chiretienne*, Ser. 4, Vol. VI. Duhem raises the possibility that Salviati's talk about an "admirable and angelic doctrine" is a jibe at Urban VIII because, according to a second hand account, attributed to Cardinal Oregio, of what Urban, then Cardinal Barberini, told Galileo after his run-in with the Inquisition in 1616. As Duhem frames the series of incidents, Barbarini essentially told Galileo what Galileo has Salviati say, hence, the speculation that Galileo is being facetious. Unfortunately Duhem ignores what I have been at pains to show is a systematic theme throughout the *Dialogue*, namely the tension between God's knowledge and man's.

WILLIAM R. SHEA

CARTESIAN CLARITY AND CARTESIAN MOTION

Descartes died an impenitent and hardened bachelor but he had a lifelong love affair with clarity and certainty and, as the years went by, he not only saw things more clearly but he grew fonder of his own certainties. A year before his untimely death in the land of "bears, rocks and ice" as he called Sweden, he wrote to Henry More, who had suggested that his philosophy could stand even if his identification of matter with extension were abandoned:

> I do not admit what you courteously concede, namely that all my views might hold even if what I wrote about the extension of matter were refuted. It belongs to the main, and to my mind, most certain foundations of my physics. I profess that no answer satisfies me in physics unless it be of the kind that is called logical necessity except, of course, concerning those things that can only be known through experience, such that there is only one Sun or one Moon around the Earth (letter of 5 February 1649, Descartes, AT, V, 275).

Such statements are not uncommon in Descartes' writings and they can be ascribed, in part, to temperamental or psychological factors. Descartes was over-sensitive to criticism and would have no truck with the give-and-take of academic debates. Those who took at face value the invitation he issued at the end of the *Discourse on Method* and sent him their objections came in for contumely greatly at variance with his much vaunted good breeding. The French mathematicians who criticized his *Geometry* were dismissed as "two or three flies" (letter to Constantijn Huygens, Descartes, AT, II, 671). To the Minim friar Marin Mersenne, who had forwarded the objections, he was even more forthright: he described Roberval as less than "a rational animal" (AT, II, 190), Petit as "a little dog" (AT, II, 267, 533), and Hobbes as "contemptible" (AT, III, 326). Jean de Beaugrand's letters were only good to be used as "toilet paper" (AT, III, 437), while Fermat's works were plain "shit", a word hardly softened by being in Latin (AT, II, 464). He expressed his contempt for his opponents by having his erstwhile servant Gillot answer their queries (AT, II, 179, 195—196, 275). He returned unread Fermat's *Isagoge ad locos solidos*, one of the great mathematical works of the seventeenth century, because he felt

J. R. Brown and J. Mittelstrass (eds.), An Intimate Relation, 23—42.

that it could only repeat what he had already said in his *Geometry* (AT, II, 495).

This overbearing and arrogant attitude was not exclusive to Descartes. Compare, for instance, the following passage from his letter to Mersenne: "I would gladly have them believe that if I am wrong about the motion of the heart, refraction, or if anything else that I have discussed in more than a couple of lines is wrong, then the rest of my philosophy is worthless" (letter of 9 February 1639, AT, II, 501), with Galileo's statement in a postil to a book by the Jesuit Orazio Grassi (who wrote under the pseudonym of Sarsi): "What do you want, Mr. Sarsi, if it was given to me alone, and to no other, to make all the new discoveries in the heavens" (Galileo 1890—1909, VI, 383, n. 13).

The age of Galileo and Descartes had long ceased to be an age of chivalry and had become the age of Spanish pride and Baroque punctilio. It is symptomatic that in 1636, the year before the publication of the *Discourse on Method*, Paris acclaimed Corneille's immensely successful play, *Le Cid*, which revolves entirely on a point of honour.

Although personal temperament and social climate must be taken into account when considering Descartes' polemics with his contemporaries, there is a more profound and interesting reason for his unbounded faith in himself and in his method. This hinges on the role or, better still, mission that he believed he was fulfilling as a philosopher. What he hoped to achieve for science was something that no modern philosopher would wish to entertain even in his wildest dreams. Since Kant, philosophers have taken it for granted that their task is to analyze scientific theories in the light of what they believe to be humanly knowable, and to locate scientific results within the realm of critical knowledge. Descartes would have considered this menial work. He had no wish to explain current scientific practice but to legislate about future scientific procedure. His goal was to formulate a general program that would yield reliable conclusions within science. As a gifted young mathematician, Descartes was fascinated early on with the clarity and certainty of mathematics. By 1619 he had found how to divide an angle into any number of equal parts as well as a method to generate a series of mean proportionals. In the *Rules for the Direction of the Mind*, which he wrote in his early thirties, he took it for granted that science, in keeping with its traditional definition, was "certain and evident knowledge" and that the paradigm cases were "arithmetic and geometry". The reason for their certainty appeared obvious to him: it lay in the simplicity of their object and the rigour of their deductions. Hence, "to seek the right path of truth, we ought to concern ourselves only with objects that admit of as much certainty as the demonstrations of arithmetic and geometry" (Rule Two, AT, X, 366, Cottingham 1985, I, 12—13).

The privileged status of mathematics rests on the conceptual purity of its object which the *Discourse on Method* describes as "nothing but the various relations of proportions". These are to be considered in the clearest way possible, namely in geometrical terms: "supposing them to hold between lines, because I did not find anything simpler, nor anything that I could represent more distinctly to my imagination and senses" (AT, VI, 20; Cottingham 1985, I, 121). Intelligibility of the clear and indubitable kind that Descartes wants is provided by embodying, or seeing, proportions in lines, the simplest kind of extension. The proper object of science is therefore extension and the comparisons of extensions. We must always ensure "that every problem has been reduced to the point where our sole concern is to discover a certain extension on the basis of a comparison with some other extension which we already know" (Rule fourteen, At, X 447, Cottingham 1985, I, 62). Gerd Buchdahl suggests that from holding that "a successful scientific treatment of nature *presupposes* its being *considered* under the aspect of extension, Descartes slides into the assertion that (material) nature *is* essentially equivalent to extension, and that this alone justifies us in postulating the existence of genuine science" (Buchdahl 1969, p. 89—90).

This is a plausible interpretation of Descartes' unconscious sliding or slithering from episteminology into ontology, but it is not how things appeared to Descartes. He saw himself as arguing that all the attributes of bodies presupposed extension, for instance that "shape is not conceivable except in an extended thing, nor motion except in extended space" (*Principles of Philosophy*, Part I, art. 53). Extension he believed to be self-evident and obvious to anyone who would but think clearly about the matter.

Descartes never wavered in his belief in the intuitive character of extension even when he subjected everything to radical doubt. In the Fifth Meditation, whose title begins, "On the Essence of Material Things", we might have expected a full account of the essence of matter, but all we get is a reminder of the clarity and distinctness that he allegedly experienced:

> I distinctly imagine quantity, or "continuous" equality as the philosophers commonly call it. That is, I distinctly imagine the extension of the quantity (or rather of the thing which is quantified) in length, breadth and depth. (Descartes, At, VII, 63; Cottingham 1985, II, 44).

The intuitive obviousness of the notion of matter is the key to understanding Descarte's concept of motion, to which we must now turn.

CLARITY OF VISION

We shall examine: (a) how Descarte's quest for clear and certain principles determined his concept of motion, (b) how this clear and distinct concept of motion was applied, and (c) why, in the end, it retained a core of unclarity that Descartes never consciously addressed.

It is clear from Descartes' published works that if people used their mind properly they would soon all be of the same mind. As the opening sentence of the *Discourse on Method* puts it: "Good sense is the best distributed thing in the word: for everyone thinks himself so well endowed with it that even those who are the hardest to please in everything else do not usually desire more of it than they possess" (Descartes, AT, VI, 2; Cottingham 1985, I, 111). The universal claim is clear but so is the biting irony, "Everyone complains of his memory, no one complains of his intelligence". We have already seen how, in his private correspondence, Descartes repeatedly dismissed his opponents as lacking in the very attribute he so generously ascribed to all mankind. When not the result of mischief (his most common explanation), he blamed error and obtuseness on habits acquired in childhood. For instance, because we played with animals when we were children, we now assume that they have thoughts and emotions like our own (Descartes, AT, V, 255—256), or because we see the sun rise and set we take the stability of the earth for granted (Descartes, *Principles of Philosophy*, Part 1, Art. 71). To a Jesuit he wrote: "The difficulty that we experience in learning science and in clearly representing to ourselves ideas that are naturally known comes from the prejudices of childhood" (letter to Mesland, 2 May 1648, AT IV, 114). The teaching of the established schools of philosophy needed no special refutation since they merely glorified childhood apprehensions with technical-sounding words like "form" and "quantity".

In the *Discourse on Method*, Descartes summarizes his own mature program under the four rules of evidence, division, order, and exhaustion. Nothing is to be assented to, unless evidently known to be true; every subject matter is to be divided into the smallest possible parts, each to be dealt with separately; each part is to be considered in the right order, the simplest first; and no part is to be omitted in reviewing the whole. In the first rule, Descartes pledged himself to assent only to what was "so clearly and distinctly" presented that he had no occasion to doubt it. Leibniz's disappointment with this rule is well known: "to say that what I clearly and distinctly perceive is true is of little use until the criteria of clarity and distinctness are made known" (Leibniz, 1965, p. 328). The four rules taken together, he adds, are not worth much more than a prescription that reads: "Take what you need, use it as you should, and you will get what you want" (*ibid*, p. 329). Leibniz was anticipated in his criticism by Gassendi and an unknown French scholar referred to as Hyperaspistes who wrote to Descartes in July 1641: "I deny that you have provided a method that is accurate enough to enable someone to know whether he perceives something clearly or not" (Descartes, AT, IV, 402). Descartes never answered this charge directly but he tackled it indirectly in a number of places. His reply took one of three forms. The first was a radical disclaimer of the need for probing the basic elements of knowledge. To Mersenne, who had asked him to comment on Herbert of Cherbury's *De Veritate*, Descartes replied that he had read the book (a rare compliment) and that it dealt with a topic upon which he had "worked all his life". He agreed with Cherbury's view that we have innate ideas but he found himself at variance with him because Cherbury "examines what is truth whereas I have never doubted what it is. It seems to me to be a notion so transcendentally clear that it cannot be ignored. We have means of examining scales before using them but we would have none to find out what truth is unless we naturally knew it" (letter to Mersenne, 16 October 1639, Descartes, AT, II, 596—597). How could we assent to someone who speaks the truth if we did not have the idea of truth! The machinery of knowledge could not get underway unless we clearly knew what truth is. The implication for Descartes is that to know truth, any truth, is to know it clearly and distinctly.

The second line of defence was equally dogmatic: even if it were the case that we cannot be absolutely certain that the deliverances of our intellect are true we should not worry about it. The most striking instance of this rebuttal occurs in Descartes' reply to the second set of objections to his *Meditations*:

> As soon as we think that we correctly perceive something, we are spontaneously convinced that it is true .... What is it to us if someone conjectures that something of whose truth we are completely convinced appears false to God or an angel, so that it is, absolutely speaking, false? *Why should this absolute falsity bother us*, since we neither believe it nor have the smallest suspicion of it (Descartes, AT, VII, 144—145; Cottingham I, 103, underlining mine).

The mind with which we are endowed is such that it automatically assents to what it clearly perceives (letter to Regius, 24 May 1640, Descartes, AT, III, 64). But even if it is in our nature to assent to what is clearly perceived, we still need criteria to avoid being rash, gullible or simply naive. It is here that Descartes introduces his third argument in the form of a page from his intellectual biography. For many years, he tells us, he conceived of gravity (literally "heaviness") as if it were a real quality present throughout the whole body but capable of being contracted to a mathematical point (Descartes, AT, VII, 442; Cottingham 1985, II, 298). In other words, although gravity was a property of the whole body, Descartes considered it as capable of acting just at one place, "exactly the way in which I now understand the mind to be coextensive with the body — the whole mind in the whole body and the whole mind in any one of its parts" (*ibid.*). He had perpetrated a category mistake by applying the innate idea of unextended mind to a feature of the world that can only be properly captured by the innate idea of extension. Once he had recognized this, it was an easy task to extricate himself from his childish ways:

> I very easily freed myself from all the doubts that my critics had put forward. First of all, I did not doubt that I "had a clear idea of my mind", since I had a close inner awareness of it. Nor did I doubt that "this idea was quite different from the ideas of other things" (Descartes, AT, VII, 443, Cottingham 1985, II, 298).

Descartes adds that what "made it especially clear" to him that his spurious idea of gravity was taken from his genuine idea of mind was the fact that he considered that gravity "carried bodies towards the centre of the earth as if it had some knowledge of the centre within itself" (Descartes, AT, VII, 442; Cottingham 1985, II, 298).

Tendency towards a specific goal or attraction towards a definite place was a characteristic of mind for Descartes. This is why he later brushed aside any suggestion that heaviness might have something to do with attraction. When Roberval made just such a proposal, he ridiculed it and claimed "that each individual part of matter in the world should have the innate property of attracting and being attracted is absurd . . . why we would have to suppose that each has a soul" (letter to Mersenne, 20 April 1646, Descartes, AT, IV, 401).

Knowledge of matter must rest, therefore, on the intrinsically obvious notions of matter and motion, or since matter is really extension, on extension and motion or, better still, since extension is indistinguishable from space, on space and motion. Space is described in the perspicuous language of geometry and different spaces can be compared in terms of their mathematical proportions. Descartes was usually content with establishing this in principle rather than attempting to work out the mathematical and mechanical details. He could not deny, however, that the correct notion of body or matter required some intellectual effort, witness his own experience. To Princess Elizabeth (of all his correspondents the one who brought out his pedagogical side to best advantage), he stated that "body, namely extension, shapes and motions, can also be known by the intellect alone, but much better by the intellect helped by the imagination". "The study of mathematics", he added, "which exercises mainly the imagination in the study of shapes and motions, trains us to form really distinct notions of a body" (letter to Princess Elizabeth, 28 June 1643, Descartes, AT, III, 692). What is interesting here is the avowal that geometry (in its Euclidean or pictorial aspect) actually helps in acquiring the notion of matter. The first thing to note is that Descartes is not saying that mathematics helps understand the *behaviour* of matter but that it assists in *conceiving* what matter is. Extension clearly known is extension clearly imagined! The second thing to note is that these geometrical bodies are, *of and by themselves*, capable of motion. This is a *crucial* consequence of Descartes' understanding of the rule of clarity and it runs counter to the Aristotelian notion of mathematics prevalent in his day. For Aristotle, mathematical entities were abstractions incapable of motion, which he considered the essential characteristic of physical objects, and he repudiated the Pythagorean identification of the geometrical solid with the sensible body (*Metaphysics*, 989 *b*, 29 ff.). Descartes approached the problem of matter purely in the light of his convictions about the realism of intuitive ideas. Because we can clearly combine points to make lines, lines to make surfaces, and surfaces to make solids, we have no reason to hesitate, once we have arrived at a solid geometrical figure, to cross the boundary into the physical world.

For Descartes, the question: How does a solid acquire motion in space? is not at all the same kind of question as, How is a geometrical figure endowed with sensible qualities? The second question, as we have seen in the case of gravity, is the very embodiment of obfuscation. Perceptible properties such as weight and colour are subjective and have no objective reality beyond the shape and motion of the matter that gives rise to them. In a radical sense, to know matter is to know that matter exists. "Where there is extension, there is necessarily body", as he puts it to Arnauld (letter of 29 July 1648, Descartes, AT, V, 224). Hence it is obvious, i.e. clear and certain, that a perfect void is impossible (*The World*, Descartes, AT, XI, 17—18; Cottingham 1985, I, 85—86). There are no spatial relations between empty places because no place is ever empty. The mere existence of spatial relation connects things together, and nothing in space can be independent. Everything exerts some effect on everything else. One cannot pick a daffodil without trembling a star!

Descartes' claim that mathematical extension (i.e. imagined space) entailed its physical reality did not go unchallenged. Gassendi objected that the extension that Descartes took for granted in his physics was a mere abstraction and only had reality in his mind. Descartes replied with *hauteur* that the objection implied that we could only behave like thoughtless monkeys or parrots. He consoled himself, he said, with the thought that his critics "linked his physics with pure mathematics", which is what he "wanted it to resemble above all". (Descartes, AT, IX—I, 212—213; Cottingham 1985, II, 275).

Mathematics, for Descartes, is a science of real quantity, a science of the quantitative aspect of the real world, not just a postulated construct. Motion is simply one of the real aspects that it describes. In the *Discourse on Method*, the "object of the geometer" is defined as "a continuous body, or a space indefinitely extended in length, breadth and height . . . that can be moved" (Descartes, A.T. VI, 36; Cottingham 1985, I, 129). To the Jesuit Ciermans he boldly proclaimed that "pure mathematics is mainly about motion" (letter of 23 March 1638, Descartes, A.T., II, 71). Cartesian bodies move, but they move geometrically. They interact not *mechanically* but *diagrammatically*. Their motion is adequately represented by a straight line, and is fully intelligible without the added notions of speed or direction. As Descartes contemplated the trajectory of a moving body on a sheet of paper, it seemed clear to him that it could be fully grasped regardless of the speed with which the distance was traversed or even the direction along which it travelled. In his *Optics*, Descartes equated this non-vectorial, geometrical, representation of motion with physical motion. This appeared so overwhelmingly clear to Descartes that two years after the publication of his *Optics* and the ensuing correspondence, he could write to Debeaune who was planning a book on mechanics:

> I should like to be able to answer you suitably concerning your work on mechanics but, although all my physics is nothing but mechanics, *I have never closely examined problems that depend on the measurement of speed.* Your way of distinguishing different dimensions in motion and of *representing them by lines* is undoubtedly the best possible (letter of 30 April 1639, Descartes, AT, II, 542, italics mine).

The notion of motion like that of extension is intuitively obvious. It need not be defined, merely inspected. Indeed "if we attempt to define things that are very simple and naturally known, such as shape, size, motion, place, time, etc., we obscure them and just get them entangled" (letter to Mersenne, 16 October 1639, Descartes, AT, II, 597).

In a passage that is reminiscent of Johnson's celebrated refutation of Berkeley, Descartes dismisses the Aristotelian definition of motion by appealing to everyday experience: "a person walking up and down a room gives a better idea of what motion is than one who says: *est actus entis in potentia prout in potentia* (*ibid.*).

This traditional definition of motion is the frequent butt of Descartes' sarcasm. He quotes it in Rule 12 of the *Rules for the Direction of the Mind* as an instance of "finding a difficulty where none exists" (Descartes, AT, X, 426). In *The World*, written a couple of years later, he cites it once more in Latin:

> For me these words are so obscure that I am compelled to leave them in Latin because I cannot interpret them . . . . By contrast, the nature of the motion I mean to speak of here is so easy to know that the geometers themselves, who among all men are the most concerned to conceive very distinctly the things they study, have judged it simpler and more intelligible than the nature of their surfaces and lives — as is shown by the fact that they have explained "time" as the motion of a point and "surface" as the motion of a line (Descartes, AT, XI, 39; Cottingham, I, 94)

But Descartes needs a double metaphysical warrant to use his notion of motion, however intuitive it may be. The first is the correspondence between the laws of nature and innate ideas; the second is the ontological grounding of these laws. Descartes had no problem with the first: the God who created the world also created minds capable of knowing it. In one of his letters to Mersenne after settling in the Netherlands, he wrote: "God is the one who established the laws of nature . . . . There is none that we cannot understand if we will but consider them for they are all *mentibus nostris ingenitae*" (letter of 15 April 1630, Descartes, AT, I, 145). The ontological grounding is laid out in chapter seven of *The World*, the first draft of the work that was to become, in a much revised and amplified form, the *Principles of Philosophy.*

Descartes states that the laws of nature (or as he prefers to call them, the laws of matter, AT, XI, 37) depend on God for their existence, and in saying this he is merely repeating what is common to all theistic philosophers of his day. Where he is original is in his belief that the laws of nature can be seen to derive from one of God's essential attributes, his constancy or immutability. Although he does not say explicitly that the notion of God's immutability is a clear and distinct innate idea, he suggests as much ("as everyone should know" AT, XI, 38, line 2), and draws the "easy" inference that God "always acts in the same way" (AT, XI, 38, line 3). From this it is an equally easy step to "two or three principal rules".

The first rule or law (Descartes uses the words indiscriminately) is:

> each individual part of matter continues always to be in the same state so long as collision with others does not compel it to change that state. That is to say, if the part has some size, it will never become smaller unless others divide it; if it is round or square, it will never change unless others force it to; if it is brought to rest in some place, it will never leave that place unless others drive it out; and if it has once begun to move, it will always continue with an equal force until others stop or retard it (AT, XI, 39; Cottingham, I, 93).

It is at this point that Descartes introduces the notion that motion is easier to understand than a geometrical line since it accounts for its genesis. A geometrical line does not change because it is not a dynamic reality, a force operating in time. Motion, for Descartes, partakes of this *timeless* quality.

The second rule reads:

> when a body pushes another it cannot give it any motion unless it loses as much of its own motion at the same time, nor can it take away any of the other's motion unless its own is increased by as much (AT, XI, 41; Cottingham, I, 94).

The first rule, therefore, postulates *the conservation of any created state of being*, the second, the *conservation of the same amount or kind of being*, namely motion. They both "follow manifestly from the mere fact that God is immutable and that, acting always in the same way, he always produces the same effect" (AT, XI, 43; Cottingham, I, 96).

This is a crucial step in the direction that will eventually lead to Newton's laws of motion, but an essential component of Newton's First Law is still missing: the rectilinearity of inertial motion. This is expressed, almost as an afterthought, in the third rule:

> when a body is moving, even though its motion for the most part takes place along a curved path . . . yet each of its individual parts tends to continue its motion in a straight line. And so their action, that is the inclination they have to move, is different from their motion (AT, XI, 44; Cottingham, I, 96).

Experiment shows this to be the case. If a stone is released from a sling "it is obvious" (*il paroist clairement*) that it will not continue to move in a circle but will fly off at a tangent to the circle. This is, from our post-Newtonian perspective, a good illustration, but its significance is obscured by what Descartes takes to be a second illustration of his third rule, namely that before the stone is released it exerts a pull at right angle to the tangential path that it will follow when it flies off. This second radial tugging away from the centre of revolution is what Huygens was to term "the centrifugal force". But in an inertial system, the centrifugal force is non physical, i.e. it arises from kinematics and is not due to physical interaction. This was a serious problem for Newtonians (see Shea, 1986, pp. 41—43) but it did not arise for Descartes since he did not consider that *direction* was part and parcel of the concept of motion. All Descartes saw was that, regardless of direction, motion tended to proceed in a straight line. There is, however, an aspect of the problem that would seem to have been almost forced upon him: rectilinear motion can only obtain in a perfect void and, according to Descartes, the void is conceptually, hence ontologically, impossible. How can a cosmos whose basic laws are incapable of realization be an ordered cosmos? But this objection never seems to have struck Descartes any more than it bothered Newton.

What Descartes considered essential was to show that the third rule can be justified on the same grounds as the first two.

This is possible because Cartesian motion is neither dynamic (involving consideration of force) nor kinematic (involving only consideration of space and time) but merely diagrammatic (involving only consideration of space). God who is eternal — above and outside time — creates and conserves all beings in the atemporality of the instant. God's "continuous action" is instantaneous, that is, it is not really continuous but quantized. He conserves things (of which rectilinear motion is one of many items) "as it is at the very instant that he preserves it" (AT, XI, 44; Cottingham, I, 96). The role played by God's instantaneous action helps to understand why Descartes felt that the *Meditations*, in which he demonstrated the existence of God, provided the rationale for his physics. To Mersenne, he wrote that it contained "all the principles of my physics" (letter of 11 November 1640, AT, III, 233), and "all the foundations of my physics" (letter of 28 January 1641, AT, III, 298). If we ask, What is being preserved in this way, i.e. in this instant? we find that Descartes' answer is purely epistemological: is conserved what can be clearly and distinctly understood to be in the instant. Now "only motion in a straight line is entirely simple and has a nature which may be wholly grasped in an instant" (AT, XI, 45; Cottingham, I, 96). This is not the case with circular motion which can only be conceived if two of its instants, or two of its parts, and their mutual relation is taken into consideration.

Descartes believes that he is contrasting rectilinear and circular motion but in the absence of any reference to force he is really comparing a geometrical line (which he implicitly reduces to a point) with a circle (a locus of co-planar points equidistant from a centre of which at least two have to be known in order to determine the nature of the circle).

Descartes cannot maintain that motion always occurs in an instant. This would fly in the face of everyday experience. But he does maintain that "whatever is in motion is determined, in each instant that can be designated while it moves, to continue moving towards some side in a straight line and never in a curve" (*Principles of Philosophy* Bk. II, art. 39).

The fascinating conclusion of this line of reasoning is drawn by Descartes in *The World*:

> According to this Rule, it must be said that God alone is the Author of all the motions in the world in so far as they exist and in so far as they are rectilinear, and that it is the various dispositions of matter that make them irregular and curved. Likewise, the theologians teach us that God is the Author of all our actions, insofar as they exist and insofar as they have some goodness, but it is the various dispositions of our wills that can render them evil (AT, XI, 46—47; Cottingham, I, 97).

There are a number of things that are arresting here. First, Descartes uses the theological doctrine of his age (God is the cause of everything) to explain not only *Being* but *Becoming*. Second, the *disposition* of the will is compared to the *direction* of motion. The analogy does not proceed from the natural to the supernatural world but the other way round: it is the perspicuity of Christian dogma that is used to shed light on the obscurities of projectile motion.

The theology to which Descartes is appealing was devised to save God's omnipotence without making Him responsible for sin. For instance, Thomas Aquinas affirmed that the physical reality of any action (say releasing an arrow) ultimately rests on God's creative power but that the intention (harming another human being) depends on man's will. The problem with this analogy is that it reintroduces the very notion of motion that Descartes was committed to extirpate from natural philosophy. Motion, conceived in this way, retains something of the Aristotelian definition that invokes potentiality. The *ontological status* of an act is distinguished from the *rectitude* or *crookedness* with which it is used. It is intelligible prior to its actualization, and it is only because of this prior intelligibility that it can be notionally (and hence, really, in Cartesian terms) distinguished (see Costabel 1982, p. 146). At this level of his thought, Descartes remains an Aristotelian.

THE ADVANTAGES OF A PURELY GEOMETRICAL NOTION OF MOTION

The drawbacks of Descartes' concept of motion has often been pointed out but we should not be blind to the important role it played in simplifying the analysis of motion. I shall briefly indicate five areas where Descartes felt that his clearer notion proved fruitful.

The first and most spectacular advantage of Descartes' concept of motion as naturally rectilinear is the solution it provides to the vexed problem of projectile motion. Indeed it dissolves the problem entirely by showing that it rests on false premisses. Since Antiquity, and more prominently from the sixteenth century onwards, natural philosophers had investigated the nature and cause of the motion of projectiles such as arrows and cannon balls once they are detached from the mechanism that sets them in flight. For Descartes, motion is conserved, and the right question to ask is no longer, Why does the the projectile continue to move? but, Why does it cease to move? In other words, what has to be explained is not why the motion goes on but why it stops.

From a Newtonian point of view, this is usually expressed by saying that we have to account not for motion but for change of motion. Descartes had not travelled that far since he did not consider a change of direction to be a change of motion but merely an alternative actualization of the same motion.

Second, Descartes believed that his notion solved the problem of the hardness or solidity of bodies by reducing it entirely to the greater or lesser state of rest of their smallest parts. The more the rest (i.e. the smaller their motion) the greater the force required to move them apart. In this way, Descartes did not have to invoke some kind of glue or some force between particles to account for solidity. Here again he could claim that earlier research had been misguided. Hardness needs no explanation beyond plain rest; what has to be explained is the internal agitation that renders some bodies less solid. Since direction is not part of the concept of motion, Descartes can affirm that a body whose internal parts are in motion will always be less resistant to breaking than one whose parts are completely stationary (*The World*, AT, XI, 13).

Third, Descartes' "quantized" explanation of motion, namely its reduction to a succession of divine instantaneous creative acts, enabled Descartes to account for sudden transitions in nature. Whereas Galileo had identified rest with an infinite degree of slowness and had argued that bodies that are accelerated pass through all the degrees of speed, Descartes neither saw the necessity nor the utility of Galileo's continuum. Although space and time are infinitely divisible, there is no ontological connection between the successive instants of a being's existence outside God's will. In the *Third Meditation*, Descartes makes this point about the self, but it has relevance to all created entities, including motion:

> For a lifespan can be divided into countless parts, each completely independent of the others, so that it does not follow from the fact that I existed a little while ago that I must exist now, unless there is some cause which as it were creates me afresh at this moment (AT, VII, 48—49; Cottingham, II, 33).

The radical discontinuity of time and the metaphysical gap between existential instants enabled Descartes to explain cases of collision in a way that he believed the continuists were unable to elucidate. His prime example is a large cannon ball encountering a very small stationary ball in full flight. Assuming that both balls are "extremely hard", we have an instance of inelastic collision, and Descartes claims that it is ridiculous to say that the small ball goes through all the degrees of speed until it is accelerated to the speed of the cannon ball (letter to Mersenne, 17 November 1642, AT, III, 592—593).

Fourth, Descartes' notion of motion proved useful in resolving ambiguities in statics. At the request of his friend Constantijn Huygens he wrote a short treatise on simple machines in 1637 (letter of 5 October 1637, AT, I, 435—447; for a revised version, letter to Mersenne, 13 July 1638, AT, II, 222—245). The basic ratios between force and resistance were known and there was nothing to add to what Archimedes had written. The ambiguity to be dispelled lay in the use of velocity and displacement as though these were simply interchangeable. For instance, in the case of the lever, since both ends move in identical times without acceleration, it is immaterial whether one uses the virtual velocities of the two weights or their virtual displacements. But this holds only for the lever and similar devices where a mechanical construction ensures that each body moves in the same time and where, because of equilibrium, there is no real but only virtual motion. Descartes' geometrical concept of notion enabled him to see (what escaped Galileo) that only the ratios of the displacement explain why the force and the resistance vary as they do: "it is not the difference of velocity which determines that one of these weights must be double the other, but the difference of space [i.e. displacement]" (letter to Mersenne, 12 September 1638, AT, II, 354). Consideration of speed, he adds in the same letter, would only obscure the issue, since it cannot be explained without explaining what weight really is and this would entail a knowledge of the whole system of the world AT, II, 355). He seems to imply that since speed is not a clear motion, it can be dispensed with in accounting for the working of nature.

Fifth, the law of refraction (what English textbooks call Snell's and French textbooks call Descartes' Law) is justified by Descartes on the assumption that light "is an action, or a virtue, that follows the same laws as local motion" (letter to Mersenne, 27 May 1638, AT, II, 143). A couple of months earlier, he had written to the same Mersenne: "Please note that I demonstrated the refractions [i.e. the sine law] geometrically and *a priori* in my *Optics*" (letter of 1 March 1638, AT, II, 31).

Matters were unfortunately less clear to Descartes' readers than they were to himself, and the abiding perplexity about his explanation of refraction can be seen in the conflicting interpretation of three distinguished historians of science. I. B. Sabra believes that Descartes was offering a proof of the law in the *Optics* (Sabra 1967, pp. 29—33); Gerd Buchdahl considers that it was rather a method of discovery (Buchdahl, 1969, pp. 141—142); and Bruce Eastwood sees it as a piece of rhetoric intended to convince artisans (Eastwood 1984, p. 483). Eastwood does not feel that the metaphysical underpinnings of the *Meditations* are necessary to understand the argument in the *Optics*: "Surely Descartes' sense of his audience", he writes, "was not so dull that he would have published a treatise requiring a knowledge of works that were unavailable to the reader" (*ibid.*, p. 483). Reading Descartes' mind is a notoriously difficult exercise, but Descartes frequently repeats in his correspondence that his physical explanations are incomplete without a grasp of the metaphysics they presuppose. For instance, he writes to his former teacher, Antoine Vatier, "all my opinions are so intimately connected and depend so much on one another, that one cannot be grasped without knowing them all" (letter of 22 February 1638, AT, I, 562).

In spite of declarations such as the one above, Descartes also maintained that what he wrote about refraction could be grasped without a prior knowledge of the nature of light (e.g. AT, II, 143, 197). To Vatier, he stated that the conclusions arrived at in his *Optics* and his *Meteors* could be "deduced in an orderly fashion from the first principles of my metaphysics" but that he had "wanted to see if the mere exposition of truth would be sufficient" (letter of 22 February 1638, AT, I, 563). Descartes believed that his notion of motion was not only true but intuitive and, indeed, true because intuitive. When people objected to his analysis, he did not believe that this indicated that he might be wrong but merely that they were blind.

It has often been pointed out that Descartes' concept of motion was inadequate and that the advantage that he so clearly saw were, at best, partial insights. In the remaining section of the paper, I shall briefly examine two of these shortcomings and show how they become intelligible when we remember that Descartes' geometrical or diagrammatic notion of motion appeared to him to possess the unassailable status of intuitive clarity and certainty.

The first case is the derivation of the law of refraction on the twofold assumptions that light is transmitted instantaneously through any

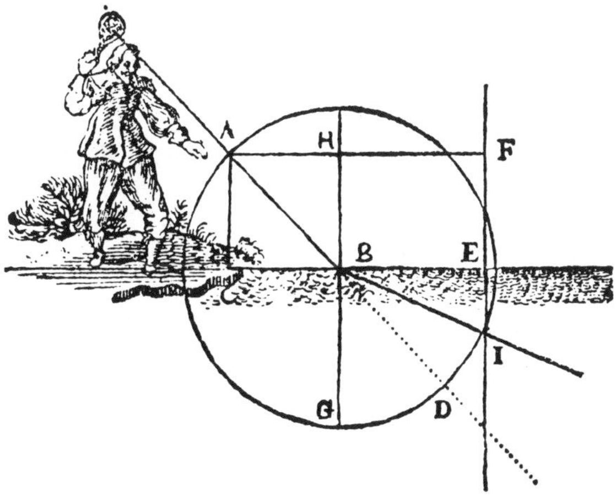

Fig. 1. (No attempt has been made at completeness, and all of the arrowed lines are merely illustrative.)

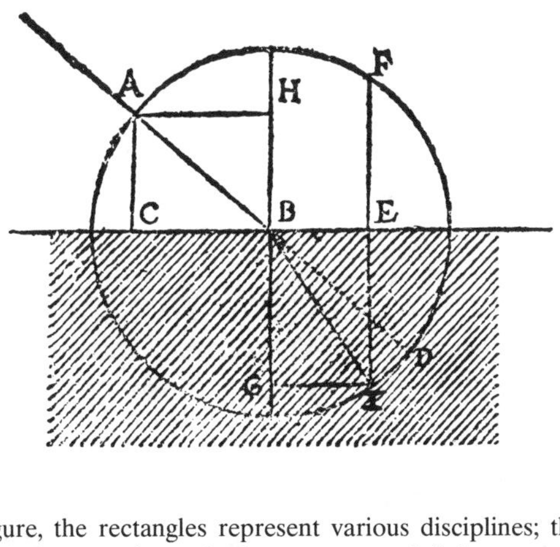

Fig. 2. (In this figure, the rectangles represent various disciplines; the thick lines the links between population genetics and other areas — such links are actually supposed to exist; and the thin lines links between the subsidiary disciplines — although such links do exist, those shown in the figure are just illustrative, they do not necessarily denote particular instances.)

medium and that the action of light follows the same laws as motion. One model that is used to illustrate the actual mechanism of the transmission of light is a tennis raquet striking a ball that moves from A to B where it hits the surface of water and is deviated towards I (see figure 1). In the case where light goes from the air at A and enters the water at B, Descartes knows from experiment that the ray will be bent along BI towards the normal (see figure 2). In order to explain this consistently with the analogy of a tennis ball, Descartes has to postulate that the ray penetrates the denser medium "with greater ease", which, as his opponents pointed out, could only mean "with greater speed", something that is precluded by his earlier assumption that light is transmitted instantaneously (*Optics*, AT, VI, 98—100). But Descartes was unmoved by the objections because he considered velocity as unessential to the intrinsic intelligibility of motion, and because the lines of motion on the diagram (i.e. the geometry of the construction) entailed the sine law of refraction.

The second case is the derivation of the laws of impact that Descartes offered in his *Principles of Philosophy.* We have already noted that motion and rest are discontinuous for Descartes and that a body starting to move does not go through all the degrees of speed as Galileo maintained. In practice, however, Descartes was faced with the fact, as he himself recognized, that "size is always opposed to speed" (*The World*, AT, XI, 51). But since matter contains no force of attraction, indeed no force of any kind, it must be wholly inert. The problem becomes acute when Descartes considers, in his *Principles of Philosophy*, what happens when a small body A strikes a larger body B at rest. It seemed logically necessary to Descartes that A cannot move B whatever A's speed. In the Latin edition of the *Principles of Philosophy*, he merely states the rule. In the French translation that appeared three years later in 1647, he explains the reasoning behind it: A cannot push B without making it go as swiftly as A itself would go afterwards, and B must resist so much the more as A comes towards it more swiftly. Thus, for example, if B is twice as large as A, and A has three degrees of speed, A can only push B if it transfers two-thirds of its speed to B. If A has 30 degrees of speed, it must transfer 20 to B; if 300, it must give 200, and so on. But since B is at rest, it resists the reception of 20 degrees ten times more than 2, and 200 degrees a hundred times more. So the greater the speed of A, the more resistance it finds in B (AT, IX—2, 90—91).

Since Descartes treats the change of motion as instantaneous, the resistance he attributed to matter had to be resistance to motion itself and not merely to change of motion. The resistance to motion that Descartes admitted cannot be reconciled with the inertness that he considered an essential property of matter. The fact that this incompatibility escaped Descartes gives us an idea of the magnitude of the conceptual change involved in the ontological identification of motion and rest. It also reveals how Descartes' allegedly clear notion of motion allowed him to side-step the whole issue.

CONCLUSION

Nature, for Descartes, "always behaves mathematically" (letter to Mersenne, 11 March 1640, AT, III, 37) and the only principles he claimed to use in understanding its behavior were those of "abstract mathematics" (*Principles of Philosophy*, Bk II, art. 64). Hence the ease with which he could use mathematical idealization and consider perfectly round balls moving along absolutely straight lines and impinging on perfectly smooth surfaces (*Optics*, AT, VI, 93–94). Whereas the entire Scholastic tradition, following Aristotle, considered this a distortion of the natural state of affairs, for Descartes it was simply a leaving aside of some merely factual details to examine the situation in the light of genuine, essential notions. Because of the ontological warrant he believed he had, abstraction was no longer an escape from the intractable irregularity of the real world into the intelligible order of Form (like Plato) but a method of grasping the essential features of the physical universe of matter. The concepts needed for the purpose are not drawn from everyday experience, however. As Descartes put it to Gassendi: "we could not recognize the geometrical triangle from the diagram on the paper unless our mind already possessed the idea of it from some other source (*Fifth Set of Replies*, *Meditations*, AT, VII, 262; Cottingham, II, 262). The key concepts are innate and present to all minds not blinded by passion or blighted by ignorance and prejudice. They are the ones for which Euclid, Archimedes and Apollonius provided a definitive grammar. It did not enter Descartes' mind that there might be more than a hyperbolical, rhetorical reason for doubting the applicability of geometry to the world of senses. The possibility of a void was ruled out of court because the very notion of mathematical extension (i.e. imagined space) entailed its physical reality. He knew that the Book of Nature was written in the language of mathematics and he never doubted that he had grasped its syntax and grammar. He spoke with the authoritative voice of clarity and simplicity, and it may well have been the tone of voice that was required to start a revolution in science.

McGill University
Montreal, Quebec
Canada

BIBLIOGRAPHY

Buchdahl, Gerd. (1969). *Metaphysics and the Philosophy of Science.*

Costabel, Pierre. (1982). *Démarches originales de Descartes Savant.* Paris: Vrin.

Cottingham, J. *et al.* (1984). *The Philosophical Writings of Descartes.* Cambridge: Cambridge University Press, 2 vols.

Descartes, René. (1897—1913, revised 1967—1974). *Oeuvres* (edited by C. Adam and P. Tannery). Paris: Vrin, 11 vols.

Eastwood, Bruce Stanfield. (1984). "Descartes on Refraction", *Isis* **75**, 481—502.

Galilei Galileo. (1890—1909). *Opere* (edited by A. Favaro). Florence: G. Barbèra, 20 vols.

Leibniz, G. W. (1960—1965, reprint of 1875—1890 edition). *Die Philosophische Schriften.* Hildesheim: Georg Olms.

Mersenne, Marin. (1933—1986). *Correspondence* (edited by C. de Waard *et alii*). Paris: Vrin, 19 vols. to date.

Sabra, A. I. (1967). *Theories of Light from Descartes to Newton.* London: Oldbourne.

Shea, William R. (1978). "Descartes as a Critic of Galileo", in R. E. Butts and J. C. Pitt eds.), *New Perspectives on Galileo.* Dordrecht: D. Reidel.

Shea, William R. (1986). "Hegel's Celestial Mechanics", in R. P. Hortsmann and M. J. Petry (eds.), *Hegels Philosophie der Natur.* Klett-Cotta.

Westfall, S. R. (1971). *Force in Newton's Physics.* London: Macdonald.

MARGARET MORRISON

HYPOTHESES AND CERTAINTY IN CARTESIAN SCIENCE[1]

1. INTRODUCTION

In his early methodological and scientific writings [the *Discourse* and *Regulae* as well as the *Optics* and *Meterology*] Descartes frequently claims that his goal is to produce a science deducible from first principles; one that would provide a level of certainty hitherto unattained. Although Descartes fails to provide in these texts the ultimate principles from which to deduce his physical explanations, he nevertheless promises to provide a complete account of his system in a later work. This is the task he set for himself in the *Principles of Philosophy*.

Cartesian scholars have argued that the *Principles* falls short of its goal of presenting a system that could provide certainty in the realm of scientific explanations.[2] Although Descartes presents first principles of physics and metaphysics that are supposedly certain he also includes a number of particular explanations that seem to be merely hypothetical or morally certain. In this paper I argue that the traditional distinction, suggested by the structure of the *Principles*, between metaphysical and moral certainty is insufficient for categorizing the various kinds of principles and hypotheses that Descartes uses. In particular there seems to be a third category — *a posteriori* certainty — consisting of claims that are argued for empirically and yet are intimately related to the higher order principles. Some specific examples of *a posteriori* certainty include the claim that light is an action and that the earth revolves.[3]

Given this analysis of certainty I conclude that one need not interpret Descartes as having failed to complete his deductive programme, or as having settled for something less than he set out to do.[4] I argue that the goal of a deductive science was compromised only to the extent that Descartes was unable to perform all the relevant experiments necessary to facilitate the deductions — not that the deductions could not *in principle* be made. And, it is in exactly the cases where Descartes feels he has provided the requisite deductions that he claims to have attained the kind of *a posteriori* certainty mentioned above. Instances where he has been unable to accomplish this comprise the domain of moral certainty.[5]

J. R. Brown and J. Mittelstrass (eds.), An Intimate Relation, 43—64.

The argument proceeds in three stages. First I discuss the importance of experience within the Cartesian programme, emphasizing its role in establishing judgments that Descartes considers metaphysically certain. Following this I examine the structure of the *Principles* particularly the relationship between the metaphysical and physical principles as well as the special interpretation of deduction that is used by Descartes.[6] With this methodological framework in place the thesis about *a posteriori* certainty is formulated. I show how it functions within Cartesian science and how it solves some of the interpretive difficulties that arise when attempting to consider Descartes' methodological and scientific writings as comprising a unified work.

2. THE ROLE OF EXPERIENCE

The arguments put forth in the *Meditations* as well as in the other works on methodology are designed to show that reason is the most reliable and critical faculty for evaluating evidence, including empirical evidence. However, once doubts about the reliability of our cognitive faculties are laid to rest Descartes willingly concedes the value and reliability of empirical evidence for establishing the metaphysical certainty of scientific claims, particularly those addressed in part two of the *Principles*. Further evidence of this is found in the *Discourse* VI where Descartes acknowledges the importance of both experiments and the kind of evidence that presents itself spontaneously to the senses (H&R, pp. 118—130). Of the many instances where Descartes appeals to the authority of experience, the most notable is the case of the mind/body dichotomy. In conversations with Burman he remarks:

> How can the soul be affected by the body and vice versa, when their natures are completely different?
>
> (R) This is very difficult to explain; but in this matter our experience is sufficient since it is so clear on this point that it cannot be denied in any way; this is evident in the case of feelings, etc. (Cottingham, p. 28).

Similarly in a reply to Elizabeth on the same point he claims that those things which pertain to the union of soul and body are known only obscurely by the understanding alone or even by the intellect in conjunction with the imagination, but they can be "known *very clearly* by the senses" (A&T, III, pp. 691—2; Fourth Replies to Objections (italics added)).

At first it seems reasonable to think that these remarks about experience are simply *ad hoc*; how can we reconcile trust in the empirical with the repeated insistence throughout the *Meditations* that reason is more reliable than the senses? But, it is possible to render these claims systematic by taking account of several subtleties in Descartes' discussion of experience.[7]

First, Descartes distinguishes between ordinary experience and experiments performed for the purposes of theory confirmation. Ordinary experience is limited to more obvious and universal phenomena which are given prior to any interference on our part with the natural occurrence of events. This contrasted with observations made in the context of theory testing. Because the latter often require very special circumstances they are often unreliable or simply unavailable.[8] When Descartes speaks of his relative distrust of experiments, and especially in his correspondence with Mersenne, it is clear that it is not the *empirical* character of the experiments that is the source of skepticism, but rather that their relative complexity makes them more liable to errors and misinterpretations than the common experience of unscientific observers. An experiment may be poorly executed — people tend to get the result they expect; it may not be repeated often enough to establish the results with certainty or, the correct result may be incorrectly interpreted.[9]

The link between experience and reason comes in the replies to the sixth set of objections:

> When we say that certitude of the intellect is much greater than certitude of the senses, that simply means that those judgments which we make in mature years as a result of some new evidence are more certain than those we make in our infancy, without any critical reflection (A&T VII, p. 438).

The crucial issue here is critical reflection, experience tempered by reason. Descartes explains, for example, that one could not possibly correct an erroneous judgment based on the visual perception of an apparently bent stick in water by reference to tactile evidence alone. One must have grounds for preferring the tactile evidence to visual evidence and this can be provided only by reason. In the final article of part I of the *Principles* we find the same contrast between senses and reason. Here Descartes claims that true philosophers should "put their trust less in the senses, that is in the ill-considered judgments of infancy than in their mature reason"; that it "does not become a philosopher to assent to things other than those which have been perceived".[10] We must be careful here not to confuse perception, as Descartes uses the term, with sense-perception. The former involves what he calls the "operation of the intellect" while the latter is thought to be only one of the diverse modes of perceiving (P I, Art. 32). Again the point is that critical reflection on experience or sensation performed by reason is always to be preferred to uncritical judgments based on observational evidence alone.

The critical function of reason is also made clear in Descartes claim that we know with certainty that material objects exist (P II, Art. 1). Although the arguments involves *a posteriori* methods, ultimately it depends on the principle that God is not a deceiver. Our senses aided by reason lead us to clearly and distinctly perceive extended matter. Furthermore, if God were responsible for presenting us with these ideas of extended matter then it would be impossible to devise any reason for not thinking him to be a deceiver. So, because we understand matter to be distinct from God and our minds, and because our reason tells us that the idea of matter comes from that which perfectly represents it, we have no further way of judging the veracity of our claims. Hence, God would be a deceiver if they were untrue.

There are of course other principles that are true *a priori*, for example, the axioms found in *Principles*, I, Art.49 (nothing is made from nothing, the principle of non-contradiction) and other metaphysical truths like the fact that a 'supremely good God exists'.[11] My point has simply been to show that metaphysical certainty is not limited to that which is the product of reason alone, there are also instances where experiential knowledge plays a role in attaining perfect certitude. Although reason is necessary to establish metaphysical certainty it is not sufficient in all cases.

Having seen that some physical principles exemplifying perfect certainty are established through the process of critical reflection on experience, the question arises whether it is possible to attain a similar kind of certainty at the level of individual explanations. To answer this question we need to first examine the relationship between the metaphysical and physical principles and their connection with hypotheses about particulars. The ways in which these different levels interact provide us with insights into Descartes understanding of the nature of deduction and its role in his architectonic.

3. METAPHYSICS—PHYSICS—PARTICULARS: THE NATURE OF DEDUCTION

For Descartes the metaphysics or first principles of knowledge given in part I of the *Principles* are logically prior to all others. They are known through the natural light of reason and are *a priori* insofar as they result from critical reflection on incorrigible data. By contrast the physical principles presented in part II involve *a posteriori* methods to the extent that they depend on judgments about the nature of objects and processes external to the self. The critical reflection of reason continues to play the major role in these judgments but more is involved than what can be concluded *a priori* on the basis of individual sensations and thought processes. In the way that the metaphysics and epistemology of part one of the *Principles* establishes the possibility of all knowledge, the mechanical philosophy, explicated by the physical principles of part two, furnishes the tools for the very possibility of *scientific* investigation. It provides an explanatory framework for the explanation of particular phenomena. It is in descending to the level of particulars that the distinction between metaphysical and moral certainty becomes evident. Unlike parts I and II of the *Principles* the contents of parts III and IV are not entitled 'principles'; their legitimacy comes from their relationship[12] to the higher order metaphysical and physical principles. As a result they are considered to be morally certain; that is, "certain to a degree which suffices for everyday life" (P IV, Art. 205). And although these hypothetical explanations provide descriptions of many natural phenomena, they may in fact be false.

Descartes often speaks of deducing the physical principles from the metaphysical ones but from an examination of the relationship between the two it becomes evident that the kind of deduction Descartes has in mind is not the kind of deductive structure exemplified by syllogistic reasoning.

It is interesting to note that the seventeenth century French use of the word deduction is associated with the notion of a narration or detailed enumeration. Deduction originates from the word "déduire" which refers not only to the verb deduce but also to the process of "drawing something out" or to "discoursing about something".

Descartes also uses the word "démontrer" interchangeably with "deduire" and uses both to refer to two kinds of demonstration; one equivalent to providing an explanation and the other referring to what we would normally call the confirmation of an hypothesis, or a proof.[13] In the former case we reason from known effects to some hypothetical cause, while in the latter we hypothesize a cause and reason toward the effects to be explained. The important point is that although both involve what Descartes calls deduction, deduction proper provides a step by step enumeration of the details whereas a demonstration need not. In the *Regulae* (X, H&R p. 32 and XI) deduction is characterized as the drawing of a conclusion from a large number of premises in such a way that the mind has to take all of them into account successively in order to reach the conclusion.[14] In other words, they must all be connected by acts of intuition which are in turn defined as the "indubitable conception of the unclouded and attentive mind which arises from the light of reason alone" (Regulae III, H&R p. 7). This process is illustrated in two different places, the first is in the *Meditations* where through an enumerative process Descartes arrives at the conclusion that God exists,[15] and in rule XII where Descartes discusses the appoach to discovering the nature of magnetism. In the latter instance we first collect all the experimental facts that have been discovered about the magnet. From this enumeration we then deduce, using the physical and metaphysical principles, which of the simple natures are adequate to explain the experimental findings; thereby eliminating other possible explanations. Having chosen an hypothesis we can claim to know the nature of magnetism *given* what is available. However, because we must rely on experimental evidence that may be incomplete we can only claim moral certainty for the hypothesis; unlike cases of metaphysical certainty where a complete enumeration of relevant alternatives can be provided and eliminated. Both these explanations or proofs are termed deductions despite the fact that they make use of a process more commonly known to us as induction.

That Descartes could not have associated deduction with syllogistic reasoning is further evident from his remark in part II of the *Discourse* (H&R, p. 91) that logic does not help one discover anything new but only provides a way of saying what must be already implicit in one's premises. And again the *Regulae* (H&R, p. 31—32) he claims that the forms of the syllogism do not help us in any way to perceive the truth about things.

Given this interpretation the claim that Descartes deduced the physical principles from the metaphysics seems relatively unproblematic. Using the same reasoning process as in the argument proving the existence of God we move from the statement of God's veracity and immutability to the guarantee that the physical principles are true, that material objects exist, and that God conserves motion and its laws (cf. n. 16 below). Hence the general principles of the mechanical philosophy are secured. But, can we apply the same method when descending to the level of particulars? The discussion of magnetism suggested that we could not, however, there may be other contexts where the method can successfully be applied.

Descartes claims in the *Regulae* that the two ways to certainty are experience and deduction; we can move only so far in the deductive schema before we must call into play experiental considerations that facilitate further steps in the demonstration. Hence the difficulty of achieving perfect certainty seems to be one only in practice; we could in principle achieve certainty even in the domain of particulars if we had at our disposal the means for eliminating all the incorrect hypotheses that experience could present us with. However, this is only a theoretical possibility and not something we are likely to achieve in ordinary scientific inquiry. Since neither experience nor deduction can lead us to metaphysical certainty outside the realm of physical principles how should we understand Descartes' goal of a deductive science? Is moral certainty the most that can be claimed for physical hypotheses and perhaps more importantly, is Descartes himself willing to rest content with this outcome? In what follows I argue that there is a certain class of physical hypotheses that are more than morally certain and it is these hypotheses that Descartes thinks exemplify certainty and the goal of science as defined in the domain of particulars.

4. A POSTERIORI CERTAINTY

In a letter to Vatier in 1638 (K, p. 45) Descartes claims that his aim in the *Discourse* (1637) is not to expound his method nor illustrate its application. Indeed the whole of his method is not contained in the work, instead it is to provide only enough information to render the *Dioptrics* and *Meteors* intelligible. Furthermore, Descartes says that he could not demonstrate the use of his method in these three treatises because it prescribed an order of research quite different from the one he thought proper for exposition:

not having here any other occasion to speak of light than to explain how its rays enter into the eye, and how they can be deflected by the different bodies that they encounter,

I need not undertake to explain its true nature. And I believe that it will suffice that I make use of two or three comparisons which will help to conceive it in the manner which to me seems the most convenient to explain all those of its properties that experience acquaints us with, and to deduce afterwards all the others which cannot be so easily observed; imitating in this the Astronomers, who, although their assumptions are almost all false or uncertain, nevertheless, because these assumptions refer to different observations which they have made, never cease to draw many very true and well assured conclusions from them (*Dioptrics*, pp. 66—71).

The one major difference between the suppositions Descartes begins with and those of the astronomers is that his are founded on principles which are ultimately deducible *a priori* from metaphysical truths. Descartes goes on to claim that his suppositions are termed hypotheses so that:

it may be known that while I consider myself able to deduce them from the primary truths which I have explained above, yet I particularly desired not to do so, in order that certain persons may not for this reason take occasion to build up some extravagant philosophic system on what they take to be my principles and thus cause the blame to be put on me (*Discourse*, H&R, p. 129).

His intention is to give some results of the new method without expressly stating the principles from which these results are obtained. Although we are invited to entertain the suppositions of *Dioptrics* and *Meteors* as hypotheses devised to explain the phenomena it would seem that a great deal more is at stake here. For, if these suppositions are deduced from primary truths they must also be certain, and it seems as though this is the favoured interpretation:

For it appears to me that the reasonings are so mutually interwoven, that as the later ones are demonstrated by the earlier, which are their causes, the earlier are reciprocally demonstrated by the later which are their effects. And it must not be imagined that in this I commit the fallacy which logicians call arguing in a circle, for, since experince renders the greater part of these effects very certain, the causes from which I deduce them do not so much serve to prove their existence as to explain them; on the other hand, the causes are (proved) by the effects (*Discourse* VI, A&T p. 76).

However, because of the rather loose sense in which Descartes understands deduction there is no reason to think that these hypotheses are metaphysically certain, even though they follow from first principles. As we shall see below, it is at this juncture that *a posteriori* certainty becomes important.

The suppositions referred to above are: (1) light is a certain power or action that follows the same laws as local motion — a consequence of the metaphysical principle that motion is the only power that can be rationally (clearly and distinctly) asserted to exist in nature (cf. *Dioptrics*); (2) that bodies such as air, water, and earth are composed of many small parts of various shapes and magnitudes (cf. *Meteors*). This latter claim of course is intimately connected with the divisibility thesis and the fact that we should consider only geometrical properties such as shape and magnitude. Another fundamental assumption is that the gaps between the small parts of bodies are filled with a very subtle and imperceptible matter — a supposition that is required if one is to deny the existence of a void. In fact, this particular claim has further implications for Descartes' hypothesis concerning the instantaneous propagation of light, which follows directly from his conception of the medium through which light is transmitted. All of these hypotheses (except for the divisibility thesis) also occur in parts three and four of the *Principles*, the parts supposedly consisting solely of morally certain explanations of particulars. However, further examination reveals that these suppositions occupy a very important place within the Cartesian hierarchy. They are not part of the principles of material objects, but nevertheless appear to be more than morally certain — thereby presenting themselves as candidates for *a posteriori* certainty.

Evidence for this interpretation of the texts is found particularly in Descartes' correspondence. In a letter to Morin (1638) Descartes remarks in defence of his method that it is not so easy to adjust one and the same cause to several different effects, unless it be the true one:

> indeed there are often effects such that one has sufficiently proved what their true cause is, if one has assigned to them one cause from which they can be clearly deduced; and I claim that those of which I have spoken in the *Dioptrics* belong to this category (A&T II, p. 199).

Here Descartes is speaking of the claim that light is a certain movement or action which passes toward our eyes through the medium of the air (and other transparent bodies) in the same manner that the movement or resistance of bodies that a blind man encounters is transmitted to his hand through his cane. This is the one cause by which Descartes wants to explain all the other effects of light. Although this hypothesis about light is not a metaphysically certain principle, by providing the basis for other deductions it achieves a status sufficient to classify it as a true cause. In the same letter to Morin Descartes claims that proving the cause by the effects constitutes what he calls *a posteriori proof* — a sufficient demonstration given the subject matter.

Although the subject matter imposes certain limitations on physical proofs, these constraints do not relegate *all* scientific inquiry to the realm of moral certainty; sufficiency given the subject matter' is to be understood differently in contexts that involve a posteriori certainty. In the case of morally certain hypotheses about particulars we need to recognize that there exists a variety of possible ways in which the phenomena could be constructed. Accumulating and eliminating the relevant alternatives would require access to all the possible empirical data. Because we are constrained by the availability of evidence at a particular time we can never be sure that we have considered all the possibilities. Although these hypotheses may in fact be false Descartes thinks he has achieved enough if the things he has written about correspond to the phenomena (P IV, Art. 204). In the case of metaphysical certainty we can enumerate the possible alternatives and using reason and experience rule out all but the seemingly correct one. Provided we apply the method of analysis and synthesis correctly and because of the limited number of possibilities, we can conclude, from the fact that God is not a deceiver, that our hypothesis is certain.[16] In contrast to moral certainty, *a posteriori* certainty applies to hypotheses which, because of their *generality*, can be considered to have a more circumscribed range of competitors that can be ruled out on the basis of experiment. These latter hypotheses are not metaphysically certain (we may not have considered *all* the relevant evidence) but on the basis of what we know now and what we are likely to know in the future it is very unlikely that they are false. Although the hypotheses discussed above concerning the nature of light, existence of subtle matter and so on involve claims about particulars they nevertheless provide a general framework, in addition to the physical principles, for the explanations of several other disparate phenomena. Their generality allows us to be more certain that we have eliminated alternatives and this together with their deductive power provides evidence for their certainty. Of course, some of the effects can be known independently of the supposed cause, as in the case of the equality of the angles of incidence and reflection; however, if these effects can *also* be deduced from the cause then so much the better.

One may object of course that this is nothing more than an application of the hypothetico-deductive method and that Descartes was quite willing to countenance the use of false hypotheses in deriving explanations. Although Descartes sometimes writes as though this were the case there is no evidence that he held this view about the claims I want to call *a posteriori* certain. In a letter to Mersenne (May 17, 1638) it is clear that Descartes wants to accord the general hypothesis about light a higher epistemic status than moral certainty:

> Now what I claim to have demonstrated about refraction does not depend on the truth of the nature of light, nor on whether its propagation is instantaneous or not but only on my hypothesis that light is an action or power, which in its propagation from place to place follows the same laws as local motion and is transmitted by means of an extremely rarefied liquid in the orifices of transparent bodies (K p. 56).[17]

The truth of this hypothesis in turn confers a degree of certainty on his conclusions about refraction. As Descartes remarks in a subsequent letter to Mersenne (Feb. 9, 1639)

> I do indeed wish that one should consider that, if what I have written on . . . refraction . . . turns out to be false, then the rest of my philosophy is worthless (A&T II, p. 501)

Although the law of refraction did not directly depend on the hypothesis of instantaneous propagation Descartes nevertheless considered the latter more than morally certain. In a letter to Beekman (Aug. 22, 1634) Descartes claims that this hypothesis is *so* certain that proof to the contrary would undermine his whole enterprise; "if someone could show it were false, I would be prepared to agree that I know nothing at all in philosophy" (A&T I, p. 307).[8] Moreover, this was not simply a rhetorical challenge; Descartes agrees to test his hypothesis and proposes a method that is considered more favourable to those who defend the finite transmission theory. Descartes suspects that the speed of light is much greater than could be measured over a short distance and so he suggests measuring the speed when light travels between the moon and the earth during a lunar eclipse. Because the instruments available were unable to measure this interval of time Descartes considers himself vindicated and remarks that his experiment "which is the one accepted by all astronomers, shows very clearly that light is not seen in a perceptible interval of time. Hence I consider this argument a demonstration" (A&T I, p. 310).[19]

A similar line of argument is put forth regarding the general assumptions that introduce the meterology. In a letter to Vatier Descartes claims that:

> what I had assumed at the beginning of my meterology I could not demonstrate a priori without giving the whole of my physics but the experiences which I deduced necessarily from it — which cannot be deduced in the same way from other principles — seem to me to demonstrate it adequately *a posteriori* (K, p. 48).

Like 'deduction', '*a priori*' also had a rather different use for Descartes than the one we normally attribute to it. Clarke (1982, pp. 109—112) points out that Descartes' use of the term a priori as closer to what we would call as "causal" rather than "independent of experience". For example, Descartes refers to his account of optical reflection and refraction in *Dioptrics* as "*a priori*" yet both explanations appeal to empirical claims about the motions of light particles and their interactions with various kinds of media or surfaces. One reason for understanding these as causal accounts rather than as independent of empirical evidence is Descartes' admission that the general claims about matter and light at the beginning of *Dioptrics* are hypotheses. Furthermore Descartes uses the term in context where empirical evidence is used as the only source of warrant for an *a priori* claim; for example, in the construction of an a priori explanation for all terrestial bodies Descartes claims that although the fixed stars

> seem to be very irregularly scattered here and there in the heavens, nevertheless I do not doubt that there is a natural order among them which is regular and determinate. And the knowledge of this order is the key and foundation of the highest . . . science . . . For by means of this science one could know priori all the diverse forms and essences of terrestial bodies whereas without it we have to content ourselves with guessing them a posteriori, and through their effects. I cannot think of anything which would help me more in coming to know this order than the observations of many comets (A&T, I, pp. 250—1).

He goes on to urge that one should, in accordance with the Baconian method and without introducing any "explanations or hypotheses" (*Ibid.*), compile an inventory of everything known about the heavens. And, having collected empirical data and formed a theory about it one can then provide *a priori* explanations about terrestial bodies.

In addition to the remarks on light mentioned above Descartes' hypotheses about the earth's revolution also seem to fall into the category of a posteriori certainty. In a letter to Mersenne (Nov. 1633) Descartes writes:

I admit that if it is false [that the earth revolves], then so are all the foundations of my philosophy because it can be clearly demonstrated from them. And it is so closely connected with all the other parts of my treatise, that I could not detach it from them without making the rest of it defective (A&T I, pp. 270—1).

Although Descartes places a great deal of emphasis on the certainty of these hypotheses he nonetheless reserves the right to argue for their truth on purely empirical grounds. In a letter to Descartes (Feb. 22, 1638) Morin points out that experience alone is unable to establish the truth of an assertion about which of several causes is the true one. In the case of apparent celestial movements one could derive the same conclusion from either of two suppositions, assuming the earth's stability or its motion. In reply, (July 13, 1638, K p. 57) Descartes emphasises that his suppositions are obtainable from higher principles which he has not yet divulged, but he also defends the truth of these suppositions on the basis of empirical evidence alone. It seems clear that Descartes wants his readers to judge from his results that he "must have some knowledge of the *general* causes on which they depend" and that he could not have discovered them otherwise (A&T II, p. 201).[20] It is in this same letter that Descartes offers a defense of the method he used in establishing his light hypothesis.[21]

Before offering a concrete example of the relationship between *a posteriori* certainty and empirical confirmation let me summarize my argument thus far. From the examples I have mentioned it seems that we can isolate a group of general hypotheses (that the earth revolves, that light is an action and that subtle matter exists) that do not occur in parts one and two of the *Principles* but play an intermediary role between the metaphysically certain principles of physics and metaphysics, and the morally certain hypotheses about particulars (for instance, that the nature of magnetism depends on screw-shaped particles). Although Descartes provides only *a posteriori* proof of these hypotheses he nevertheless claims on several occasions that if they are false his entire programme is in vain. These hypotheses are *constrained* by higher order physical principles but are argued for empirically and considered by Descartes to be conclusively demonstrated given the subject matter. In a letter to Mersenne (May 27, 1638) Descartes writes:

You ask whether I consider what I have written about refraction (in the Dioptrics) to be a demonstration; and I believe it is, at least in so far as it is possible to give a demonstration in this matter without having previously demonstrated the principles of physics and metaphysics (which I hope to do one day, but which hitherto has not been done), and in so far as any question of mechanics, or of optics, or of astronomy, or any other matter that is not purely demonstrated. But to require of me geometrical demonstrations in a matter which depends on physics is to demand that I achieve impossible things .... For I could demonstrate that the definition of the center of gravity, which Archimedes has given, is false, and that such a center does not exist .... As for Ptolemy and Vitellion, they have suppositions that are much less certain, and nevertheless one should not for this reason reject demonstrations they have deduced from them (A&T II, pp. 141—2).

Although Descartes claims that geometrical demonstrations are impossible in the realm of physics he nevertheless claims in the *Principles* [II, Art. 64] that there are certain common notions from which we can derive claims about substance that are so certain that they must be "*considered* as a mathematical demonstration" (italics added). These are the physical "principles". In the domain of physics proper, where hypotheses and explanations about particular phenomena are given, metaphysical certainty is unavailable. It would seem that the best we can do is to isolate, within this group, the candidates for *a posteriori* certainty and show how they are deducible from first principles. Because of his rather loosely construed sense of deduction Descartes is able to claim success in this task; a feat that renders his assumptions more certain than those of his predecessors. Hence based on the demands of the physical and metaphysical principles the general hypotheses about light, subtle matter, etc. could not be otherwise.

Because of the nature of *a posteriori* certainty one is tempted to see the difference between it and moral certainty as one of degree rather than kind. However, although these general hypotheses are not "principles" they have a status that is inconsistent with claims made by Descartes on behalf of the morally certain hypotheses. The latter may be false (P IV, Art. 204) while the former could not be otherwise (Art. 206). In light of this is more reasonable to see *a posteriori* certainty as differing from metaphysical certainty in degree and moral certainty in kind. So, although the empirical nature of the subject matter imposes some limitations on physical proofs, these constraints do not relegate all of scientific inquiry to the realm of moral certainty.

Adherence to the goal of a deductive science is evident from Descartes' insistence throughout his writings and correspondence on the interdependence of his first principles and specific applications of them. If moral certainty was the goal of science then the role of the physical and metaphysical principles could be drastically downplayed.[22]

Finally, it remains to show *how* these general hypotheses are empirically confirmed; how their *a posteriori* certainty is established. In order to do this I shall discuss Descartes' account of the nature of light, a classic example of the combination of a priori and a posteriori elements in Cartesian methodology.

5. EMPIRICAL CONFIRMATION: A POSTERIORI CERTAINTY IN CONTEXT

In much of Descartes' correspondence concerning his theory of light a distinction is made between material and formal light rays. The latter are composed of mathematical straight lines whereas material rays are rarely, if ever, mathematically straight. Although particles of the transmitting medium are not arranged in rows that are mathematically straight, the geometric path of light (the pressure of the movement of the second element) is along lines that are perfectly straight. Descartes draws an analogy between the subtle matter of this transmitting medium and water flowing in a river. Although the water is always pushed by the same force and moves in the same way, it flows in straight lines in some places and turns in others. The same is true of light rays. We think of them as always moving in the same way, traveling in mathematically straight lines even though their actual movement through the pores of bodies deviates from a strictly linear path.[23] The difficulty with this analogy is that light is not in fact the actual motion that Descartes describes, but rather, a tendency. Hence, the action Descartes calls light consists of an instantaneously transmitted *tendency* to move but is thought, at the same time, to obey the laws of local motion. These notions of tendency and rectilinear motion of light have their foundations in Descartes' third law of motion which states that when a body is set in motion it always tends to move in a straight line even though it may be forced to move otherwise. It is this tension between the material and mathematical or formal aspects of explanation that prompts descartes to abandon abstract geometry for a more concrete kind of explanatory framework, the aim of which is to explain phenomena in a more natural way (Letter to Mersenne, July 27, 1638, A&T II, p. 198).

Descartes experimental confirmation for his light hypothesis takes the form of an argument by analogy. He ues a variety of different models to represent different aspects of light; for example, a blind person's perception of physical objects using a walking stick is compared to our perception of light while the linear transmission of light is compared with wine leaking from a vat. Because the *true* nature light is describable using certain formal principles based on the laws of motion, Descartes feels the need to concentrate on the *material* aspect for which complete and (metaphysically) certain proofs could not be given. We call upon experience to provide us with the proper analogies that depict the phenomena to be explained. These analogies, such as the tennis ball model used in the discussion of refraction in *Dioptrics*, use actual motion as the model for light because in order to argue empirically for a particular hypothesis we must make use of actual instances rather than the abstract notion of a tendency. Once we explain particular properties of light such as refraction on the basis of the hypothesis that light is an action, and explain its transmission by assuming instantaneous propagation, we have provided the required *empirical* evidence for these two hypotheses that are in turn deducible (in the sense discussed above) from higher level physical principles.

Descartes defends the analogical method despite the differences between models and the reality they are designed to explain. It is through the use of this method that we can "submit our explanations to the test of the senses, just as I always attempted to do" (A&T II, p. 366). In fact, he places so much reliance on this method that he remarks:

> I claim that they [models and analogies] are the most appropriate way available to the human mind for explaining the truth about questions in physics; to such an extent that, if one assumes something about nature which cannot be explained by any analogy, I think that I have conclusively shown that it is false (To Morin, Sept. 12, 1638, A&T II, 368).

Despite the advantages Descartes is careful to distinguish the physical-mathematical, from the purely heuristic aspects of the method. The model of light used in *Le Monde* and the *Principles* is the physical-mathematical model. It is used to analyse the transmission of mechanical force by means of the collision of bodies moving in a straight line, a situation that could not actually occur in his universe of vorticies. Its heuristic counterpart is used in *Dioptrics* to derive the laws of reflection and refraction. Here the model of instantaneous transmission of a light impulse is conceived in terms of a moving ball colliding with an obstruction. Again the difficulty with this model is that light is not an actual motion. Both models are related in such a way that in the heuristic model the velocity of the ball, representing *force* of impulse, corresponds to constant *velocity* in the physical-mathematical model.[24] Force is resolvable into magnitude, direction and directional components while velocity is resolvable into speed, determination and directional components. The media of transmission in the heuristic model are replaced by the empty space of the ideal geometric motion in the physical-mathematical model.

Even in the case of the heuristic model Descartes is concerned that his account be mathematizable. He writes to Mersenne (Oct. 28, 1640) that

> It is for this reason that I assumed in my Dioptrics that the surface and the ball are perfectly hard, and that the ball has neither weight, nor size, etc. in order to make my demonstration mathematical (i.e. exact) for I am well aware that the reflection of an ordinary ball never occurs at exactly equal angles (A&T III, p. 208).

This emphasis on mathematization is in many ways the cornerstone of Descartes' method; the ability to set hypotheses in a mathematical framework provides them with increased confirmation and certainty.

In a letter to Plempius (Dec. 20, 1637) (cf. n. 21) Descartes remarks that his comments on the nature of light in *Dioptrics* are not "principles" but conclusions proved by everything that comes later. He goes on to point out that the only principles from which he derives these conclusions are the axioms on which geometers base their demonstrations. The difference is that in Descartes' case these axioms are not abstracted from sensible matter, as they are by the geometers, rather they are "applied to various experiences which are known by means of the senses and are indubitable" (A&T I, p. 476). By abandoning abstract geometry one can apply mathematical methods to the study of nature, thereby achieving certainty through both empirical confirmation and mathematical demonstration.

An obvious objection to my account is that a great deal of the evidence for it is drawn from Descartes' correspondence and from scattered remarks in *Dioptrics*, *Meterology* and the *Discourse*, rather than from specific claims made in the *Principles*. There is however a reason for this. It has to do with the fact that Descartes has very little to say about his methodology in the *Principles* despite the fact that Descartes had promised in it an exposition of his method in its complete form. Thus we have no choice but to attempt a reconstruction of that method from his remarks made in the earlier and incomplete texts and correspondence. It seems likely that Descartes felt that the synthetic unity he intended to display in the *Principles* would speak for itself; that once people saw the entire system of physical and metaphysical principles together with explanations of particular phenomena, in one coherent whole, the method and its advantages would become obvious.

In addition, the evidence cited in the correspondence and elsewhere about the nature of certainty provides a context within which we can understand Descartes' remark in Art. 206 of the *Principles*. Having stated that his physical explanations are at least morally certain he claims that it is scarcely possible that some of the *general* hypotheses about the world and the earth could be understood in a manner other than as he has explained them. For example, as mentioned above instantaneous propagation was a consequence of Descartes' account of the medium through which light was transmitted. This, in turn, was determined by his theory of matter. Given the metaphysical certainty of the general principles in parts one and two, their relationship to these empirically confirmed hypotheses about light, subtle matter, and the earth's revolution, together with the fact that these hypotheses can be given a mathematical demonstration, one can hardly understand their place within the Cartesian hierarchy as examples of moral certainty:

> Besides, there are, even among natural things, some which we judge to be absolutely and more than morally certain; basing our judgement on the Metaphysical foundation that God is supremely good and by no means deceitful, and that, accordingly, the faculty which He gave us to distinguish the true from the false cannot err when we use it correctly and perceive something clearly with its help. Such are Mathematical demonstrations; such is the knowledge that material things exist; and such are all evident demonstrations which are made concerning material things. These reasonings of ours will perhaps be included among the number of these absolutely certain things by those who consider how they have been deduced in a continuous series from the first and simplest principles of human knowledge. Especially if they sufficiently understand that we can feel no external object unless some local movement is excited by them in our nerves . . . for once these things have been accepted, it will scarcely seem possible for all the rest, at least the *more general things which I have written about the World and the Earth*, to be understood otherwise than as I have explained them (italics added).[25]

These general things are the hypotheses that are *a posteriori* certain.

6. CONCLUSIONS

I have argued that the distinction between metaphysical and moral certainty is insufficient to categorize the hypotheses presented in Descartes' scientific writings. An intermediate category, what I have termed "*a posteriori* certainty", is required. Given their relationship to the physical and metaphysical principles the hypotheses that belong to this category are considered certain yet they are argued for empirically. In a letter to Mersenne (Mar. 11, 1640) Descartes writes:

> As far as physics is concerned, I believed that I knew nothing at all if I could only say how things might be, without being able to prove that they could not be otherwise (A&T III, p. 39).

The impossibility of the general hypotheses about the earth and the world being understood other than as he explained them, shows that Descartes takes them to be certain. By contrast, the hypotheses which are morally certain are consistent with the higher order principles but Descartes is quite willing to concede the possibility of their error.

Given the evidence for this new class of *a posteriori* certain hypotheses it seems reasonable to conclude that moral certainty cannot be the ultimate aim of scientific activity. The goal of science is not defined in relation to particulars nor can it be. Instead, the quest is for metaphysical and physical principles together with general hypotheses, similar to the light hypotheses, that will provide certainty that is empirically grounded but surpasses the certainty sufficient for the purposes of everyday life.[24]

Dept. of Philosophy
University of Minnesota
Minneapolis, Mn 55455
U.S.A.

NOTES

[1] The following texts and abbreviations will be used throughout: A&T — *OEuvres de Descartes*, Charles Adam and Paul Tannery (eds.) 12 vols, Paris, 1897—1913; H&R —*The Philosophical Works of Descartes*, Elizabeth Haldane and G. R. T. Ross (trans.), 2 vols., Cambridge, 1931; K —*Philosophical Letters*, Anthony Kenny, (ed. and trans.)

Minnesota, 1970; P — *The Principles of Philosophy*, R. Miller and V. Miller (trans.) Reidel, 1983.

[2] On this point see Garber (1979).

[3] It is important to note that there are also several statements in parts I & II of the *Principles* that involve a posteriori methods and are nevertheless considered to be *metaphysically certain*, for example, that material objects exist and that the mind is united with the body. I shall discuss these below and show how they differ from the kinds of claims that I have categorized as *a posteriori* certain.

[4] Garber suggests this.

[5] I argue below that the difference between moral and *a posteriori* certainty is one of kind and not degree while the difference between metaphysical and a posteriori certainty is one of degree.

[6] Clarke (1979) was the first to provide a detailed analysis of this use of deduction and how it might affect the historian's interpretation of Descartes' programme.

[7] Again Clarke (1982) provides an excellent discussion of the role of experience in Descartes' thought. My discussion borrows much from his.

[8] See for instance the discussion of experiment and observations in the *Discourse* (H&R pp. 106—130) and the *Regulae* (H&R pp. 35—49), especially the rule concerning simple natures and the obvious where Descartes claims that booklearning is not as close to truth as common sense reasoning. See also a letter to Mersenne, April 20, 1646 (A&T IV, p. 392).

[9] Cf. the introduction to the *Passions of the Soul*, A&T XI, p. 319 as well as Discourse VI and a letter to Mersenne dated April 20, 1646, A&T IV, p. 392.

[10] Of course divine authority is always to be preferred to perception. Descartes claims that trust must be placed solely in divine authority rather than in our own judgement, except of course in matters where divine faith teaches us nothing (P I, Art. 76).

[11] Two rather different attitudes toward the interplay of experience and reason are evident in Descartes' remarks about the certainty attributed to his laws of motion. In part II Art. 52 of the *Principles* Descartes concludes his discussion of the laws of collision by claiming that the things he has discussed require no proof because they are "obvious in themselves". In the Miller and Miller translation (1983, p. 69) it is pointed out that the final sentence in the French text reads, "And the demonstrations of this are so certain that even if experience were to appear to show us the opposite, we would nevertheless be obliged to place more trust in reason than in our senses". The same text provides an expanded version of the laws. Immediately following in article 53 he seems to have reconciled the difficulty and acknowledges that experience often seems to contradict these laws but interprets the problem as one of practice rather than principle. Because each body is always surrounded by others contiguous to it and because bodies are seldom perfectly solid it is difficult to perform calculations that would determine the truth of his laws.

[12] The explicit nature of this connection will be discussed later in the paper. For now suffice it to say that the relationship is one of consistency.

[13] Clarke (1982) has an extended discussion of this point in an appendix pp. 207—210.

[14] Descartes characterizes a special type of deduction, dealing with inference from a relatively large body of evidence from disparate sources, as induction; an example is the procedure used for inquiring into the nature of the magnet. For additional discussion of

Descartes' rather loose sense of deduction see Garber (1979) and especially Clarke (1979).

[15] Garber (1979) provides a nice illustration of this process. In the argument for the existence of God one provides an enumeration of the possible sources of the idea of God and through a process of elimination it is concluded that only God could have caused the idea; hence, God exists. Because the possible alternatives have been eliminated the conclusion is considered to be certain. As I argue above, this form of argument by enumeration is quite different in the case of the magnet. In the latter instance we have an enumeration of all the experimental evidence from which we deduce the explanation or cause. Experience and reason coordinate the process but because we can never be sure that we have provided a complete enumeration we fail to attain metaphysical certainty.

[16] A similar case can be made for the argument establishing the existence of material objects. The only possible sources of our ideas of material bodies are that God caused them, that I caused them or that they were caused by the bodies themselves. Using the appropriate methods (critical reflection on reason) I am able to rule out all but the correct conclusion that the ideas we have of sensible objects are caused by those objects. See Garber (1978) on this point. It is important to note that it is not simply our use of reason or God's divine veracity that *proves* material objects exist. As Miller & Miller point out (P II p. 39 n. 1) the fact that we cannot verify or deny this belief by use of reason is what would make it a deception if it were untrue.

[17] Descartes refers to the *hypothesis* that light is an action because in this context he has not provided the principles of physics from which it can be deduced.

[18] It is important to note here that Descartes did not think that light travelled from its source in an instant. Light did not involve actual motion, instead it was thought to be a tendency. For more on this point see Pendergast (1975) and Burke (1965).

[19] Recall from the discussion in section two above, 'demonstration' is used to denote an explanation or a proof and is often used interchangeably with 'deduction' (cf. Letter to Morin, July 13th, 1638 as well as the discussion in the *Discourse* VI).

[20] I want to stress the importance of this notion of a "general" cause for it is only these that can be legitimately called *a posteriori* certain. (Sometimes the principles are referred to as general causes but in these cases Descartes explicitly states that he is referring to his "principles"). I point out below that further evidence for this can be found at the end of part four of the *Principles*, Art. 206. There Descartes claims that it scarcely seems possible that the more general things he has written about the world and the earth (in parts three and four) could be understood to be other than he has explained them.

[21] That Descartes' claim about the nature of light is not one of his principles is made explicit in a letter to Plempius dated Dec. 20, 1637 where he writes that: "the things which I say in the first chapters of (*Dioptrics*) about the nature of light . . . are not my principles as you seem to object, but rather conclusions which are proved by everything that comes after" (K, p. 43).

[22] That this is not the case is evident from Descartes' remarks at the end of the *Principles*.

[23] This analogy is discussed by Descartes in *Dioptrics*, A & T II, pp. 24—30. See also Burke (1965).

[24] This point is not original with me but I am unable to locate the reference.

[25] It is important to note here that what was written about the world and the earth refers to parts III and IV of *Principles*.

[26] I would like to thank Dagfinn Follesdal and Paul Forster for valuable comments and suggestions. The idea for this paper originated out of a graduate student project at the University of Western Ontario. During that time I benefited greatly from the guidance and friendship of Robert Butts.

BIBLIOGRAPHY

Adam, Charles and Tannery, Paul. (1913). *Oeuvres de Descartes*, Cerf, Paris.

Burke, John G. (1965). "Descartes on the Refraction and the Velocity of Light", *American Journal of Physics* **XXXIII**, 390—400.

Clarke, Desmond. (1979). "Physics and Metaphysics in Descartes Principles", *Studies in History and Philosophy of Science* **10**, 89—112.

Clarke, Desmond. (1982). *Descartes Philosophy of Science*. University Park: Penn State Press.

Descartes, René. (1911). *The Philosophical Works of Descartes*. Elizabeth Haldane and G. Ross (trans. and ed.), Cambridge: Cambridge University Press.

Descartes, René. (1965) *Discourse on Method, Optics and Meterology*. Paul Olscamp (trans.). Indianapolis: Bobbs-Merrill.

Descartes, René. (1970). *The Treatise on Light*. Michael Mahoney (trans.). Princeton: Princeton University Press.

Descartes, René. (1970). *Philosophical Letters*. Anthony Kenny (trans. and ed.), Minneapolis: University of Minnesota Press.

Descartes, René. (1983). *The Principles of Philosophy*. R. Miller and V. Miller (trans.), Dordrecht: Reidel.

Garber, Daniel. (1978). "Science and Certainty in Descartes", in *Descartes: Critical and Interpretative Essays*, Michael Hooker (ed.), Baltimore: Johns Hopkins University Press.

Prendergast, Thomas. (1975). "Motion, Action and Tendancy in Descartes' Physics", *Journal for the History of Philosophy* **13**, 453—62.

Sabra, A. I. (1981). *Theories of Light from Descartes to Newton*, Cambridge: Cambridge University Press.

HOWARD DUNCAN

DESCARTES AND THE METHOD OF ANALYSIS AND SYNTHESIS

I believe that Descartes used a variant of the ancient method of analysis and synthesis in his scientific and philosophical works. Some have argued this general interpretive point before, but there has been little agreement about what this method involves.[1] I hope to make a contribution here. A considerable amount of evidence exists supporting the general thesis. Some of it is indirect, displaying that Descartes sought just the sort of axiomatically structured scientific theories that the ancients used the method of analysis and synthesis to establish. Some of it is more direct, to be found in his explicit descriptions of his method, usually in the philosophical works, and in some striking accounts of his own scientific reasoning.

This method was exceptionally well-suited to aiding scientists develop axiomatically structured theories in ancient times. The ideal of a theory was, perhaps, that described by Aristotle in *Posterior Analytics* and illustrated by Euclid's *Elements*.[2] A scientific theory consisted of two main parts, the undemonstrable principles and the propositions demonstrated from these. The fundamental principles, whose epistemic credentials were unquestioned, were of two sorts: principles common to all sciences no matter what their subject matter, which were called common notions or axioms; and principles that were specific to the science in which they appeared. These were called theses and included definitions which made no existence claims and hypotheses which did make existence claims. Hypotheses were in some cases rather suspect and were given the name 'postulate', as in Euclid's fifth postulate about parallel lines. The suspicious were asked to suspend their doubts, however, because the results demonstrated using these postulates would be so secure and valuable as to finally convince the doubter that they must be true despite initial appearances. The remaining propositions of the science were to be demonstrated from these principles. The truth of the demonstrated propositions was secured because the truth of the principles was assumed and the inferences used in the demonstrations were all non-ampliative and therefore truth-preserving. Thus, if a science could be put into axiomatic form, its epistemic integrity was assured.

J. R. Brown and J. Mittelstrass (eds.), An Intimate Relation, 65—80.

The method of analysis and synthesis was a tremendously effective tool for developing sciences in this form. The difficulty in doing so with logic alone is easy to see. If one possesses a set of fundamental principles, an infinite number of valid strings can be deduced from them, but most will be irrelevant or trivial. Further, the time consumed in generating these deductive strings would be extraordinary. This method is to assist the deductive work. The description from Pappus of Alexandria remains very helpful. This is taken from the *Collection* of 340 A.D.:

Now, analysis is the path from what one is seeking, as if it were established, by way of its consequences, to something that is established by synthesis. That is to say, in analysis we assume what is sought as if it has been achieved, and look for the thing from which it follows, and again what comes before that, until by regressing in this way we come upon some one of the things that are already known, or that occupy the rank of first principle. We call this kind of method 'analysis', as if to say reduction backward. In synthesis, by reversal, we assume what was attained last in the analysis to have been achieved already, and, setting now in natural order, as precedents, what before were following, and fitting them to each other, we attain the end of the construction of what was sought. This is what we call 'synthesis'.[3]

There is some scholarly disagreement over how to interpret this passage, but the general idea is clear enough and is all that we need for understanding Descartes' methodology.[4] Analysis and synthesis differ most noticeably in the 'logical direction' of the reasoning, analysis proceeding backwards and synthesis proceeding forwards, the direction most natural to us. Synthetical reasoning is more familiar and is what most of us think of when we think of what proofs look like: the deduction of a theorem from a set of accepted premises. But as every mundane student of logic and mathematics knows, finding the appropriate logical pathway from premises to theorem is a difficult task indeed. Herein lies one of the values of analysis.

In analysis, one begins with a conjecture that a proposition is demonstrable from the established set of first principles; again, these principles are unprovable, and analysis plays no role in determining them. One then asks what are the presuppositions of the conjecture, from what could this proposition be deduced? The presuppositions are traced backwards until one reaches a proposition known to be true, for example, a first principle or previously proven theorem, or until one reaches a proposition known to be false, the negation of a first principle or previously proven theorem. This step signals that the analysis is complete. If the analysis ends in a true proposition, it can be concluded that the conjecture is provable as a theorem and if it ends in a false proposition, it can be concluded that the conjecture is not a theorem. Following a successful analysis, a synthetic proof may be given by retracing the steps of the analysis in the opposite direction from principles to theorem. If the proof is convertible, the steps in the synthesis will be precisely those taken in the analysis, but in the reverse order. Clearly the synthetic arm of the method is the easier. It is in analysis that the hard work of science was done.

Analysis was a method for discovery. It was not, however, a mechanical procedure, but required considerable skill and insight, not to mention hard work. One must be careful in specifying what is discovered by analysis. As it was used by the ancient geometers, the method helped to discover which were the appropriate premises from which a theorem could be demonstrated. The theorems themselves resulted, methodologically speaking, from an act of creative insight. The framework of established principles within which analysis was carried out served to orient and guide the backwards reasoning in the direction likely to yield a proof. They also led the reasoning towards propositions of greater epistemic security which were needed to secure the integrity of the theorem. And the principles served as criteria for recognizing when enough analysis had been done; without this context of accepted truths, analysis could continue forever. Thus, analysis is a method of reasoning which helps one to discover the premises from which a proposition can be deduced, and which is used within a context of accepted truths which direct the reasoning along the most fruitful pathways, which signal the completion of analysis, and which enable one to evaluate the epistemic credentials of the original conjecture.

Descartes did not incorporate the method of the ancients unadulterated, but there can be no doubt that his is a close variant. There were many reasons for modifying the method, among them its use in natural science rather than in pure mathematics alone, and absorbing it into his general and multi-faceted philosophical programme. Taking the method over from the mathematical into the scientific arena involved using it to facilitate explanations of phenomena as well as proofs of theorems. Embodying it in his general philosophical system meant embellishing the process of seeking presuppositions in quite specific ways. This we shall see shortly. The clearest direct textual evidence that Descartes adopted analysis and synthesis is found in *Rules for the Direction of the*

Mind. This text is invaluable because it describes not only the specific methodological rules that Descartes thought ought to be followed, but also provides a general methodological context and a kind of metaphysical justification for using analysis and synthesis within this context. Descartes describes this context as mathesis universalis (universal mathematics), and the underlying metaphysics is the mechanical philosophy which he inherited, at least in spirit, from Galileo. Descartes tailors analysis and synthesis quite specifically to reside comfortably with these other of his commitments.

Early in *Rules for the Direction of the Mind*, Descartes explains that theories ought to be axiomatically formed to secure their epistemic credentials. With many writing at the time, Descartes reacted to the excessive speculation of the Scholastics with a rather conservative view of science and its method. Anything that deserved the name 'science' must be certain and evident knowledge, and he believed that only axiomatically structured theories could meet this condition. The certainty of Cartesian science rested upon its intuited first principles; the derivative propositions were to be made evident by exceptionally strong deductive links with the first principles. Ideally, extended deductive chains could be intuited and thereby possess immediate certainty; failing this, however, each step in a chain must be intuitable, even if the entire chain cannot be comprehended in a single act of cognition. This picture of scientific theories is highly restrictive because of its demands on the logical relations between first principles and derivative propositions, but Descartes considered this a necessary antidote to the scientific morass of the past.

The first principles of the metaphysical foundations of Descartes' physics describe what he calls simple natures: extension and its modes. All natural bodies are thought by Descartes to be various combinations of the simple natures, and thus explaining natural phenomena is showing what are the combined simple natures and how they come to be combined in that way. An explanation is to represent the causal relations between the simple natures in a deductive chain from first principles to the phenomena. The method of analysis and synthesis Descartes took to be the best way to carry out this sort of explanatory project.

The fourth of the rules for the direction of the mind contains Descartes' first general description of his method, and it conveys quite clearly that his model is the ancient method of analysis and synthesis.

He refers explicitly to Pappus and to the use of a kind of analysis by the ancient geometers whom, he thought, begrudged the method to posterity. (Pappus' writings were lost for many centuries and were recovered apparently during the Renaissance and translated into Latin only in 1589).[5] In this rule, he displays a firm conviction that the fundamental methodological problem is to find the appropriate logical pathways necessary in proofs and explanations. For example, he says,

> If we are ignorant of some one of the things we are able to know, this ignorance can . . . be attributed to our failure to notice a path which would have lead us to such knowledge.

Further, in the same paragraph, he writes,

> Our method correctly explains . . . how deductions are to be discovered so that we reach the knowledge of everything.[6]

In the fifth rule, Descartes advances beyond these general remarks to a description of method that can only be understood as a version of analysis and synthesis.

> Method consists entirely in the order and arrangement of those things upon which the power of the mind is to be considered in order to discover some truth. And we will follow his method exactly if we reduce complex and obscure propositions step by step to simpler ones and then try to advance by the same gradual process from the intuitive understanding of the very simplest to the knowledge of all the rest.[7]

Analysis and synthesis is a method for discovering logical relations between propositions, for arranging propositions in a way that allows demonstration from first principles. In Euclid's geometry, a relatively complex theorem was conjectured, and its presuppositions traced in the logical direction of the established first principles, which were simple and epistemically prior to any theorem. Descartes' "step by step reduction" of complex to simple propositions is analysis, backwards deduction. Notice, too, that the advance from the simplest to the more complex is by "the same gradual process", namely step by step deduction, now in the forwards direction, the process of synthesis. Descartes says that this rule contains the sum of all human endeavour. We can be confident, then, interpreting Descartes' method as at least a version of the ancient method of analysis and synthesis. What remains is to articulate some details; here reside the variations.

The all too common caricature of Descartes' method as a priori deductions from first principles like "I exist", synthesis without analysis as it were, is certainly wrong. So too is attributing to him a deep suspicion of the senses. Some commentators who recognize that Descartes used analysis and synthesis expressed this suspicion of the senses by saying that Descartes analyzed thoughts, not phenomena.[8] Yet, in a striking passage from the discussion of rule 5, he argues that the sort of rationalism once fashionable to attach to Descartes is a hopeless approach for discovering truth.

> This is the way all astrologers act who hope to be able to predict the results of heavenly motions without knowledge of the heavens, and without even observing their motions accurately. This is the way many do who study mechanics apart from physics, and rashly invent new instruments for initiating motion. So also do those philosophers who neglect experience and believe that truth will arise from their own heads as Minerva did from that of Jove.[9]

The demand for experiential grounds is a crucial way to avoid speculative excess in science; the epistemic credentials of a science will be enhanced if it is rooted in experience. This is the very epistemic point made by Galileo and Newton: we do not invent nature.

In general, when analyzing experiential data, one seeks the causes of of the phenomena under study, tracing those causes back to the simple natures. The simple natures provide a directional guide to the analysis and stand as criteria for the completion of the analytical work; once one has driven an explanation to the point of the recognized simple natures, the analysis is complete. Thus one regards the phenomenon as a complex effect and tries to reduce it to its simple causes. It is specified in advance what sorts of things will be the simple causes, and one directs the search for the cause of the specific phenomenon in their direction, just as one tries to trace the logical presuppositions of a geometrical theorem towards the previously established first principles of the geometry. This procedure would find an analogous use in proofs of physical laws, such as the laws of refraction. I will limit myself here, though, to the use of analysis and synthesis in explanation.

This is to say that the analysis of phenomena occurs within a framework of previously established principles of scientific explanation. For now, I will take it for granted that Descartes has indeed determined a set of basic explanatory principles which include the theory of matter as geometric extension, the fundamental tenets of the mechanical philosophy including the laws of motion, at least the rudiments of a theory of perception, especially of vision, a general epistemology and attitudes about the limits of human cognition, and others additional to these. I assume all this for now because if I can postpone discussing how Descartes justifies these contentious claims, my methodological story will be easier to tell.

Descartes mentions many other features of the analytical tracing of phenomena to their causes. These include devices to assist the procedure: dubbing things as absolute and relative within their context, the absolutes being simpler and causal, the relative being more complex and resultant; an extensive enumeration of the relevant empirical data; and considerable practice with easy problems and in analogous activities such as weaving, whose requirements of order will transfer to the more intellectual arts. Perhaps more interesting, however, are his prescriptions on using hypotheses.

Descartes realized as well as most the formidable problems in discovering causes that do not present themselves to our gaze. Yet the mechanical philosophy demanded that explanations be driven to the level of unseen mechanical causes. Thus, he had to make room within his method for hypotheses and in a way that did not degrade the epistemic merits of the results. Descartes first alerts us to the need for hypotheses in Rule 8, where he discusses how one might solve the problem of the anaclastic curve, that shape of a lens that causes parallel rays to refract to a single point. After having told us that the problem cannot be solved within mathematics alone or by purely experimental means, he explains that a mathematics informed by physics and following the procedures of analysis will come upon the correct solution. This requires reducing the problem step by step, ultimately to that of the nature of illumination — for Descartes, a form of natural power. Descartes admits that no understanding of illumination itself is available, and instructs the researcher to frame a hypothesis about illumination using what is known about other natural powers as a guide. With this hypothesis, he thought, a solution to the problem should be forthcoming.

The recommendation that we appeal to hypothesis to explain natural phenomena has deep implications for Descartes' epistemology of science. It is no doubt startling to find this advice in the context of a rule that advocates stopping an investigation when one's store of clear and distinct ideas has become depleted. However, from the point of view of methodology, this recommendation is really no surprise at all given the long heritage of hypothesis in the methodological tradition that Descartes followed. Indeed, in Rule 12, Descartes describes his use of hypotheses as the same as found in geometry and astronomy. His description is reminiscent of the postulates of ancient axiomatics, principles that do not command the assent of the reader or student:

> It will be enough for me to explain as briefly as possible which manner of conceiving our powers of recognizing objects will be most useful to my plan. You need not believe that the facts are thus, unless you choose; but what is to keep you from following these suppositions, nevertheless, if it appears that they are in no way opposed to the truth about these matters, but only make everything clearer? It is no different than when in geometry you make certain assumptions about quantity, by which the force of the demonstrations is in no way diminished.[10]

While Descartes recognizes the increased risk of appealing to hypothesis in explanation, he claims that so long as they are invoked in the way of geometry and astronomy, there will be no significant compromising the epistemic worth of the explanation or of the science that uses them. This means that Descartes' use of hypotheses is much more strictly controlled than what he took to be their use in earlier sciences.

We learn more about Descartes' attitude toward hypotheses in his Rule 12 hypothetical account of perception. The explanation is but sketched, but the idea is clear: perception is caused by external bodies impressing their shape upon our organs of sense by striking them. This is said not only of touch where such an account has some plausibility, but also of the other senses. Thus, the colours of vision are caused by the impression of shapes upon the opaque parts of our eyes. Descartes draws the common analogy with the wax and seal, but in order to deny that his explanation is intended merely analogically; it is to be understood literally, even though it has the status of an hypothesis. His mechanical account of perception follows Galileo's denial of the existence of secondary qualities and the attendant Scholastic theory of causal action as the communication of qualities between bodies. But further, the elements of the explanation are restricted to Cartesian simple natures, extension, shape, and motion. This has epistemic force for Descartes:

> And it helps a great deal to conceive all things in this manner, since nothing is more susceptible to examination than shape, for it is both felt and seen. Moreover, that no falsity follows from making this assumption rather than any other is demonstrated by the fact that the concept of shape is so common and simple that it is involved in every perception.[11]

We ought to understand colour, not rashly as some entity unique in its determinations, not as an occult quality, but as itself a shape, and the differences between colours as similar to the differences between such shapes as these:

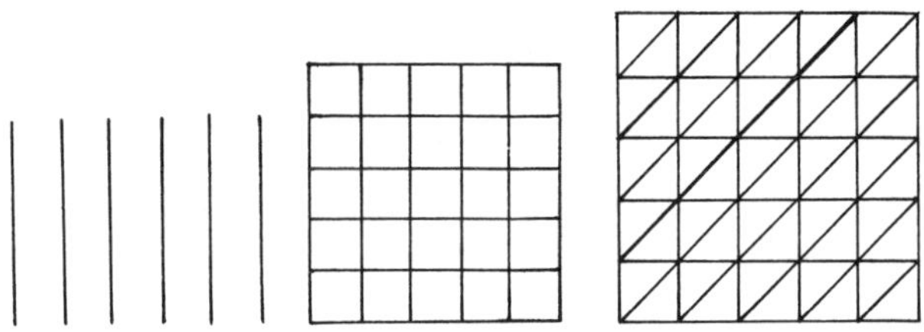

Though many explanations must ultimately be hypothetical, not any hypothesis will do. Legitimate hypotheses are by way of the simple natures alone.

Hypotheses are indispensable for Descartes' use of analysis and synthesis; without them, one could not explain a phenomenon as an effect of specific unseen simple natures. Hypotheses are inventions, methodologically speaking; there is no logic productive of them, nor is the method of analysis alone able to generate them. However, it would be wrong to suggest that they are invented out of the blue, in a manner like Popperian bold hypothesizing.[12] For Descartes, inventing hypotheses is to be very strictly controlled, though not to the point where the controls uniquely determine the hypothesis. Descartes' controls go beyond conformity with empirical data, though this is especially important. Hypothetical explanations are to some extent confirmed by a good fit with experimental results, he thought. But the hypotheses must, just as importantly, be expressed in terms of simple natures and conform to the laws governing them; they must conform to the metaphysical first principles of the Cartesian mechanical philosophy; and therefore they must be representable mathematically as demanded by the mathesis universalis. Others, notably Laudan,[13] have regarded Descartes as adopting a form of the method of hypothesis; I want to emphasize, however, that his appeal to hypothesis must be understood within the context of the method of analysis and synthesis.

Descartes' description of his reasoning about the phenomenon of the rainbow is one of the clearest expressions of analysis in his scientific writings. As such, we should expect to see him doing some of these things:

(1) His reasoning will be directionally guided by a set of metaphysical first principles and scientific results already established; these may include physical hypotheses.

(2) These first principles and scientific results will function, too, as criteria for the completing of an analysis.

(3) Analysis will begin by enumerating empirical data and, so far as possible, will reduce these data to simple natures. This reduction will provide a clear and distinct description of the phenomenon which represents it much more simply than originally, and in terms by which it can be explained as a causal effect of simple natures in lawlike combination.

(4) Once the analysis is complete, the phenomenon can be explained by giving a deduction of it from the established propositions of the science. The synthetical explanation may be confirmed experimentally, which confirmation will carry over to the analysis and any hypotheses it uses.

Descartes explains the phenomenon of the rainbow in the eighth discourse of the *Meteorology*.[14] This discourse appears within a well-articulated framework of metaphysical first principles, the results of the *Optics* and the *Geometry*, and the previous seven discourses of the *Meteors* which concern the nature of such things as vapours, winds, clouds, and precipitation. This framework guides the direction of the reasoning about the rainbow, leading it, as it were, from the empirical data towards the fundamental principles of explanation. In the language of rule 6, these principles stand as the absolutes to which the data must be reduced if they be understood. These stand, too, as Descartes' criteria for success in his analysis and signal that a synthetical explanation of the rainbow is available.

Descartes opens the discourse by noting that the rainbow is the most appropriate subject for illustrating the power of his method, it not yet having received any remotely adequate explanation.[15] He immediately begins to enumerate empirical data, citing the occurrence of rainbows in fountains in addition to those appearing high in the sky. This observation has immediate results for solving the problem because it leads the problem one step toward the simple natures that are the goal of the analysis. Because rainbows appear high in the sky and in fountains, he concludes that the sky plays no causal role and can be ignored, and that the phenomenon can be reduced to causal relations between rays of light, drops of water, and our organs of vision. This use of eliminative induction appears frequently in Descartes' reasonings, and while he has no specific name for this form of inference, he recognized its reductive powers. Discovering that the phenomenon concerns drops of water is of great moment in solving the problem, partly because Descartes already knew quite a lot about water drops; he had previously shown them to be spherical, and thus he can utilize the geometry of the sphere, something, too, of which he had considerable knowledge. We see in this one inference a move to understand the problem under the rubric of already established scientific principles.

To better understand the causal influence of a water drop upon rays of light, Descartes constructs a very large drop from a transparent glass flask filled with water. He then positioned this drop so that sunlight would fall upon it. If the angle formed by the line DE from the bottom of the drop to the eye and the line EM from the eye following the direction of the sun's rays was near to 42°, Descartes found that the bottom portion of the drop would appear bright red. Enumerating further data, he found that this red colour appeared no matter what his distance from the drop or his lateral position with respect to it, so long as this angle remained very near to 42°. A slight varying of the angle resulted in the colour shifting from red to yellow, blue, and the others in the spectrum; a larger deviation from this angle presented no colours at all at this point on the drop. Similar effects were observed towards

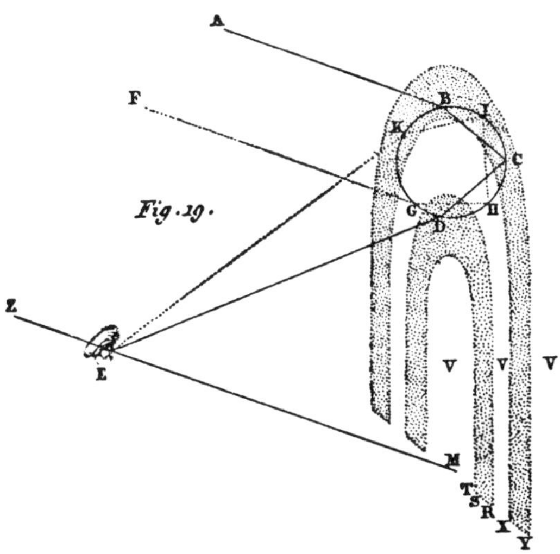

the top of the drop if the angle formed by the counterpart lines, KE, EM, was very near to 52°. These effects could be generalized for those positions that would be occupied by water drops in which full rainbows, both primary and secondary, appear.

It is important to notice that Descartes has said nothing so far about the action or transmission of light. The analysis has been restricted to the relative positions of the sun, the water drop, and the observing eye. Again, enumerating empirical data takes precedence over explanation in the early stages of the investigation. This, of course, is in the interest of minimizing risk. Yet, this enumeration is for the purpose of clearing the way for a causal explanation; it helps to reduce the number of plausible causal avenues to explore. The data that Descartes has enumerated so far strongly suggest that the cause of the rainbow phenomenon will involve the pathway of the light from the sun to the water drop to the observer. Discovering that it is only a highly restricted range of angles that accompany the production of colour leads Descartes to attempt to trace the rays of light in their travels through the water drop. Descartes' reasoning here is not just the method of hypothesis and test nor merely an exhibition of orderly thinking, but it follows a specific pattern of refining descriptions of the phenomenon in terms closer and closer to the mathematizable simple natures; this is required by his method before offering any explanation. Indeed, this descriptive project is essential to the explanatory.

Had Descartes' method been one merely of inventing and testing hypotheses, he might have conjectured at an early stage that rainbows result from refracting light, and then described the refractions *a priori* using his theory of refraction and the geometry of spheres. This he does not do, though he might have. However, his analytical procedure requires a more cautious approach of reducing the data step by step towards the simple natures. Accordingly, *he discovers by experimentation*, by a primitive form of ray tracing, that the production of rainbows involves refraction.

After this, examining in more detail what caused the part D of the ball BCD to appear red, I found that it was the rays of the sun which, coming from A toward B, were curved as they entered the water at point B, and went toward C, whence they were reflected toward D; and there, being curved again as they left the water, they tended toward E. For as soon as I put an opaque or dark body in some place on the lines AB, BC, DC, DE, this red colour would disappear. And even if I covered the whole ball except for the two points B and D, and put dark bodies everywhere else, provided that nothing hindered the action of the rays, ABCDE, the red colour nevertheless appeared.[16]

Descartes' largely empirical ray tracing not only simplifies the phenomenon considerably, but allows him to bring in the full force of his geometrical optics; representing the phenomenon as a combined reflection and refraction of light rays by the medium of the water drop reduces it to a problem of universal mathematics, solvable by the combined resources of geometry, physics, and optics.

The final step in this analysis occurred upon Descartes' examining the effects of a prism upon light rays. This illustrates yet another aspect of analysis, its need for creative insight. Nothing in the method compelled Descartes to examine the prismatic effect; this was his insight alone, and a happy one it was. Following steps similar to those used to examine the giant water drop, Descartes discovers that the colours of the rainbow are caused by a single refraction, not a double refraction, and not by reflection at all, for there is no reflection in the pathways traced through the prism. *This is the major discovery of the analysis: the rainbow phenomenon is a case of the refraction of light.* Note that in this analysis, Descartes has achieved, primarily, a precise description of what the phenomenon is. This description is much simpler than it was initially, given in terms more akin to the simple mathematical natures which must be used in explanations. Further, having reduced the phenomenon to one of refraction, the explanatory problem is essentially solved, the analysis complete. Descartes had previously developed a theory of refraction that he could use to explain the rainbow, once it was recognized that the rainbow is but a case of the refraction of light. This is obvious in the text, for immediately following this step in the analysis, Descartes gives a synthetical explanation of the colours of the rainbow using his account of refraction from the *Optics*. It was a signal that the analysis was complete that he had reduced the problem to an example of this general phenomenon for which he had already established an explanation.

Recall that, in the context of explaining phenomena, analysis is a method for discovering appropriate explanatory principles. These principles must both yield an explanation and satisfy strong epistemic demands. By discovering that the rainbow phenomenon is a case of refraction, Descartes had driven his analysis to this point. It was now determined that he must explain the colours of the rainbow as an effect of the refraction of light by water droplets in the air. Not only is he in a position to do just this because of the results of the Optics on refraction, but the explanation will meet the epistemic requirements that he imposed to avoid the problems of Scholastic science. The account of refraction appealed to only the sanctioned principles of the mechanical philosophy. Thus, his explanation of the colours of the rainbow in terms of the differential rotation of particles of light, too, meets these requirements.

Much could be said about Descartes' synthetical explanation of the rainbow, but here I wish only to point to the hypothetical nature of the explanation. In the Optics, the theory of refraction is general and does not itself countenance the generation of colours. Rather, it describes the mechanical actions of refracting media on the motion of light particles. Colours are generated, of course, only when these light particles assault human retinas. Thus, when Descartes explains specifically the production of colour by referring to a differential rotation of these particles, he is not merely deducing this from the theory of the *Optics*. He is invoking an hypothesis, but subject to constraints imposed by the general theory of refraction. However, the hypothetical nature of the explanation runs much deeper than this. We have already seen Descartes describe his mechanical theory of perception as hypothetical; further, the theory of refraction, too, is an hypothesis owing to its reliance upon a hypothetical account of light and its transmission. Of his account of light, Descartes says this in the first discourse of the *Optics*:

> Thus, not having here any other occasion to speak of light than to explain how its rays enter into the eye, and how they can be deflected by the different bodies they encounter, I need not undertake to explain its true nature. And I believe that it will suffice that I make use of two or three comparisons which help to conceive it in the manner which to me seems the most convenient to explain all those of its properties that experience acquaints us with, and to deduce afterwards all the others which cannot be so easily observed; imitating in this the Astronomers, who, although their assumptions are almost all false or uncertain, nevertheless, because these assumptions refer to different observations which they have made, never cease to draw many very true and well-assured conclusions from them.[17]

Descartes makes this sort of defence of the hypothetical explanation of the rainbow by appealing to what amounts to its experimental confirmation.

> And in all of this, the explanation accords so perfectly with experience that I do not

believe it possible, after one has studied both carefully, to doubt that the matter is as I have just explained it.[18]

I believe that we must consider this defence of hypothetical explanation as an analog to the ancient attitude towards geometrical postulates. Postulates were existential principles whose truth was assumed but not obvious, but whose truth would gradually be assented to once the demonstrated results were generated. In the case of physical hypotheses, if their demonstrative results were confirmed by experience, this, together with their meeting Descartes' basic epistemic and metaphysical constraints, would speak highly in favour of their truth. Descartes' use of hypotheses, therefore, falls within the scheme of the ancient methodology.

Of course, this defence of hypothetical explanatory principles is fallacious, amounting to the fallacy of affirming the consequent. And, Descartes' use of hypotheses may be disastrous for his epistemology of science, especially if he intends his sciences to be certain in any strong sense. However, it would be a mistake to regard Cartesian method as liberal, as allowing nearly unbridled use of explanatory hypotheses. The spirit of his method is fundamentally conservative; it is a strong reaction against what he thought to be a scandalously loose method in Scholastic science and philosophy. Hypothesis he thought to be unavoidable in explanatory science, but he took considerable care in shackling the hypotheses to specific experiential data and to metaphysical principles directly applicable to empirical phenomena. For Descartes, our clues to understanding the unseen must be taken from the observed nature of things; anything less would be wanton. Cartesian analysis and synthesis is a mark of methodological restraint.

Gloucester, Ontario
Canada

NOTES

[1] One can get a good idea of the range of interpretations from Paul Olscamp's introduction to Descartes' *Discourse on Method, Optics, Geometry, and Meteorology*, tr. P. Olscamp, Indianapolis, Indiana: Bobbs-Merrill, 1965; "Newton and the Method of Analysis" by Henry Guerlac in *Dictionary of the History of Ideas*, ed. Philip Wiener, New York: Scribner, 1973; chapter 7 of *Descartes' Philosophy of Science* by Desmond Clarke, Manchester: Manchester University Press, 1982; and "Descartes' Project for a Mathematical Physics" by Stephen Gaukroger in *Descartes: Philosophy, Mathematics and Physics*, ed. Gaukroger, Totowa, New Jersey: Barnes and Noble, 1980.

[2] Aristotle, *Posterior Analystics*, tr. G. R. G. Mure, in *The Basic Works of Aristotle*, ed. R. McKeon, New York: Random House, 1941; Book I, ch. 2.

[3] Pappus of Alexandria, *Book 7 of the Collection*, tr. A. Jones, New York: Springer Verlag, 1986, p. 82.

[4] Some players in this debate are Jaakko Hintikka and Arpad Szabo. See Szabo's *The Beginnings of Greek Mathematics*, Dordrecht: Reidel, 1978; *The Method of Analysis* by Hintikka and Unto Remes, Dordrecht: Reidel, 1974; and *Theory Change, Ancient Axiomatics and Galileo's Methodology*, ed. Hintikka, D. Gruender, E. Agazzi, Dordrecht: Reidel, 1981. I have some comments in "The Euclidean Tradition and Kant's Thoughts on Geometry", *Canadian Journal of Philosophy*, March, 1987.

[5] The Latin translation was Commandino's. He translated the Greek terms 'analysis' and 'synthesis' as 'resolution' and 'composition'. For my purposes, I will consider the analytical-synthetical method as the same as the method of resolution and composition. Many in Descartes' day did the same.

[6] Descartes, *Rules for the Direction of the Mind*, tr. L. Lafleur in *Descartes: Philosophical Essays*, Indianapolis, Indiana: Bobbs-Merrill, 1964, p. 157.

[7] Descartes, *RDM*, p. 163.

[8] cf. Guerlac's "Newton and the Method of Analysis".

[9] Descartes, *RDM*, p. 163.

[10] Descartes, *RDM*, p. 188f.

[11] Descartes, *RDM*, p. 189.

[12] cf. Olscamp's introduction to Descartes' *Discourse on Method, Optics, Geometry, and Meteorology.*

[13] See Larry Laudan's, "The Clock Metaphor and Hypothesis", in his *Science and Hypothesis*, Dordrecht: Reidel, 1981.

[14] Descartes, *Meteorology*, Olscamp's translation.

[15] There can be little question that Descartes borrowed heavily from prior explanations of the rainbow, both in gaining the refraction account itself, and in some of the details of the analysis including constructing an artificial water drop. This was done by Theodoric of Freiberg in the 13th century, and Descartes' account so closely parallels that of Freiberg that one cannot help but believe that Descartes was familiar with his explanation. See A. C. Crombie's classic, *Robert Grosseteste*, London: Oxford, 1962, ch. IX.

[16] Descartes, *Meteors*, p. 334.

[17] Descartes, Optics (Olscamp translation), p. 67.

[18] Descartes, Meteors, p. 338.

THOMAS M. LENNON

PHYSICAL AND METAPHYSICAL ATOMISM: 1666–1682

My topic is an episode in what I take to have been the most interesting philosophical debate in the latter half of the seventeenth century. The debate was between the Cartesians and the Gassendists, and the episode concerns the divisibility of matter. The Gassendists plumped almost by definition for an ontology of atoms and the void, while the Cartesians of course were almost invariably plenum theorists who argued the infinite divisibility of matter. My thesis is that those who argued atomism in the period 1666–82 were moving away from atomism as a physical theory toward a metaphysical theory that found its most important expression in Locke. Arguing the case for Locke is far beyond my resources here; instead, I shall discuss the relatively minor figures who prepared the way for him.

Physical atomism is the thesis that the infinite divisibility off matter is a physical impossibility. The laws of physics are such that the integrity of certain quantities of matter cannot be upset. Thus, one might hold, for example, that while matter is conceptually divisible to infinity, only God could effect its division beyond a certain point. It seems to me, however, that physical atomism in this period is itself a conceptual impossibility, at least given the conception of homogeneous matter universally accepted by those who were scientifically *au courant* in this period. The principal attraction of the concept of universal homogeneous matter was that according to it the same set of physical laws is sufficient and necessary for the explanation of all material states. If matter is physically divisible at one level then it should be physically divisible at all levels, no matter how microscopic. Within the eye of a mite we should find worlds containing, if not mites with eyes containing further worlds, then at least nothing essentially different from our own divisible world. Size can make no difference.

Metaphysical atomism is perhaps a more intelligible thesis. One expression of it is the denial that there are internal relations; it is the claim that the existence of a thing and its attributes are independent of everything else. We might say that it is the ontological version of twentieth-century logical atomism. But this statement of the view, that

J. R. Brown and J. Mittelstrass (eds.), An Intimate Relation, 81–95.

things are independent, is not quite sufficient, for then with Descartes's definition of substance, any substance ontology including Descartes's would be atomist. Another way to put it might be to say that not everything that can exist does exist. The obvious image, historically very important indeed, is of discrete particles of matter randomly dispersed in the void. The discreteness suggested by this image, however, can only be a physical accident. Instead, a stronger account is required and thus we have to think of the individuals in non-material terms and as located in logical space. What this comes to in its extreme form is a binary system of bare particulars and bare non-existence. This is by way of obvious contrast with the Cartesian monist picture of space necessarily filled with matter identical to it, with every part or bit of it being necessarily what it is because of its essential mathematical structure.

The early Gassendists were no more than physical atomists. For the purposes of doing physics, they said, let's assume that beyond a certain point matter will in fact never be divided. Such was the attitude of Pierre Petit of Montluçon, who took an unusually sophisticated instrumentalist view of the whole matter.[1] Despite certain anticipations of metaphysical atomism noted below, it seems to me that Gassendi's too was nothing more than physical atomism.[2] Even so, Gassendi's atomism is not without interest by way of contrast with the Cartesians. The following sketch, especially of the Cartesian contrast is exceedingly schematic, but I hope it will suffice as background to the case I shall present in more detail.

Gassendi's aim of course was to rehabilitate the atomism of Epicurus by correcting its theologically unacceptable features. The principal corrective Gassendi makes in Epicurean atomism is to regard atoms as created. He also reduces the number of atoms and their shapes to a finite (if incomprehensibly large) number, which thus makes atoms by themselves incapable of explaining the order in the world. Finally, while weight (*pondus, seu gravitas*) is for Gassendi as for Epicurus a property of atoms, this notion is given a radically new explanation which rejects the picture of atoms falling primordially downward through the void. Instead, an atom which would otherwise be inactive is created with "a natural and internal faculty or force" which it cannot lose and by which, unless it is inhibited by a similarly grounded motion in some other atom(s), it moves itself and moves others.[3] Thus the law of inertia, for the first publication of which Gassendi has been given credit, is true for him only of objects at the observational, or at most the molecular level, where bodies at rest remain at rest only as a result of exactly equal and opposed motorforce. The upshot of created, quantitatively and qualitatively finite, and intrinsically, if not natively, inactive atoms is an argument for God as a provident creator.

The contrast with the Cartesian picture is sharp and dramatic. For Descartes there is but a single real individual, *res extensa*, which is capable of (and, according to the *Discourse's* fable version of the world, actually assumes) an infinite number of shapes. That is, the one intelligible object of the *mathesis universalis* can be perceived as an infinite number of sense-instantiated corporeal things, which can be regarded as the objects of mathematical theorems. Motion is an essential feature of corporeal things only insofar as it individuates them; it is not a feature at all and *a fortiori* not a natural force of the real. Regardless of the ontological status of motion, however, the law of inertia obtains at all levels of moving objects and quantity of motion is conserved in all instances of collision. Finally, though God has His purposes, they are irretrievably hidden from human view and play no role either in physics or metaphysics. The only intelligibility is the mathematical intelligibility of what God creates from one moment to the next. Malebranche later altered this picture somewhat. What are corporeal things for Descartes are for him representations of corporeal things, which things are real, multiple and mobile. But is just this departure that he had difficulty in sustaining in a way that preserves both Cartesian and Christian orthodoxy. He also admits teleological explanations, but only with respect to a restricted range of phenomenon that does not argue the existence of God. In short, Malebranche essentially preserved the Cartesian contrast with Gassendism.

Ironically, the metaphysical atomist challenge to Cartesianism first came, not from the Gassendists, but from within the Cartesian camp itself. Gerauld de Cordemoy (1626–1684) was a practicing lawyer who was an *habitué* of the leading *salons* (Mme. de Bonnevaux) *conférences* (Rohault) and *académies* (Bourdelot) of the 1660's. As suggested by some of the circles he frequented, he was regarded as a Cartesian and in fact was among the anointed Baillet lists as attending the Descartes funeral jamboree of 29 June 1667.[4] His principal work appeared in 1666 entitled, *The Distinction between Body and Soul in Six Discourses, intended as a clarification of physics.* The review of it for the *Journal des Savants* of 7 June of that year thought that its author "generally followed the principles of Descartes, but (that) he was not so attached to them that he did not relinquish them in certain places."[5] *In certain places* indeed, for Cordemoy argued in no uncertain terms an ontology of atoms and the void. His Cartesian critic Desgabets credited him with being the first to advance metaphysical arguments, however faulty, for the ancient doctrine he espoused.[6]

Cordemoy announces right at the outset of his first *Discourse* that "bodies are extended substances". And, "1. Since there are many bodies, the extension of each of them must be limited, this limit being what is called shape. 2. As each body is but a self-same substance, it cannot be divided — its shape cannot change and it is so necessarily continuous that it excludes every other body, which is called impenetrability."[7] He goes on to argue against the Cartesians' divisibility of matter that (a) the indefinite divisibility they assert really means that it is infinitely divisible, and this is inconceivable; (b) a body at rest among others at rest but different from them in fact is conceivable but on Cartesian grounds is not conceivable; (c) on Cartesian grounds there can be no fixed shape for any body, each of whose parts *ad infinitum* would be acted upon by a different body *ad infinitum*; (d) the Cartesians anyhow assume atomism in their actual explanations. He thinks the Cartesians assert the divisibility of body because they confuse body with *matter*, which they take to be a substance, but which in fact is only an aggregate of bodies.[8]

Cordemoy argues that because individual bodies are substances, i.e. things independent of all other things except God, unoccupied space is not inconceivable as the Cartesians think. It may be that a vase emptied of all its contents would be crushed by the matter surrounding it unless its contents were replaced. But to say that its sides would *ipso facto* touch without any external pressure, is no argument, he thinks. For no given body depends upon any other.[9] Each body, including those composing the edges of the vase, is an individual extended substance and not a mode of extension, Descartes's single material substance. This obviously anti-Cartesian thesis of Cordemoy's is reinforced for him by his perfectly Cartesian occasionalism. No body is able to move another, he argues;[10] *a fortiori* no body can cause another to exist. Indeed, these two operations come to the same thing. To create bodies is to create them in a certain place, i.e. with certain relations among them; and to move them is to recreate them with different relations. Both operations belong only to God.

Cordemoy's book was sent by Clerselier to Dom Robert Desgabets

(1610–1678), who responded with a letter whose sectarian reaction of indignation rather typifies intramural debate among Cartesians in the period.[11] To extirpate the atomist heresy so supportive of Gassendi's camp, Desgabets undertook a refutation of Cordemoy's principal theses. For one thing, according to Desgabets, his atomism is based on a false conception of the continuum and infinite divisibility. The alleged atoms have extension, hence parts, and thus are divisible at least by God. In every one of them we can conceive an entire world and so on to infinity, argues Desgabets, using the imagery of what is found in a grain of sand, the eyes of mites, and so on — imagery extensively used by Malebranche and others in the period, and made all the more compelling by contemporary discoveries in microscopy. In addition, as Gassendi had argued that infinite divisibility was a mathematical concept that was not physically applicable, so Desgabets accuses the atomists of confusing mathematical points with the real, or at least of trying to construct the latter from the former, which in his view cannot be done.[12] Instead, individual material things are modal beings produced by us when we perceive the single extended substance.[13] Desgabets thus responds to atomism by construing individuation in the idealistic terms I identified at the outset here as Cartesian. A similar response is made to Cordemoy's advocacy of the void, a response which is worth developing since it bears directly on the issue of the reality of individuals.

We have a natural inclination, according to Desgabets, to form ideas of genera, species and "other universals", which he thinks can be true. But what happens is that we have become so accustomed to such ideas that we begin to treat them as real, existing outside individuals, and to regard them in the fashion of Plato as eternal, ungenerable and incorruptible essences. This is what has happened in the case of the void. It is a reified abstraction — it is only real, i.e. material, extension considered in general, just as human nature is nothing but particular men conceived perfectly.[14] Desgabet's criticism is hard to make out in orthodox Cartesian terms, for Descartes' space/extension is, while created, nonetheless an essence and one the idea of which logically precedes all ideas of individually extended things. This is the doctrine of *Meditations* III according to which "there is manifestly more reality in infinite substance than in finite, and therefore [we have in us] the notion [*perceptionem*] of the infinite before that of the finite" — a doctrine which was made so much of by Malebranche in the explication of his notion of intelligible extension.[15] If anything, individually extended things are the abstraction on this view.

Another critic of Cordemoy, Pierre-Sylvain Régis, argued at least vaguely along the more orthodox Cartesian line.[16] He distinguished between body, which is indivisible, and quantity, which is infinitely divisible. "There is no one who does not know that there is a substance extended in length, breadth and depth called *body*, for besides the fact that its existence is demonstrated in metaphysics, its idea is so comprised in all those that the imagination can form that necessarily either we know it or we never imagine anything."[17] Size "in itself" as a necessary result of extension is an essential property of body; a given size, quantity, is not. "Quantity is nothing else but body itself considered as such or such according to size."[18] The distinction between quantity and body is only one of reason, but this distinction is sufficient for them to have different properties. (It is the kind of distinction there is between numbers and things numbered, which have different properties.) (Infinite) divisibility (into proportional parts) is an essential property of quantity, but not of body. If body were divisible, then "since all division brings a change to the thing divided", its essence would be changed, which is contrary to reason.[19] Thus, after division each part still has the whole essence of body. (The properties of having some shape or other and of impenetrability are also essential properties of actually divided quantity). "The failure to distinguish body and quantity is the paralogism of those who following Epicurus maintain that atoms are indivisible because they are substances."[20]

Régis now gives what I take to be a fair approximation of Descartes's view as outlined above. "From this doctrine it follows, first, that having some quantity or other [*la quantité indeterminée*, as Regis perhaps misleadingly puts it] is of the essence of particular bodies . . . and second, that quantity is not an interior mode of body, but an exterior mode consisting in a certain manner in which body is conceived in relation to a given size. Whence it follows that quantity is not distinct from body by a formal or modal distinction, but by a distinction of reason, such as is found between substance and exterior modes".[21] Now, this notion of an exterior mode is what Descartes called an extrinsic attribute. In replying to Caterus, for example, he says that it is an extrinsic attribute to be objectively in the understanding, i.e. to be thought of. Régis explicates the notion as typically relational and as depending on how things are conceived.[22] The upshot is that, while Régis distinguishes, as

Desgabets had not, between modal being and beings of reason, the individuation of particular extended things still depends on *us*.

I'll turn now to a more orthodox atomist theory, as presented by the best-known continental proponent of atomism in the period, François Bernier (1620–1688). This student of Gassendi sought to promulgate the views of his teacher in his *Abregé de la philosophie de Gassendi*. This work appeared in two editions of seven and eight volumes in 1678 and 1684, although the first two volumes had appeared in 1674 and again in 1675.[23] It consists of translation, paraphrase, explication and, occasionally, tacit emendation of Gassendi's views as Bernier's own views departed from those of his teacher. Bernier's presentation of atomism, however, seems to me essentially to follow Gassendi's physicalist theories. For Bernier's own views we must look to a work of his called *Doubts about some of the main chapters of the Abregé*, which he published as an independent work in 1682 and then again in 1684 as part of the *Abregé* of that year. He raises substantive questions concerning Gassendi's views on space, time, motion, collision and other important topics, but he claims not to question the basis of Gassendi's system, for "[he does] not think that one can reasonably philosophize on any system other than that of atoms and the void". The *Doubts* of 1682 contain a section, curiously lacking in the 1684 edition, called "Whether the response I give here to the great and ordinary objections made against the existence of atoms is acceptable". In this section Bernier does not question or reject Gassendi's views as he does throughout the rest of the work. Rather, he provides arguments that go beyond them. For Bernier, perhaps guided by Cordemoy, with whom he was later to be associated by Dourtous de Mairan,[24] was edging toward metaphysical atomism.

Bernier's treatment is notable for its presentation of atomism as both a physical theory about the arrangement and behaviour of matter and as a metaphysical theory about why ultimately it exhibits the arrangement and behaviour it does. The section begins with two arguments "which among others demonstrate the view of atoms".[25] The first argument, which is found in many previous authors, is that infinite divisibility is incompatible with the finitude of things. What is composed of infinite parts must itself be infinite; the finite therefore cannot have infinite parts. The second argument Bernier draws specifically from Lucretius; only if there are absolutely solid, hard and impenetrable bodies with more or less void between them can the relative hardness and softness of bodies be explained. To these Bernier later adds a third argument: only if there are indivisible bodies can there be the resistance called impenetrability. The second argument, concerning hardness, seems more clearly physical than metaphysical, while the other two, concerning finitude and impenetrability, are ambiguous. In Bernier's consideration of objections to atomism, however, the metaphysical comes to prevail.

The first objection is one that Desgabets had raised against Cordemoy, *viz.* that parts of an alleged atom can be distinguished which are thus really distinct and separable. But for Bernier, "the sign, reason or cause of the separability of any body whatsoever is that it give way, not resist, not be perfectly hard",[26] and if all bodies were such, there would be no hardness whatsoever. He does not quite state it, but the generalized version of his argument is that there would be no *body* whatsover. That is, if all bodies were perfectly soft there would be no difference between them and the void. Thus, in order for there to be separable bodies there must be something that is inseparable — something whose integrity is assured, for example, by its conceivability apart from everything else.[27] So what we are talking about here are not real but conceptual parts. This is an explication, it seems to me, of Leibniz's cryptic argument later in the *Monadology*:[28] if there are composites then there must be (absolute) simples. A problem this dialectic raises among others is to explain how it is that parts that are inconceivable apart are nonetheless known as different. I shall return to this, at least briefly, toward the end.

Bernier argues further that impenetrable solid bodies cannot be considered as composed only of perfectly contiguous but separable parts, for if there are no continuous, inseparable parts, there are no finite bodies, which can have only a finite number of parts. The worry it seems to me is not that otherwise bodies would be too big, *viz.* infinitely large because of their infinite number of parts, but again that there that there would not be any bodies at all. A condition for the physical separability of bodies is the metaphysical integrity of other bodies, without which there would not be bodies at all. This gives the sense of the physical considerations Bernier cites. The point is that the physical explanation at the macro-level depends on metaphysical explanation at the micro-level. Thus he argues that ties or hooks of the sort to which Gassendi had appealed are of no use here (Bernier would not have had to look to Malebranche, nor Newton to either of them, for the obvious objection that the coherence of ties still needs explaining) and that motion, to which the Cartesians appeal, is of no use either: "for it is not the idea of motion or of rest that gives us the idea of separability; let diamond powder be as moved and disturbed as you wish, each individual grain will be no less hard and difficult to divide".[29]

The conclusion of Bernier's *Doubt* underlines the metaphysical character of his atomism. Here he raises the conundrum of whether God can divide an atom. "It is true that nothing opposes divine omnipotence, that the human understanding often sees only obscurely what does not imply contradiction, and that thus it must not inconsiderately determine what is or is not possible; so that we can only say here that if God [were to divide an atom] He would do so in a way no less supernatural, nor less inconceivable than the penetration [of bodies], and consequently in a way beyond the consideration of physics, which considers things only according to the ordinary course nature".[30] Bernier does not draw the comparison, but were God to divide an atom He would do something that was metaphysically no less impossible than if He were to separate a thing from its qualities.[31]

The conceivability of the different parts of an atom, it seems to me, is no more problematic for the atomist than the conceivability of the distinction between a thing and its shape or between its size and its shape should be for the Cartesian. The problem is that what is distinguishable should be distinct and separable. Thus for Berkeley, instantiated qualities such as this blue are ultimately the only physical individuals, which I think is the real significance of his well-known claim that primary qualities vary as do secondary. Thus, too, the use to which Hume puts his famous principle about distinguishability — even if he cannot get clear on just what it is that is no longer further distinguishable. (Is it such items as colour or shape, or is it things that have colour and shape?)[32] These subscribers to what we might call phenomenalist atomism still require that there be qualitative differences that are *given*. As we shall see momentarily, however, in our period the view was already emerging that only *numerical* differences are given. But how does this address the problem of the conceptual distinguishability of atoms into parts?

The distinction above, between (unobservable) micro- and (observable) macro-levels cannot be construed in the way Bernier thought as the distinction between metaphysical and physical accounts. This was argued as a *reductio* of atomism by Desgabets in his criticism of

Cordemoy, who as we have seen also relied on considerations of metaphysical independence in plumping for atoms and the void. Desgabets argued that Cordemoy's claim that each body *qua* substance cannot be divided was true only formally; materially it is false. Desgabet's distinction here is not altogether clear; it seems to be between the thing as nominally defined and the thing as having the constituents necessary for it to satisfy that definition. The constituents can exist apart, but not as such, as when a part is removed from a watch. With this distinction he replies to Cordemoy's contention that contrary to Cartesian doctrine a body at rest amidst others at rest is at least conceivable. According to Desgabets, bodies may be formally distinguished independently of motion in the way in which an army is independently of its members.[33] The entity formed by two drops, for example, is formally indivisible; but this is not to say that one drop cannot be separated from another. This shows in what sense Cordemoy's alleged atoms are indivisible. But it also shows, according to Desgabets, that unless they confuse mathematical and physical points, atomists have no reason for thinking that atoms must be imperceptibly small.[34] Drops could become indefinitely large.

Desgabets is right, it seems to me, that with respect to the kind of atomism that emerges in Cordemoy and above in Bernier the size of atoms is an irrelevant consideration. This is not the crude physicalist atomism of Gassendi, whose atoms were in principle imperceptibly small, or of Boyle whose view of an atom, one of the *minima* or *prima naturalia* as he called them, was that " 'tho' it be mentally divisible, yet, by reason of its smallness, and solidity, nature scarce ever actually divides".[35] For the problem is not that because of the size of atoms we cannot find knives thin and small enough to cut them. Instead it is a kind of atomism that at this historical juncture is groping its way toward an account of numerical differences in terms of bare individuals which have *no* properties and therefore no size at all.[36] The problem of the conceptual divisibility of atoms into parts is thereby resolved. Like size, shape, motion, and all other qualitative designatioins, divisibility is (as the problem suggests) purely conceptual. This is the metaphysical atomism sketched at the outset, according to which things just are (as opposed to the void). Such properties as things are said to possess are a function of our perception of them. To use the traditional language, they are accidents, which here means they are only phenomena or the appearances of things. This, I believe, is the view that emerges more clearly in Locke.

With this view, finally, we can see how an answer might go in response to two problems we have raised: (1) Bernier's problem of how it is that parts that are inconceivable apart are nonetheless known as different and (2) Berkeley's and Hume's problem of how it is that the colour and shape of a thing, which are inconceivable apart from the thing, are conceivable apart from each other. Both are answered by the most general form of Bernier's argument concerning separability: all prediction may depend ultimately on us, but every predication requires a metaphysically integral subject. Of any subject we can (within causal constraints) conceive anything we want, including that it has different parts; but that subject *qua* subject is metaphysically indivisible. (And what is true of parts is true *mutatis mutandis* of color and shape.) To this extent Desgabets's distinction between material and formal divisibility rather advances the cause of the metaphysical atomists. The difference between his view and theirs is twofold. First, the atomist (formal) definitions of things will be merely nominal, while the Cartesian definitions will be real. This difference is due to a second difference, *viz.* in what is here defined. For the realist Cartesians it is the single *res extensa*, the essence whose mathematical structure grounds a system of internal relations. Régis gets this exactly right in distinguishing between the (merely) conceptual entity he calls *quantity* which may change, and *body* which could not change without a change of essence. (This could happen only if, for example, God were to falsify Euclidean geometry.) For the nominalist Gassendists on the other hand, what is defined is one from a multiplicity of bare particulars, for there is no essence which exists apart from such things, or of which they are a part, or in which they participate. The real essence of a thing, as the atomists put it, just is that thing. So a thing cannot be essentially divisible or indivisible, or anything else. Classification of things is entirely a matter of how we classify them. This view too begins to emerge more clearly in Locke as it had already emerged in Hobbes.

To conclude, I would like to draw attention to an historical connection between the issues we have just examined and a concept that is central to Robert E. Butts's *Kant And The Double Government Methodology*. For it emerges that Kant was not the first, even in the early modern period, to have conceived the true philosopher as the physician of the soul. In a letter of 23 June 1676, Tschirnhaus[37] raised the problem of individuation with Spinoza, arguing that the variety of things cannot be deduced a priori from the concept of extension[38] and that, as Descartes thought, motion is also required. According to

Tschirnhaus, what is true in mathematics is true here, *viz*. "that from the definition of anything considered in itself, at most one property can be deduced";[39] for the deduction of further properties, the definiendum must be related to other things. Thus its uniformity as a curve can be deduced from the circumference of a circle, but any further properties require that it be related, for example, to the radii drawn from its center or to two or more intersecting lines. This is an example of Tschirnhaus's general method of definition, which requires a fixed part, e.g. an immobile point, and mobile part, e.g. a moving point that may define a figure. The application of this method to physics, especially in Cartesian terms, is tolerably obvious.

It is clear that Tschirnahus rejects Gassendi's ontology of atoms and the void as falsely extrapolated from what we are able to merely to imagine, although that ontology represents an advance over the obscurities of Fludd. Tschirnhaus clearly views Descartes as representing a further advance over Gassendi (*omnium optime explicavit materiam universi per impenetrabilem extensionem*), but it less clear whether Tschirnhaus's method actually yields the Cartesian position as outlined at the outset above, for his concept of the fixed element in physics as impenetrable may significantly differ from the Cartesian *res extensa*. In any case, the significance of these issues for Tschirnhaus is less ontological than epistemological and ethical, since his ultimate concern lies with the happiness made possible by knowledge of true physics liberated from the imagination. Tschirnhaus thus anticipates the pragmatic turn given philosophy by Kant,[40] even if his anthropology remains bound to the physiological standpoint, in "the Cartesian fashion", as Kant put it in contrast to his own properly pragmatic standpoint.[41] More importantly, however, Tschirnhaus conceives of his physics as the true medicine of the mind, thus restoring the physician model of the philosopher exploited previously by the Stoics[42] and shown by Butts to be so crucial to our understanding, later, of Kant. Tschirnhaus calls his book, indeed, *Medicina mentis*.[43] Finally, Tschirnhaus's main influence seems to have been on Wolff,[44] which would have been sufficient for a connection with Kant and thus with Butt's work. But there is a more direct and more important one, *viz.* through Leibniz, for it was Tschirnhaus who during Leibniz's *hibernas mirabilis* of 1675–6 brought him his first information concerning Spinoza's *Ethics*.[45] Indeed, Gerhardt published a conversation between them in which Tschirnhaus describes the last part of Spinoza's *Ethics* as "de . . . beatitudine seu perfecti hominis idea, de Medicina mentis, de Medicina corporis".[46]

University of Western Ontario
London, Ontario
Canada

NOTES

[1] "This philosophy of atoms is quite useful although self-contradictory, shape and indivisibility of bodies being unable to subsist together either in physics or mathematics." *Dissertations academiques* (Paris, 1674). Petit was active much earlier than the publication date of his book. In 1646 he introduced Pascal to the Torricelli experiments.

[2] I think it fair to say that Gassendi's atomism generally takes the form of a crude physical theory invoked to explain certain phenomena. Thus, for example, he supposes on the basis of his observation of the formation of salt crystals that gross bodies are "by some sort of necessity" composed of similarly shaped bodies and these of others until atoms of the same shape are reached. [to Peiresc, 6 July 1635; cited by B Rochot, *Les travaux de Gassendi sur Epicure et sur l'atomisme 1619–1658* (Paris, 1944) pp. 66–7.] Rochot, however, draws attention [pp. 174ff.] to one point in the *Syntagma* where Gassendi, in response to difficulties with respect to the divisibility and size of atoms, may be offering a rather more sophisticated theory than anything he resurrected from the ancients. I am none too certain of the interpretation of *Syntagma*, *Opera* I, 268a, but it may go something as follows. In the *minimum visibilium* there must be perceptible parts, e.g. the top, the bottom, each of the sides, that yet are not further *visibilia* and do not upset the *minimum* as a *minimum*. Similarly, atoms may have *conceptual* parts that account for their differences in size but that are themselves not separable as further atoms.

[3] *Syntagma*, *Opera Omnia*, I 273a–274a.

[4] For more on Cordemoy, *V.* the introduction by P. Clair and F. Girbal to his *Oeuvres philosophiques*, (Paris: P.U.F., 1968).

[5] *Ibid.* p. 33.

[6] Incipient metaphysical arguments were not the only non-physical arguments being advanced for atomism. Maignan earlier (1653) had argued atomism on theological grounds. Creation is possible, he thought, only if the continuum is composed of determinate parts, which as the object of creation, must be simple. *V.* Prost, *Essai sur l'atomisme et l'occasionalisme dans la philosophie cartésienne* (Paris, 1907) pp. 59–60. The idea seems to be that a limited creation must terminate in limited individual objects. An unlimited creation, for example Cartesian extension, presumably would be a rival divinity, although I don't know that Maignan ever made this charge as such.

[7] *Oeuvres*, pp. 95–96.

[8] *Oeuvres*, p. 98. This raises the question as to who it was that Spinoza was attacking for having conceived corporeal substance as an aggregate of bodies. For Tschirhaus's

role in relating to Leibniz Spinoza's relation to Descartes on this, *v.* H. A. Wolfson, *The Philosophy of Spinoza*, (Cleveland: Meridian, 1958; first ed. 1934) vol. I, pp. 268—70. We shall return to these important connections below.

[9] *Oeuvres*, pp. 103—4.

[10] For the details of his argument, irrelevant here, *v.* my "Occasionalism and the Cartesian Metaphysic of Motion", *Canadian Journal of Philosophy*, Sup. Vol. 1974, pp. 29—40.

[11] The letter remains unpublished in the library of Epinal. It is noted by Rodis-Lewis in *Analecta Cartesiana*, III p. xxxi. Excerpts from it appear in Prost, p. 156. ff.

[12] Interestingly, when Desgabets replies to Cordemoy's contention that the infinite divisibility of matter is inconceivable, it is the mathematical notion to which he appeals: "geometers know perfectly well what they [mean by infinite divisibility]." Prost, fn. 1 pp. 166—7.

[13] *V.* Supplément à la philosophie de Monsieur Descartes, ed. J. Beaude, *Analecta Cartesiana*, 2, pp. 243—4 & *passim.*

[14] Prost, p. 172.

[15] HR I, p. 166. Descartes is referring to the perception of God, but his doctrine applied to the perception of extension as well, as Malebranche showed.

[16] Arguably, the best-known Cartesian of his time, Régis (1632—1707) published his *Système de Philosophie* (Paris) only in 1690 but, according to Bayle, had written it some eighteen years before and delayed its publication due to political problems.

[17] *Système*, pp. 279—80.

[18] *Système*, p. 280.

[19] *Système*, p. 282.

[20] *Système*, p. 282.

[21] *Système*, p. 283.

[22] *Système*, p. 3 Alas, he gives shape as an example of an *internal* mode, i.e. as being conceived *in* the substance rather than as depending on something outside it. Perhaps he can be made consistent on this; but perhaps not.

[23] The publishing history and composition of this work are complicated. I discuss them in an article for the forthcoming *Ueberweg.*

[24] *v. Malebranche: Oeuvres Complètes*, (Paris: Vrin, 1961—   ) XIX, p. 902.

[25] *Doutes . . . sur quelques-uns des principaux chapitres de son Abregé* (Paris, 1682) p. 28.

[26] *Ibid.* p. 136.

[27] As it was for Descartes; *v. Principles* I, 60; which shows why for him the whole of extension is a single entity: none of its parts can be conceived apart from the whole.

[28] Para. 1—2.

[29] *Doutes*, p. 142.

[30] *Ibid*, pp. 143—4.

[31] *V.* below. This of course was generally how the miracle of the Eucharist was described. Gassendi early on came close to elaborating a doctrine of bare individuals, i.e. metaphysical atoms, with respect to the problem of the Eucharist. The essence of quantity he took to be "external extension", the tridimensionality perceived by the eyes that is commensuratc with place. To distinguish "internal extension" from this as consisting of the relations among non-spatial or entitative parts is either to make no distinction or to introduce something which cannot have any relation at all among parts. But Gassendi thought the notion of internal extension incoherent ("how can there be a

relationship between the parts where there is no relationship at all between up and down or any other position" — i.e. apart from absolute space). Instead he opted for a fideistic response to the problem: "since this mystery surpasses the capacity of the human mind no matter how you consider it, it is best to refer anything miraculous and inconceivable in it to divine will and omnipotence". *Exercises against the Aristotelians*, 2, 3, 10–11; *v. The Selected Works of Pierre Gassendi*, trans. C. B. Brush, (New York: Johnson, 1972) pp. 58–64.

32 "Whatever subjects are different are distinguishable, and . . . whatever objects are distinguishable are separable by the thought and imagination. And we may . . . add, . . . these propositions are equally true in the *inverse*." *Treatise*, Selby-Bigge (ed.), p. 18. But consider: "Tho' a particular colour, taste, and smell are qualities all united together in this apple, 'tis easy to perceive they are not the same, but are at least distinguishable from each other." *Ibid.* p. 2. This is Hume's view when discussing the simple-complex distinction; when he discusses the distinction of reason he says: "we consider the figure and colour [of marble] together, since they are in effect the same and undistinguishable". *Ibid.* p. 25.

33 Prost, p. 67.

34 Prost, p. 159 ff.

35 *Origin of Forms and Qualities* (ed. of 1725) vol. 8, p. 209.

36 More's atomism represents a curious half-way house in this regard. His "perfect parvitudes" or "minima corporalia" have no shape — as the infinitely large has no shape, so the infinitely small has none. An infinitely small cube in fact is a contradiction for him. But they do have extension because finite extensions are composed of them. *The Immortality of the Soul* (London, 1659) BK I, ch. 6; preface, sec. 3. Nor are they bare individuals: "the subject, or naked Essence of a thing, is utterly unconceivable to any of our faculties." BK I, ch. 1, Axiom viii. All things have one from each of two pairs of "immediate properties": penetrability/impenetrability; discerpability/indiscerpability.

37 Ehrenfried Walther von Tschirnhaus, Freiherr von Kieslingswalde und Stoltzenberg (1651–1708). For more on this fascinating inventer of both the Tschirnhaus Transformation in mathematics and, perhaps, the Meissen method of glazing china, *v.* L. W. Beck, *Early German Philosophy: Kant And His Predecessors* (Cambridge, Mass.: The Belknap Press of Harvard University Press, 1969). p. 189 ff.

38 *An ex solo Extensionis conceptu rerum varietas a priori possit demonstrari . . . . Spinoza Opera*, ed. C. Gebhardt (vol. IV, p. 334. *V.* also his letter of 5 May and Spinoza's reply of 23 June; *ibid.* pp. 332–3.

39 *Ibid.*, p. 335.

40 R. E. Butts, *Kant And The Double Government Methodology: Supersensibility And Method in Kant's Philosophy of Science*, (Dordrecht; D. Reidel, 1984) p. 14.

41 *V.* Butts, p. 285.

42 *V.* Cicero, *Tusculanae Disputationes*, IV, x, 23.

43 Amsterdam, 1687; 2nd ed., Leipzig, 1695. Repr. Hildesheim, Georg Olms, 1964. For Tschirnhaus's discussion of the atomist issues of concern to this paper, *v.* esp. p. 177 ff.

44 Beck, p. 190.

45 *V.* G. W. Leibniz: *Philosophical Papers and Letters*, trans. ed. L. E. Loemker [Dordrecht: D. Reidel, 2nd ed., 1969] fn. 1 p. 166.

46 Quoted by H. A. Wolfson, vol. II, p. 263.

KATHLEEN OKRUHLIK

THE FOUNDATION OF ALL PHILOSOPHY: NEWTON'S THIRD RULE*

The recent revival of the interest in scientific realism and the status of theoretical entities makes it appropriate to look once more at the debate surrounding Newton's third "Rule of Philosophizing". The questions at issue concern when (if ever) one is justified in making inferences from observation reports to claims about unobservable entities and what constraints (if any) must supplement the requirement of hypothetico-deductive adequacy in making such inferences.

The rule first appeared in the second edition of the *Principia*, published in 1713. From that time until about 1964, it was generally thought to be on a par with the other three rules; that is to say, it was viewed as a purely methodological dictum. Though the rule was sometimes singled out for criticism, complaints were of a methodological sort. They questioned the efficacy of the rule as a guide to scientific procedure. Beginning in the 1960's, the third rule was seen to be fraught with ontological and epistemological commitments and to be intimately bound up with Newton's atomism. A pre-requisite for this late-blooming recognition was a prior appreciation of the depth of Newton's belief in the real existence of atoms.

This essay looks first at some traditional interpretations and evaluations of the rule. Special attention is paid to the criticisms of William Whewell because some of the distinctions he draws are akin to ones I shall invoke in disagreeing with parts of J. E. McGuire's treatment of the rule. The second and the third sections examine the rule in the new light cast on it by Maurice Mandelbaum's and McGuire's interpretations of it as a rule of transduction. In these sections the intensification/remission criterion will be ignored as far as is possible. This criterion will form the subject matter of the fourth part, in which the question of its significance will be closely examined. Is it meant to distinguish primary from secondary qualities as Mandelbaum maintains; and, if so, is McGuire's larger claim justified? Is it true that this rule makes possible the identification of the primary with the essential characteristics of matter? It is on this issue that I shall disagree most strongly with McGuire.

J. R. Brown and J. Mittelstrass (eds.), An Intimate Relation, 97—113.

Not until the final section will the application of the rule to the case of gravity be discussed. We must decide whether gravity is the sort of thing which can be justly universalized under the auspices of this precept. It is necessary also, to examine the implications of our answer for its status as a primary or secondary, essential or non-essential characteristic of matter.

The main body of the third rule reads as follows:

> The qualities of bodies which admit neither intensification nor remission of degrees, and which are found to belong to all bodies within the reach of our experiments, are to be esteemed the universal characteristics of all bodies whatsoever.[1]

HISTORICAL BACKGROUND

Ralph Blake's evaluation is typical of the traditional view:

> The Third Rule seems to be no more than a reformulation of the first two with respect to another set of conclusions. It seems obvious that in bringing before our minds these notions of the simplicity and uniformity of nature, Newton has in mind not so much any general theory of the matter as the utility of these principles for the defense of his own system against various forms of objections.[2]

E. A. Burtt, in *The Metaphysical Foundations of Modern Science*, asks: "Is not this a highly speculative assumption of the Cartesian sort, that it is legitimate to generalize *ad infinitum* the qualities discovered in the small realm of our experience or is it perhaps a purely methodological postulate?"[3] Burtt opts for the latter alternative, saying that the rule is nothing more than a combination of the experimental principles of nature's uniformity. He bases this interpretation on the first elaborating sentences of the rule:

> For since the qualities of bodies are only known to us by experiments, we are to hold for universal all such as universally agree with experiments; and such as are not liable to diminution can never be quite taken away. We are certainly not to relinquish the evidence of experiments for the sake of dreams and vain fictions of our own devising; nor are we to recede from the analogy of Nature which is wont to be simple and always consonant to itself.[4]

Alexandre Koyré, also, emphasizes the rule's demand for experimental warrant and treats and it chiefly as a polemic against the Cartesians.[15]

Like the three commentators just cited, William Whewell regards the rule as a methodological postulate or "rule of prudence" — but as one of the most disreputable sort. He views the rule chiefly in the light of its role in the universalization of gravity, but his criticisms are able to stand independently of that particular application. Whewell's contention is that "the assertion of the universality of any properties of bodies must be grounded upon the reason of the case and not upon any arbitrary maxim."[16] No amount of experience can completely justify the universal extension of any quality of matter; for what is at issue here is not a relation of facts, but of ideas. Only a rational insight into the *necessity* of a quality legitimates our attribution to it of universality. Experience is required to bring about that insight, but the necessity itself is in no way derived from experience. Rather, there comes a point at which one realizes that matter *could not exist* without having certain characteristics: this is to say, one realizes that these characteristics are *essential* to matter. It is the essential nature of the characteristic which warrants universalization, not experience. Newton is wrong, according to Whewell, to maintain as he does in the elaboration of rule three that it is experience which teaches us that all bodies are extended, hard, impenetrable, mobile, and possessed of inertia. This is because we are never justified in saying that we *find* something in experience unless we could conceivably find the contrary:

> But who can conceive our finding by experience some bodies which are not extended? It appears, then that the reason given for the Third Rule of Newton involves a mistake regarding the nature and authority of experience. And the Rule itself cannot be applied without attempting to decide, by the casual limits of observation, questions which necessarily depend upon the relation of ideas.[7]

Whewell is not merely arguing that the third rule does not justify the universal extension of gravity. Rather he maintains that it does not legitimate *any* such extension; it is simply a bad rule. Though the non-gravitational properties mentioned by Newton may be legitimately predicated of all bodies whatsoever, according to Whewell, it is on account of their essential character rather than because of their grounding in experience.

Whewell's criticism is important in this context because he introduces much of the conceptual apparatus required for later discussion. I shall maintain in section three that — regardless of the validity of Whewell's criticism — he did correctly understand Newton's intent vis-à-vis essences and that McGuire misconstrues Newton's intention in this same regard. Newton *was* maintaining as Whewell contends he was, that we universalize qualities of matter without benefit of any insight into the essence of matter. We have no warrant which transcends experience on Newton's account.

THE PROBLEM OF TRANSDUCTION

In his 1964 book, *Philosophy, Science, and Sense Perception*,[8] however, Maurice Mandelbaum offered a new and very different interpretation of the third rule. He argued convincingly that this rule is heavily laden with ontological and epistemological import and furthermore, that it was via this postulate that Newton was able to resolve an apparent tension in his thought. On the one hand, Newton demanded that all scientific laws be "deduced from the phenomena" — i.e., be empirically derived. On the other hand, however, Newton spoke often of corpuscles which were not only unobserved, but perhaps in principle non-observable. (Because of their structure, the smaller corpuscles would be unable to interact with light. It would become stifled and lost within their vacuities.[9]) Moreover, these non-observable entities often played an explanatory role in Newton's thought. Such a role might seem to be in conflict with Newton's much touted empiricism. This conflict had not been studied extensively in the past, according to Mandelbaum, because few commentators took Newton's atomism seriously. Instead, his natural philosophy had been portrayed in an overly-positivistic light — probably as a result of excessive emphasis on the "Hypotheses non fingo" aspect of his methodology. His interpreters and critics had thought that he was interested only in the establishment of *laws* which allowed extrapolation from one set of observed phenomena to another. Few had attributed to him any interest in the *mechanism* via which the laws might be explained. If Newton sometimes spoke of atoms or of corpuscles, these were understood as mere constructs facilitating various extrapolations.

More recent scholarship, however, has tended almost unanimously to the view that Newton was firmly convinced of the real existence and causal efficacy of atoms. This new interpretation has thrown into sharper focus what Mandelbaum calls the problem of transdiction and J. E. McGuire, the problem of transduction. (For the sake of consistency, I shall use the word "transduction" throughout.)

In general, the problem of transduction is that of making inferences from observed phenomena to entities which are in principle non-observable. There are obviously many faces to this problem in philosophy; but we are concerned here with just one: How can Newton use data from experience to justify descriptions of in principle non-observable corpuscles? Is there any way in which he can justify his beliefs regarding the ultimate constituents of matter without in the same breath violating his requirement of experiential warrant?

Before proceeding, an important distinction must be made. Though Newton demanded that all scientific propositions be *derived* from experience, he did *not* maintain as would some of his successors that explanation must be in terms of the phenomenal. Ultimate units of explanation need not be units of experience. Nothing in Newton's work prohibits non-observable entities as units of explanation. So atomism is not ruled out *a priori*. The problem that remains, however, is this: How can we legitimately make inferences to non-observable atoms while preserving experiential warrant? Which properties of the phenomenal realm may be legitimately extended to atoms?

It was Mandelbaum's contention that the third rule provides precisely the legitimation needed. In the first place, we can attribute no properties to atoms which are not derived from experience: the properties must be properties of *observed* entities. If Mandelbaum made much ado about the primacy of the empirical derivation of atomic properties, it was in order to contrast Newton's attitude with a more modern emphasis on confirmation and, especially, predictive power. For more recent philosophers of science, a theory's origin — the method of its derivation — is relatively unimportant; what counts is explanatory and predictive power. Mandelbaum took great pains to point out that regardless of the explanatory power of a hypothesis, it was unacceptable to Newton unless tied very intimately into the phenomena. Derivation was more important than confirmation — or perhaps more accurately, the method of derivation was a very important consideration in determining how well confirmed any theory might be.

If the first earmark of a true scientific law was for Newton its empirical derivation, then the second was its universalizability. True scientific laws were those applicable throughout nature — that is, even to cases of a different sort from those for which they were first formulated. Newton's demand that a law be capable of being "rendered general by induction" is, in effect, a demand for universalizability.

Furthermore, this capacity for universal extension is evidence of the *truth* of a law. Consequently, theories about atomic entities must concern only universalizable properties.

Now Mandelbaum's conclusion is undoubtedly a correct interpretation of Newton's requirement in the third rule. I do not think, however, that his effort to derive that requirement from a more general characterization of Newton's methodology is legitimate. Surely Newton does not require that all laws be universalizable in the strong sense which Mandelbaum described. In this case, *all* laws would be applicable to the atomic realm; and they manifestly are not. Furthermore, "universalizability" is a tricky criterion. The only way to get mileage out of it here would be with a very full-blown notion of consilience; and I am not aware that Newton had any such notion. It would seem instead that the "universalizability" requirement is peculiar to cases of transduction and that it is not a requirement upon all scientific laws for Newton.

Mandelbaum's conclusion was that "Characteristics which are invariably associated with experienced objects are also characteristics of all objects whatsoever."[10] This maxim specifies the conditions under which transduction is permissible. Experiential warrant is preserved in moving from the observed to the non-observable if and only if the properties being attributed to non-observable entities are properties "invariably associated with experienced objects." It is important to note here that Mandelbaum attached little significance to the intensification/remission criterion. For him, it was just an amplification of the requirement that properties be associated with all experienced objects: If a property were subject to remission, it would be reasonable to assume it altogether absent in some objects. J. E. McGuire makes much more of the intensification/remission criterion, giving it priority over the requirement of empirical derivation.

Mandelbaum also argued that the maxim quoted above provides a hitherto unrecognized means of distinguishing the primary from the secondary qualities of matter. He supported this conclusion largely by pointing out that those properties which pass the third rule's test are in fact those which appear on traditional lists of primary qualities. There was little attempt made at conceptual elucidation of this co-incidence. McGuire's efforts in this direction are more ambitious. These efforts are to be found largely in his essay "Atoms and the 'Analogy of Nature': Newton's Third Rule of Phisolophizing".[11] The goal there is to integrate several aspects of Newton's thought: his atomism, his belief in the analogy of nature, his doctrine of primary and secondary qualities, and his theories regarding the transmutability of various sorts of matter. This allows us to locate the universalizability criterion in the larger context of Newton's natural philosophy. Mandelbaum saw Newton's alleged requirement of universalizability as a reflection of his belief that nature is always consonant with itself; that is, as a reflection of the analogy of nature. On Mandelbaum's account, however, the analogy itself remained an isolated justificatory principle. McGuire's efforts at integration are very useful, and the following discussion is largely derived from his work.

ATOMS AND THE ANALOGY OF NATURE

There are three tenets of Newton's atomism which we ought to examine briefly. These tenets allow us to see clearly how bound up Newton's atomism was with the rest of his metaphysics.

First, atoms are the ultimate units of explanation. The whole may be explained in terms of its parts, but not vice-versa. Supra-atomic phenomena are not randomly emergent, but are determined by atomic properties and configurations. This is a pre-requisite for the possibility of a unified science; for if this relationship did not obtain, the macro- and micro-realms would be governed by entirely different sets of laws. Newton once flirted briefly with the possibility of a world in which this part-whole relationship did not obtain[12] (i.e., with one in which atomic and phenomenal laws were entirely unrelated.) but quickly rejected this notion as a blatant violation of the analogy of nature. The doctrine of the analogy of nature teaches that there are no discontinuities, no radical breaks in nature. Instead, various levels of being merge gradually into one another in such a way as to form a continuum. Whether plenitude is actual or potential, it is a reflection of God's perfection. The constancy of the ultimate particles, which is to be discussed shortly, mirrors God's immutability.

The point here is this: The fact that macro-phenomena can be explained in terms of atomic properties and configurations guarantees for Newton the analogy of nature. If gross phenomena were randomly emergent, the analogy would be violated. Instead, the properties of the elementary corpuscles determine the nature of the universe which is built up out of them.

An interesting and important consequence of this relationship is this:

If the "original properties" of atoms had been other than they are, then the laws at the macro-level would be different:

> . . . it may also be allowed that God is able to create particles of matter of several sizes and figures, and in several proportions to space, and perhaps of different densities and forces, and thereby to vary the laws of nature and make worlds of several sorts in several parts of the universe. At least, I see nothing of contradiction in all this.[13]

This belief in the possible existence of other worlds will play a role in my disagreement with McGuire on essential properties. Newton differed radically from Descartes on this issue. For Descartes, only the present laws of nature are really possible. Conceptual considerations regarding extension make other laws impossible on his account. No such conceptual limitations are present in Newton's model. We must *discover* empirically which are the original properties; no amount of conceptual analysis can give us the answer because there are many possible worlds governed by different fundamental laws.

Consequently, "Why?" questions about observed entities can be answered in terms of elemental properties and configurations. But "Why?" questions about the atoms themselves are ill-formed. They can be answered only by saying; "God made it so in the beginning."

The second tenet of Newton's atomism to be canvassed here is closely bound up with his adherence to the primary/secondary distinction, just as the first tenet is inseparable from his belief in the analogy of nature. Newton held that the atoms themselves are immutable and that all change and variety in the world is dependent upon atomic motions and configurations rather than upon any change in the atoms themselves. All change is with respect to form.

We said earlier that a unified science is possible only if there are determinate relations among various levels of matter. Here something even more basic is at stake. No science whatsoever is possible unless there are patterns in nature; certain elements must be invariant through space and constant through time. This constancy is for Newton the bedrock of rationality. It makes the universe a cosmos rather than a chaos. A world with no universalizable truths would be a chaos, and science would have no place within it. This is not to say that there are elements of explanation which are necessarily universalizable, but that necessarily there are elements of explanation which are universalizable. (We shall see that McGuire mistakenly attributes the former position to Newton.)

Atomism locates these universalizable elements in the atomic realm. This is because all variety and all change in the phenomenal world is a result of change in the *configuration* of atoms themselves. The classic and very familiar statement of this view is in Query 31 of the *Optics.*

All these things being considered, it seems probable to me that God in the beginning formed matter in solid, massy, hard, impenetrable, movable particles, of such sizes and figures, and with such other properties, and in such proportion to space, as most conduced to the end for which he formed them; and that these primitive particles being solids, are incomparably harder than any porous bodies compounded of them; even so very hard as never to wear or break in pieces; no ordinary power being able to divide what God himself made in the first creation. While the particles continue entire, they may compose bodies of one and the same nature and texture in all ages; but should they wear away, or break in pieces, the nature of things dependent on them would be changed. Water and earth, composed of old worn particles and fragments of particles, would not be of the same nature and texture now with water and earth composed of entire particles in the beginning. And, therefore, that Nature may be lasting, the changes of corporeal things are to be placed only in the various separations and new associations and motions of these permanent particles; compound bodies being apt to break, not in the midst of solid particles, but where these particles are laid together, and only touch in a few points.[14]

This passage bears directly on both of the first two tenets at issue here. First, it explicitly says that if the nature of the atoms were changed, "the nature of things depending on them would change", confirming that atoms are the ultimate causal principles. Secondly, it unambiguously asserts the immutability of atoms and attributes all change to "the various separations and new associations and motions of these permanent particles."

The second tenet, supplemented by the first, sets the stage neatly for the primary/secondary distinction. It is traditionally held that only primary qualities have causal efficacy. Secondary qualities are caused by primary qualities, but do not resemble their causes. Perceived primary qualities exactly resemble their causes. The first tenet, then, which locates ultimate causal efficacy in the atoms allows us to see why the primary qualities must inhere in atoms rather than in bodies *qua* bodies. The second traditional characterization of the primary/secondary distinction confirms this conclusion. This characterization says that primary qualities and they alone are inseparable from matter. All that is configuration-dependent is subject to change and alteration, according to Newton. Only the properties of the atoms themselves are immutable.

Because they are not configuration- or motion-dependent, they are proper to all bodies regardless of the arrangement of their corpuscles. Because they inhere in the atoms themselves, these properties cannot be phased out by any amount of re-arrangement. They are not form-dependent; they are inseparable from matter.

So Newton's brand of atomism nicely accommodates the now-traditional primary/secondary distinction, although this distinction is patently untenable for anyone who locates the ultimate units of explanation in phenomenal entities rather than in unobservable atoms.

The third tenet of Newton's metaphysics to be canvassed here concerns the homogeneity of Newton's atoms; and the aspect of his metaphysics to which I'd like to relate this tenet is his belief in the transmutability of matter. In the first place, Newton's atoms differ from their twentieth-century counterparts in that there are not different sorts of atoms proper to various substances. An atom of iron is for Newton like an atom of gold. (In the *Opticks*, he does sometimes admit *differentiae* such as size and shape; but he does not seem to consider these important.) This homogeneity is absolutely essential to Newton's enduring belief that different sorts of matter (and even light) can be transformed into one another through continuous degrees. When one traces the history of Newton's transmutation hypotheses, it becomes clear that remnants of it remain even in Newton's most mature work. Now, if there were radically different sorts of atoms each proper to a specific sort of matter, transmutation would be manifestly impossible. No amount of re-arranging of corpuscles could yield a substance composed of different *sorts* of corpuscles.

In his earlier work, prior to formulation of the primary/secondary distinction, Newton could talk as if *all* qualities of matter might be changed through continuous degrees — even eliminated by moving through all gradations to degree zero. Later, though, he had to exempt the primary qualities for obvious reasons. It is worth noting here that when Newton re-interprets the analogy of nature in terms of atomism, it ceases to be a hierarchical principle as it was for the neo-Platonists and becomes instead an egalitarian principle: All atoms are created equal, have the same properties, and are governed by the same laws.[15]

It is clear that the transmutation hypothesis is closely tied to Newton's belief in the analogy of nature, as well as to the homogeneity of atoms. The logical relationships among the various aspects of Newton's metaphysics are far more numerous and complex than those traced out here.

The object here has been merely to spell out in a fairly simple-minded way some of the metaphysical interrelations that ground the methodology of the third rule.

In conclusion, there is very good evidence for Mandelbaum's and McGuire's contentions that the third rule is ontologically-loaded with atomism, that it is intended to provide a method for making valid transductions, and that it separates the primary from the secondary qualities. I have avoided here as far as is possible discussion of the intensification/remission criterion because its most important role in this debate involves a still larger claim advanced by McGuire. He argues that the third rule does not only provide a sieve for separating the primary from the secondary qualities, but that it also allows us to identify the primary with the essential qualities. This claim is made in a 1968 article entitled "The Origins of Newton's Doctrine of Essential Qualities".[16]

PRIMARY vs ESSENTIAL PROPERTIES

The chief value of this essay lies in the fact that it shows that the intensification/remission criterion is not of the same origin as the rest of the third rule. What McGuire calls the First Draft of the rule does not mention this criterion. Its emphasis, instead, is on the requirement of experiential warrant for properties to be universalized. This, as we have seen, is the aspect emphasized by Mandelbaum who virtually ignores the intensification-remission criterion. McGuire, on the other hand, gives short shrift to universalizability and is intrigued by the "more philosophical" Second Draft of the rule. It is here that Newton first says that only qualities not subject to intensification and remission can legitimately be predicated of non-observables. McGuire interprets this as a *conceptual* requirement which Newton wishes to integrate into his empirical methodology. He attributes its more philosophical character to Newton's association with Locke around 1690.

One of McGuire's chief aims in the paper was to trace the history of the intensification/remission phraseology from Aristotle through the Scholastics and through Dibgy to Newton. He concludes that:

> What Newton does in effect is to appropriate the traditional distinction between qualities and substance, express it in the language of the latitude of forms, and then apply it to the general conception of matter more or less acceptable in his time. As we

have seen, his conclusion is that only certain qualities are intrinsic to it. He refers to these as being such as are not liable to diminution and can never be taken away.[17]

Bear in mind that McGuire believes, on the basis of his historical analysis, that the limitations imposed by the intensification/remission criterion are *conceptual*, not empirical. This is most evident in the following passage which I quote rather than paraphrase because the exact wording is very important.

(Newton) is not merely using his criterion as a means of categorising primary and secondary qualities — this in itself would not imply that the primary are the essential qualities — but his is also arguing that it is because certain properties *cannot* be held to intend and remit that they are essential. Thus, the two lists of qualities, in connection with Newton's argument, purport to distinguish those which a body *must* have from those which it *might* have under certain conditions.[18]

Now, if Newton's intent is what McGuire says it is, then it follows inexorably that those qualities which do not intend and remit are indeed the essential ones. For, whereas the primary qualities are as a matter of fact universal, essential properties are those without which a body can neither be nor be thought of. Essentiality *is* largely a conceptual matter.

My contentions are that McGuire is quite mistaken about Newton's intent, that the intensification/remission criterion does not impose conceptual constraints, and — most importantly — that Newton does not believe that conceptual considerations can shed much light on the composition of matter.

My disagreement with McGuire can be located in his introduction of unwarranted modal considerations in the passage just quoted. He says there that "it is because certain qualities *cannot* be held to intend and remit that they are essential" and, furthermore, that the third rule distinguishes qualities "which a body *must* have from those it *might* have under certain conditions". The underscored words (emphasis is McGuire's) all pertain to conceptual possibility or necessity — notions alien to the third rule as I understand it. McGuire's point seems to be that there is a sort of conceptual insight to the effect that it is inconceivable that matter should lack certain properties and that it is this insight which licenses universal attribution of these properties. This interpretation is supported by McGuire's assertion that ". . . it became impossible for Newton that bodies, existing as phenomena independent of us, could be otherwise conceived".[19]

But Newton's argument seems to be precisely the opposite of what McGuire says it is. He is insisting that we have nothing more than *experiential* evidence for the universality of any property. The light of nature gives us no insight into essences. Newton says in the General Scholium of the *Principia*:

> ... what the real substance of anything is we know not. In bodies we see only their figures and colours, we hear only the sounds, we touch only their outward surfaces, we smell only the smells and test the savors, but their inward substances are not to be known either by our senses or by any reflex act of our minds ...

The entire thrust of the third rule is toward Newton's conclusion that our evidence for the universality of gravity is of precisely the same sort as our evidence for impenetrability, extension, etc. In each case, says Newton, the evidence is *experiential*, not conceptual. There is no evidence for universality which transcends experience. The elaboration of rule three says:

> We in no other way know the extension of bodies than through our senses, nor do these reach it in all bodies; but because we perceive extension in all that are sensible, therefore we ascribe it universally to all others also. That abundance of bodies are hard we learn by experience; and because the hardness of the whole arises from the hardness of the parts, we justly infer the hardness of the undivided particles, not only of the bodies we feel, but of all others. That all bodies are impenetrable, we gather not from reason, but from sensation. That all bodies are movable and endowed with certain powers (which we call the inertia) of persevering in motion or in their rest, we only infer from the like properties observed in the bodies we have seen. The extension, hardness, impenetrability, mobility, and inertia of the whole result from the extension, hardness, impenetrability, mobility, and inertia of the parts; and hence we conclude that the least particles of all bodies to be also extended, and hard and impenetrable, and movable and endowed with inertia. And this is the foundation of all philosophy.[20]

Conceptual analysis can only tell us that *some* qualities must be universalizable; it can tell us nothing about *which* qualities are universalizable. Newton's very interesting discussion of the possible divisibility of corpuscles[21] is evidence for this interpretation. Though he clearly believed that atoms are indivisible, he argues that should even a single corpuscle be proved infinitely divisible we would have to assume all matter to be infinitely indivisible. No amount of thought can establish with certainty whether divisibility or indivisibility is proper to the corpuscles. Conceptual analysis along the lines discussed above can tell us only that the properties of atoms, *whatever they are*, must be universalizable. My complaint is that McGuire confuses the following two requirements:

$$\Box[(\exists x)\,(x \text{ is universalizable})] \quad (1)$$
$$\exists x[\Box(x \text{ is universalizable})] \quad (2)$$

Newton opts only for (1) while McGuire attributes (2) to him.

Two final observations should be made before leaving this topic. First, if Newton meant what McGuire says he meant, then Whewell can have no complaint with the third rule (except that it is inapplicable to gravity). Whewell's oft-repeated dictum is that necessity implies universality, but not vice-versa. If the intensification/remission criterion *did* pick out essential (necessary) qualities, it would be perfectly legitimate to extend these universally. If it *did* concern a relation of ideas rather than of facts, then the third rule would be a good one rather than an "arbitrary maxim". Whewell is right, however, in saying that Newton does not intend the intensification/remission criterion as a relation of ideas, but of facts. *Contra* McGuire, it is an empirical test. Consequently, the tension between the "empirical" and "conceptual" components of the third rule which he tries to resolve are non-existent. McGuire says, for instance:

> In the General Scholium (Newton) seems to suggest that we cannot know the essence of matter, but in the Third Rule and the Queries, he states that *all* matter is extended, and so forth. Still, we must not too hastily conclude that these statements are contradictory.[22]

The preceding analysis makes it clear that there is not even apparent contradiction here. Newton believes we must universalize without benefit of insight into the essential make-up of matter.

IS GRAVITY AN ESSENTIAL PROPERTY OF MATTER?

Despite the fact that recent attention has focused on the third rule's applicability to atoms, its original *raison d'etre* was the universalization of gravity, both in the large and in the small. (Distant planets, as well as atoms, are in some sense non-observable.) Newton's only specific invocation of the rule occurs at proposition 6 corollary 2 of the third book of the *Principia* where he says that because gravity is a property of all bodies within the reach of our experiments, it is to be affirmed of all bodies whatsoever. He had set the stage for this invocation in the final paragraph of the rule. There he had written that the evidence for a principle of mutual attraction is at least as good as that for extension, impenetrability, mobility, hardness, and inertia. It is present in all bodies we have ever observed. We are, therefore, justified in attributing it to all bodies whatsoever.

It is a curious fact that although the third rule appears to owe its existence largely to Newton's needs in universalizing gravity, there is some question as to whether gravity even falls under the criteria it sets up. Much of the difficulty stems from equivocal use of the word "gravity". Most of the time Newton uses it to refer to a *force*; that it, to something with casual efficacy. When it suits his needs, however, he uses it to refer to weight or *pondus* — i.e., to the *effect* of the aforementioned cause. Now, the difficulty is this: *Pondus* fits some of the rule's requirements; others are met by gravity-as-force. Neither concept alone appears to fit the rule exactly.

First, isn't gravity subject to intensification and remission? Clearly *weight* is. So the only candidate here is gravity as a force of mutual attraction. What is constant here is its proportion to quantity of matter within any given latitude. Newton clearly had this in mind when formulating the third rule. He soft-peddled the notion somewhat in the published Principia; but from early sketches of the rule, it is clear that the invariance of the proportionality is what guided Newton's formulations of the rule. Furthermore, within the *Principia* itself, Newton sometimes uses the invariance of proportionality to distinguish gravity from other forces, such as magnetism.[23]

Secondly, can gravity legitimately be called a "quality of bodies"? Clearly, gravity-as-weight can, but we have already seen that it fails to satisfy the intensification/remission criterion. So what about the *force* of gravity? Is it a quality of bodies? In what sense can a force operating between two bodies be said to be a quality of either one? In order to justify this locution, we might begin to think of gravity as force or possibly even an *attracting power* residing within each particle. This was the source of Newton's difficulties with his contemporaries who thought that he was making an attracting power essential to matter. Part of Newton's ploy in the third rule had been to give gravity a sort of respectability-by-association. He wanted to show that, in certain respects, gravity was right up there in a class with extension, impenetrability, etc. This ploy backfired, however. Instead of thinking that Newton wished to put gravity in this class on the basis of empirical well-foundedness alone, the public, naturally enough, assumed that he wanted to group the properties on the basis of essentiality.

Even Cotes, in first drafting his Preface to the second edition of the *Principia*, called gravity an essential property of matter. Clarke corrected him and Cotes ultimately wrote that:

> Either gravity will ultimately have a place among the primary qualities off all bodies, or Extension, Mobility and Impenetrability will not. And the nature of things will either be rightly explained by gravity of bodies or it will not be rightly explained by their extension, mobility and impenetrability.[24]

Calling gravity "primary" rather than "essential" did not placate Newton's readers, however. So in the third edition of the *Principia*, Newton appended the following disclaimer to the third rule:

> Not that I affirm gravity to be essential to bodies; by their *vis insita* I mean nothing but their inertia. This is immutable. Their gravity is diminished as they recede from the earth.

Now, this is a perfect red herring; for it refers to *weight*. And no one thought Newton was trying to make weight an essential property. As we have seen, the only way to make gravity fit the third rule's criteria is by regarding it as a force residing within corpuscles. It is this conception which is at issue, not weight.

Newton wanted to leave open to himself the option that gravity might not be an ultimate cause, that it might be explained by something else. But properties which are primary and original are supposed to be ultimate units of explanation. It would appear that the only way to maintain consistency here is by saying that the third rule provides a necessary but not sufficient test for primary status. A less than satisfactory solution.

Despite all these difficulties in fitting gravity to the third rule, it must be noted it does enjoy very special status in Newton's system. Other forces are subject to intensification and remission in a way that gravity is not. They are dependent upon the *form* of matter in a way that gravity is not; proportionality with respect to quantity of matter is not constant. It is not difficult to see why Newton wished to put gravity in a class with the primary qualities.

One can only conclude, however, that the foundation of all philosophy is somewhat shaky.

University of Western Ontario
London, Ontario
Canada

NOTES

* An earlier draft of this essay was written in 1975 (at about the time I first met Bob Butts) and has been circulating since then. Because the essay deals with both William Whewell and the methodological heritage of Newton (two of Bob's favorite topics), it seemed the appropriate piece to publish in his honor. A new introductory paragraph has been added to indicate why I think the issues raised are especially timely in 1989.

[1] Isaac Newton, *Mathematical Principles of Natural Philosophy*, hereafter *Principia* Motte-Cajori Translation. Page numbers refer to the Great Books edition, p. 270.

[2] Ralph Blake, "Isaac Newton and the Hypothetico-Deductive Method", in *Theories of Scientific Method: The Renaissance Through the Nineteenth Century*, Edward H. Madden (ed.). (Seattle: 1966) p. 143.

[3] Edwin Arthur Burtt, *The Metaphysical Foundations of Modern Science*, (Garden City, New York: 1954) p. 219.

[4] *Principia*, p. 270.

[5] Alexandre Koyré, *Newtonian Studies*, (Chicago: 1965) p. 267ff.

[6] William Whewell, *Philosophy of Discovery*, (New York: 1971 reprint) p. 194.

[7] *Ibid.* p. 196.

[8] Maurice Mandelbaum, *Philosophy, Science, and Sense Perception: Historical and Critical Studies*, (Baltimore: 1964).

[9] Isaac Newton, *Opticks*. Book II, Part IIII proposition 7.

[10] Mandelbaum, *op. cit.*, p. 84.

[11] J. E. McGuire, "Atoms and the 'Analogy of Nature': Newton's Third Rule of Philosophizing", in *Studies in the History and Philosophy of Science* **I**, No. 1, pp. 3—58.

[12] *Ibid.*, p. 45.

[13] *Opticks*, query 31.

[14] *Ibid.*

[15] This point has also been made independently by Ernan McMullin in *Newton on Matter and Activity*, Notre Dame, 1978, pp. 13—21.

[16] J. E. McGuire, "The Origin of Newton's Doctrine of Essential Qualities', *Centaurus* (1968), **XII**, No. 4, pp. 233—260.

[17] *Ibid.*, p. 250.

[18] *Ibid.*, p. 251.

[19] *Ibid.*, p. 254.

[20] *Principia*, Book III, Third Rule, p. 270.

[21] *Ibid.*

[22] McGuire, "The Origin. . . ." p. 253.

[23] *Principia*, Book III, Proposition 6, Corollaries 1, 2, and 5.

[24] *Principia*, Preface to Second Edition. Quoted in Koyré, *Newtonian Studies*, p. 159.

WILLIAM HARPER

CONSILIENCE AND NATURAL KIND REASONING

(*in Newton's Argument for Universal Gravitation*)

1. STRENGTH *vs* SECURITY

In his ongoing debate with Clark Glymour and other scientific realists, Bas van Fraassen (e.g. 1983 pp. 165—168, 1985 p. 247 pp. 280—281 pp. 294—295) has often appealed to a widely accepted doctrine that *strength* and *security* are conflicting virtues which must be traded off one against the other. In limiting his commitment to only the empirical adequacy of a theory van Fraassen claims to be simply more cautious than his realist opponents. Certainly, as he delights in pointing out, a theory T cannot be more probable than its empirical consequences E. For Probability is monotone with respect to entailment, so $P(T) \leqslant P(E)$ if T entails E. If security is measured by any function which, like probability, is monotone with entailment, then it would seem that the kind of trade-off between strength and security van Fraassen appeals to is unavoidable.[1]

Upon hearing some of these debates it occurred to me that some celebrated cases from the history of science which involve unification under a theory provide *prima facie* counterexamples to the opposition between strength and security. Examples such as the unification of Galileo's terrestrial mechanics and Kepler's laws of planetary motion by Newton's theory of universal gravitation seem to be ones where accepting the stronger unified theory provides more security than would be obtained by accepting only the weaker hypotheses alone.

Imagine an acceptance context wherein the theoretical commitments include the approximate truth of Kepler's Laws of Planetary Motion and the approximate truth of Galileo's terrestrial mechanics. If we choose the approximations in a reasonable way then these commitments will be entailed by Newton's account of the system of the world in *Principia* Book III. Now consider a rival planetary hypothesis. Let it be a version of Tycho Brahe's system which has just the relation to Kepler's system that Tycho's original proposal had to Copernicus' system. According to such an hypothesis the six known primary planets

J. R. Brown and J. Mittelstrass (eds.), An Intimate Relation, 115—152.

all orbit the sun (just as they do for Kepler), but the sun (in its annual motion) orbits the earth. We now know that in so far as Kepler's system gives a correct account of the relative (annual) motions of the solar system from a reference frame on the sun (that is not rotating with respect to the fixed stars), so too will this Brahean system be a correct account of the relative (annual) motions of the solar system from a reference frame fixed at the centre of the earth (and not rotating with respect to the fixed stars). Now consider some data — the absence of stellar parallax. This data rightly counts against Kepler and for the Brahean account, since on the Keplerian account we have the diameter of the earth's orbit (two AU's) as a base for generating parallax.[2]

Now imagine an acceptance context in which the theoretical commitments include those of Newton's account in *Principia*. This is a context in which the theoretical commitments are stronger — the original commitments are entailed by this much richer account. Nevertheless, it seems clear that these commitments may reasonably have been regarded as more secure in the sense of more immune to revision. Once these richer commitments are accepted, Newton's dynamical argument at Proposition XII (using a centre of mass calculation to find an approximately inertial frame) can be used to support the Keplerian hypothesis against its Brahean rival (Stein 1967, pp. 177—184).[3] Given Newton's dynamical argument the proper response to the absence of stellar parallax data was to reason that the fixed stars are very much farther away than they had been thought to be.

Sometimes it is rational to hang on to one's theory in the face of putative counter-evidence. Our little thought experiment suggests that security in the sense of such immunity of our theoretical commitments, to revision in the face of putative counter-evidence can sometimes be legitimately increased by moving to accept a stronger theory. What it shows is that if only the weaker theoretical commitments are accepted the data about absence of stellar parallax might reasonably lead to rejection of Kepler (in favour of Brahe); but if the stronger Newtonian theory is accepted it would be legitimate to protect the theory by putting the blame on initial conditions (the fixed stars are much farther away than they had been thought to be).

This idea of immunity to revision in the face of hypothetical (or actual) evidence is one of a number of closely related conceptions of security that have begun to attract attention as philosophers of science have become aware that high probability is not the only conception of security we need to do justice to real life scientific reasoning.[4] I shall argue that a closer look at Newton's unification of terrestrial and celestial motion phenomena in *Principia* supports the suggestion that some such alternative conception of security does play a critical role in the evolution of science. I shall also argue that Newton's reasoning in support of universal gravitation is a vivid illustration of non-empiricist aspects of scientific practice that are quite widespread, reasonable, and perhaps even essential to science as we know it. As far as I know, no contemporary work by philosophers of science deals as adequately with some of these important but neglected aspects of the practice of science as the methodological work by William Whewell to which Robert Butts has for so long been calling our attention.

2. THE COLLIGATION OF FACTS INTO GENERAL PROPOSITIONS

In his analysis of the process of induction (Chapter V) of Book III of his *Novum Organon Renovatum* Whewell tells us,

> "... the colligation of ascertained Facts into general Propositions may be considered as containing three steps, which I shall term *the selection of the idea, the construction of the conception, and the determination of the magnitudes* (Butts 1968, p. 211).

He also tells us

> "*These three steps correspond to the determination of the* Independent Variable, *the* Formula, *and the coefficients, in mathematical investigations*, or to *the* Argument, *the* Law *and the* Numerical Data *in a Table of an Astronomical or other* Inequality (Aphorism XXXV, Butts 1968, pp. 210—211).

Malcolm Forster (1988) has recently interpreted Whewell's notion of induction as colligation of facts by explicating how these three steps work in a number of examples. One of these is the colligation of tidal data into the phenomenon that heights of highest tides vary sinusoidally with the angle from solar conjunction of the moon, so that spring tides (highest high tides) correspond to the two syzygies (0° = 360° and 180°) while neap tides (lowest high tides) correspond to the quadratures (90° and 270°). The selection of the idea is the selection of the angle of the moon from solar conjunction as the independent variable with respect to which to represent the approximately fortnightly cycle of spring and neap tides. The construction of the conception is the selection of a certain curve (a cosine curve) as that of the best fitting forumula representing the variation of heights of high tides with respect to this independent variable. Finally, the determination of the magnitudes is the determination of values of such coefficients of this curve as its amplitude and (of particular interest to Newton as we shall see) the numerical ratio between its high and low values.

It should be emphasized that what we are talking about here is as Whewell put it in our first quotation, "the colligation of ascertained facts into *general* propositions". The output of the three-step procedure is a generalization — the empirical law of the variation of neap and spring tides with the moon's angle from solar conjunction. It is this general proposition that counts as the primary tidal *phenomenon* which Newton (*Principia*, Book III, Proposition XXIV) explained by the joint action of the moon's and sun's gravity on our seas.

According to Whewell we extract such a general proposition from the mass of particular facts by imposing our idea and conception on the data.

> Thus in each inference made by induction there is introduced some General Conception, which is given, not by the phenomena, but by the mind (Butts 1968, p. 141).

He points out that the task of finding an illuminating idea (independent variable) and an appropriate conception (formula) requires sagacity and often also great effort and even luck. Anyone who doubts this should consider the difficulties faced, and the problems overcome, in Kepler's great struggle to master the motion of Mars.[5] He called this his war on Mars. It ended triumphantly with his successful extraction of the elliptical orbit law and the law of areas for Mars from the mass of data made available by Tycho Brahe's extensive and careful line of sight observations. Kepler's extraction of these two laws for Mars, his extension of them to the other planets and finally his extraction of his harmonic law some years later are all examples cited by Whewell (Butts 1968, pp. 130, 141, 142) as colligations of facts into general propositions.

It is important to note that once the idea is selected and an appropriate formula constructed the magnitudes of the coefficients are fixed (up to appropriate limits for error) by the data. Having settled upon an elliptical orbit with the sun at one focus the data fix the magnitudes of the greatest and least semi-axes and the period of the orbit, therefore they settle such interesting magnitudes as the average centripetal acceleration of the planet.

The idea or independent variable depends on what we choose to consider. Some choices are more illuminating than others and we shall defer discussing what Whewell has to say about evaluating such choices. For now I want to say something about Whewell's account of special methods of induction available in quantitative investigations to constrain the construction of an appropriate formula once the independent variable has been chosen. Whewell offers a very interesting account of how methods such as curve fitting, and the methods of means and least squares facilitate finding generalizations that might otherwise lie hidden among the errors and secondary effects which infect the data (Chapter VI, Book III of *Novum Organon Renovatum* (Butts 1968, pp. 223–237). Forster (1988) suggests that a proper appreciation of how such methods can average out errors (especially in large data sets such as the 13 000 observations extending over 19 years Whewell cites [Butts 1968, p. 233] for Lubbock's first investigation of the tides at London) will lead us to interprete Whewell's controversial claim that the result of a colligation can be 'more true than the individual facts themselves' (Butts 1968, p. 227) in a way that makes it both interesting and plausible.

Forster (1988) also explains how Whewell's method of residues can be used to separate out effects of component forces. On Forster's account Kepler's laws are true descriptions of that part of the motions of the planets due to the gravitation of the sun. Forster argues quite plausibly that when one considers them as generalizations revealed by averaging over a great mass of data the Keplerian effects are there to be found, even though mutual perturbations among the planets are also there to be found in the data.[6] Perturbations such as those resulting from the mutual action of Jupiter and Saturn on each other near conjunction can be revealed when the method of residues is applied to colligate laws about the various deviations from the Keplerian average values.[7]

I may have left the impression that data do not put any very strong constraint on our choice of formulae. That this impression is wrong should be clear from Forster's example of the tides. Once the independent variable is fixed any appropriate formula has to fit the general trend correlating spring tides with syzygies and neap tides with quadratures. Our choice of formula may be a conception we impose to go beyond the data, but it is a conception we construct to fit the data. Our scope for free invention in constructing this conception is significantly constrained by the data even if the data do not uniquely determine the correct formula.

3. WHEWELL's TESTS OF HYPOTHESES

According to Whewell, our choice of a formula in a colligation is not *just* a construction made to fit some given data. It is also an hypothesis subject to empirical tests. Whewell devoted a section of his chapter on scientific induction (Section III, Chapter V, Book II, NOR) to tests of hypotheses. Here are the subsection headings for this section from his table of contents (NOR, pp. XV–XVI):

Section III *Tests of Hypotheses*
Art. 10 True Hypotheses Foretell Phenomena
Art. 11 Even of Different Kinds — Consilience of Inductions
Art. 12 True Theories Tend to Simplicity
Art. 13 Connexion of the Last Tests

These headings suggest that Whewell distinguishes three tests of hypotheses, but that he sees a special connection between the last two of them.

3.1. *True Hypotheses Foretell Phenomena*

After setting the stage with a discussion of hypotheses which may be useful even though they turned out to be false (in article 9), Whewell opens article 10 with the following introduction to his first test.[8]

> 10. Thus the hypotheses which we accept ought to explain phenomena which we have observed. But they ought to do more than this: our hypotheses ought to *foretell* phenomena which have not yet been observed; at least all phenomena of the same kind as those which the hypothesis was invented to explain. For our assent to the hypothesis implies that it is held to be true of all particular instances. That these cases belong to past or to future times, that they have or have not already occurred, makes no difference in the applicability of the rule to them (Butts 1968, p. 151).

Whewell also devoted one of his aphorisms on induction to this first test

> It is a test of true theories not only to account for, but to predict phenomena (Aphorism XII, Chapt V, Butts 1968, p. 138).

He illustrated the application of this test by noting a number of examples of useful theories which eventually failed to meet it. These include what he called the "Epicyclical Theory of the Heavens", Newton's hypothesis of Fits of Easy Transmission and Easy Reflection and the doctrine of Phlogiston (Butts 1968, pp. 151—152).

I want to consider how this test applies to Kepler's so-called 'vicarious theory' of the orbit of Mars. This theory was Kepler's intitial attempt.[9] It is a classic example of an hypothesis that was useful, but discovered to be false. Kepler has left us a detailed account of the episode (in the second part of *Astronomia Nova*). We know exactly what data were used to construct it, what data verified it and what it could be relied upon to predict. We also know how Kepler refuted it.

For our purposes it may be sufficient to point out that the theory specified an eccentric (off-centre) circular orbit with an equant point (point about which equal angles are swept out in equal times) on the aphelion-line (line from the sun through the centre. It meets the circle on the far side at the aphelion, the point most distant from the sun. On the near side it meets the circle at the perihelion, the point closest to the sun). The equant point was to be on the side of the centre opposite the sun so that the planet would move fastest when it was closest to the sun. These specifications can be regarded as the conception in a Whewellian colligation. Kepler included two key geometrical parameters among the magnitudes to be fixed by the data. These were the distance between the sun and the centre as a fraction of the radius (the eccentricity) and the distance between the sun and the equant point as a fraction of the radius. Two more parameters, the direction from the sun of the Aphelion and the mean anomaly (time from aphelion as a proportion of the planet's period) of any one observation of the direction of Mars from the sun would complete this theory's specifications of the variation of Mars' direction from the sun with time (Stephenson 1987, p. 42).

The angular distance of the projection onto the plane of the ecliptic of the line from the sun through Mars from the point on the zodiac marking the vernal equinox is called its heliocentric longitude. About every 780 days Mars and the sun are in opposition (exactly on opposite sides of the earth). At such a time the observed location of Mars against the fixed stars in the zodiac fixes its heliocentric longitude. Kepler had twelve such observations to work with (ten from Tycho and two more he made himself). He was able to use four of them to fix the geometrical parameters (see Stephenson 1987, pp. 42—43). He had mean-anomalies for all so he was able to complete the theory when he managed to specify the heliocentric longitude of Mars' aphelion (Small, pp. 185—187). He then used the eight additional opposition observations to test the theory. Kepler concluded from these tests (the observed heliocentric longitudes were well scattered about the zodiac) that the theory could be relied upon to predict heliocentric longitudes to within about two minutes of arc, which was about the accuracy of the observations.

He went on to give two different refutations (Small, pp. 189—196, Stephenson, pp. 45—49) of the theory each of which used two observations, one near perihelion and one near aphelion, to constrain the ratio of the sun-aphelion distance to the sun-perihelion distance (Wilson, p. 92, diagram on p. 102). This contradicted the value of the eccentricity parameter (ratio of sun centre distance to radius) that had been fixed by the original four longitudes. Kepler was able to show that if the eccentricity were fixed to be consistent with the sun-Mars distances then any way at all of fixing the equant point would make for errors of up to eight minutes in some longitudes. This was sufficient for Kepler to reject this type of theory. The original version, however, continued to be a useful hypothesis for predicting longitudes. Indeed, Kepler continued to use this "vicarious" theory to make such predictions long after he had given it up as a candidate for the true account of Mars' orbit.

When we apply Whewell's first test to this episode we can surely count the four heliocentric longitudes used to fix its parameters as phenomena it was invented to explain. Perhaps, the eight heliocentric longitude observations used to verify it can count as successful predictions. They established its usefulness as a predictor of heliocentric longitudes. What about the solar-Mars distances that refuted the theory? If they are also phenomena of the same kind as those it was invented to explain then we can say that (after some initial success) the vicarious theory failed Whewell's first test. Suppose, however, we consider only the four longitudes used to construct the theory as the phenomena it was *invented* to explain and count only other heliocentric longitudes of Mars as phenomena of the same kind. On this way of settling what are to count as phenomena of the same kind as those it was invented to explain the vicarious theory did not fail Whewell's first test. It was a reliable predictor of heliocentric longitudes. In addition to the observations used to explicitly verify it, we can count later successful predictions when Kepler continued to rely on it to calculate longitudes.

We see here that Whewell's prediction test depends on how we specify which data are to count as phenomena of the same kind. Whewell's other examples show a similar relativity of the application of this first test to how data are sorted into kinds of phenomena. What he says about the Epicyclical Theory has an especially close parallel with the episode of Kepler's "vicarious" theory.

For example, the *Epicyclical Theory* of the heavens was confirmed by its *predicting* truly eclipses of the sun and moon, configurations of the planets, and other celestial phenomena; and by its leading to the construction of Tables by which the places of the heavenly bodies were given at every moment of time. The truth and accuracy of these predictions were a proof that the hypothesis was valuable, and, at least to a great extent, true; although as was afterwards found, it involved a false representation of the structure of the heavens (Butts 1968, p. 151).

In both cases the successful predictions are of line of sight observations, and in both cases the problem is a false representation of the structure of the heavens, specifically wrong relative distances.

In all of these examples Whewell cites as cases where an initially useful hypothesis failed this first test, the original regularities the hypotheses were actually constructed to fit continue to hold, at least more or less, even today. Whewell never provides a case where such regularities just stopped holding, as would be the case, for example, if the general trend correlating spring tides with syzygies and neap tides with quadratures were to begin to go radically wrong. Nor does he show us cases where we have discovered that we were radically mistaken from the beginning about such putative empirical regularities.

If we stick with a narrow interpretation of what count as phenomena of the same kind, then each of these examples is a case where the rejection of the hypothesis resulted from a failure to predict phenomena of a kind different from those it was invented to explain. This would make the rejection a failure of Whewell's second or third tests rather than this first test. Some ambiguity about this on Whewell's part may be shown by the fact that he discusses exactly these same cases as examples where it is the third test which fails to be met — what this third test demands is convergence to simplicity and unity as the hypothesis is generalized to explain phenomena of kinds different than those it was constructed to fit (Butts 1968, pp. 156—157).

We can see that Kepler's elliptical orbit unifies both heliocentric longitudes and sun-Mars relative distances so that it is superior to the vicarious theory. This may suggest that the vicarious theory fell because it was displaced by this superior rival, but such a suggestion is incorrect. Kepler rejected the vicarious theory before he had developed any hypothesis that could successfully account for both heliocentric longitudes and relative distances. Kepler was committed to finding a theory that could give a true account of the motions of the heavenly bodies among themselves. He intended the vicarious theory to account for relative distance phenomena as well as the heliocentric longitudes that were used to construct it.

A tradition in astronomy going back, at least, to Ptolemy treated astronomical theories as geometrical devices for predicting eclipses, eclipse magnitudes, and geocentric line of sight locations of heavenly bodies with respect to the fixed stars. These are the phenomena that Whewell (quoted above) pointed out that the epicycle theory could correctly predict. On this mathematical tradition the only constraints on an astronomical theory were that it *save these phenomena* and provide convenient methods of predicting them. Kepler can be regarded as a pioneer in a new physical-astronomy in which the aim is to give a true account of the motions of the heavenly bodies and not just to predict the geocentric line of sight data.[10]

Whewell offers some interesting remarks about the role played by advances in measurement in making new phenomena count as kinds that an adequate theory must account for (Butts 1968, pp. 156, 157). Tycho Brahe's observations made it possible to give much more accurate measurements of relative distances by triangulation than had been available when the line of sight angles were less closely fixed by the data. The use of the telescope and especially of the micrometer to give, by Newton's day, really precise data about relative visual angles occluded by bodies made an even more impressive increase in the availability of accurate relative distance data (van Helden 1985, pp. 118—156). Similarly, Whewell points out (Butts 1968, p. 157), it was the use of the balance in chemistry that made available the data about weights that led to the rejection of the phlogistan doctrine (though here, it appears, the rejection came only after the superior rival oxygen theory was developed).

Most of Whewell's discussion of his first test was devoted to its use to falsify hypotheses; however, in the last paragraph before moving on to discuss the much greater force of the evidence in favour of an hypothesis provided by its predicting phenomena of a different kind from those it was designed to fit, Whewell does acknowledge that successful predictions do count as positive evidence in favour of an hypothesis. This suggests that he would regard the accumulating mass of data points which the tidal correlation continues to fit as a continuing increase in its evidential support.

3.2 (*Predicting*) *even Phenomena of Different Kinds – the Consilience of Inductions*

Whewell introduces his second test of hypotheses with the following passage, which opens article 11, Chapter V:

> 11. We have here spoken of the prediction of facts *of the same kind* as those from which our rule was collected. But the evidence in favor of our induction is of a much higher and more forcible character when it enables us to explain and determine cases of a *kind different* from those which were contemplated in the formation of our hypothesis. The instances in which this has occurred, indeed, impress us with a conviction that the truth of our hypothesis is certain. No accident could give rise to such an extraordinary coincidence. No false supposition could after being adjusted to one class of phenomena, exactly represent a different class, where the agreement was unforeseen and uncontemplated. That rules springing from remote and unconnected quarters should thus leap to the same point, can only arise from *that* being the point where truth resides.
>
> Accordingly the cases in which inductions from classes of facts altogether different have thus *jumped together*, belong only to the best established theories which the history of science contains. And as I shall have occasion to refer to this peculiar feature in their evidence, I will take the liberty of describing it by a particular phrase; and will term it the *Consilience of Inductions*.

This test also has one of Whewell's aphorisms on induction devoted to it.

> The Consilience of inductions *takes place when an Induction, obtained from one class of facts, coincides with an Induction, obtained from another different class. This consilience is a test of the truth of the Theory in which it occurs.*
> (Aphorism XIV, Chapter V NOR, Butts 1968, pp. 138–139)

Let us put aside, for the present, Whewell's claims about the forcible character of the evidence consilience provides, and look to these passages to find out what consilience is.

The first part of this passage suggests that an induction has the virtue we are trying to explicate

> when it enables us to explain and determine cases of a *kind different* from those which were contemplated in the formation of our hypothesis.

The main emphasis, however, appears to be on a coming together of two or more different inductions and the term that actually gets introduced in the definition at the end is "consilience of inductions". The Aphorism, similarly, characterizes the coming together of different *inductions*. Here is how Malcolm Forster explicates what these passages say about the sort of coming together that counts as a consilience of inductions:

> Here we must recall Whewell's insistence that every colligation of facts adds a new *conception* to the facts. Now, if we consider two separate domains of inquiry, then the colligation of facts within one domain will introduce conceptions unconnected (or not thought to be connected) with the conceptions introduced in the second domain. When these different inductions, i.e. colligations of facts, "jump together", Whewell means that the magnitudes independently measured within separate domains agree with one another, or are connected by some law-like regularity (i.e. connected by some formula). (Forster 1988, p. 74)

On this explication the consilience of inductions takes place when a higher level induction colligates the results of two or more lower level inductions.

Forster's claim that each case of the consilience of inductions is a higher order induction is, surely, a correct explication of Whewell's view, as the following passage indicates:

> For when the theory, either by the concurrence of two indications, or by extension without complication, has included a new range of phenomena, we have, in fact, a new induction of a more general kind, to which the inductions formerly obtained are subordinated, as particular cases to a general proposition (Butts 1968, pp. 159—160).

The phrase "concurrence of two indications" refers to consilience, while the phrase "by extension without complication" refers to the tendency toward unity and simplicity which is the virtue Whewell discusses in his third test of hypotheses. Here we see that each of these two virtues is realized by a higher order induction to which the lower level inductions it brings together are subordinated.

Forster's contention that by "consilience" of two or more inductions Whewell means just that their magnitudes have been colligated by a higher order induction — that, in consequence, any such higher order colligation counts as a consilience of inductions — is perhaps more controversial. Whewell tells us that the cases colligated have to be of different kinds. According to Forster, the absence of the higher-order colligation shows that they count as different kinds at least in the respects which are relevant to the case. This line is attractive even if it means admitting as examples of consilience many higher order colligations which do not have some of the impressive characteristics of Whewell's favorite examples.

Kepler's third law is a functional relationship among the magnitudes measured from the several applications of his first law to planets orbiting the same primary. According to this Law the ratio R^3/T^2 is constant, where R is the major semi-axis of the elliptical orbit and T is its period.[11] On Forster's account this shows a consilience of the different inductions colligating each of these planets' motions into its own elliptical orbit.

Whether or not this is something Whewell would count as a consilience of inductions depends upon whether or not we can find an appropriate respect in which these separate elliptical orbits can be regarded as phenomena of different kinds. I think we can reasonably follow Forster's suggestion here and take the absence of knowledge of any such functional relation among them as is specified by the third law as an exactly appropriate respect in which they were regarded as of different kinds before and as of the same kind afterward. If this is viable, then, on Forster's interpretation, consilience exhibits exactly what Larry Laudan suggested is its real virtue.

> The real strength of such an hypothesis is usually that it *shows that events previously thought to be of different kinds are, as a matter of fact the 'same' kind of event.* (1971, p. 374)

This historical relativity of what count as appropriately different kinds of phenomena is in line with Whewell's grand historical view of induction on which inductions are represented as stages in hierarchically organized tables representing the progress of science toward the unification of different kinds of phenomena. It suggests that the sorting of phenomena into kinds relevant for any stage in the development of science is determined by the colligations which have been achieved by that stage.

3.3. *True Theories Tend to Simplicity*

Here is how Whewell introduced his third test of true hypotheses:

> 12. In the preceding Article I have spoken of the hypothesis with which we compare our facts as being framed *all at once*, each of its parts being included in the

original scheme. In reality, however, it often happens that the various suppositions which our system contains are *added* upon occasions of different researches.

This being the mode in which theories are often framed, we have to notice a distinction which is found to prevail in the progress of true and false theories. In the former class all the additional suppositions *tend to simplicity* and harmony; the new suppositions resolve themselves into the old ones, or at least require only some easy modification of the hypothesis first assumed: the system becomes more coherent as it is further extended. The elements which we require for explaining a new class of facts are already contained in our system. Different members of the theory run together, and we have thus a constant convergence to unity. In false theories, the contrary is the case. The new suppositions are something altogether additional; — not suggested by the original scheme; perhaps difficult to reconcile with it. Every such addition adds to the complexity of the hypothetical system; which at last becomes unmanageable, and is compelled to surrender its place to some simpler explanation (Butts 1968, p. 155).

This third test is one which applies to the process of theory construction. Whewell tells us that as the various suppositions are added in the course of its development a true theory tends toward greater simplicity and unification, while a false theory degenerates into complications until it becomes so unmanageable that it is compelled to surrender its place to some simpler explanation.

Robert Butts has long argued that one ought to take seriously Whewell's contention that his account of induction is to be construed as an account of discoverer's induction (1968, p. 20; 1973, pp. 56—57, 66—70). Whewell's discussion of this third test suggests that Butts has been on the right track in looking to the context of discovery for illumination of these views. In his important paper on Whewell's logic of induction (Butts 1973, p. 68) he called attention to an interesting passage at the beginning of section 13 where Whewell discussed the close connection between consilience and the explanatory virtues of convergence toward simplicity and unity exhibited in successful theory building.

13. The last two sections of this chapter direct our attention to the two circumstances, which tend to prove, in a manner which we may term irresistible, the truth of the theories which they characterize: — the *Consilience of Inductions* from different and separate classes of facts; — and the progressive *Simplification of the Theory* as it is extended to new cases. These two characters are, in fact, hardly different; they are exemplified by the same cases. For if these Inductions, collected from one class of facts, supply an unexpected explanation of a new class, which is the case first spoken of, there will be no need for new machinery in the hypothesis to apply it to the newly-comtemplated facts; and thus, we have a case in which the system does not become

more complex when its application is extended to a wider field, which was the character of true theory in its second aspect. The consiliences of our Inductions give rise to a constant convergence of our Theory towards Simplicity and Unity.

Butts argues that this close connection Whewell claims between consilience and the theory-building virtues of simplicity and successive generalization shows that consilience is more a pointer toward explanatory excellence than a measure of either entailment content or corroboration (1973, p. 69).

Butts also suggests (1973, p. 68) that for Whewell the concepts of consilience, simplicity, and successive generalization mean the same thing. Whewell does tell us that Consilience and Simplicity are hardly different, and that they are exemplified in the same cases; but he does not tell us that they mean the same thing. I think Forster (1988) has shown that Butts' main point — the importance of the context of theory building as the arena where consilience operates — can be even better served by distinguishing consilience from these explicitly explanatory virtues and then explicating the last sentence of the passage — where Whewell tells us that the consiliences of our inductions give rise to a constant Convergence of our theory towards Simplicity and Unity.

According to Forster the bottom-up character of Whewell's philosophy of scientific discovery is the key to understanding how it is that the consilience of inductions gives rise to convergence toward simplicity and unity.

However, Whewell's emphasis is on the philosophy of scientific discovery, whereas the top-down emphasis of hypothetico-deductivism obscures the most important component of this process — the *discovery* that two, or more, coefficients invented to explain diverse phenomena actually represent a common "cause". This feature is given its proper emphasis in the "bottom-up" approach to the philosophy of science.

Whewell has provided roughly the following picture of science. At the phenomenological level a colligation of facts will generally introduce the means of measuring formula coefficients, which may be instances of variable quantities when viewed more globally. These 'theoretical' variables may themselves enter into higher-level colligations, which may then introduce further coefficients and so on. This hierarchical procedure will eventually end when the coefficients introduced are no longer variable — as in the case of fundamental physical constants — or when the coefficients introduced are easily interpreted as conversion factors between different scales of measurement. (Forster 1988, pp. 76—77)

On this view the bottom-up character of Whewell's philosophy of discovery is shown by the step by step progression through the history of science toward increasing simplicity and unification. These are what

Butts (1973, p. 68) calls the steps of successive generalization exhibited in Whewell's inductive tables. Each step is guided from below by correlations colligated among the magnitudes measured in lower level colligations.

Forster's remarks continue with a simple illustration from laboratory physics of the way consilience can facilitate convergence toward simplicity and unity.

Examples are everywhere; e.g. consider the magnitude of 'mass' as measured from balance phenomena or from spring phenomena. In cases like this, the observed correlation between independent measurements is *explained* in terms of an *identity* relation: The independently determined values are approximately the same (after being converted to a common scale) *because* they are measurements of the *same* thing, e.g. of 'mass', rather than measurements of the different properties of 'spring mass' and 'balance-mass'. The importance of these explanations is that they give rise to a simplification and unification of our previous theoretical committments (e.g. by replacing 'spring-mass' and 'balance-mass' with 'mass') (1988, p. 77).

In later remarks Forster expands upon this theme in a specific comment on the passage from Whewell we have been considering.

Consilience is not the same as simplicity; the former merely gives rise to a *convergence towards* the latter. The goal — of simplicity and ontological unification — is only *achieved* once the observed consilience is deemed to be non-accidental and is subsequently *explained* in terms of an identity. Consilience is the evidence for accepting the identity. After that the theory can be further tested by its deductive consequences, but it is the *discovery* of the identities and unity in nature that is the admirable part of the process (Forster 1988, p. 79).

On this view the consilience of inductions gives rise to the convergence toward simplicity and unity by providing the evidence for introducing an identity. The agreement between the separate values is explained by a common cause — they are measurements of the same magnitude.

We have already pointed out that, on Forster's view, Kepler's third law is, itself, an example of consilience even if it doesn't directly illustrate the level of the explanatory virtues of convergence to simplicity and unity associated with Whewell's favorite examples where consilience is exhibited in grand unifying theories. The functional relationship colligated in Kepler's third law tells us that the ratio R^3/T^2 is constant for all of the planets orbiting the sun. On Forster's view this gives rise to convergence toward simplicity and unification when these approximately equal values can be explained by introducing an identity so that each can be regarded as a separate measurement of the same magnitude. This is just what Newton did. The equal ratios R^3/T^2 (suitably adjusted for scale) calculated from these various orbits can be understood as separate measurements of what Newton (Definition VI, Book I) called the absolute quantity of a common inverse square centripetal force holding all these planets in their orbits. He went on to establish that these could be understood as measurements of the mass of the sun. This magnitude can be regarded as the common cause of the ratio's characterizing the orbits of all the planets.

4. NEWTON'S ARGUMENT FOR UNIVERSAL GRAVITATION

4.1. *Phenomena*

Newton begins by citing six phenomena. The first specifies that Jupiter's moons satisfy Kepler's Law of Areas (by radii to the centre of the primary they describe areas proportional to the times of description) and Kepler's Harmonic Law. The second specifies that the moons of Saturn also satisfy these two laws. The last, Phenomenon VI, specifies that Kepler's Law of Areas is satisfied by our moon as well. The third, fourth, and fifth are of special interest to us.

Phenomenon III

That the five primary planets, Mercury, Venus, Mars, Jupiter, and Saturn, with their several orbits, encompass the sun.

Phenomenon IV

That the fixed stars being at rest, the periodic times of the five primary planets, and (whether of the sun about the earth, or) of the earth about the sun, are as the 3/2th power of their mean distances from the sun.

Phenomenon V

Then the primary planets, by radii drawn to the earth, describe areas in no wise proportional to the times; but the areas which they describe by radii drawn to the sun are proportional to the times of description.

Two points, rather immediately, strike an attentive reader. First, these phenomena do not specify that any of these orbits are ellipses. Secondly, they leave open our Brahean alternative to Kepler's account. Whether the earth orbits the sun, as Copernicus and Kepler specify, or the sun orbits the earth, as in the Brahean hypothesis, is not decided by what Newton takes as phenomena.

We saw that Whewellian colligations are general propositions colligated from the data by imposing a conception on them. Each of these phenomena cited by Newton is such a general proposition giving what purports to be an empirical law of the relative motions of a satellite (or satellites) with respect to its (their) primary. Another feature we noted about Whewellian colligations is that they are constrained to fit the data they colligate, even if they go beyond what these data logically imply. Newton discusses the evidence for each phenomenon. I think these discussions make it clear that the idealizations introduced by the conceptions which generalize on the data are fairly well within what would count for Whewell (or even contemporary statisticians) as reasonable estimates of fit.[12]

Not just any colligations count as phenomena. I think we can say, roughly, that only those colligations that can be reasonably regarded as presumed common-knowledge among appropriate experts are to count. Moreover, this presumption is defeasible in that corrections are allowed for. Professor I. B. Cohen (conversation, November 1987) suggested to me that G. E. L. Owen's (1961) interesting discussion of Aristotle's *Tithenai Phainomena* can illuminate the role played by Newton's phenomena. According to Owen (pp. 170—173) Aristotle's *Tithenai Phainomena* are the common conceptions on the subject accepted by all, or most, or by the wise. They are not rock hard data, but starting points that can be corrected as the investigation proceeds.

4.2. *Deductions from Phenomena*

In Proposition I, Book III, Newton infers that the moons of Jupiter (and Saturn)

(a) are held in their orbits by forces directed to the centre of their primary

and

(b) that these centripetal forces vary inversely with the square of the distance from their centre,

from the phenomena that they satisfy (a′) Kepler's Law of Areas and (b′) Kepler's Harmonic Law. In Proposition II he infers by similar arguments that the five primary planets are also held in their orbits by centripetal forces that vary inversely with the square of the distance from the centre of their primary. These are the paradigmatic examples of Newton's practice of inferring forces from phenomena.

(a) *Equivalence Theorems and Bootstrap Confirmations.* In each case the inference is backed up by theorems establishing equivalences between values of a phenomenal magnitude (say the behaviour of the rate at which areas are being swept out by radii to a centre as increasing, constant, or decreasing) and corresponding values of a theoretical magnitude of interest (say the direction of a deflecting force as off centre in the direction of tangential motion, centripetal, or off centre in opposition to the tangential motion). The assumptions used in the proofs allow for a range of possible values for the phenomenal magnitude, and equivalences with corresponding values of the theoretical magnitude are proved to hold over the whole range.

These equivalence theorems show that Clark Glymour (1980, pp. 203—225) was correct when he argued that Newton's deductions of forces from phenomena satisfy the requirements for bootstrap confirmations. A bootstrap confirmation is an inference where an instance of a theoretical claim is deduced from data together with theoretical background assumptions. The basic requirement is that the background assumptions be compatible with alternative data which would have led to a contradiction of the instance of the theoretical claim that was deduced from the actual data. Newton's deductions of forces from phenomena, however, are especially interesting special cases of bootstrap confirmations. Not only are the background assumptions compatible with alternative values of the phenomenal magnitude which would have contradicted the theoretical claim that was actually inferred, as bootstrap confirmation requires, but these background assumptions also entail the theorems which give the equivalences between the various values of the phenomenal magnitude and corresponding values of the theoretical magnitude over a whole range of alternative values which they leave open. This suggests that (given the background assumptions) the values of the phenomenal magnitudes can be regarded as *measuring* the corresponding values of the theoretical magnitude just as readings on a thermometer measure temperatures. Given Newton's theorems the motions of orbiting bodies can be regarded as measurements of directions of deflecting forces, and for cases where these are centripetal forces the motions can measure the variation of those forces with power of distance.

(b) *The Dynamical Significance of the Law of Areas.* Theorems I and II of Book I give the two directions of the equivalence between centripetal direction of a force deflecting a body from uniform rectilinear motion and constant rate for sweeping out areas by radii drawn to that centre, while Corollary I of Theorem II, Book I gives the relevant equivalences for increasing and decreasing rates. These theorems show that motions satisfying Kepler's Law of Areas with respect to a centre moving inertially measure the centripetal direction of a force deflecting the body from uniform rectilinear motion with respect to that centre. Theorem III extends this result to centres, howsoever moved, so that motions of a body satisfying Kepler's Law of Areas with respect to the centre of another body measure a force compounded of a centripetal force together with all the accelerative force by which that other body is impelled.

This shows that the dynamical significance Newton finds for Kepler's Law of Areas is not limited to idealizations where the centre counts as in inertial motion and the only force operating is the centripetal force deflecting the body. Rather, as Forster points out, Newton shows how relative motions satisfying Kepler's Law of Areas let us infer centripetal component forces even in the presence of other unknown accelerating forces.

This virtue of the theorems backing up Newton's deductions of centripetal forces from phenomena suggests that Forster's emphasis on the use of residues in large masses of data to make room for decomposing effects into components from which to infer component causes may not always be necessary. Such deductions of component centripetal forces from appropriate relative motions will work even when there are no multiple data points corresponding to repetitions of the motion. This is important for Newton's applications to motions of comets where there might be just a few observations on a single fly past of the sun. Newton's equivalence theorems make it possible to infer a component centripetal force from *even a single case* of a motion where equal areas are swept out in equal times by radii to a centre.

4.3. *Newton's Mathematical Theory of Forces and Motions*

Howard Stein (conversation) once called my attention to the fact that Newton places the laws of motion together with the definitions in a separate section before the actual beginning of Book One. Newton actually tells us (Scholium to the Laws translation [1987] by I. B. Cohen and Anne Whitman) that these principles are "accepted by mathematicians and confirmed by many kinds of experiments." They are assumptions we readers are expected to agree to before we are to even begin working our way through his rich investigation of the motions of bodies. These laws together with Euclidean geometry, and the lemmas on first and last ratios (which do for Newton what we would do with the calculus) are the general background assumptions which make possible Newton's deductions of forces from motion phenomena.

We saw that the deductions from phenomena we considered were special bootstrap confirmations because they were backed up by equivalence theorems that let us measure the forces from the phenomena. Here we see that they are special because the background theory is special — the very general assumptions of Newton's mathematical theory of forces and motions. In "On the nature of the truth of the Laws of Motion" (Butts 1968, pp. 79—100) Whewell held that the laws of motion were necessary truths even though they also had empirical elements. Butts (1965, especially pp. 169—173) has offered an interesting account of the interplay of empirical and *a priori* elements in Whewell's account of how we come to have a clear intuition of the necessity of the laws of motion to the application of the idea of cause construed as force.

Consider the situation with kinematics. The vector composition of motions according to Galiliean transformations were explicitly taken by Kant and Hertz as *a priori*. Moreover, I doubt that many scientists or philosophers from Newton's day through most of the 19th century would have regarded them as empirically refutable. Nevertheless, when Poincare developed an alternative conception that could take over their role in making intelligible the composition of motion and forces and Einstein used it to formulate special relativity theory, few had any difficulty understanding certain experiments as empirical refutation of Galilean kinematics. Putnam (1983, p. 95) coined the nice term "contextually *a priori*" for the special status I claim was enjoyed by kinematics. In the earlier context, when no alternative conception was available to make intelligible the compositions of motions, the Galilean transformations were treated as immune to any empirical refutation. When an alternative conception was developed enough to count as a serious rival the context changed and the earlier conception became open to empirical test.

4.4. *Maintaining Simplicity in the Face of Hard Tests: Proposition 3 and the Lunar Precession Problem*

In Proposition 3 Newton argued that the moon is held in its orbit by an inverse square centripetal force. That the force is centripetal is deduced from Phenomenon VI by the Areas Law theorem. The argument to establish the inverse square variation was more problematic. What made it a problem was the well established precession of the lunar apsides of about 3° forward every revolution.

Newton appealed to Corollary 1 of Proposition XLV, Book I, which establishes equivalences between precession of aphelion for a body orbiting under a centripetal force and the variation of that force with power of distance. A stable elliptical orbit corresponds exactly to an inverse square variation. Precession forward measures variation that falls off faster than the inverse square, while precession backward measures variation that falls off slower than the inverse square.

The precession of our moon is about 3° forward every revolution which would measure variation inversely as the 2 and 4/243 power rather than the inverse square variation. Newton argued that this lunar precession can be neglected. For one thing the variation is approximately equal to the inverse square. More importantly, the precession is to be neglected because it is due to the action of the sun (as he promised to show later). Finally, to the puzzlement of more than a few, he explicitly appealed to the moon test agreement with Huygens' value that results from the inverse square variation as a reason for neglecting the lunar precession.

The precession theorem is a one-body idealization. It only takes into account the centripetal force, but Newton used it as his main tool to solve for the gravitational effect of a third body on an orbit. His method was to calculate the centripetal (say earth-moon axis) component of the action of the third body, then compose this force with the basic inverse square centripetal force to give a new total centripetal force that no longer would vary inversely as the square. In Corollary 2 of Proposition XLV he developed a formula for calculating precession from the centripetal component of a foreign force as a fraction of the main central force.

When Newton applied this method with the correct value for the ratio of the centripetal component of the sun to the action of the earth on the moon he could only account for about one-half of the observed precession. When Clairault, d'Alambert and Euler all worked on the problem in the 1740's with no more success than Newton had, Clairault and Euler were led to propose modifying the basic inverse square law (Waff 1976). When Clairault used a calculation that took into account not just the centripetal but the transverse component of the action of the sun he was probably trying to conclusively falsify Newton's theory, much as Kepler falsified his vicarious theory (Waff 1976, p. 175). To his great surprise, he found that this accounted for the lunar precession.

4.5. *Unification, the Moon Test and the Tides*

(a) *The Moon Test.* Newton's famous moon test showed that the inverse square force on the moon (calculated from the centripetal acceleration of its orbit) would, at the surface of the earth, equal the force of terrestrial gravity (measured by Huygen's seconds pendulum). In Proposition IV, he appealed to his first two Rules of Reasoning,

Rule 1 No more causes of natural things should be admitted than are both true and sufficient to explain their phenomena.

Rule 2 Therefore, the causes assigned to natural effects of the same kind must be, so far as possible, the same (Cohen I. B. and Whitman A. 1987).

to argue from this agreement in measured value to the claim that the centripetal force on the moon is identical with terrestrial gravity.

Newton's calculation of the centripetal acceleration of the moon used the simplifying geometrical assumptions that the moon orbits the centre of the earth (rather than the common centre of gravity of the earth-moon system) and that the earth was a perfect sphere (rather than oblate with a larger semidiameter at the equator). Yet he put great weight on the resulting close agreement with Huygens' value in the moon test. He made the key inference to identify the centripetal force on the moon with gravity, and he even appealed to the agreement with Huygens' as an additional excuse for neglecting the lunar precession in his inference to the inverse square variation of the force on the moon in Proposition 3.

Forster sees the role of Newton's first two rules here as that of endorsing a Whewellian policy to identify magnitudes when this is appropriately supported from below by a consilience showing agreement in their *independently* measured values. On this view Newton should have kept his evidence for colligating the inverse square variation of the force on the moon completely independent of the resulting agreement with Huygens. Moreover, he should have made sure that the simplifying assumptions were not giving an accidental agreement before regarding the consilience as independently established. Forster (1988, pp. 86—87) explicitly worrries that the moon's orbit cannot provide an appropriately independent measurement of its centripetal acceleration from celestial phenomena until it can be backed up with an independent measurement of the Moon's mass so that its contribution to the joint force can be factored out by using the common centre of gravity.

Newton's use of the rules to put so much weight on the agreement, even when its warrant from below had not been established, shows that his practice of theory construction is more free-wheeling than the bottom up procedure Forster admires.[13] I think it is clear that, once the initial agreement has been shown, the task of showing that it will hold up as simplifying assumptions are removed is not one of attempting to generate enough force of evidence to carry the inference to the identification through, rather it is simply making sure that the new unified theory can account for the data, so that the unification will not be undone by a failure to satisfy the first test. For Newton the Rules provide the weight for the identification. The consilience is needed only to license permission to give the unification a chance to get off the ground.

(b) *The Tides*. Newton did provide a way to measure the mass of the moon so as to show that the agreement in measured values in the moon test would hold up when the simplifying assumption that the moon's orbit is centred on the centre of the earth is replaced by the more correct assumption that they orbit a common centre of gravity. Newton's way of measuring the mass of the moon was to appeal to the tidal phenomenon which Forster gave as his first example of a Whewellian colligation. Newton used the ratio between high and low values calculated from the coefficients of the equation colligating the correlation of spring and neap tides with syzygies and quadratures to measure the ratio of the moon's mass to that of the sun from their relative contributions to this tidal phenomenon.

Consider the effect one body (say the moon alone) would have on the seas of a water-covered earth. On the side toward the moon the water is pulled away from the earth, because it is nearer to the moon than the solid part of the earth is. On the side away from the moon there is a similar increase in the depth of water, because the solid earth is closer to the moon and so is pulled away from the water on that side. The basic effect of the moon's gravity on a sea-covered earth is to cause nearly equal water bulges on the near and far sides of it.

Now consider the additional effect of the sun's gravitation. At conjunction and equally at opposition the sun's force adds maximally to that of the moon to increase the basic effect. This is why the spring tides (highest high tides) occur at both syzygies (at 0° and at 180°). At quadratures (90° and 270°) the action of the sun is maximally opposed to the stronger action of the moon (its basic effect here would be to take water away from the sides on which the moon is making it deeper). This is why the neap tides (lowest high tides) occur at quadratures.

Newton was not satisfied with this merely qualitative causal explanation. He wanted to be able to use the tidal ratio to calculate the ratio of the mass of the moon to the mass of the sun. Such a calculation can be given from the theory. The basic differential effect of the moon between the water at a point B and the centre of the earth C when the moon is at A directly over head is

$$L = GMm/(\overline{AB})^2 - GMm/(\overline{AC})^2$$

Similarly, for the sun

$$S = GMs/(\overline{A'B})^2 - GMs/(\overline{A'C})^2.$$

When it is directly over head at A'. If we had a perfectly unimpeded water flow, an untilted earth with the moon's orbit and the earth's equator in the plane of the ecliptic then we would have

$$\frac{\text{(Spring height)}}{\text{(Neap height)}} = \frac{L + S}{L - S}$$

so that we could start from the tide ratio and solve the equations for the ratio of Ms to Mm.

For cases where the earth is in equinoxes we can calculate the differential effect at a given latitude, taking into account the inclination of the lunar orbit by averaging over the various combinations of greatest and least angles from zenith of the sun and moon at syzygies and quadratures. Philip Catton (1988) was able to show that such a calculation for latitude 50° would give Newton's 9:5 ratio, when it was further corrected by three corrections Newton actually applied in his own calculation. Newton's tidal data were taken from equinoxial syzygies and quadratures (Cajori, p. 479) and were from about latitude 50°. One of the corrections is for tidal delays which could be established by local observations. The other two are corrections for true special features exhibited by the lunar orbit when the earth is near equinoxes. We can go from Newton's data back through the equations to recover the true ratio of the mass of the moon to that of the sun.

(c) *Explanation.* Newton established for the first time a new mathematical ideal for causal explanations. One should develop one's mathematical force model sufficiently so that when it is applied to explain a phenomenon one can establish from the model the sort of equivalence theorems that back up the deductions of forces from phenomena. This very strong sort of explanation requires that the relevant features of the magnitudes that explain the phenomena are also *measured* by the phenomena they explain. Here we see a realization with Newtonian component forces of Cartwright's (1983) ideal that causal explanations require that the cause be measured by the effect it explains.

4.6. *Generalization Using the Rules of Reasoning*

In Proposition V Newton extended the identification of orbital force with gravity to the moons of Jupiter and Saturn and to the Planets. His argument for this consisted in noting the inverse square centripetal nature of the accelerative measures of these forces (accelerations of satellites of the same primary depend only on their distances not on their respective sizes) and applying Rule 2. These inverse square centripetal forces are of the same kind as the force on the moon, therefore they should, as far as possible, be assigned the same cause — gravitation of the satellites to their respective primaries.

Whewell (Butts 1968, p. 332) argues that the problem of deciding when effects are of the same kind renders applications of Rule 2 vacuous.

> Are the motions of the planets of the same kind with the motion of a body moving freely in a curvilinear path, or do they not rather resemble the motion of a floating body swept round by a whirling current? The Newtonian and the Cartesian answered this question differently. How then can we apply this rule with any advantage? (p. 332).
>
> It is only when it appears that comets pass through this plenum in all directions with no impediment, and that no possible form and motion of its whirlpool can explain the forces and motions which are observed in the solar system, that he [the Cartesian] is compelled to allow the Newtonian classification of events of the same kind (p. 333).

Whewell suggests that Newton's claims to deduce inverse square centripetal forces from the phenomena of orbital motion (as well as his use of Rule 2 in Proposition V to identify these as gravitational forces) depend on his being able to refute the Cartesian alternative hypotheses of vortices. This would make Newton's internal criticism of Cartesian vortices in Book II an essential part of the argumentation supporting Newton's deductions from phenomena in Book III.

Rule 4 makes it clear (as does the famous remark about hypotheses in the general scholium) that Newton thought that mere hypotheses can be ignored because they are not deduced (or inferred by general induction) from phenomena.

> Rule 4 In experimental philosophy, Propositions gathered from phenomena by induction should be considered either exactly or very nearly true notwithstanding any contrary hypotheses, until yet other phenomena make such propositions either more exact or liable to exceptions (Cohen I. B. and Whitman A. 1987).

We have seen that Newton's deductions from phenomena depend upon background assumptions and upon idealizations about what phenomenon a certain effect is to count as. In the case of the centripetal force on the moon the very precession theorem (Corollary I, Proposition XLV, Book I) connecting variation of force with power of distance to apsidal motion would allow inverse variation with the 2 and 4/243 power of distance to be deduced from the observed apsidal motion, unless the assumption of no interfering forces can be undercut. Relative to that assumption, the theorem shows that the phenomenal magnitude would count as a measurement of this alternative to the inverse square variation. This is why Newton must make plausible the claim that the apsidal motion of the moon is due to an interfering force.

I assume that Newton thought that the Cartesian hypothesis was not backed up by any similarly detailed mathematical equivalences between vortical magnitudes and specifications of relative motions that could count as phenomena.[14] I see no reason why the reasonableness of this opinion should depend on his actually having carried out his internal criticisms of Cartesian theory in Book II. These criticisms may perhaps be viewed as an attempt to show that no plausible vortical magnitudes could have values equivalent to Keplerian orbital phenomena. Even if Newton had not provided these arguments, however, he would have been reasonable in regarding the burden of proof on the Cartesians to provide such equivalences if their proposals are to count as serious rivals rather than mere hypotheses that can be safely ignored.

In the first part of Corollary I, Newton extends the power of gravity to planets without moons.

> There is, therefore, a power of gravity tending to all the planets, for, doubtless, Venus, Mercury, and the rest, are bodies of the same sort with Jupiter and Saturn.

Here there are no already established inverse square centripetal forces to go on, so there are no similar natural effects that can be used to apply Rule 2. Instead, the inference is carried by the assertion that these planets are bodies of the same sort as the others. In Corollary II, Newton infers that the gravitational forces to all planets are inverse square forces. Again, no rule and no additional facts are cited. The idea seems, again, to be that the gravitational forces toward the planets without moons are the same kind of forces as those towards planets with moons and therefore that they too should be inverse square.

Rule 4 tells us to regard as true those propositions inferred by a induction from phenomena, and Newton tells us

> This rule we must follow, that the arguments of induction be not evaded by hypotheses (p. 400).

This makes it clear that it is even propositions like Corollary II and the first part of Corollary I, which are not deduced or inferred by Rule 2 but only inferred from phenomena by induction, that get protected from evasion by mere hypotheses.

We can use the fourth rule as a clue to illuminate some of the puzzles about the "vera causa" clause in Rule 1. There has been considerable discussion over what role, if any, the requirement that we admit only causes which are *true* should play in applications of Rule 1 (e.g. Whewell, Koyre). If we take Newton at his word, Rule 4 provides the answer. We are to regard as true those propostions inferred by general induction from phenomena (as well, of course, as those actually deduced or otherwise more directly inferred). This suggests that Forster's Whewellian interpretation of the "vera causa" condition in Rule 1

> Whewell's reading of Newton's "true causes" (verae causae) as those "causes" that are found to be identical with the causes of other phenomena corresponds with our understanding of Newton's Rule 1 and reinforces the previous interpretation of what Whewell meant by consilience (Forster 1988, p. 80)

would be rejected by Newton. The extension of the power of gravity and its inverse square variation to planets without moons such as Venus and Mercury in Corollaries I and II of Proposition five is not backed up by any consiliences of independently measured forces.

In Proposition VI Newton argued that the motive quantity of the centripetal force of gravity of a planet on a body is proportional to the inertial mass of that body. In Proposition VII he argued that the absolute quantity of the gravity of any body is proportional to its inertial mass. This stage in his argument involved an appeal to rule three to extend gravity and its proportionality to inertial mass to all planets. It also involved the generalization to universal gravitation. These generalizations are even more radical departures from Forster's bottom-up ideal.

Forster's common cause realism may be less restrictive than the rather extreme bottom up interpretation I have put upon it. This is suggested by the warm remarks he makes toward Newton's third Rule of Reasoning (1988, p. 94).

To complete this story, we should emphasize that, once discovered to hold within a certain domain, the law is automatically extended universally to apply in all future instances and *for all bodies* whatsoever. There is an inductive generalization as traditionally construed, but it is a generalization *constrained* by the results of experimentation. Newton's third rule of reasoning is as follows: "The qualities of bodies, which admit neither intensification nor remission of degrees, and which are found to belong to all bodies within the reach of our experiments, are to be esteemed the universal qualities of all bodies whatsoever." In particular, the *power* of gravity measured by the gravitational mass of a body is found to be invariant over all independent determinations for all applications and for all times so far examined, so this property should be extended to other bodies and to other times. So, we should extend the previous measured value for the mass of the cup to apply to future times, and we assume that all as yet undiscovered bodies obey the same laws (as was instrumental in the discovery of Neptune). However, the preconditions placed on this inductive inference — conditions demanding the consilience of the coefficients so far measured — protect this form of inference from the usual counterexample to enumerative induction (such as the chicken that is fed every day 99 times in a row and then killed on the 100th day). As Whewell says, it is the *intermediate* step of adding a new conception — in this case the 'mass' concept — that enables us to *verify* the truth of our law in terms of the consilience of inductions. It is this form of verification (Whewell's second and third tests of hypotheses) that the naive cases of enumerative induction lack (Forster 1988, p. 94).

Whewell (Butts 1968, pp. 333—334) did not at all share Forster's enthusiasm for Newton's third rule. As Butts (1970) has pointed out

It is central to Whewell's philosophy of science that we recognize that observation, even experimentally regularized observation, can confer on propositions about facts only a probable universality (Butts 1970, p. 144).

Forster has largely avoided talking about Whewell's views on necessity and conceptual change (except in so far as colligation involves imposing a conception). These are aspects of Whewell's philosophy of induction that Butts has drawn to our attention (Butts 1976). I think the issue of inferences to universality Forster confronts in this passage require him to deal with these more problematic aspects of Whewell's thought.

5. NATURAL KIND REASONING

I think Newton's generalizations by induction, and his applications of Rule 3, are examples of Natural Kind inferences such as the inference from a value of the specific gravity of one or a few samples of a given alloy of metal to the same specific gravity for others. Such inferences abound in science. We measure the charge of one or a very few electrons and are quite sure all other electrons have the same charge. They are cases where we generalize even when there are no significant masses of data points on which to use means, least squares and other statistical techniques to support the inference. I have argued (1986) that such super-inductive inferences are legitimate when made relative to a natural kind conception that specifies which features are essential to the kind in question. Thus, we can infer specific gravity but not shapes for samples of an alloy.

If we accept a conception into the role of specifying a feature ø which allows such super-inductive inferences for a kind x then we have to accept the enabling conditional,

If any x has ø then every x has ø,

as certain; or we will not be able to infer anything more than probable universality. So it is, I think, that when Newton tried to use Rule 3 to establish that gravity applied universally to all bodies and that it universally varied directly with inertial mass and inversely with square of distance, he could not avoid making these features essential to a natural kind conception for bodies, without giving up his inference to universality.

I also think that it is a part of our regular practice in science to regard these essential properties, on which we base such inferences to universality, as necessary. Kripke and Putnam are surely correct in their explication or how we reason about identities and natural kind terms. Therefore, if I am correct about the fact that we do make natural kind inferences and thus treat some conceptions as explicating natural kind essences, it follows that (contrary to the view of most empiricists) science does seriously involve modalities.

These scientific natural kind conceptions I have been talking about do not have some features Newton worried about when he resisted calling gravity essential to bodies. He was worried making gravity essential would preclude any further investigation into its causes. In particular he wanted to keep open the possibility of a mechanical explanation of gravity; nevertheless he was sure that gravity really existed and operated by the universal laws he had discovered. I think Newton's discussion in *De Gravatatione* (Hall, A. R. and Hall, H. B., 1962, pp. 89—156) where we can generate true and universal mathematical laws of motion phenomena without any metaphysical account of their ultimate nature can be regarded (in part) as an argument for introducing mathematical natural kind conceptions of just the scientific sort I am talking about. We don't stop making natural kind inferences with respect to charge on an electron when we explore deeper explanations of their nature. That their nature has such and such features essentially does not preclude further investigation in which we search for an explanation of these features.

Secondly, as I have indicated, scientific natural kind conceptions are subject to overthrow when an alternative conception becomes available which is developed enough to count as a serious rival. The status these natural kind conceptions enjoy when they are accepted is at most only contextually apriori. They are regarded as absolutely certain and, perhaps, even immune to any merely empirical refutation, but they are not incorrigible. Once a serious rival becomes available these conceptions lose their invulnerability and are open to refutation by being tested against the rival.[15]

There is a great deal of similarity here with Whewell's program. Not just any colligations count as natural kind conceptions, but those colligations Whewell characterizes as theories of causes do.

> Inductive truths are of two kinds, Laws of Phenomena, and Theories of Causes. It is necessary to begin every science with the Laws of Phenomena; but it is impossible that we should be satisfied to stop short of a Theory of Causes (Aphorism XXIV).

On Forster's common cause interpretation we begin to have a theory of causes when we can unify by identifying magnitudes so that several lower level colligations can be regarded as measurements of the same thing. On the view I am advocating these measurements do not count as causal explanations unless the equivalences grounding them are based on a natural kind conception. We do not treat them as really causal until they are based on conceptions we are prepared to universally generalize upon. This is why there is a point to Whewell's distinction between causal theories and laws of phenomena.

We have seen the close connection between Newton's use of his third rule and the respectable scientific practice of making natural kind inferences. Newton's other rules also make good sense in the context of natural kind reasoning (Harper 1983, 1986). The first two rules function to endorse a policy to introduce unifying conceptions wherever possible, while the 4th rule expresses the resilience of an accepted kind conception. It can be corrected by phenomena, but it will not be undercut by mere hypotheses. Until an alternative conception is developed enough to rival capacity of the original theory to illuminate the phenomena it colligates that alternative hypothesis should not be allowed to undercut the theory.[16]

What makes an alternative count as a serious rival, rather than as a mere hypothesis to be ignored? When we think of the role of a natural kind conception as that of making intelligible the phenomena in its range by providing causal explanations, the answer is obvious. A rival has to be a plausible candidate for developing into a theory that will provide alternative explanations for enough of these phenomena to count as a better account of the causal powers underlying them. This is why the Brahean hypothesis became a non-starter once Newton's dynamical explanation of Kepler's phenomena became available. Any hypothesis to restore it would have to come up with a rival treatment of Newton's dynamical explanation of motions including, for example, the force of the sun to move the tides.

What makes unification and simplicity contribute to security is that each identification of coefficients as measurements of a common cause adds a new range of phenomena that must be explained by any rival conception that would replace the conception grounding the causal explanation. Each supposition that is deduced from a phenomenon in the old theory must be maintained as true or approximately true, or the phenomenon must be challenged, or the background assumptions of the equivalence theorems must be undercut. Moreover, if the rival theory is to explain such a phenomenon it must provide for new equivalence theorems which are more defensible than the old ones and which allow the relevant coefficients of its alternative causal supposition to be measured by that phenomenon. Thus it is, that as the network of deductions from phenomena supporting a theory becomes more extensive and more closely knit the theory becomes more secure.

Forster's recommendation that each move to identify magnitudes as measurements of the same common cause be first backed up by consilience established by independent measurements may be a more restrictively bottom-up requirement than we want to impose on theory building. His moderate "common-cause" realism, which counts as real only those causes which have been measured by more than one effect, may also be less generous than we would like. Nevertheless, his positive view — that unifications which are supported by successful over-determination of a common cause by independent measurements provide powerful evidence in favor of a theory — accurately represents both common sense reasoning and scientific practice.

We can agree with Forster that this common sensical as well as scientific practice of attributing evidential force to independent measurements of a common cause is a reasonable one. We have seen that new (successful) independent measurements add new phenomena for which a rival must account. Indeed, we have seen that such new measurements can increase support, even though the theory being supported may be regarded as already certain. The increased support adds to security by legitimately stiffening resistance to revision.[17]

NOTES

[1] I say "it would seem to be unavoidable" rather than that "it would be unavoidable" because it is not clear that conceiving the new theory would not lead to a new prior probability function (perhaps by actually modifying the relevant algebra of propositions over which degrees of belief are defined) that would legitimately make the new theoretical commitments more secure (even more probable) than the old ones were. Some such proposal was suggested by L. J. Cohen in his insightful comment (1967) on Mary Hesse's (1967) interesting attempt to explicate Whewellian consilience within the framework of confirmation theory.

[2] Brahe actually used this point to argue against the Copernican hypothesis. His measurements were good enough to detect stellar parallax greater than 1′ of arc. This implied that if the Copernican hypothesis were correct the fixed stars would have to be much farther away than any of what were regarded as reasonable estimates of this distance (van Helden, 1985, p. 51).

There was also a problem about the implications for the sizes of the stars. It was generally agreed that the apparent diameter of a third magnitude star was about 1′ so that the Copernican hypothesis implied that even third magnitude stars would be at least as large as the earth's orbit with major stars comparably larger (van Helden 1985, p. 51). Galileo helped defend the Copernican hypothesis against this latter implication by pointing out that the stars' apparent diameters are much smaller when the sparkling rays surrounding them are removed by observation through a telescope or in twilight (van Helden 1985, pp. 66—67).

Stellar parallax was not actually observed until Bessell detected a shift of Alpha Centauri against more distant stars in 1838 (Hansen 1973, p. 20) long after the Newtonian revolution had settled the issue in favour of Kepler rather than Brahe.

[3] We can specify an approximation under which the Keplerian account follows from Newton's and the Brahean account is ruled out by Newton's. Newton applied an hypothesis that the centre of the system is at rest. Kepler's centre (the sun) is much more nearly unaccelerated than Brahe's centre (the earth).

[4] Other closely related ideas are Skyrms' (1980) concept of *resilience*, ideas about *relative importance of information* used by Gardenfors (1984, 1988) and by Spohn (1988), and concepts of *robustness* which have been appealed to by Rosenkrantz (1977), Wimsatt (1981) and Glymour (1988).

[5] Kepler's *Astronomia Nova: Tradita Commentariis de motibus Stellae Martis*, Heidelberg 1609 gives his own account of this struggle. N. R. Hanson (1973) *Constellations and Conjectures* (pp. 253—282) and C. R. Wilson "How Did Kepler Discover His First Two Laws?", *Scientific American* 226 (1972) pp. 92—106 are two quite accessible and illuminating accounts. The most impressive and helpful account I found was in B. Stephenson (1987, pp. 39—49).

[6] Howard Stein suggested to me that this is somewhat oversimplified since the so-called "secular inequities" don't average out, but rather accumulate.

[7] Forster makes this point a cornerstone of his cogent answer to skeptical arguments by Cartwright (1983) and Ellis (1965) about the existence of component forces.

[8] There is a discrepancy between Whewell's text and table of contents. In the text (NOR, pp. 83—85) he includes article 9 which is entitled. "Hypotheses may be useful though inaccurate" in Sect. III, while his table of contents (NOR, pp. XV) located this article in Sect. II on the use of Hypotheses.

[9] I am leaving aside Kepler's work on the inclination of Mars' orbit to the ecliptic (Wilson 1972, pp. 94—95; Small 1894, pp. 145—184; Stephenson 1987, pp. 21—39).

[10] Stephenson (1987, pp. 1—4 and especially pp. 21—49) makes a compelling case that Kepler was motivated throughout his work on Mars (and indeed in all his work) by an attempt to give an account that would fit with plausible reasoning about physical causes of the motions. His discussion makes it clear that there is much more going on in Kepler's move from the vicarious theory through various stages to eventually settle on the elliptical orbit than can be accounted for by simply noting the availability to him, in Brahe's data, of greatly improved information about relative distances.

[11] Herbert Simon's group have developed computer programs for discovery of quantitative empirical laws [The Bacon Programs (Langley *et al.* 1987, pp. 65—170)] that will generate Kepler's third law from the data. There is, perhaps, a significant difference between Kepler's war on mars, where causal reasoning led eventually to successful

colligation of the elliptical orbit (as well as the law of areas) and the more purely computational search for higher order symmetries that led to the third law.

[12] Here I am disagreeing with R. Laymon's (1983) suggestion that it was Newton's regular practice to count as phenomena idealizations that were beyond reasonable error limits on the measurements available. I think Laymon has failed to distinguished between those empirical generalizations Newton explicitly cited as phenomena and the special mathematical idealizations appealed to in proofs of theorems, in what I. B. Cohen (1980, pp. 52—153) has called "Newton's mathematical style".

My investigations suggest that Newton devoted some effort to, and on the whole did quite well at, keeping the idealizations he explicitly cited as phenomena within reasonable limits of error on the measurements available to him. For example, when we analyse the data in the table corresponding to the evidence for the Harmonic Law of Jupiter's moons, we find that the distances calculated from the periodic times differ from the means of the cited astronomical measurements by respectively 0.74, 1.1, 1.28, and 1.02 standard deviations. Newton goes on to give additional values for these distances calculated from Pound's measurements (Cajori, p. 402). If these are added in, the differences from the new means are respectively 0.324, 0.471, 0.525 and 0.372 standard deviations.

[13] Though I shall not take them up here, Mary Hesse's interesting discussions (e.g. 1967, 1974) of the role of analogical reasoning in consilience illuminate important aspects of this more free wheeling character of Newton's practice of theory construction.

[14] Howard Stein pointed out to me that Huygens and Leibniz (major defenders of vortical theories) accepted Newton's deductions of centripetal forces from phenomena, and did not take issue with Newton's argument for Propostition V. This suggests that Whewell's objection is even less cogent than I have allowed. Indeed, much work by vortical theorists following the publication of *Principia* can be regarded as an attempt to make vortical theory recover more or less similar versions of Newton's own deductions from phenomena.

[15] According to Quine's (1970) compelling picture, science is a piecemeal process of introducing and improving theories to replace prescientific natural kind conceptions based on innate ideas of similarity. Quine (pp. 21, 22) sees explicit scientific theories as rendering the modal associations of the old natural kind conceptions redundant. I see our scientific theories as taking on the modal features of prescientific natural kind conceptions as they take over their role in making ranges of phenomena intelligible. I think natural kind reasoning is unavoidable. We cannot get outside the process Quine describes. I do not think Cartesian doubt is an epistemic stance that any of us can actually take. What we can do is to try to improve our natural kind conceptions. This, I suggest, is just what science is doing for us.

[16] When we consider the failure of the vortical hypothesis to support equivalences that could seriously rival those Newton provided we can see that this is not as Van Fraassen (1985, p. 265—266) would have it, an excessively conservative principle. Nor is it, as he would also have it (van Fraassen 1985, pp. 265—266), limited in its descriptive application to the superceded scientific practices of an earlier age. As I see it, this rule and the others are reasonable policies for introducing and revising scientific theories

construed as natural kind conceptions. The history of science and its developing situation today reveal that these policies have been and continue to be a regular part of our scientific practice.

[17] The same point can be made probabilistically if we agree with Brian Skyrms (1979) that science aims at not just high probability but a *resiliently* high probability. The resiliency of an hypothesis h is defined relative to a set of conditions c_i, each of which is consistent both with h and with not-h.

$$\text{Res}\,(h) = \text{Min}\,\{P\,(h/c_i)\text{: all the } i\text{'s}\}.$$

Let the c_i's be various possible data to be added by future observations to those already colligated by two or more phenomena which have been successfully unified by an hypothesis h_u which takes each of the phenomena as independent measurements of the same common cause (say the mass of the Sun). Let h_j be an hypothesis about coeffiencts of one of these phenomena separately (say the ratio R^3/T^2 for Jupiter's orbit in some appropriate units). Let c_i be some bad data for the hypothesis h_j (say new observations which would put R^3/T^2 far away from the previously colligated value). If our conditional probabilities are sensitive to reasonable statistical estimates of fit in relevant data sets then

$$P_1(h_j/c_i) < P_2(h_u/c_i),$$

where P_1 is the original prior and P_2 is the new epistemic probability after the unification, since the relevant data base for h_u includes all the data supporting h_j together with all the data supporting the other phenomena that have been successfully unified as independent measurement of h_u. Moreover, it sould be clear that some such worst case for h_j will be worse for it than any worst case for h_u, just because the set of data supporting h_u is so much larger.

BIBLIOGRAPHY

Baigrie, B. S. (1987). "Kepler's Laws of Planetary Motion Before and After Newton's Principia: An Essay on the Transformation of Scientific Problems", *Studies in History and Philosophy of Science* **18.2**, 177–208.

Butts, R. E. (1965). "Necessary Truth in Whewell's Theory of Science", *American Philosophical Quarterly* **2.3**, 1965.

Butts, R. E. (1968). *William Whewell's Theory of Scientific Method*, Pittsburgh.

Butts, R. E. (1970). "Whewell on Newton's Rules of Philosophy", in Butts and Davis, pp. 132–49.

Butts, R. E. (1973). "Whewell's Logic of Induction", in Giere, R. N. and Westfall, R. S. (eds.), *Foundations of Scientific Method: The Nineteenth Century*, Bloomington, Indiana University Press.

Butts, R. E. (1977). "Consilience of Inductions and the Problem of Conceptual Change in Science", in Colodny, R. G. (ed.), *Logic, Laws and Life*, Pittsburgh, University of Pittsburgh Press, pp. 71–88.

Butts, R. E. and Davis, J. W. (1970). *The Methodological Heritage of Newton*

Cajori, F. (1962). *Sir Isaac Newton's Mathematical Principles of Natural Philosophy and his System of the World*, Berkeley and Los Angeles, University of California Press.

Cartwright, N. (1983). *How the Laws of Physics Lie*, Oxford, Clarendon Press.
Catton, P. (1988). "Disputing 'Newton and the Fudge Factor' ", Manuscript University of Western Ontario.
Cohen, I. B. (1971). *Introduction to Newton's Principia*, Cambridge, Harvard University Press.
Cohen, I. B. (1980). *The Newtonian Revolution*, Cambridge, Cambridge University Press.
Cohen, I. B. and Koyré A. (1972). editors *Isaac Newton's Philosophiae Naturalis Principia Mathematica*,
Cohen, I. B. and Whitman A. Translators (1987). *Isaac Newton's Mathematical Principles of Natural Philosophy*, Forthcoming, Cambridge, Massachusetts: Harvard University Press, and Cambridge, England: Cambridge University Press.
Cohen, L. J. (1967). "An argument that confirmation factors for consilience are empirical hypotheses", in Lakatos, I. (ed.), *The Problem of Inductive Logic*, North Holland, pp. 247—250.
Earman, J. (ed.). (1983). *Testing Scientific Theories*, University of Minnesota Press.
Ellis, B. D. (1965). "The Origin and Nature of Newton's laws of Motion", in Colodny, R. G. (ed.), *Beyond the Edge of Certainty*, New York, Prentice Hall.
Forster, M. R. (1988). "Unification, Explanation, and the Composition of Causes in Newtonian Mechanics", *Studies in History and Philosophy of Science* **19.1**, 55—101.
Freidman, M. (1983). *Foundations of Space-Time Theories*, Princeton.
French, A. P. (1971). *Newtonian Mechanics*, MIT Press.
Gardenfors (1984). "Epistemic Importance and Minimal Changes of Belief", *Australasian Journal of Philosophy* **62**, 136—157.
Gardenfors (1988). "Causation and the Dynamics of Belief", in Harper, W. and Skyrms, B. (eds.), (1988).
Glymour, C. (1980). *Theory and Evidence*, Princeton.
Glymour, C. (1988). "AI for Statistical and Causal Modelling", in Harper, W. and Skyrms, B. (eds.).
Godin, G. (1972). *The Analysis of Tides*, Toronto.
Hanson, N. R. (Humphrey's W. C. ed.). (1973). *Constellations and Conjectures*, Reidel, Dordrecht.
Harper, W. (1986). "Kant on the *a priori* and Material Necessity", in Butts, R. E. (ed.), *Kant's Philosophy of Physical Science*, pp. 239—272.
Hesse, M. (1968). "Consilience of Inductions", in Lakatos (ed.), *The Problem of Inductive Logic*, North Holland.
Hesse, M. (1971). "Whewell's Consilience of Inductions and Predictions", *Monist* **55.3**, 520—524.
Hesse, M. (1974). *The Structure of Scientific Inference*, Macmillan.
Koyré. (1973). *The Astronomical Revolution*, Cornell University Press.
Langley *et al.* (1987). *Scientific Discovery*, MIT
Laudan, L. (1970). "William Whewell on the Consilience of Inductions", *Monist* **55.3**, 368—391.
Laudan, L. (1970). "Reply to Hesse", *Monist* **55.3**, 525.
Laymon, R. (1983). "Newton's Demonstration of Universal Gravitation and Philosophical Theories of Confirmation", in Earman (ed.), (1983). pp. 179—199.
Owen, G. E. L. (1961). "Tithenai ta Phainomena", in Moravcsik (ed.), *Aristotle*, New York, Doubleday Press, 1967. pp. 167—190.

Putnam, H. (1983). *Realism and Reason*, Cambridge, Cambridge University Press.
Quine, W. V. (1970). "Natural Kinds", in Rescher, N. (ed.), *Essays in Honor of Carl G. Hempel*, Reidel, Dordrecht. pp. 5–23.
Rosenkrantz, R. (1977). *Inference, Method and Decision*, Reidel, Dordrecht.
Russell, J. L. (1964). "Kepler's Laws of Planetary Motion 1609–1666", *British Journal for the History of Science* **2.5**.
Shea, W. R. (1980). Nature Mathematized, Reidel, Dordrecht.
Skyrms, B. (1980). *Causal Necessity*, Yale Press.
Small, R. (1804). *An Account of the Astronomical Discoveries of Kepler*, Madison, The University of Wisconsin Press.
Spohn. (1988). "Ordinal Conditional Functions: A Dynamic Theory of Epistemic States", in Harper, W. L. and Skyrms, B. (eds.), (1988). pp. 105–134.
Stephenson, B. (1987). *Kepler's Phsycial Astronomy*, Berlin, Springer Verlag.
van Fraassen, B. (1980). *The Scientific Image*, Oxford: Oxford University Press.
van Fraassen, B. (1981). "Theory Construction and Experiment: An Empiricist View" in *P.S.A. 1980*, Asquith P. D. and Giere, R. N. (eds.), vol. 2, pp. 663–78. East Lansing, Michigan: Philosophy of Science Association.
van Fraassen, B. (1983). "Evidence and Explanation", in Earman (ed.) pp. 165–177 (1983).
van Fraassen, B. (1985). "Empiricism in the Philosophy of Science", in Churchland, P. M. and Hooker, C. A. (eds.), *Images of Science*, University of Chicago Press. pp. 245–308.
van Helden, A. (1986). *Measuring the Universe*, University of Chicago Press.
Waff, C. B. (1976). *Universal Gravitation and the Motion of the Moon's Apogee: the Establishment and Reception of Newton's Inverse Square Law, 1687–1749*, PhD Dissertation, John Hopkins University, University Microfilms, Ann Arbor.
Whewell, W. (1842). *The Philosophy of the Inductive Sciences (Founded Upon Their History)*, London, John W. Parker and Son.
Whewell, W. (1858). *Novum Organon Renovatum*, London, John W. Parker and Son.
Wilson, C. A. (1972). "How did Kepler Discover his First two Laws?", *Scientific American* **226**, 92–106.
Wimsatt, W. (1981). "Robustness, Reliability and Overdetermination", in Brewer, M. and Collins, B. (eds.), *Scientific Inquiry and the Social Sciences*, San Francisco, Jossey-Boss, pp. 124–163.

FRANÇOIS DUCHESNEAU

LEIBNIZ'S 'HYPOTHESIS PHYSICA NOVA': A CONJUNCTION OF MODELS FOR EXPLAINING PHENOMENA

Commentators like Hannequin,[1] Dugas,[2] and Gueroult[3] have tended to consider the *Hypothesis physica nova* (1671)[4], dedicated to the Royal Society, as a mere appendix to the *Theoria motus abstracti*, which Leibniz had sent that same year to the Académie des sciences in Paris. Both texts were intended as shows of skill on behalf of the young German philosopher in the area of natural philosophy, at the time he was planning his diplomatic mission to France. When Oldenburg, secretary of the Royal Society, received the *Hypothesis*, he requested Leibniz to send him a copy of the *Theoria* so as to cast more light on his theses in physics.[5] Also Wallis, who had been mandated to examine these for the Royal Society, did not fail to consider the *Theoria* as the true ground for the *Hypothesis* and to agree that the object of physics cannot be analyzed without resorting to reasons in geometry. Leibniz himself connected the physical construction, which is real yet exact, with the geometrical construction, which is both imaginary and exact.[6] He was convinced that everything in the physical world obeys the laws of *phoronomia elementalis*. However, these abstract laws do not suffice to account for the determination of material parts to circular motion nor for the effects resulting from mass. Is therefore the *Hypothesis* an attempt to correct the untoward consequences of an abstract geometry of conatuses? Is it a kind of vast ad hoc hypothesis to save the phenomena in view of abstract inadequate reasons?

Such views, akin to those of Gueroult, would outdo the case. Leibniz never pretended that phenomena could be deduced from abstract reasons. The physical construction need rely on a complex set of theoretical postulates, but part of these at least would outstretch the compass of the geometry of conatuses. Yet, were the principles heterogeneous, it would not mean that we end up with an artificial apparatus of reasons like those which Leibniz denounced in *phoronomia experimentalis*, because then piecemeal inferences from empirical data are

J. R. Brown and J. Mittelstrass (eds.), An Intimate Relation, 153—170.

blended with mixed constructions.[7] Hannequin follows closely on the texts, and he suggests that Leibniz's is an integrated system in which metaphysical and physical considerations harmonize with the analytic statements of the conatus theory. However, Hannequin believes that the abstract reasons rule over the system deductively. Analyzing the implications of the concept of motion, one would show the coherence of the geometry of conatuses with the original principles of the system of nature. What is lacking in this otherwise faithful account of the *Hypothesis* is an assessment of the epistemological features of Leibnizian physical constructions. Our proper objective is to fill up this gap.

On this aspect of Leibniz's endeavor, the letter to H. Fabri of 1676 gives interesting information.[8] In his correspondence with Pardies, Fabri had attempted to dissociate his case from that of the corpuscularians and for that purpose he had criticized the systems of both Gassendi and Descartes. In an additional attack on Hobbes's *De corpore*, he took objection to Leibniz's treatise. Leibniz initiates his defense by showing what he kept from Aristotelian science, how he differed from atomists and from Gassendi in particular, and why he could not adopt Descartes's methodology. His admiration goes to Aristotle's contributions to "civil science" and to the natural history of animals. He is ready to incorporate and reinterpret some developments in the *Physics* concerning principles, motion, continuum, soul. He objects to Democritean natural philosophy on the ground that incorporeal substances ought to be admitted, that it is impossible to derive motion from body, that naturally indivisible corpuscles are inconceivable, that any theory about *simulacra* is extravagant. He notes with interest that Gassendi drew a significant distinction between extension and the essence of material realities, thus starting on phenomenalizing the *extensum*. His position regarding Descartes is far more radical. Leibniz denies that he had adequate knowledge of the Cartesian theory when writing the *Hypothesis*. He did not yet master the mathematical techniques of the Cartesian school. Also his purpose was not, contrary to Descartes's, to frame hypotheses that pointed to mere possibilities without allowing physics to reach anything real.

The methodological stand Leibniz favors consists in "deducing" more complex phenomena from simpler ones which have been previously investigated. This makes it possible to discard fictitious hypotheses (*fictitiae hypotheses*) from natural philosophy.[9] Indeed, the postulated causes are then homologous with those phenomena referred to in explaining observed effects. The transition from causal models to resulting effects is secured by means of such laws as those one derives from geometrical phoronomia. Causal models refer to the simplest and most general phenomena instantiated in nature; their capacity for deductively generating more specific phenomena is considerable.[10]

Two demonstrative ways from effects to causes are indeed possible. In some instances, from a mere collection of effects one can identify the causal process which they issue from. This Baconian approach is subject to difficulties when effects can be explained by reference to a plurality of possible causes. Secondly, the demonstrative way can be vouched for when causes express themselves directly in the field of the observable or can be derived through unproblematic analyses and inferences. If physics is to be raised to demonstrative certainty, there will need to be explanations of more specific phenomena by more general ones: the latter are meant to be the simplest elements at the phenomenal level. This type of explanation depends upon the availability of conceptual analytical tools. The 1671 developments take the geometrical conatus theory to be of that kind, as well as rules for algebraic summation which can be applied to motion indivisibles. In view of this jointly phenomenal and analytic approach to hypotheses, the *Hypothesis* points in the direction of an original research programme.[11]

How can science develop under these conditions? Its point of support resides in experiences. These may have already been achieved, or analysis itself suggests achieving them. The proper function of analysis is to unify the known facts or empirical elements in a chain of sufficient reasons. The conceptual ground of analysis consists, as we suggested, in abstract elements provided by geometry, in particular, motion indivisibles or conatuses. In 1676, after his stay in Paris, Leibniz seems to add to these the specific laws of physical mechanics, such as the laws of Wren or Huygens. These laws formalize more directly the sensible data of motion in the phenomenal world. But then, Leibniz does not yet question the fact that in principle these empirical laws might conform with his mathematical system for the summing up conatuses. In any event, by means of analysis, a chain of sufficient reasons can allow for a demonstrative rendering of specific phenomena, provided one has already proceeded to the required empirical inventories and experimental verifications. The principles of scientific knowledge consist in notions about the fundamental motions of the physical world to which the analytic means of geometry and mechanics can be applied. The consecutive derivation of phenomena from such principles may allow us to uncover the combinatorial order implied in these phenomena; and so one will know "what is the nature of the elements and by which combinations everything is formed".[12]

One of Leibniz's preoccupations is with the logic of discovery. In 1676, he protests the strategy of raising arbitrary hypotheses, as Descartes among others did. He proposes instead a process that is more in accordance with the practice of mathematical analysis. One need inventory the available data for solving the problem and specify the experiments to be performed so as to account demonstratively for the phenomena and substitute certainty for mere probability. Indeed, the inventory could have been achieved haphazardly (*bona fortuna*), but the demonstrative process can always be obtained by methodical search. Such a process may increase our knowledge significantly and prevent waste of time and effort. These views of 1676 rest on Leibniz's growing confidence in the analytic methodology which mathematics exemplifies. This analytic potential of mathematics had formed a major object for consideration during his stay in Paris, but the same methodological trend was immanent in the *Hypothesis*. This is well expressed in this passage:

> From these few elements, I believe one may understand what a large field is open to us for accurate philosophizing without hypotheses, provided from now on experiments (which diligent men have noted or which true analysis has shown should be undertaken) be connected with geometric elements and mechanical laws; and I do not doubt that with due application something certain can be demonstrated concerning the system of things and the principal motions which take place around us: hence afterwards the various phenomena of particular things will be explained and our dominion over nature will be secured.[13]

The phenomenalist and analytic methodology Leibniz puts forth is probably the main distinguishing element in a comparison drawn with Descartes. Commentators, like Mouy[14] and Dugas, have tended to overlook this difference because the succession of themes is similar for both Descartes and Leibniz. Indeed, Leibniz's criticism of Cartesian methodology and physics will expand mostly in the period when he is working out his dynamics (1676–1686). However, the 1671 project is already quite original in its methodological aspect. If Hobbes had influenced the conatus theory, the construction plans for physics in the *Hypothesis* represent a revised empiricist strategy, better structured and fitter for drawing demonstrative chains of sufficient reasons.[15] In his letter to Hobbes,[16] Leibniz had stated his reservations concerning the English philosopher's theory of cohesion, and he had opposed the causal reduction of motion to a mere order of material changes. Conspiring motions as a ground for the formation of discrete masses in extension and the global economy of phenomena outride the capacity of Hobbes's theory. In the *Hypothesis*, Leibniz tried precisely to account for such requisites through a process of empirico-analytic construction. We shall now try to describe and analyze this process.

While excluding any preliminary recourse to a corpuscular theory, Leibniz postulates a minimum of prerequisites for the generation of the physical phenomena out of diversified motions in homogeneous space. Insofar as concerns the planetary system, a simple model contains the solar sphere, the earth and the "mass at rest" of the ether, a subtile fluid occupying the intermediary space.[17] The parts of the sun and earth rotate about the axis of their respective globes. These celestial bodies have resulted from a primordial differenciation in ether, which could only come about through conspiring circular motions. For the penetration by conatuses issuing from the ether would be total, were it not for a sui generis determination to motion causing the dissociation of celestial bodies from the ether. Apart from the gyratory motion of solar particles because of axial rotation, one must suppose a specific motion of these parts causing the emission of light and the resulting mechanical effects. These effects have been wrongly assigned by Torricelli and Hobbes to gyration proper: were it the case, projection along the tangent would exhaust the solar matter and the required differenciation among the ether particles would not be secured. Therefore, Leibniz supposes a large number of circular motions in the sun so that according to the angles of convergence of these motions the resulting paths follow the direction of external bisectors.[18] Determinations to motion in the ether are modified accordingly: this phenomenon corresponds to the emission of light. In the letter to Fabri, Leibniz appeals to arguments that are both notional and empirical[19] to justify indefinite transmission in all direction in such fluid and resistance-free medium (=imponderability). Based on the analogy of sight experience, one may grant any sensible point a capacity of transmitting light, and hence identify a fluid matter of the heavens, in which conatuses of light expand without impediment. Due to the analytic implications of the *plenum* notion, one needs to admit that all those motive determinations compensate mutually but one need also suppose a functional heterogeneity between these various determinations underpinning differences among the masses in the diaphanous ether.

Indeed, Leibniz partakes with Descartes in the view that the main mechanical effects in the physical world can be derived from the nature and action of light. But his theory excludes the corpuscular models the author of the *Principia* had set up. Let us suppose the rotation of the earth, a sphere initially formed with matter whose density holds the middle between air and earth, namely of the order of density of water. Light emissions produce impacts in the fluid mass of the earth: hence the various significant phenomena arise. *Ex hypothesi*, the earth moves on its axis from west to east; the subtile ether undergoes the motive determinations of light and presses on our globe; there follows an apparent motion from east to west, which we can take as a real motion if we consider the earth as motionless on its axis. Such is the universal motion out of which the explanation of all phenomena is to be drawn.[20]

Under this pattern, Hannequin is right in pointing out that the letter to Fabri sketches an hypothesis about the formation of the terrestrial sphere, when Leibniz writes: "Fluids surrounded by heterogeneous fluids, gather in the form of a round drop".[21] Due to the various internal motions in fluids, the convergence of heterogeneous fluids disrupts the overall equilibrium. The denser fluid gathers naturally under the volume and shape that will secure the greater resistance to change caused by external impacts. Hannequin does not perceive anything essential in that addition, which would not provide sufficient reason for the functional heterogenity of fluids and the resulting figures.[22] This criticism does not hold. After all, Leibniz tries to account for the structural conditions of gravity when it operates in all points of the sphere along a centripetal tendency. If one considers axial rotation alone, the conatuses of light emissions should produce pressure following a ratio related with cylindrical surface. This problem, which arose also in Descartes's vortex theory, meets with a solution that Leibniz deems adequate, by means of the bubble theory. Proposition 7 in the letter to Fabri develops the argument pertaining to that theory, which was also central in the *Hypothesis*.

The bubble theory is built on empirical analogies and on the analytic requirements arising from the conatus theory. The phenomena suggesting it are boiling and fusion, whether they result from heat or from the mechanical action of light emissions. The analytic rule implies attributing conspiring inner motions to the spherical structures of the denser fluids. Subtile fluids penetrating dense structures increase the endogenous tendency to resist, which their "hard crust" opposes to surrounding fluids. Glass blowing suggests an experimental process that conforms to this mechanism: the glass bubble results from the circular motion of fire and from the rectilinear motion of air as it is insufflated in the cane; both motions act jointly on the vitrescible fluid matter. "Similarly, from the circular motion of the earth and the straight one of light, there arise bubbles."[23] The analogy is to be understood by means of an inversion in hypotheses: the earth is supposed motionless as the vitrescible fluid matter; it would be acted on by the ether in circular gyration and subject to impacts of light perpendicular to the axial diameter. Yet Leibniz seemed to acknowledge the need for a determination to spherical shape. This determination would depend on those motions which, within a mass of fluid parts, are occasioned by the multidirectional impacts of a surrounding fluid. The hollowing of those bubbles, the increase of their endogenous conatus to react would result from such a combination of circular and rectilinear motions. As the matter stands, one glimpses here at the composedness of material elements, since heterogeneous conatuses produce the bubble structure. The synthesis of these dispositions issues from the gyration of bubbles about their own centre.

Leibniz borrows from the bubble notion the main tool for raising a physical theory.[24] If we suppose an original fluid state of the earth, the terrestrial particles will possess a dominant conspiring motion along the Equator, due to the cohesion resulting from axial rotation; and the same will happen for the successive parallels. The parts located on the same parallel benefit from a cohesion pertaining to conspiring rotative motions. From one parallel to the next, cohesion will be, if not null, very weak. Hence, the ether particles moved by the light emissions penetrate the sphere along the lines of lesser resistance and combine with the parts of the generic terrestrial fluid to form bubbles. As a result, a sum of individualized elements of various sizes and densities compose the successive layers of the sphere, with infinite gradation in the constitutive materials.

Also, as noted by Hannequin,[25] a Leibnizian bubble has no similarity with a Democritean atom: first, because it is subject to many transformations following surrounding motions, secondly, because bubbles comprehend other bubbles, and these latter yet others of a lesser size, without our being able to mark out the end of that regress. And so, one may take bubbles to represent individualized and complex elements with nested structural determinations. Interestingly enough, Leibniz connects with this concept of the element, observations about microscopical organizations, which micrographers like Kircher and Hooke had collected.[26]

Indeed, this theory about integrated bubbles representing nesting microcosms will provide, in the correspondence with Herzog Johann Friedrich,[27] a material basis for the analogy to be drawn between the circle of influence of the *mens* and a convergence point for motions in the phenomenal world. But this is a metaphysical view forecasting the concept of a finite substance or monad at later stages of Leibniz's philosophy.

From the point of view of scientific construction, the characteristic feature of this theory rests in the cohesion of the explanatory system restricted to the four elements: ether, air, water, and earth, insofar as ether, according to the geometrical properties of conatuses, achieves the functional diversification of the large "masses" of bubbles forming the other elements; indeed, within those masses there occurs a diversification of bubbles *ad infinitum* in terms of density, size, and specially dynamic properties (due to conspiring internal motions in connection with ether impacts). Leibniz discards the hypothesis of an even more subtile ether which would form a fifth element.[28] On the one hand, no phenomenon suggests the need to appeal to such a postulate; on the other hand, the structural conditions of the physical universe require a type of element that can generate gravity and the other analogous properties of bodies, without having to depend on properties similar to those that need explanation. Indeed, the structure of the ether and its action form an abstract theoretical construct. If the latter is coherent and suffices to explain the sensible effects, why would one need to regress further in the order of hypotheses, contrary to Occam's principle? It seems preferable to segment the masses of the other elements in an infinite variety, according to the modalities of ether action on the material constituents of bubbles. As stated by Hannequin:

> Ce qu'il faut noter, c'est l'effort de Leibniz pour dériver des actions de l'éther, soumis exclusivement aux lois du mouvement abstrait, et cette nature et cette densité [des éléments matériels], et en même temps cette division en bulles qui va à l'infini et qui, en supprimant leur continuité et en les faisant 'interrupta', confère enfin aux corps cette puissance proportionnelle à leur grandeur de résister au mouvement qui est ce que les modernes entendent par la masse.[29]

In line with this theory, Leibniz attempts to work out a model to explain the phenomenon of weight (*gravitas*) and analyze its effects. This model would be a drawing pattern for building a system of the phenomena that could be extended into more specific explanations.[30] The proposed model relies on the equivalence of hypotheses on circular motion and rest between the earth and ether.[31] Let us suppose once again the earth motionless and surrounded by an ocean of ether which would move from east to west. This ocean does not only enfold the globe; through an infinite network of conduits, it exerts its moving action in the interior of masses and of the bubbles they consist in. Any heterogeneous body which opposes the free circulation of the subtile ether particles, provokes a reaction: the ether particles push back towards the centre of the earth any body or part of a body whose mass cannot be dissociated or dispersed through subtilization to a degree similar with ether. The centripetal motion of heavy bodies seems to correlate with the spherical surface of the ether layer which they disturb. Leibniz then states a false but logical law on the increase of gravity in a double ratio to the diameter of the sphere of circulation of ether. Moreover, this law seems to account for the accumulation of successive impacts in the effect of fall.[32] So the rate of increase in velocity would decrease with the distance gone through in the process of fall. This error is indeed perfectly coherent with the notions of the *Theoria motus abstracti*, since the effect of mass is discarded and we only deal with conatuses that are subject to the algebraic summing up of impacts which decrease as the square of the diameter. Gravity measures the effort of ether in restoring the uniformity of its motion broken up by the insertion of bodies with a different density; the letter to Fabri adds that this effect is due to the same cause that produces the spherical figure of fluids contained in less dense media; such a cause operates through conspiring motions of the inner parts under the impact of surrounding parts.[33]

The explanation of elasticity builds up in a parallel way: it is a general phenomenon caused by the disruptive and expansive action of the ether, when it tends to reduce to the same subtilty as its own, those heterogeneous material parts comprised in its circular flow.[34] Here, Leibniz undertakes to analyze phenomena pertaining to the physical economy and not merely to the abstract geometry of motion as amended by auxiliary hypotheses on the formation of the planetary system. For instance, Leibniz relates with elasticity conceived as a flow of ether in more or less cohesive bodies, the physically assignable laws of reflection and refraction whether about light or mechanical effects. The laws of impact of Wren and Huygens, which the *Theoria motus abstracti* could not account for, seem to him to be compatible with the basic mechanical hypothesis, once the conception of elastic action and reaction has been established.[35] In the same way, one can give a theoretical representation of causes for the oscillatory motions and isochronism of pendula, for the dioptrical and catoptrical effects, and even for a manifold of more specific physical effects, for instance the accelerated and then decelerated motions of projectiles in the atmosphere. Air is indeed the basic support for elastic phenomena in the sensible world. It is important to note the discontinuous structure of material parts, which alone justifies the effects of mass and the reaction of bodies to impulse by ether parts. Discontinuity is achieved by the conspiring gyratory motions of bubbles, which represent the "hard" elements in the physical world; and reaction depends on the dispersive power of ether particles disturbing endogenous gyration. At least, that interpretation of sensible data is allowed, granted the relativity of our cinematic representations.[36]

Starting from these basic theoretical concepts, Leibniz aims at connecting the statics of fluids with a coherent representation about the elementary structure of bodies interacting because of gravity and elasticity.[37] Then, the endeavour will extend to phenomena described through specific aspects. The specific aspects can indeed be analyzed down to associations of sensible qualities, which in turn refer to subtile, though properly imperceptible, motions.[38] Yet Leibniz sets himself the task of reconstructing a general model that could link these phenomena with gravity and elasticity beyond sensible appearances and seemingly inassimilable motions.

This applies particularly to so-called sympathetic and antipathetic motions, that is to say magnetism and electric attractions on the one hand, chemical reactions on the other. The mechanism of magnetic attractions gets explained through a variant of gravity.[39] The pressure of ether on the parts of the globe repels solid corpuscles towards the centre of the earth, but, because of axial rotation, when the centripetal motion of these corpuscles is impeded by the impact on denser parts, there occurs a deflection of "magnetic" matter towards the poles along the meridian lines. In short, as made clear about the spherical consolidation of denser fluids in the letter to Fabri,[40] the motion of magnetic declination happens through the convergence of centripetal conatuses (connected with the forming of drop-like structures) and conatuses of gyration along the parallel lines. Along the same line of thought, magnetization gets explained by a porosity of iron, congruent with the flow of ether particles along meridian lines. Similarly, electric attraction is accounted for by the dynamic of exchanging fluid bodies so as to reach a stage of equilibrium relative to the circulation of ether.

The general model drawn up by Leibniz does not only bring together electromagnetic and gravitational effects, which is an interesting programmatic thesis, but the same model tends to connect chemical phenomena with the explanation of elasticity by means of disruptive ether flows. The effects deriving from elasticity are then varied to infinity through the combinatorial relating of *containers* with *contents*, of bubbles with their endogenous motions under the impact of ether particles. A frame is thus set for collecting a plurality of chemical phenomena under analytical concepts provided by the geometry of conatuses. Article 46 of the *Hypothesis* is a significant statement in this regard: there Leibniz justifies his model in view of the general conciliation it achieves between abstract rules of motion and the integrated economy of natural processes:

> In order that the reader struck by a slight appearance of contrary experiment do not disturb at once the entire harmony, since most of the time experiments as I showed in the case of motion, disagree plainly at first sight with the inner principles of things, and since they cannot be reconciled with them except by the considerable artifice of a universal economy, with the admirable wisdom of the Creator enfolding the origin of things; it must be shown a priori briefly that our hypothesis is slightly more than an hypothesis. For, without bubbles and containers (*vasa*), the most subtile corpuscles cannot be constrained. There must therefore be two types of bodies: containers and contents or small contents (and I would not deny that some escape out of bubbles, even though perhaps these also consist in smaller bubbles), solids and liquids, bubbles and masses.[41]

As noted by Hannequin,[42] one must consider in this instance the full scope of Leibniz statements concerning the composition of elementary structures and the state of dynamic equilibrium which the system of bubbles tends to produce. Masses or the so-called indetermined principles of composition, correspond with the typical elements taken in their extensive quantity without further differenciation: these are ether, air, water, and earth. The conspiring motions inside these masses generate spherical structures which diversify and get endowed with dynamical properties in connection with impacts by ether particles. Masses determined in bubble structures form then the contents of those "vasa" which relate to the motions and densities of surrounding fluids. The relation of inner densities and motions to external ones can be "ordinary" when there occurs an equitable compensation of effects: this state then tends to subsist and restore itself if disturbed. Major changes can nevertheless happen, breaking the equilibrium between the mechanical actions and reactions in the range of low amplitudes: then the conatus relationships between surrounding fluids and bubble components produce abrupt transfers in ether circuits: a phenomenal effect suggesting explanatory analogies consists in permutation between density and rarefaction states. Precisely, Leibniz attempts to annex various concepts from previous and contemporary chemistry to such a physical model: for instance the "deflagration" resulting from some struggle between fire and water, the fermentation known to the ancients, the sulfur and nitre of Paracelsians, the various principles called upon by the moderns, such as acids and alkalis. In order to promote a physical model concerning agents and processes of chemical reaction, attempts such as Boyle's, jointly empirical and mechanicist, seem to develop along the line drawn in Leibniz's programme, as witnesses his reference to the *Sceptical Chemist* (1665). The terminological variations among chemical doctrines reflect probably the complexity of the phenomena that have to be described from experience or discovered through experiments. Leibniz suggests that attempts be made to account for the system of reactions through a representation of the presumed generic mechanism. After referring to density and rarefaction to account for motions generating ebullition, Leibniz concludes:

> The same can be said about all the other reactions, diversified in their origin solely because of the size and number of bubbles, their position and figure, and the degree of exhaustion or compression.[43]

The rather extensive developments in the *Hypothesis* about the variations in bubble characteristics result in a table of combination for the elementary properties of bodies. This is a significant attempt at setting a theoretical cipher for unravelling specific processes which could be inventoried through experience alone. This representation is subject to a principle of architectonic order: a host of diversified effects are generated by a most limited system of causes, factors, or elements acting as general sufficient reasons. Indeed, such a fiction requires some a posteriori justification which has to be gained through empirical analysis; yet without this type of fiction conformable to the general analogy of facts, the phenomena under observation would remain illegible. The hypothetical framing of this table produces a possible schematism under cover of an architectonic idea.[44]

The resulting schema is probably the first reasoned table of the chemical elements according to a general physical hypothesis. That was the programmatic objective Leibniz had set for this presentation. Did he not aim then at giving a simple set of interconnecting concepts to link the physical theory with experience through conciliating the methodologically acceptable principles which belonged to partially antagonistic theories? This schema for theorizing would probably allow a progressive unfolding of scientific knowledge and its applications.[45]

This Leibnizian schema manifests its "theoretical functionality" mostly in the critique of corpuscularian physics. Yet Leibniz will not question the maxim of the moderns that everything in the physical phenomena is to be explained mechanically. In fact, he agrees that there is a perfect continuity between explaining phenomena by geometrico-mechanical reasons and building "machines of the art" which exemplify the laws of motion operating in the case of duly specified material structures, for instance clocks built according to the functional properties of gravity or elasticity.[46] But proper to Leibniz's model is the aptness to produce the theoretical sufficient reasons for the mechanisms involved. There are three possible grounds for the property of increasing the power of bodies, meaning for instance the lifting considerable weights with reduced forces being applied: the distance of bodies from the terrestrial centre of gravity, the *impetus* produced by falling, and finally the *nisus* which certain types of things possess because of their inner arrangement and intrinsic convergent motions: for example, the blowing force of gunpowder or the moving force of animal muscles. The first two grounds are clearly accountable for through the concepts of weight and elasticity; the third one should also be by means of a more indirect analysis focusing on a complex interplay of "micro-systems". Now in this Leibnizian procedure, one tends to spare the use of Democritean properties: antitypy or primary density, elementary figures for such and such types of solid particles or atoms; vacuum is also discarded to the benefit of some relative ratio between the dense (*densum*) and the rare (*exhaustum*): this ratio is derived from the sole primary tenet, namely that motion generates structures which differentiate because of their relative discontinuity. Leibniz builds on the perfect conformity of his basic hypothesis, which avoids postulating unanalyzable physical properties, with the theory of conatuses.

Yet justifying the theoretical concepts he adopts goes along with denouncing the relative inaptness of corpuscularian physics in offering any "ultimate" account for empirical properties. Leibniz criticizes Boyle for imagining helixes in order to explain the elastic force of air, as it exceeds that of water. It is easy to get rid of Boyle's model if one hypothesizes a greater agitation of the air bubbles due to dispersive impacts by ether particles; hence gyratory motions make it more difficult for air to penetrate narrow tubes, while the water particles, denser and less compressible, are less capable of resuming the specific form of their "mass" in its cohesive disposition. So, Leibniz shows that he can avail himself of a model, which Boyle lacked, to study capillarity in narrow tubes. The followers of Gassendi, and this applies to Boyle, as well as Descartes, entertained a false conception of density as the cause of cohesion. This conception made them assume the existence of specific figures of the elementary corpuscles, whether these were to be considered as atoms or not. For Leibniz, assuming these figured corpuscles or, which gets to the same, vortex motions of solid particles with various sizes, is tantamount to creating pure fictions void of explanatory power because they cannot "harmonize" with a coherent and systematic analysis of the phenomena. On the contrary, one should try to combine the resources of a uniform and coherent geometrico-mechanical model for all phenomena with specific inferences based on the observation of those phenomena which seem to express the inner structure of the bodies (bubbles for instance). Construed in this manner, the Leibnizian hypothesis should not fail when accounting for the sufficient reason of natural realities and processes. On the one hand, one has to abide by a rule of simplicity in postulation, on the other hand, one has to set up a network of explanations patterned after the specific phenomena without superfluous ad hoc hypotheses. On that account, the contrast between the corpuscularian approach and his own is well stated by Leibniz:

> I have always believed that what we are told about variously figured atoms, vortices, ramifications, hooks, globules, and such other arrangements, is appropriate for entertaining the mind, remote from nature's simplicity and from all experiences, and sterile, so much so that there is no question of being able to connect it manifestly with the phenomena. Our Hypothesis, on the contrary, unifies the separate and flowing

corpuscles by means of bubbles; it deduces the motions and effects of bubbles and finally of all apparent objects from the sole universal motion of the universal system, and so originating from the first and abstract principles, raising itself on the basis of the most essential experiments of chemists, applying to the case of the most simple phenomenon of gravity and elasticity an explanation drawn from the total state of our globe, it connects theory and mechanical observation with great clarity and harmony.[47]

En résumé, this paper has aimed at describing and analyzing the Leibnizian model for empirico-scientific theorizing which the *Hypothesis physica nova* (1671) sketched. Leibniz never really revised it in the later expressions for his physical theories. Evidently, the *Theoria motus abstracti* and the initial formulations of the conatus theory were abandoned with the advent of the dynamics. On the other hand, the *Hypothesis* was essentially programmatic: it provided only some directive ideas on how to secure a system of scientific explanations in dealing with the indefinite inventory of phenomena. The specific models Leibniz had cast, for instance in connection with the bubble analogy, were probably not marking any significant progress on the models of corpuscularian philosophy he was challenging on the basis of their sterile adhocness. Yet, the *Hypothesis* is worth re-examining insofar as it embodies the provocative view that scientific theorizing depends on an analytic process, which endeavours at systematizing the inventory of specific phenomena through an economy of models: these models are to be cast and progressively reframed in the light of architectonic norms. These norms spell out the necessity of harmonizing abstract mathematical or quasi-mathematical patterns for explanations with a systematic networking approach to the various phenomena. Observational and experimental data are to be considered as more or less provisional expressions for tentative conceptual links in a complex system of sufficient reasons.[48]

Université de Montréal
Montréal, Québec
Canada

NOTES

[1] Arthur Hannequin, "La première philosophie de Leibniz", in *Etudes d'histoire des science et d'histoire de la philosophie*, Paris: F. Alcan, 1908, I, pp. 17—224.

[2] Martial Gueroult, *Leibniz: Dynamique et métaphysique*, 2nd ed., Paris: Aubier-Montaigne, 1967, chap. 2. Les premières conceptions de physique, pp. 8—20.

[3] René Dugas, *La mécanique au XVIIe siècle*, Neuchâtel: Ed. du Griffon, 1954, chap. 14. La pensée mécanique de Leibniz, pp. 460—466.

[4] G. W. Leibniz, *Die philosophischen Schriften*, hrsg. von C. J. Gerhardt, Hildesheim: G. Olms, 1965, IV, pp. 177—219.

[5] Letter of Oldenburg to Leibniz, 14 April 1671, in G. W. Leibniz, *Mathematische Schriften, hrsg.* von C. J. Gerhardt, Hildesheim: G. Olms, 1971, I, p. 17.

[6] François Duchesneau, "The Problem of Indiscernibles in Leibniz's 1671 Mechanics", in K. Okruhlik and J. R. Brown, *The Natural Philosophy of Leibniz*, Dordrecht: D. Reidel, 1985, pp. 7—26. Cf. P. IV, p. 235: "Triplex constructio est: Geometrica, id est imaginaria, sed exacta; Mechanica, id est realis sed non exacta; et Physica, id est realis et exacta."

[7] Cf. P. IV, p. 240.

[8] P. IV, pp. 244—261.

[9] P. IV, pp. 248

[10] Leibniz confirms these various points thus, cf. P. IV. p. 248: "Nimirum tentandum putabam, an non phaenomena naturae difficiliora ex aliis quibusdam phaenomenis manifestis atque exploratis deduci possent. Hoc enim praestito patebat frustra causas possibiles assumi pro veris, dum ipsae verae atque certae causae in promptu essent . . . Volui harum causarum tam potentium tamque late fusarum [the question is about the motions of celestial bodies and the transmission of light] consequentias scrutari adhibitis Mechanices legibus."

[11] P. IV, pp. 248—249: "Credidi tamen excitari posse ingeniosores hoc exemplo, ut imposterum quoad ejus fieri posset sine fictitiis Hypothesibus Philosophiam naturalem tractare conentur, assumtis causis, quas revera in natura esse constaret . . . Ad novam enim et ni fallor veriorem de rerum natura ratiocinandi viam homines vocavi."

[12] P. IV, p. 257.

[13] P. IV, p. 256.

[14] Paul Mouy, *Le développement de la physique cartésienne, 1646-1712*, Paris: J. Vrin, 1934, pp. 218—29.

[15] Hannequin seems to have been justified in linking Leibniz's and Hobbes's approaches in physics. However, one should be more careful in drawing strict analogies between the *Hypothesis* and the *De corpore*, since there are significant methodological differences involved, cf. Hannequin, *op. cit.*, pp. 133—135.

[16] Letter of Leibniz to Hobbes, 13/22 July 1670, P. I, pp. 82—85.

[17] P. IV, 1, p. 181.

[18] P. IV, 7, p. 182.

[19] P. IV. 1—2, p. 249.

[20] P. IV, 10, p. 184: "Atque hic est ille Universalis motus in globo nostro terr-aqu-aereo, a quo potius, quam atomorum figuris aut ramentorum ac vorticum varietatibus, res sunt repetendae."

[21] P. IV, 7, p. 251.

[22] Hannequin, *op. cit.*, p. 111.

[23] P. IV, 11, p. 184.

[24] Cf. P. IV, 12, p. 184: "Hae jam bullae sunt semina rerum, stamina specierum, receptacula aetheris, corporum basis, consistentiae causa et fundamentum tantae varietatis, quantum in rebus, tanti impetus, quantum in motibus admiramur: hae si

abessent, omnia forent arena sine calce, avolaretque gyratione densorum expulsus aether, ac terram nostram mortuam damnatamque relinqueret. Contra a bullis, gyratione circa proprium centrum firmatis, omnia solidantur et continentur. Quae ratio est etiam, quod fornicata, ea quam admiramur firmitate polleant, cur vitra rotunda in experimentis Elasticis subsistant, alterius figurae dirumpantur."

[25] Hannequin, *op. cit.*, p. 115.

[26] P., IV, 43, p. 210: "Sciendum est enim, ut praeclari illi Micrographi, Kircherius et Hookius, observavere, pleraque quae nos sentimus in majoribus, lynceum aliquem deprehensurum proportione in minoribus, quae si in infinitum progrediantur, quod certe possibile est, cum continuum sit divisibile in infinitum, quaelibet atomus erit infinitarum specierum quidam velut mundus, et dabuntur mundi in mundis in infinitum."

[27] Cf. Letter of 21 May 1671, P., I, pp. 52—53: "Dann auch meine Demonstrationen gegründet sein auff der schwehren doctrina de puncto, instanti, indivisibilibus, et conatu; dann gleich wie Actiones corporum bestehen in motu, so bestehen Actiones mentium in conatu, seu motus, at sic dicam, minimo vel puncto; dieweil auch mens selbsten eigentlich in puncto tantum spatii bestehet, hingegen Ein Corpus einen platz einnimbt. Welches ich, nur populariter davon zu reden, daher klarlich beweise, dieweil das gemüth sein musz in loco concursus aller bewegungen, die von den objectis sensuum uns imprimirt werden."

[28] P. IV, 49, p. 203.

[29] Hannequin, *op. cit.*, pp. 116—117.

[30] P. IV, 15, p. 186: "Porro has bullas, haec vitra varie intorta, figurata, glomerata esse, facile cogitatu est, ad tantum rerum apparatum producendum, de quo mox in origine specierum, nunc totius systematis affectionem, id est gravitatem praeoccupemus: ac merito quidem, cum gravitas plerorumque in globo nostro extraordinariorum motuum causa, aut certe clavis sit, eorum etiam, qui in speciebus privatim exeruntur, et danda sit Physico opera, ut ad mechanicas rationes, quippe simplicissimas, quoad ejus fieri potest, omnia reducantur."

[31] P. IV, 2, p. 181.

[32] P. IV, 18, p. 186: "Cum enim turbent circulationem, expelluntur; non sursum, nam eo magis turbabunt (quia superficies sphaericae crescunt in duplicata ratione, non in eadem cum diametris ratione; ac proinde sectionum quoque in idem corpus agentium inaequalitas major evenit) ergo deorsum, id est descendent. Hinc porro incrementum impetus ob novam ubique inter descendum in qualibet aetheris liberi aut liberioris, quam rei illius ratio fert, parte impressionem."

[33] P. IV, 10 p. 251.

[34] One should note that, according to Leibniz, heavy bodies are never without some degree of elasticity and that the most elastic bodies, approximating the subtilty of ether, are nevertheless subject to some degree of descent towards the centre of the globe.

[35] P. IV, 22, p. 190.

[36] P. IV, 21, p. 188: "At corporum sensibilium alia plane facies: omnia enim dura sunt motu quodam intestino in se redeunte; omnia discontinua sunt, unde caeteris paribus plus efficit moles; omnia Elastica sunt, seu compressa ac mox sibi relicta, ab aetheris gyratione in statum priorem restituuntur."

[37] P. IV, 23—29, pp. 191—195.

[38] P. IV, 30, p. 195.
[39] P. IV, 33*—34*, p. 197—199, 17, p. 254.
[40] P. IV, 7—8, p. 251.
[41] P. IV, 46, pp. 202—203.
[42] Hannequin, *op. cit.*, p. 127 n. 3
[43] P. IV, 40, p. 201.
[44] P. IV, 51, p. 204: "Et hic certe Hypothesin condituro, nisi temerarius haberi affecta, subsistendum est; specialior enim applicatio ab experientia pendet. Credidi tamen semper admirabilem Conditoris sapientiam ita res instituisse, ut paucis multa gerantur. Unde si somniandum esset, dicerem . . ."
[45] P. IV, 54, p. 206: "Sufficit causam omnibus motibus explicandis suffecturam reddidisse, sufficit ex simplicissimis et liquidissimis et intellectu facillimis, ad hanc usque experientiae portam volatiles, alioquin et usui vitae atque analysi practicae inconciliabiles Theorias deduxisse; sufficit ea attulisse, quae sectae omnes, salvis domesticis opinionibus, ferre possunt."
[46] P. IV, 58 pp. 210—211.
[47] P. IV, 57, p. 209.
[48] Work on this paper was supported by a research grant from the Social Sciences and Humanities Research council of Canada.

RICHARD T. W. ARTHUR

RUSSELL'S CONUNDRUM: ON THE RELATION OF LEIBNIZ'S MONADS TO THE CONTINUUM

INTRODUCTION

In his influential book on Leibniz (1900), Bertrand Russell often let his new-found anti-Hegelianism come between him and his subject, attacking doctrines held by the likes of Bradley and MacTaggart in the belief that he was attacking one of their main sources. This is particularly true of his chapter on Leibniz's treatment of the continuum, where after rather wildly accusing Leibniz of courting "the essentially Hegelian view that abstraction is falsification",[1] he proceeds to the charge that "his whole deduction of Monadism from the difficulties of the continuum, seems to bear a close analogy to a dialectical argument" (110). This is intended as a hostile criticism, as Russell makes clear, explaining that in calling the argument "dialectical" he means that Leibniz infers the reality of monads "from premisses admittedly false, and inconsistent with each other". He charges that Leibniz's argument, "in obtaining many reals, assumes that these are parts of matter — a premiss which it is compelled to deny in order to show that the reals are not material".[2]

Russell presents Leibniz's argument as follows: "The general premiss is: Since matter has parts, there are many reals. Now the parts of matter are extended, and owing to infinite divisibility, the parts of the extended are always extended. But since extension means repetition, what is repeated is ultimately not extended. Hence the parts of matter are ultimately not extended. Therefore it is self-contradictory to suppose that matter has parts. Hence the many reals are not parts of matter. (The argument is stated almost exactly in this form in G.VII.552.)"

I doubt whether many commentators today would be prepared to accept such a "dialectical" reading of Leibniz's reasoning. But putting aside for now the this mooted Hegelian logic, it is clear that Russell has uncovered an important problem for the interpretation of one of Leibniz's most central concerns. For it is hard to see how monads, which Leibniz describes as mind-like units containing force and perception, could be *derived* from difficulties with the notion of continuous matter; or how their existence contributes towards a resolution of these

J. R. Brown and J. Mittelstrass (eds.), An Intimate Relation, 171—201.

difficulties. Yet Leibniz is adamant that his doctrine of substance is a solution to these difficulties, the only exit he can find from the "labyrinth of the continuum".[3] This problem has tended to dog Leibniz's interpreters from Euler through McGuire, and few have been able to avoid finally falling into the same kind of difficulty as Russell. For no matter how subtle their attempts to explain the relation between monads and the continuum, the commentators I shall consider here all end up subscribing to some version of the view that monads are actual parts of extended material bodies.

This, then, is Russell's conundrum: if monads are not parts of continuous bodies, it is difficult to see how Leibniz could have supposed that their introduction would solve the problems of the continuum; whereas if they are, his position seems to lapse into inconsistency.

I cannot, within the confines of this paper, present a full exposition of Leibniz's solution to the problem of continuum, which would involve tracing both its origin in the words of Suárez, Fromond and others, and the respective influences of Hobbes, Descartes, Galileo, Aristotle and Plato; and also showing how his views on substance, mind and soul developed in harmony with his progress in the mathematics of the infinite, and issued in his mature dynamics.[4] What I want to tackle here is altogether more modest: namely, to identify the extent to which expositions of Leibniz's solution have been vitiated by a misidentification of monads as matter's *actual parts*, a misreading which, far from leading out of Leibniz's labyrinth, takes us straight into the deadend of Russell's conundrum.

1. EULER: ATOMS OF SUBSTANCE

The most straightforward misinterpretation of how monads are supposed to ground the continuum is the *naive atomist* one, which reads Leibniz's "atoms of substance" as *indivisible parts of matter*: as the analogue in the continuum of (material) substance to indivisibles in the mathematical continuum. Although it is not a position any self-respecting commentator could hold today, its consideration will set the stage for an appreciation of the difficulties to come.

This was how Leonhardt Euler understood the monadology, although in fairness to him what he had in mind likely has more to do with the philosophy popularized by Christian Wolff than Leibniz. Consequently, in one of the more ironic episodes in the history of philospophy, Euler set about attacking this "solution" to the problems of the continuum in his *Letters to a German Princess* of 1761 — with arguments very reminiscent of Leibniz's own. For Euler supposed monads to be *parts* of bodies that result from a "limited division" (Euler 1843, pp. 48—49), "ultimate particles which enter into the composition of bodies" (p. 39). Naturally he had little difficulty in making hay with this straw man, drawing out the contradiction between this idea of a limited division and the fact (acknowledged by Leibniz) that matter is infinitely divisible, and therefore divisible without limit.

Leibniz, of course, was adamant that although the reality of any phenomenal body is constituted by an infinite aggregate of monads, these monads, having no extension, are not *parts* out of which any extended body can be *composed.*

> But, strictly speaking, matter is not composed of constitutive unities, but results from them, since matter or extended mass is nothing but a phenomenon founded in things, like a rainbow or mock-sun, and all reality belongs only to unities . . . Substantial unities, in fact, are not parts but foundations of phenomena. (To de Volder, 30/6/1704: G.II.268)

Euler's interpretation is also flatly contradicted by Leibniz's repeated assertions that phenomena are not only infinitely divided, *they have no smallest part* (G.II.268, G.II.305, GM.II.157, etc.); and that monads, far from being such smallest parts of the bodies, are presupposed in as small a part of a body as we want to consider (G.II.135, G.II.301, G.II.305, G.II.436, G.II.329).

So much for the naive atomist reading. Although it had the merit of clarity in suggesting a plausible connection between Leibniz's monadology and his infinitesimal calculus, it is evident from the above objections that no such simple connection is to be expected, but something considerably more labyrinthine.

2. RESCHER: IDEAL AND REAL WHOLES

But before proceeding to more elaborate interpretations of the relation of monads to the continuum, it is worth enquiring how in the face of this uncompromising preliminary conclusion, there could remain any justification for interpreting monads as actual parts of extended phenomena. The answer is that the tendency to interpret monads this way has been fostered largely by Leibniz's frustratingly terse statements of his solution. These usually consist in the remark that in ideals the whole is prior to its parts, which are merely potential or *indeterminate*, whereas in actuals the parts are real and *determinate*, and prior to the whole; and that confusion about the continuum could be avoided if we would only heed this distinction between determinate and indeterminate parts, and refrain from looking for actual parts in ideal wholes, and vice versa. Here it is natural to interpret these "actual parts" of the real whole as the monads that give rise to a phenomenal body, as the following passage would suggest:

It is also obvious from what I have said that in Actuals there is only discrete Quantity, namely the multiplicity of monads or simple substances, a quantity which is in fact greater than any given number for any aggregate that is sensible or corresponds to phenomena. But continuous Quantity is something ideal, which pertains to possibles and to actuals considered as possibles. The continuum, that is, involves indeterminate parts, whereas in actuals there is nothing indefinite — indeed, in them any division that can be made, is made. Actuals are composed as is a number out of unities, ideals as a number out of fractions: the parts are actual in the real whole, not in the ideal whole. In fact we are confusing ideals with real substances when we seek actual parts in the order of possibles, and indeterminate parts in the aggregate of actuals, and we entangle ourselves in the labyrinth of the continuum and inexplicable contradictions. (To de Volder, 19/1/1706: G.II.282)

Thus just as any number can be divided into parts in innumerable different ways, so that it has no parts that can be identified as *the* parts out of which is composed — no determinate parts — so too, for the same reason, no continuous quantity such as time or Cartesian matter can have determinate parts either. But it is axiomatic for Leibniz that anything actual must have such determinate parts, just as integers are made up out of a determinate number of units. If therefore follows that continuous quantities, such as time and Cartesian matter, must be *entia rationis* or ideal entities.

But passages such as these only seem to push us inexorably toward a more comprehensive atomism. For if continuity is to be found only in ideal wholes, it seems that everything real must be merely discrete. But in this case how does Leibniz's position constitute a *solution* to the problem of continuum?

Nicolas Rescher explains things as follows. Leibniz, he contends, locates the source of the difficulties of the continuum in the "failure to distinguish between the ideal or phenomenal and the real or actual"

(1967, p. 111). For it is not case that "both the indivisible constituent and the continuum to which it belongs [can] both at once be real":

In mathematics the continuum, the line, is real and the point is merely the ideal limit of an infinite subdivision. In metaphysics only the ultimate constituents, the monads, are actual, and any continuum to which they give rise is but phenomenal. This is the Leibnizian solution of the paradoxes of the continuum. (Rescher 1967, p.111)

Despite its obvious neatness, however, this interpretation does not tally with what Leibniz says. In the first place, with regard to the mathematical side of the solution, a point for Leibniz (after 1676) is always an *endpoint* of a line segment, and is never the ideal limit of an infinite subdivision.[5] Infinitesimals, on the other hand, are *not* indivisible, and are in any case fictions rather than parts, as we shall see later. But more importantly, Leibniz is explicit that the continuum or line, like any mathematical object considered in the abstract, is ideal, and (*contra* Rescher) no such ideal entity can be regarded as *real* since it is not *composed* of its parts:

Properly speaking, the number 1/2 in the abstract is an entirely simple relation, in no way formed by the composition of other fractions, although in numbered things there is equality between two quarters and a half. And one may say as much of the *abstract* line, composition being only in *concretes*, or masses of which these abstract lines mark the relations. And this is also the way in which mathematical points occur, which are again only modalities, that is extremities. And as everything is indefinite in the abstract line, one takes into account there everything that is possible, as in the fractions of a number, without troubling oneself over the divisions actually made, which designate these points in an entirely different manner. (Remarks on the Objections of M. Foucher (1695): G.IV. 491)

And conversely, on the metaphysical side, any phenomenal whole resulting from an aggregate of monads ("secondary matter" or "body") must be well founded or *real*, and cannot therefore have the indeterminate parts characteristics of the continuous: "But in real things — i.e. bodies — the parts are not indefinite" (To de Volder, 6/30/1704: G.II.268); "in actuals there is nothing indefinite — indeed in them every division that can be made is made . . . the parts are actual in the real whole" (To de Volder, 1/19/1706: G.II. 282). Consequently such a real phenomenal whole is *discrete* rather than a continuum: "Matter is not continuous, but discrete . . . The same holds for changes, which are not continuous". (To de Volder, G.II.245)

So the analogy and contrast between mathematical and metaphysical on which Rescher found his interpretation falls through: points are mere modalities in a (continuous) ideal whole, whereas monads are actuals which give rise to a (discrete) phenomenal one. Where Rescher goes wrong here, apparently, is in his initial supposition that there is a dichotomy between ideal or phenomenal, on the other hand, and real and actual on the other. But as we have seen, the phenomenal whole that results from an aggregate of simple substance is *real* in the sense that it is actually divided into determinate parts; monads cannot "compose a continuum" (Rescher 1967, p. 110), a continuum being merely an abstract entity, not a real phenomenon.

3. MCGUIRE: THE IDEAL, THE PHENOMENAL AND THE ACTUAL

J. E. McGuire, on the other hand, is prone to no such conflation of the phenomenal with the ideal in his provocative article on the development of Leibniz's philosophy of substance and continuity (McGuire, 1976). He carefully distinguishes a *trichotomy* of the ideal, the phenomenal and the actual (p. 309). Space, time and motion are *entia rationis* (p. 307); they are arbitrarily divisible, though not composed of parts (p. 309).[6] Space and time are *resolvable* into points or instants, but neither divisible into nor composed out of these, since points and instants are mere modalities (309—310). Well-founded phenomena, on the other hand are extended aggregates, and as such presuppose a plurality of entities from which their extension results. Thus they are *resolvable* into units of substance. They are also *divisible* into actual parts, and *composed* of these parts. The only *actuals* are the substantial unities into which phenomena are resolved. Being simple these actual substances "can be neither composable, nor resolvable, nor divisible" (310).

Since simple substances (monads) are the only actuals, however, there appears to be no way to avoid inferring that they are the actual parts of the phenomenal wholes. McGuire duly draws this conclusion (his scare quotes presumably being a tacit acknowledgement of Leibniz's insistence that monads can *not* be parts of the phenomena): since "extended things have a plurality of coexisting 'actual parts'" (306), and they have this coexistence "in common even with things that are not extended" (To de Volder, 23 June 1699: G.II.183), it follows that "the 'actual parts' of extended things are non-extended substances" (306).[7] Similarly, he claims, phenomenal changes 'involve a succession of extraneous 'parts' . . . these 'parts' are in reality the diverse states of substances that exist at one time but not at the next" (306—307).

Now of course, we are right back in Russell's conundrum, and the challenge is to see how the above claims can constitute any sort of solution to the problems of continuum. McGuire attempts this by allowing the phenomenal wholes an *apparent* continuity. He reasons that since extended things are resolvable into simple substances, which are discrete, this means they are "in reality a plurality of substances that *appear* to be continuous and extended" (306): their continuity is simply an appearance. Likewise phenomenal change: McGuire interprets Leibniz's "the same holds for changes, which are not truly continuous" as meaning that "the perceptual continuum is not truly continuous, but comes into existence in a series of discrete chunks" (311).

Thus McGuire"s trichotomy is a rigid partition. "The actual cannot be continuous as it is simple and indivisible . . . Being composable, resolvable and divisible the perceptual continuum cannot be truly continuous. Yet it is not actual, since these notions [i.e. composability, resolvability and divisibility] do not apply to what is truly simple . . ." (311). One might say that where Rescher saw Leibniz's solution as consisting in a kind of *apartheid*, where indivisible and continuum were not both allowed reality in the same region (Mathematics or Metaphysics), McGuire extends this model to allow for the separate status of well-founded phenomena. But with true continuity restricted to its homeland in mathematics, and actuality to the ruling elite of monads and their states, phenomena come out (uncomfortably like the Cape Coloureds) as neither one nor the other.[8]

Yet if this is indeed Leibniz's solution, then, as McGuire admits, "it bristles with difficulties" (311), some of which he proceeds to outline. In the first place, contrary to Leibniz's claim that his *law of continuity* applies everywhere, on this model it seems properly applicable only to space, time and motion (311). Indeed, Leibniz seems to contradict himself on this score in three distinct contexts. For he claims (i) that "continuity is found in . . . extension . . . and in fact in all natural changes" (Leibniz, 1715); but this seems incompatible with what he said to de Volder about matter and change being "not truly continuous" (G.II.278). He also asserts (ii) that "the change is continuous in each created monad" (*Monadology*, §10: G.VI.608); and (iii) that there is a continuity of kinds of monad, each differing infinitesimally from the next (To de Volder, 3 April 1699: G.II.168; *New Essays* 50: G.V.49) — assertions that seem impossible to reconcile with his stipulation that the actual cannot be continuous. For on McGuire's interpretation, this stipulation would rule out any "continuity of forms"; the difference between different kinds of monads would always be discrete. It would also entail that a monad could not be regarded as a continuous temporal whole, so that there could be no question of continuous monadic activity. Further, if it were such a continuous whole, then it would be divisible temporally into *indeterminate parts*, which is quite impossible for anything actual.[9]

In the face of all these difficulties Leibniz's introduction of monads seems to constitute not so much a solution to the difficulties of the continuum, as a source of new ones. This should at least make us wary of the adequacy of McGuire exposition. Indeed, on further reflection there are problems with its adequacy to deal even with the problems it does appear to resolve. For if, as McGuire says, it is only "the *perception* of continuity" (312, my italics) that is created by "the repetition of discrete but simultaneous states" (312) or the aggregation of an infinity of discrete substances, then presumably this would be all there is to Leibniz's solution: actuals, monads and their states, are discrete, but they give rise to the appearance of continuity. Thus when Leibniz said that extension and change are continuous, he meant only an *approximate* continuity, and not a true one. But how are we to understand the approximation here? One can imagine a many-sided regular polygon appearing as a continuous circle if there were sufficiently many discrete and very small sides. But monads are not discrete in this sense: this is precisely the sense of monad as atom of matter that we saw to be wrong in considering the interpretations of Euler and Rescher. Thus the appearance of continuity must be an effect created in the percipient by the action of the monad: but then why shoudn't the created appearance be truly continuous? Why does Leibniz need to claim that phenomenal wholes are made up of discrete parts at all — why not claim that they are continuous appearances produced by discrete substances?

Secondly there is the whole question of the applicability of the mathematics of the continuous to the real world. If the actual is really discrete, then surely continuous geometry would not be applicable to it without error? In answer to this difficulty, McGuire tells us how "Leibniz believed that there was no inconsistency between coherent notions and the actual determination of phenomena", that ideal notions such as continuity are "applicable to phenomena in so far as they are consistent, and lead to correct empirical results" (312). But how can the application of the continuous to the discrete be consistent? Again, this seems little more than a verbal solution. It does not seem to do justice to Leibniz's own explanations or how continuity applies to the actual, which invokes the same wording he uses in justifying the infinitesimals of his differential calculus. For instance, in the continuation of the above quoted passage from his letter to de Volder, Leibniz says:

> But the science of continua, i.e. of possibles, contains eternal truths, which are never violated by actual phenomena, since the difference is always less than any assignable given difference. (To de Volder, 19/1/1706: GT.II.282)

Intermezzo

These difficulties seem rather daunting. But there is a clue in the wording of the above passage that can help us begin to unravel them. This is Leibniz's reference there to *actual phenomena*, an expression that appears to violate McGuire's strict trichotomy of Leibniz's ontology into the actual, the phenomenal and the ideal. The inconsistency, however, is only apparent, for the sense of 'actual' here is given by its contrast with the merely possible; and indeed this is a sense the word often carries in Leibniz writings. There is, for instance, the actual world that God decides in his infinite wisdom to create, as opposed to all the possible worlds that he could have created, each one of which contains "actuals" in the other sense — simple substances and their states — as well as the various phenomena founded in them. Thus one can have possible as well as actual substances, and actual as well as possible phenomena.

But the recognition of this other sense of 'actual' immediately opens up the possibility that this is the sense of 'actual' in the expression "actual parts": not parts that are substances, but actually existing parts as opposed to parts into which phenomena might have been divided. Perhaps this is the "Ariadne's Thread" we need to escape Russell's conundrum and lead us out of Leibniz's labyrinth.

That this is the correct interpretation of Leibniz's "actual parts" can be confirmed by further inspection of the passages in which the phrase occurs. The two passages quoted above are cases in point. For in each of them, the actual parts are mentioned in the context of an *actual division*. In the letter to de Volder of 19/1/1706, Leibniz writes "in actuals there is nothing indefinite — indeed, in them any division that can be made, is made" (G.II.282). And in his Remarks of 1695, the points marking the possible divisions of an abstract line are contrasted with the "the divisions actually made, which designate these points in an entirely different manner" (G.IV.491) Now since only phenomenal bodies can be divided (simple substances are indivisible), this means that Leibniz's "actual parts" must be parts of actually divided phenomenal bodies. This is confirmed more explicitly in the following passages:

But in real things, that is, bodies, the parts are not indefinite (as they are in space, a mental thing), but are actually assigned in a certain way, as nature actually institutes the divisions and subdivisions according to the varieties of motion, and . . . these divisions proceed to infinity . . ." (to de Volder, June 30th, 1704: G.II.268).

No body is so small that it is not actually divided into parts which are excited by different motions; and therefore in every body there is actually an infinite number of bodies. (*A Specimen of Discoveries* (c. 1686), Parkinson 1973, pp. 85—6)

This discovery by itself does not take us very far. For if actual parts are phenomenal bodies, it is still not clear what role monads have in the resolution of the problem of the continuum. In order to get any further, we will have to go beyond a mere textual analysis of Leibniz's mature pronouncements and delve into the historical origins of his views. If we are to discover the basis for the above assertions, we have to ask some more pointed questions. In particular, (i) what is the origin of Leibniz's claim that matter is "actually divided"? and (ii) what is the basis for the claim that this division proceeds to infinity?

4. ACTUAL DIVISION

A clue to the answer of the first of these questions may again be found in Leibniz's wording, this time in the two passages just quoted: "nature actually institutes the divisions and subdivisions according to the varieties of motions"; every body is "actually divided into parts which are excited by different motions". The implication is clear: it is *motion* that is responsible for the actual divisions of matter. That is, different actual parts of matter are distinguished by their differing motions; adjoining parts not in mutual relative motion are not actually divided from each other, and are thus (potential) parts of the same actual part.

Although this view may seem strange to us today, in the latter part of the 17th century it represented the orthodox Cartesian position.

Leibniz knew this well. Indeed, in 1669 when he was still optimistic that Aristotle's physics could be completely reconciled with the new mechanistic philosophy of Descartes and Hobbes, he saw this construal of the actual division of matter as a key to their reconciliation. This was to be achieved by reinterpreting the principles of Aristotelian physics solely in terms of magnitude, figure and motion, the one qualification to this otherwise pure mechanism being that mind is understood to be the principle of all motion.[10] Thus Leibniz identifies Aristotle's *primary matter* with Descartes' *subtle matter*, since each is continuous and homogeneous, and has no divisions in it prior to motion.[11] That is, matter in itself is undifferentiated, and has only an indefinite quantity: "For so long as it is continuous, it is not cut into parts, and therefore does not actually have boundaries in it" (A.VI.ii.435; Loemker 1969, p. 95). Form, on the other hand, being "nothing but figure", is the boundary of a body, and in order for there to be a variety of such boundaries in matter "a discontinuity of parts is necessary" (ibid). The operative notion of continuity here is explicitly Aristotelian:

> For by the very fact that the parts are discontinuous, each of them will have separate boundaries (for Aristotle defines continuous things as *ὧν τὰ ἔσχατα ἕν*, those whose boundaries are one). (A.VI.ii.435; Loemker 1969, pp. 95—96)

Now it is crucial to realize that on this account of continuity, as Leibniz points out, there can be *two different kinds of discontinuity*: the first kind arises when parts are actually pulled apart, leaving a vacuum; and the second when the parts are displaced relative to each other, but remain contiguous: "For example, two spheres, one included in the other, can be moved in different directions and yet remain contiguous, though they cease to be continuous" (ibid). Leibniz argues that, since vacua cannot be created by the "supernatural" annihilation of parts of the plenum, matter could only be discontinuous in the first sense if it were actually created this way. "But if it is continuous in the beginning" — the view he obviously favours — "forms must necessarily arise through motion" (*ibid*).

Thus here we can see the origins of Leibniz's determinate/indeterminate parts distinction, and his view that mass or secondary matter is discontinuous. For whereas *primary matter* — that is, matter as it is in itself, devoid of form — is not divided into determinate parts, and is therefore *continuous*, this is not true of *secondary matter*[12] or *mass*, which is actually divided into parts by its various motions. These parts are fully determinate, each having its own boundary. But consequently they are merely *contiguous* with each other. Thus "mass is not truly continuous", but is rather *discontinuous* in the second sense above. Only a homogeneous, undifferentiated whole such as primary matter can be truly continuous.

5. ACTUALLY INFINITE DIVISION

This brings us to the second question posed above: what is the basis for Leibniz's claim that the actual division of matter proceeds to infinity?

Most commentators seem to have assumed that by this claim he simply means that matter, like any continuum, is infinitely *divisible*, and that it is from this infinite divisibility that Leibniz derives the infinite plurality of its parts. Thus Russell: "owing to infinite *divisibility*, the parts of the extended are always extended", "if what appears as matter is a plurality, it must be an infinite plurality. For whatever is extended, *can be* divided ad infinitum" (1900, pp. 110, 108); Rescher: "an actual whole, though *divisible* into actual parts, cannot be resolved into actual ultimate constituents" (1967, p. 108); and McGuire: "plurality is actually infinite, as there is no possible or actual limit to the *divisibility* of the extended" (1976, p. 307). But these claims run roughshod over the distinction between the determinate parts into which matter is actually divided and the indeterminate (or merely possible) parts into which it could be divided, so that this cannot be the way Leibniz derives the plurality of matter's actual parts. For the point of his contrast between contiguity and true continuity is that whereas in a continuum all divisions are possible, in reals there is some one particular division. Of course, anything that is actually divided is *a fortiori* divisible. But the converse does not hold: infinite divisibility does not entail infinite division. For an abstract line is divisible into parts in an infinity of different ways, but anything real is divided in only one way — say, into thirds or quarters, or into some other partition. Infinite divisibility concerns the infinity of possible such partitions, but it does not entail the infinity of any one partition.

This distinction, of course, is precisely the one employed by Aristotle in his treatment of the continuum, by means of which he is able to deny the actual infinite, and so avoid the paradoxes of the infinite first raised by Zeno of Elea.[13] An abstract line, on his account, is *potentially* infinitely divisible, but is actually divided only into a finite number of determinate parts. Thus there is never anything but finite number in actuality: the only infinite is the potential infinite, which pertains to the possibility of division of the continuum, and the infinity of potential parts into which it can be divided; there is no actual infinite. But in conscious to this, Leibniz asserts even in his earliest writings that matter is divided into an *actual infinity of parts*.

Clearly, then, Leibniz's espousal of the actual infinite does not have its origin in Aristotle, despite the obviously Aristotelian pedigree of his distinction between determinate and indeterminate parts. Rather, as we might have suspected given Leibniz's attempt to reconcile the old and new physics discussed above, it appears to have its source in Descartes. On the Cartesian scheme, as we saw, actual division is effected by motion, and it is motion of a special kind that results in actually infinite division. The argument for it is given by Descartes in his *Principles* (1644) as a corollary to his argument for the possibility of the circulation of matter through unequal spaces in a plenum. Such circulation can occur without the creation of a vacuum, he argues, so long as the matter travels proportionately faster through the narrower space, and vice versa. This cannot happen, however, if all the corpuscles of matter are of a fixed shape and size. Some corpuscles may be so; but if circulation is to occur, then at least some part of the fluid matter in which they are floating must be able to adjust its shape by infinitely gradual degrees; "and for this to happen," Descartes reasons, "all imaginable parts of this piece of matter — in fact, innumerable parts — must be to some degree displaced from their positions relative to each other; and this displacement is actual division" (*Principles*, II, 34: 1644, p. 60). Consequently,

There is an infinite or indefinite division of matter into small parts; and the number of these is so great that, however small a part of matter we may imagine, we must conceive it as undergoing actual division into still smaller parts. (*ibid*)

Leibniz seldom misses a chance to praise this argument. In his published critique of the *Principles* he calls it "most beautiful and worthy of his genius" (Loemker 1969, p. 393). But there he regrets that Descartes "does not seem to have sufficiently pondered the importance of this last conclusion", viz. the actually infinite division itself. This is a sound criticism. For if a part of matter is actually infinitely divided, it would (at first sight) seem that it must have actually infinitely small parts. But this would require Descartes to give an account of the composition of his continuous matter out of the *indivisibles* whose existence he categorically rejected — or at least to explain how this conclusion could be avoided.

Such, at any rate, was Leibniz's estimation of Descartes's argument in *Pacidius Philalethi*, his unpublished dialogue on continuity and motion written in October of 1676.[14] This dialogue is particularly significant for the understanding of Leibniz's views on continuity, not only for the exposition he gives there of a good part of his "labyrinth of the continuum"; but also because it contains the first elaboration of his mature interpretation of actually infinite division, which remains a constant factor in his solution to the labyrinth from then on. But as he indicates there, he did not settle on this interpretation before going through several profound changes of mind. Without attempting to trace in detail the actual development of Leibniz's views on this point, we may learn a good deal by comparing three distinct positions he took on actual parts and the continuum in the 1670's, culminating in the position reached in the *Pacidius*.

The first of these is the account Leibniz gave in his *Theory of Abstract Motion* of 1671, where he came closest to the atomistic view ascribed to him by Euler. In this work he held that the continuum is actually infinitely divided and contains *indivisibles*, or *extensionless points* of non-finite but non-zero magnitude.

There are actual parts in the continuum, . . . and these are actually infinite, for the indefinite of Descartes is not in the thing but in the thinker . . . *There are indivisibles or unextended entities*, otherwise neither the beginning nor the end of motion or body could be conceived. (A.VI.ii.264; Loemker 1969, p. 139)

Here the allusion to Descartes hints at the Cartesian origin of Leibniz's doctrine of the actual infinite. But the argument for indivisibles is based on a combination of Hobbesian physics, in which every point of matter is endowed with a *conatus* or endeavour, together with Leibniz's understanding of Cavalieri's Geometry of Indivisibles, so that these points and conatuses are interpreted as indivisibles with diverse magnitudes but no extension.[15] An immediate advantage of this scheme, as Leibniz explained in an earlier letter to Hobbes, is that the interpretation of endeavours as beginnings of motion enables one to give an account of the cohesion of bodies, that is, of how the actual parts of matter unite to form a continuous body:

I should think that the conatus of the parts towards each other would itself suffice to explain the cohesion of bodies. For bodies which press upon each other are in a conatus to penetrate each other. The conatus is the beginning; the penetration is the union. But when bodies begin to unite, their boundaries or surfaces are one. Bodies

> whose surfaces are one, or τὰ ἔσχατα ἕν, are according to Aristotle's definition not only contiguous but continuous, and truly one body, movable in one motion. (To Hobbes, July 1670: Loemker 1969, p. 107).

Thus the continuity of a body is attained only by virtue of a sustained "effort" or conatus to penetrate. What Leibniz does not reveal to Hobbes, however, is his mentalistic reading of conatus; it is for him something that is only interpretable on the model of the human mind. Thus a body can only remain united, can only exist, if it contains at every point within itself something quasi-mental at every moment. But bodies are not themselves true minds, they are only "momentary minds". For although there may be many contrary conatūs in a given body at the same time, these cannot last longer than a moment, since they will be immediately composed into a new conatus. In contrast, a mind is able to retain such contrary conatūs through its memory, and it is on the basis of such contrasts that perception is founded.

But this conatus physics fails, and is regarded by him with mild embarrassment after he "became a mathematician" under Huygens' tutelage in Paris. The second of Leibniz's positions on actual parts and the continuum I want to consider occurs in the notes he made there in February 1676. Here we find him regretting that "it does not seem possible to explain the origins of solids from the motions of fluids alone", and instead appealing directly to minds in order to explain the cohesion of solid bodies (A.VI.ii.473; Loemker 1969, p. 157). Nevertheless, he still maintains that "matter is divided into perfect points or into all the parts into which it can be divided", a result that "seems to follow from a solid in a liquid" (as in Descartes' argument). For "every perfect liquid is composed of points, because it can be dissolved into points, as I prove by the motion of a solid within it" (ibid, p. 158).[16] Only now, these "infinitely small points or bodies less than any assignable ones" cannot be considered as united by their mutual endeavours to penetrate, but will rather be separated by "inassignable metaphysical voids". Consequently, it would follow that

> a perfect fluid is not a continuum, but discrete, or a multitude of points. From this it does not further follow that the continuum is composed of points, since liquid matter will not be a true continuum, even though space will be; whence it is again clear how great the difference between space and matter. Matter alone can be explained by a plurality (*multitudine*) without continuity . . . . So matter is a discrete entity, and not a continuum; it is merely contiguous, and is united by motion or by some mind. (A.VI.ii.474; Loemker 1969, p. 158).

The wording in this passage is very suggestive of some of Leibniz's later dicta on continuity, and consequently helpful in resolving some of the difficulties we noted above. Of particular import is Leibniz's use of the word "discrete" in this passage, which pertains not to monads (as McGuire and others believed of the later formulations), but to the parts into which the fluid is divided. Matter is said to be discrete because its parts — here, its infinitesimal parts — are actually separated from each other by their differing individual motions. But since the voids that separate them are "inassignable metaphysical voids", the parts of matter may be said to be physically next to each other; that is, matter is *contiguous*, rather than truly continuous. Thus we can already discern in this early position the basis for Leibniz's mature statements that matter is *discrete* — in the sense that it is divided into actual, discriminable parts, each contiguous to the next — and yet that no assignable part of space is devoid of matter.

The passage is also illuminating about Leibniz's argument for the plurality of the parts of matter. For in this passage, at any rate, Leibniz does not derive plurality from the continuity of the extended, as Russell and McGuire would have it. On the contrary, he is explicit that (secondary) matter "can be explained by a plurality without continuity", since the plurality of its parts has already been established by the actual division argument.

But these conclusions are phrased by Leibniz quite tentatively, and depend on an interpretation of "inassignables" as actually infinitely small parts; that is, on an interpretation of the infinitesimals of his newly discovered differential calculus as *infinitely small actuals*. This in turn depends on the hypothesis of infinite number, or more precisely, infinite collections. As Leibniz writes: "there must be infinite number if a liquid is really divided into parts infinite in number. But if this is impossible, it will follow too that a [perfect] liquid is impossible. Since we see that the Hypothesis of infinities and infinitesimals is clearly consistent and succeeds in Geometry, this also increases the probability that it is true" (A.VI.ii.475; Loemker 1969, p. 159).

This tie between infinitesimals, infinite number and the idea of a perfect fluid brings me to the third of Leibniz's position on actual parts and the continuum, that advanced by him in his *Pacidius Philalethi* a bare eight months later. For by then he had found a new interpretation of actually infinite division which did not result in infinitesimal points, and consequently did not commit him to the existence of actually infinite number. His inspiration for this appears to have derived from his work on *infinite series*, his model being a converging infinite series such as 1/2 + 1/4 + 1/8 + 1/16 + ... For such a series is a finite whole made up of an infinite number of finite parts; it has no limiting or infinitieth term.[17] But even though the series is open-ended and does not actually attain its limiting value, every one of its terms can easily be determined by the law of the series, as can its sum, which is (in this case) 1. Thus the series is not a completed collection, but rather a *distributive* whole, whose unity is determined by the series' law. To say that it has a sum of 1 is to say that it can be made as close as desired to 1 by taking a sufficient number of finite terms; or, equivalently, that the difference between its sum and 1 is inassignable.

This allows for a different interpretation of how discontinuous matter can fill a continuous space. For instead of conceiving matter as an infinite collection of inassignable parts, it can be conceived distributively as an actual infinity of finite parts, each smaller than the last. Despite its discreteness, it can be regarded as filling continuous space, since the gap between matter and space is inassignable — it can be made arbitrarily small by taking a sufficient number of ever smaller parts.

Thus in the *Pacidius* Leibniz proposes what will turn out to be his final interpretation of Descartes' argument for actually infinite division: instead of reaching a limit in indivisible points, it is reinterpreted by him as an open-ended division, each part of matter being further divided into further parts without limit. Consequently he now rejects the idea of a "perfect fluid", matter that is "resolved into a powder (so to speak) consisting of points:"

> On the contrary, every liquid has some tenacity, so that even when it is torn into parts, all the parts of the parts are not so torn, but merely formed for a while, then transformed; and so there is no dissolution all the way down into points, even though every point is distinguished from every other by its motion. (Leibniz 1676, p. 615).

On this new conception of matter, there is no absolute distinction between solid and liquid. Rather every body is "everywhere pliant", and has "a certain unequal resistance to bending". Thus even though matter is divided to infinity at any given moment, its parts or cells resist such division with a kind of inherent *elasticity* — they are "merely formed for a while and then transformed" — resulting in a different infinite partition at each assignable instant:

If a perfectly fluid body is assumed, a finest division or division into minima cannot be denied; but a body that is indeed everywhere pliant, though not without a certain unequal resistance to bending, still has cohering parts, although these are variously opened up and folded together. Accordingly the division of the continuum must not be considered to be like the division of sand into grains, but like that of a sheet of paper or tunic into folds, so that even if the folds are infinite in number, there are [always] some smaller than others, and for this reason a body is never dissolved into points or minima . . . Although some folds are smaller than others down to infinity, bodies are always extended and points never become parts, but always remain mere extremities. (*ibid*)

This new conception of actually infinite division has two crucial consequences for Leibniz. On the one hand, it means that body or secondary matter, being actually divided into discrete contiguous parts, is not truly continuous according to the Aristotelian definition. Indeed, since every part of every piece of matter is further divided, it follows that there is no principle of unity in matter itself. At each instant, any given body is the sum of all its parts. But this sum is an infinite sum which, like an infinite series, is never completed. It is not a true whole, or even a collection of them, but a distributive whole. Nevertheless, it is perceivable as a whole insofar as the "law of its series" is grasped by a perceiving mind. Therefore body, in so far as it is material, does not contain a principle of spatial unity in itself, and is consequently merely phenomenal.

Secondly, and even more importantly for Leibniz, body in so far as it is material, does not contain any principle of temporal continuity either. Since "there is no part of time in which some change or motion does not happen to every single part of a body", it follows that "no motion stays the same through any space or time however small" (Leibniz 1676, p. 622). Therefore between any two points in time there will be an infinity of different partitions of matter, but no continuity between them. There is no continuity of motion or change, just different partitions at different instants. For this reason there is no state of change in material bodies either, "but only an aggregate of two states old and new; and so there is no state of action in a body, no moment can be assigned at which it acts" (1676, p. 623). He continues:

Thus action in a body cannot be conceived except through a certain aversion. If you really cut to the quick and inspect every single moment, there is no action. Hence it follows that proper and momentaneous actions belong to those things which by acting do not change. (*ibid*)

One of the most intriguing features of the *Pacidius* is that there is no explicit mention in it of Leibniz's belief that it is necessary to introduce minds as principles of unity of bodies, despite the fact that he had long been convinced of this necessity. I am inclined to believe that this suppression does not represent a change in his position, but is explicable in terms of his purpose in composing the dialogue. If, as seems likely, he composed it with a view to drawing Spinoza into discussion, it would have been good policy to try to gain the latter's agreement on the lack of continuity and unity of phenomena before becoming embroiled in a controversy over the nature of substance.

There is in fact a hint that Leibniz had not given up on his idea of true substances underlying and underwriting the "actionless" phenomenal bodies. For although the explicit conclusion that he draws in the *Pacidius* is the Occasionalist one that "what moves and transfers the body is not the body itself, but a superior cause which by acting does not change, which we call God" (1676, pp. 623—4), the more rigorous conclusion of the argument is previously stated in terms of substances in the plural: "it follows that proper and momentaneous Actions belong to *those things* which by acting do not change" (p. 623), an indication that if each body had some principle of action in it, this would also be sufficient to account for the unity and temporal continuity of substance.

This hint is important in understanding how the account we have been discussing evolves into his mature position.

6. UNITS OF SUBSTANCE

I have indulged in this rather long digression on Leibniz's evolving views on the physics of fluids and actually infinite division not only for the clarification it brings to what Leibniz means by "actual parts", but also because of the perspective it gives us on the relation of Leibniz's monads to the continuum. For as we saw in the first three sections of this paper, if we ignore this physical context it is very hard to see the relevance of these mind-like units of substance to the difficulties of the continuum, and Leibniz's motivation for introducing them is fraught with mystery. But in the preceding considerations concerning the phenomenality of matter and change taken by themselves, we can already discern some of Leibniz's motivations for his mature doctrine of substance; and in his positing of minds to account for the cohesion and self-continuance of bodies, we can see the bare bones of his mature solution. To avoid prolonging my historical narrative, some of the salient features of the emergence of his doctrine of substances may be briefly summarized as follows:

At any instant, matter is actually divided into parts any one of which is further divided. Although all these parts are constantly folding and unfolding over time due to their constant motion, some retain their integrity for longer than others by virtue of their greater resistance to division, which results from an intrinsic elastic force. These are called bodies. There is no absolute distinction between solid and liquid bodies: the more solid a body, the better its parts cohere. Yet neither the spatial unity of bodies not the continuity of their changes is explicable in terms of the material principles of magnitude, figure and motion, nor is their elastic force. Consequently, insofar as any body remains a true unity, such as the body of a living being does, there must be in it some non-material principle of unity which accounts for its self-identity through time, its ability to act whilst remaining the same, and this principle must also be what is responsible for its intrinsic elastic *force*. Otherwise, as Leibniz later explains to Arnauld,

> It will be of the essence of a body to be phenomenon deprived of all reality, like an ordered dream; for phenomena themselves, like a rainbow or a heap of stones, would be wholly imaginary, if they were not composed of entities with a genuine unity. (To Arnauld, 30 April 1687: G.II.97)

These non-material principles are, of course, the descendants of the material points endowed with conatus that Leibniz had earlier placed in bodies to account for their cohesion; which, after the failure of his conatus physics, were replaced by minds proper as principles of motion and unification of solid bodies.[18] Although these minds were suppressed in the *Pacidius Philalethi* in an apparent reversion to the Occasionalism Leibniz had espoused in his letters to Thomasius, his reference in the dialogue to "things which by acting do not change" suggests that he had not abandoned his commitment to mind-like substances everywhere in matter. By 1686 his commitment to these non-material principles is firm. These he now calls *souls* (and will later call *monads*); and body combined with a soul is what he calls *corporeal substance*: "either there are no corporeal substances and bodies are mere phenomena which are true and consistent with each other, such as a rainbow or a perfectly consistent dream, or in all corporeal substances there is something analogous to the soul, which old authors called a form or species" (*A Specimen of Discoveries* (1686), Parkinson 1973, p. 81).

What is more, Leibniz is quite explicit that he was led to postulate these souls as unifying principles because of the lack of unity of a body that is actually infinitely subdivided. In the continuation of the passage just quoted from *A Specimen of Discoveries* he writes:

> For that is not one substance or one being which consists merely of an aggregation, such as a heap of stones, nor can beings be understood where there is no one true being. Therefore, . . . since it is to be held as proved that every body is actually subdivided into other parts, . . . the reality of a corporeal substance consists in a certain individual nature; that is, not in mass, but in a power of acting and being acted on. (Parkinson 1973, p. 81)

A fascinating consequence of this line of reasoning is that there must be monads *everywhere*. For since every part of matter is further divided, and each of these parts is either united by some monad into a corporeal substance or is an aggregate of such corporeal substances, it follows that every part of matter, however small, contains an infinity of corporeal substances and therefore an infinity of monads. Thus for any monad contained in one part of matter, one can find another monad situated in another part of matter arbitrarily close to it. In other words, the difference in the points of view of the two monads will be inassignably small.

This suggests a solution to a second of the three problems raised by McGuire concerning the application of Leibniz's Law of Continuity (pp. 177—8 above). In answer to the first, we have already seen how phenomena such as extended matter and motion, although they are truly continuous only when considered abstractly, can approximate continuity arbitrarily closely owing to their actually infinite division, just as an irrational like π can be approximated arbitrarily closely by a series of rationals. Similarly here we see that since any difference in point of view will result in a different representation of the universe — a different monadic state or "form" — the arbitrarily small differences of monadic points of view result in an approximate "continuity of forms". On the other hand, McGuire's third problem concerning the continuity of change in monads is an altogether more complicated matter, and it would take me too far afield to sketch an explanation here.

There are, of course, other developments in Leibniz's philosophy between 1676 and 1686 which reinforce his belief in the phenomenality of mere bodies and their motions, and the need for an underlay of non-material substances. Not least of these is his growing realization as to the unreality of space:

> For when I formerly conceived space as a real immobile place, endowed only with extension, I could define absolute motion as a change of this real space. But gradually I began to doubt whether there existed in nature such an entity as we call space . . . Whence it seemed to follow that what is real and absolute in motion does not consist in what is purely mathematical, such as change of neighborhood or situation, but in motive power itself; and if there were no such thing, this would mean that absolute and real motion would be abolished. (Leibniz 1688, p. 580)

Added to the considerations in the *Pacidius* about the absence of an action or state of change in matter, these considerations reinforce the need for something in matter to function as the source of its motion. But the resources of a purely mechanistic physics are spent: situation is merely relative, and the space or extension that matter is supposed to fill is, as something truly continuous, only an ideal entity. Apart from motion, the only other modification that extension might take on is that of shape or figure. But in Leibniz's eyes the phenomenality or imaginary character of shape is also established by the argument from the actual division of matter:

> For from the fact that no body is so small that it is not actually divided into parts which are excited by various motions, it follows that no determinate shape can be assigned to any body, nor is an exact straight line, nor a circle, nor any assignable figure of any body found in the nature of things, though in the derivation of an infinite series certain rules are observed by nature. And so shape involves something imaginary, and no other sword can cut the knots we weave for ourselves by our imperfect understanding of the composition of the continuum. (Parkinson 1973, p. 81)

Thus none of the material properties of body is sufficient to explain its apparent substantiality. Extension, situation, shape and motion are all phenomenal effects of something substantial in matter, they are all secondary qualities. Because of the actually infinite division of matter consequent on its internal motions, and the changes of these from one moment to the next, strictly speaking no body has a fixed shape or a continuous motion over time. And yet we know there are real motions and shapes in bodies, so that the soul or monad which is responsible for the unity of each corporeal substance must also be responsible for the series of changing modifications which it undergoes. Thus again it must have the character of a primitive force, it must be a principle of activity:

> And so the essence of body is not to be located in extension and its modifications, namely shape and motion, for these involve something imaginary, no less than heat and colour and other sensible qualities. It is to be located in the power of acting and resisting alone . . . (Parkinson 1973, p. 82)

To sum up: mere bodies are doubly phenomenal, both in respect of their spatial unity, and in respect of the unity through time of the series of changes they undergo. For each body is spatially discontinuous, a mere aggregate of discrete contiguous parts that can constitute only a perceived whole, unless it is united by an organizing principle, the soul of a living being; and the series of shapes and motions it manifests has a temporal unity only insofar as they are coherent effects of a continuously acting primitive force.

Correspondingly, Leibniz has two ways of arguing for monads from the difficulties of the continuum. Because of its actual division into parts, any body is a plurality, so that if it is not purely phenomenal, it must consist of a plurality of substantial unities. These cannot be material atoms, since every part of matter is further subdivided. Thus there must be non-material atoms, or atoms of substance in every part of matter, however small, which account for the spatial integrity of bodies. But the argument from actual division also establishes that the shapes and motions of any part of matter do not remain constant from one instant to the next. Therefore, assuming that matter is not purely phenomenal, whatever is substantial in it is something other than extension, figure and motion, and yet accounts for the unity and continuity of its changes through time, and the reality of its motion. Thus there must be some principle of activity in every body, however small, which accounts for this temporal integrity of bodies and their motions.

Now if only human beings contain such organizing principles, then all other bodies are mere phenomena, and we are left with a kind of Malebranchian phenomenalism. But the above arguments clearly incline Leibniz to grant that other bodies contain souls, that there are corporeal substances in all bodies or parts of matter however small.[19] This leads Leibniz to construct his "hypothesis of concomitance", so that the shapes and motions of bodies are not just constituted as phenomena by the perceiving monad, but result from the agreement among all the representations of its constituent monads. As Leibniz says in one of his last statements of his position: "aggregates themselves are nothing but phenomena, since everything apart from the ingredient monads is added by perception alone, by the very fact of their being perceived together" (To Des Bosses, May 28, 1716: G.II.517).

In one sense, then, corporeal substances are Leibniz's units of substance: they are the unities of body and soul from which everything real is composed. But on a deeper metaphysical level, bodies apart from souls or monads are but the effects of aggregates of these monads: all corporeal properties such as extendedness, elasticity and anitypy, and all physical forces, are but well-founded phenomenal effects. In this more rigorous sense, monads are the only substances, since they are "absolutely without parts" and are the only entities which can really act. They are "the absolute first principles of the composition of things, and as it were the ultimate elements into which substantial things can be analysed" (*New System*, Parkinson 1973, p. 121).

7. CONCLUSION

Having gained some understanding of the origins of Leibniz's theory of substance in the difficulties of the continuum, we are now in a position to solve Russell's conundrum. Russell, we recall, claimed that Leibniz derived the existence of a plurality of monads from the infinite divisibility of matter by identifying them as the actual parts into which matter can be divided. He therefore accused Leibniz of inconsistency in his derivation of the Monadology from the difficulties of the continuum, charging that his argument, "in obtaining many reals, assumes that these are parts of matter — a premiss which it is compelled to deny in order to show that the reals are not material" (1900, p. 110, fn. 1). Similarly McGuire argued that since "extended things have a plurality of coexisting 'actual parts', . . . this means that extended things are *phenomena bene fundata*, and that they are simply a plurality of coexisting and active substances," so that "the 'actual parts' of extended things are non-extended substances" (McGuire 1976, p. 306).

Despite his occasional mentioning of actual parts in the same breath as monads,[20] however, Leibniz denies that monads are parts of matter, claiming that they are "foundations of phenomena", substances whose actions give rise to phenomena. But if monads are not parts, then how can they be argued for by arguments couched in terms of parts? This is the conundrum.

Russell, as we have seen, saw this as no mere verbal difficulty, but an outright inconsistency in Leibniz's thought. In a nutshell, his account of Leibniz's "Hegelian" deduction runs as follows: Matter is divided into an infinity of extended parts, and therefore has an infinity of real constituents. But "since extension means repetition, what is repeated is ultimately not extended. Hence the parts of matter are ultimately not extended . . . Hence the many reals are not parts of matter". (Russell 1900, p. 100). He claims that Leibniz's argument for his monads "is stated in almost exactly this form" in the following passage:

> Everybody agrees that matter has parts, and is consequently a multiplicity of many substances, as would be a flock of sheep. But since every multiplicity presupposes true unities, it is evident that these unities cannot be matter, otherwise they would in turn be multiplicities, and by no means true and pure unities such as are finally required to make a multiplicity. Thus the unities are properly substances apart, which are not divisible, nor consequently perishable. (Reply to Foucher: G.VII.552)

But in this passage Leibniz argues only that "every multiplicity *presupposes* true unities", i.e. is *resolvable* into them; and that since every part of matter is divided into further parts, it follows that every part of matter is a multiplicity which presupposes unities. He does not assert that these unities *are* the parts into which matter is divided.

As we have seen above, Leibniz does not argue that this infinite plurality of matter's parts follows as a simple consequence of the infinite divisibility of extension, as Russell alleges. For this would only establish matter's divisibility into an infinity of potential parts, and would not preclude material atoms from being the substantial unities. Rather, as we saw in the preceding historical excursus, Leibniz argues for the infinite plurality of the actual parts of matter on the basis of its actual division, an actually infinite division that results from the differing motions of its various parts. But this same argument also establishes that matter cannot be composed of atoms, since each of its parts is further subdivided. Moreover, even the shapes and motions of the parts do not remain constant from one instant to the next. So, assuming that matter is not purely phenomenal, whatever is substantial in it must be something other than extension, figure and motion, and must therefore be non-material. Moreover, on account of the infinite division of matter, this non-material principle must be found in each and any of its parts, however small.

These considerations, as I sketched above, issue in Leibniz's mature doctrine of substance. The non-material unit of substance found in any of matter's parts is the *monad*, or simple substance. An infinite aggregate of monads gives rise to the phenomena of extendedness, antitypy etc., properties which do not belong to any of the constituent monads taken singly, but result from an agreement among their states or perceptions. The resulting phenomenal whole is what Leibniz calls variously *mass, secondary matter*, or *body.* Although it is well founded, and is in this sense a *real* phenomenon, it does not have a true unity unless it is united by a dominant monad. A body which is "made into one machine by the dominant monad" is what Leibniz calls a *corporeal substance*. Thus "*body* is either corporeal substance, or a mass composed of corporeal substances" (NE.722: G.VII.501).

These considerations provide us with all we need for a successful resolution of Russell's conundrum. A simple substance, having no extension in itself, is *not* the same kind of thing as an aggregate of monads, since the latter is accompanied by the phenomenon of extendedness: thus monads cannot be actual parts of secondary matter or body. But every such body is actually divided into determinate or actual parts. These parts, therefore, are the same kind of things as the whole which they compose: they themselves are *bodies*. That is, they are infinite aggregates of monads accompanied by the phenomenon of extendedness, each such body being either a true whole united by a dominant monad (a corporeal substance), or an aggregate of such corporeal substances which is merely perceived as a whole.

As a concrete illustration we may consider Leibniz's example of the flock of sheep. The flock itself, as Leibniz mentions in his reply to Foucher quoted above (G.II.552), is a mere aggregate of corporeal substances or animals, namely the sheep themselves. It is what he calls elsewhere merely a *unum per accidens*, not a *unum per se*. But each sheep, being a corporeal substance, has a body which has determinate parts — its eyes, stomach, etc. These parts are not corporeal substances, however, since they have no animating monad. But they too are actually divided into constituent parts, so that if we continue dissecting the sheep, we will find that the eye is composed of living cells, that the stomach lining is composed of bacteria and other microscopic organisms which possess a real unity, corporeal substances. And given that the subdivision proceeds to infinity, it follows that however small a part of the sheep's mass we take, it will be composed similarly of other corporeal substances. Thus any body or corporeal substance is composed of determinable parts, which can always be subdivided until we reach other far smaller corporeal substances out of which it is composed.

This analysis is confirmed by Leibniz's own words in his further (unpublished) remarks on Foucher's objections on the same topic:

> In realities, where only divisions actually made enter, the whole is only a result or an assemblage, like a flock of sheep. It is true that the number of simple substances in any mass, however small, is infinite; for beside the soul, which makes the real unity of the animal, the body of the sheep, for example, is actually divided, i.e. is an assemblage of invisible plants or animals or plants, similarly composite except for what makes *their* real unity; and though this goes to infinity, it is plain that all in the end depends on these unities, the rest, or the results, being only well-grounded phenomena. (G.IV.492)

It is in this sense, then, that Leibniz can write without inconsistency that although simple substances are "prior" to assemblages or bodies, these bodies have actual parts, namely other bodies, which will always be extended corporeal substances or aggregates of these. Phenomenal body is divided into actual parts, and these parts are further divided without limit; but always some of these parts will be the bodies of corporeal substances, real unities, that are united by a dominant monad.

Thus Russell's attempt to foist a bogus Hegelianism onto Leibniz's logic fails. For Leibniz does not assume as a premise that monads are parts of matter only to deny it in his conclusion. Indeed, Leibniz never explicitly asserts that monads are actual parts of matter. In fact his position is perfectly consistent: it is that any part of extended matter, however small, *presupposes* a multiplicity of true unities of substance or monads.

Middlebury College
Middebury, Vermont

NOTES

Acknowledgement. When I developed the initial insights which form the heart of this paper in the summer of 1983, I was out on a limb career-wise, teaching Applied Mathematics at the University of Western Ontario. During this time, Robert Butts's advice and encouragement of my research on Leibniz was invaluable, so it is a great pleasure for me to dedicate the fruits of my labour to this festschrift in his honour.

[1] I call this allegation wild because Leibniz never deviated from the view that abstract ideas contain truth, particularly the mathematical idea of continuity. As he writes to de Volder in a passage quoted by Russell (246): "But the science of continua, i.e. of possibles, contains eternal truths, which are never violated by actual phenomena, since the difference is always smaller than any assignable given difference." (G.II.282). Indeed, in his *New Essays on Human Understanding* Leibniz explicitly repudiates the view ascribed to him by Russell: "an abstraction is not an error, provided we know what we are ignoring is really there. This is the use made of abstractions by mathematicians when they speak of the perfect lines they ask us to consider, and of uniform motions and other regular effects . . .", Parkinson (1967), p. 159.

[2] Russell then adds sarcastically: "Those who admire these two elements in Hegel's philosophy will think Leibniz's argument the better for containing them" (1900, p. 110).

[3] See in particular his account in the *New System* (1695), Parkinson (1973) pp. 116, 120—121; and his reference in the *Theodicy*, Preface, ¶8, (Leibniz 1951, p. 53); G.VI.29.

[4] I intend to present just such an exposition in a forthcoming work, tentatively titled *Leibniz's Labyrinth of the Continuum.*

[5] The same mistake is to be found in McGuire's otherwise very penetrating analysis of Leibniz's solution to the continuum: "mathematical points . . . are mere modalities, or the limit of a process of subdivision" (McGuire 1976, pp. 307—8).

[6] Actually, McGuire distinguishes geometrical lines, "which are divisible into finite parts", from the notion of the spatial continuum, which "is different from the geometrical, since it involves the ideal of a possible order of coexistence. As such, it does not involve actual distances that are extended and presuppose division into parts." A similar attempt to distinguish between space and geometric extension can also be found in Martial Gueroult's "Space, Point and Void in Leibniz's Philosophy" (Hooker 1982, pp. 284—301). As I mentioned in a previous publication (1985, p. 310), I am unconvinced by it; but since nothing concerning continuity hangs on this disagreement, I have ignored their distinction here.

[7] McGuire's actual wording here is that "extended things have a plurality of coexisting 'actual parts' *that they have* 'in common even with things which are not extended'" (306), implying that it is the actual parts they have in common. This is not supported by Leibniz's text, according to which it is *coexistence* that extension "has in common even with things which are not extended".

[8] This is not to deny that well-founded phenomena do indeed have a "separate status" with respect to their reality. Leibniz writes to Arnauld that "mock-suns and perhaps even well-ordered dreams will be able one day to lay claim to reality, unless very precise limits are set to this *droit de bourgeoisie* which is to be accorded to entities founded by aggregation" (30th April, 1687: Parkinson 1973, p. 71).

[9] McGuire also claims to find a number of difficulties concerning the application of *time* to what is actual and real, stemming from the unreality of relations (312, 313—318). I believe these difficulties to be spurious, as I have argued in my (1985), where I present a construal of Leibniz's theory of time based on the relations of ground containing and compatibility holding between pairs of monadic states.

[10] See his letter to Jakob Thomasius, April 1669, in Loemker (1969), pp. 93—104: G.I.15—27; G.IV.162—174.

[11] Cf. *On Primary Matter*, G.VII.259—260: "Aristotle's primary matter is the same as Descartes' subtle matter. Each is divisible to infinity. Each lacks form and motion in itself, each acquires forms as a result of motion. Each receives its motion from a mind."

[12] This distinction between primary and secondary matter is Scholastic in origin, as Dan Garber points out in his (1985). There he quotes Eustacius: "Primary [matter] is said to be that which, before all else, we conceive as entering into the composition of any natural thing; regarded as lacking all forms . . . Secondary [matter] is said to be that very primary [matter], not, however, bare, but endowed with physical actuality [i.e. forms]. (Eustacius a S. Paulo, *Summa Philosophiae Quadripartita* . . . , Cambridge, England, 1648: p. 119)."

[13] The relevant passages off Aristotle's writings on the continuum have been collected together in translation in Appendix A of Kretzmann (1982), pp. 309—321; many of the articles in this book should also be consulted, as should Richard Sorabji's excellent and extended account of the progress Aristotle made in his treatment of the continuum, its shortcomings, and its historical importance (1983). There are also two extremely stimulating accounts of its contemporary significance: a book by José Benardete which deserves to be much better known, (1964), esp. pp. 18—27; and a provocative new article by Feyerabend, "Some Observations on Aristotle's Theory of Mathematics and of the Continuum", pp. 219—246 in his (1987).

[14] Charinus, one of the two characters speaking for Leibniz in the dialogue, comments: "He [i.e. Descartes] ought to have at least explained how in this case matter is not resolved into a powder, so to speak, consisting of points, when no point would appear to be left attached to anything, since each moves by itself with a motion different from the motion of any other." (Leibniz 1676, p. 614). I was first alerted to the significance of this dialogue by Nicholas Rescher, who gives several excerpts in his (1967). With his encouragement, I made a complete translation of it myself some years ago, which I hope soon to publish.

[15] I would hesitate to accuse Leibniz here of outright inconsistency in upholding indivisibles yet denying minima (points with no magnitude), as for instance Earman does in his (1975), p. 241. First one would have to take into account the scholastic doctrine of signs (cf. Loemker's note in his 1969, p. 145). That this theory is not inconsistent has been shown in an interesting article by Steve Thomason (1982).

[6] A "perfect liquid" is roughly synonymous with Descartes' "subtle matter". In *Pacidius Philalethi* Leibniz defines it as a liquid "any part of which, however small, could be separated from any other given part" (1676, p. 613).

[17] As Leibniz writes to Bernoulli in 1698: "Suppose all the subdivisions of a line, 1/2, 1/4, 1/8, 1/16, 1/32 &c., actually existed. To infer from this series that an infinitieth term absolutely exists would be an error, for I think nothing more follows than that there exists an assignable finite fraction as small as you please . . ." (Quoted from Benardete 1964, p. 19).

[18] For an interesting analysis of the role of minds in Leibniz's early physics, see Daniel Garber's (1982).

[19] It will be evident that I am ignoring any changes in Leibniz's position with regard to the reality of bodies or of corporeal substances. But with the exception of his

correspondence with Des Bosses, where he toys with the idea of a *substantial chain* which would unite corporeal substances into a really continuous spatial whole, I do not believe that there are any important changes in his philosophy of substance. Although I do not have the space to defend my heresy here, I take the position (implicit in my argument in this article) that there are always corporeal substances for Leibniz — indeed, an actual infinity of them — from the time of his solution of the continuum problem in the early 1680's till his dying day.

[20] For instance, in a letter to Remond, having declared that in his view "the whole universe of Creatures consists only in simple substances or Monads, and in Assemblages of them," and that "the Assemblages are what we call bodies", Leibniz writes: "in the real the simple is prior to the assemblages, the parts are actual, and precede the whole" (To Remond, July 1714: G.II.622). Here Leibniz's talk of "actual parts" immediately after referring to assemblages of simple substances gives the impression that it is the monads themselves that are the actual parts of phenomenal bodies. But the continuation of the passage makes clear that the parts are the parts into which phenomenal bodies are divided.

BIBLIOGRAPHY

Arthur, Richard. (1985). "Leibniz's Theory of Time", in Okruhlik and Brown, pp. 263—313.

Benardete, José A. (1964). *Infinity: An Essay in Metaphysics*, Oxford: Clarendon Press.

Bos, H. J. M. (1974—5). "Differentials, Higher-Order Differentials and the Derivative in the Leibnizian Calculus", *Archive for History of Exact Sciences* **14**, 1—90.

Couturat, Louis (ed.). (1903). *Opuscles et fragments inédits de Leibniz*, Paris: Felix Alcan.

Descartes, René. (1644). *Principles of Philosophy*, Vol. VIII in Adams, Charles and Paul Tannery (eds.), *Oeuvres de Descartes*. Paris: Libraire Philosophique J. Vrin, 1973.

Earman, John. (1975). "Infinities, Infinitesimals and Indivisibles: The Leibnizian Labyrinth", *Studia Leibnitiana*, Band VII/2, pp. 236—251.

Euler, Leonhardt. (1843). *Letters of Euler on Different Subjects in Natural Philosophy Addressed to a German Princess*, David Brewster, (ed.), New York: Harper and Brothers, vol. 2 of 2 vols.

Feyerabend, Paul (1987) *Farewell to Reason*, London/New York: Verso.

Garber, Daniel. (1982). "Motion and Metaphysics in the Young Leibniz", in Hooker, pp. 160—184.

Garber, Daniel. (1985). "Leibniz and the Foundations of Physics: The Middle Years", in Okruhlik and Brown, pp. 27—130, (1985).

Gerhardt, C. I. (ed.). (1875—90). *Die philosophische Schriften von Gottfried Wilhem Leibniz*, 7 vols, Hildesheim, New York: Georg Olms Verlag, 1978 (cited as G.II.268 etc.).

Gerhardt, C. I. (ed.). (1849—55). *Leibnizens Mathematische Schriften*, Berlin and Halle (cited as GM.II.157 etc.).

Hooker, Michael (ed.). (1982). *Leibniz: Critical and Interpretive Essays*, Minneapolis: University of Minnesota Press.

Kretzmann, Norman (ed.). (1982). *Infinity and Continuity in Ancient and Medieval Thought*, Ithaca: Cornell University Press.
Leibniz, Gottfried Wilhelm. (1676). *Pacidius Philalethi*, my translations of the text in Couturat (1903), pp. 594—627.
Leibniz, Gottfried Wilhelm. (1686). *Phoranomus; or On Power and the Laws of Nature*, my translations of extracts given by Gerhardt in *Archiv für Geschichte der Philosophie* **1**, 566—581.
Leibniz, Gottfried Wilhelm. (1923—00). *Sämlichte Schriften und Briefe*, ed. Preußische Akademie der Wissenschaften, Darmstadt, Leipzig (cited as A.VI.ii.264 etc.).
Leibniz, Gottfried Wilhem. (1951). *Theodicy*, trans. E. M. Huggard, London: Routledge and Kegan Paul.
Loemker, Leroy E. (1969). *Gottfried Wilhelm Leibniz: Philosophical Papers and Letters*. Dordrecht: D. Reidel.
McGuire, J. E. (1976). "'Labyrinthus Continui': Leibniz on Substance, Activity and Matter", in Machamer, K., and Turnbull, R. G. (eds.) *Motion and Time, Space and Matter*, Columbus: Ohio State University Press. pp. 290—326.
Okruhlik, Kathleen and James Robert Brown (eds.). (1985). *The Natural Philosophy of Leibniz*, Dordrecht/Boston: D. Reidel.
Parkinson, G. H. R. (1973). *Leibniz: Philosophical Writings*, London: J. M. Dent & Sons Ltd.
Rescher, Nicholas. (1967). *The Philosophy of Leibniz*, Englewood Cliffs, N.J.: Prentice-Hall, Inc.
Russell, Bertrand. (1900). *A Critical Exposition of the Philosophy of Leibniz*, Cambridge: Cambridge University Press.
Sorabji, Richard. (1983). *Time, Creation and the Continuum: Theories in Antiquity and the Early Middle Ages*, Ithaca: Cornell University Press.
Thomason, S. K. (1982). "Euclidean Infinitesimals", *Pacific Philosophical Quarterly* **63**, 168—185.

NICHOLAS RESCHER

THE PHILOSOPHERS OF GAMBLING

0. PREFACE

When I wrote this essay for you, Bob — actually, during your Pittsburgh visit of 1987 — I certainly did not labor under the delusion that you have much of an interest in gambling. (So far as I know, you do not gamble at all.) But I did have in mind your long-standing interest in the era from Descartes to Newton, the starting-point of so much that is important in the development of European philosophy. It is in your role as an historian of ideas and as a connoisseur of strange twists in the development of philosophical issues that I offer this essay to you.

1. INTRODUCTION

In his informative study of the origination of the mathematical theory of probability, Ian Hacking maintained that "the decade around 1660 is the birthtime of probability".[1] And he set out a convincing case for regarding the period 1650—1670 as the time that saw the mathematical taming of chance in the emergence of a calculus of probability at the hands of mathematicians like Pascal, Fermat, and Huygens, who concerned themselves with problems of the division of stakes in gambling. The aim of the present discussion is to indicate that the immediately preceding period 1610—1650, the unsettled era of the Civil War in England and the Thirty Years' War on the Continent, witnessed a new concern on the part of philosophers with gambling, gaming, and the role of chance, accident, fortune and luck in human affairs. In this way, the intellectual climate of a *philosophy* of chance paved the way to the development of the *mathematics* of chance. Only after initially attracting the attention of moralists and philosophers, who resumed the ruminations of moralists and philosophers, who resumed the ruminations about fate and fortune inherited from classical antiquity,[2] did intellectual preoccupation with matters of chance pass into the hands of the mathematicians who revolutionized thought on the subject by developing that "calculus of chance" which we ourselves characterize as the theory of probability.

J. R. Brown and J. Mittelstrass (eds.), An Intimate Relation, 203—220.

In substantiating this historical perspective, it is helpful to consider the ideas of three very different thinkers working in three mutually distant corners of Europe: Gataker, Gracián, and Pascal.

2. THOMAS GATAKER (1574–1654)

Born in London and educated in Cambridge, Thomas Gataker was for a time preacher to the society of Lincoln's Inn.[3] He was a versatile scholar and influential Puritan divine who was nevertheless one of the forty-seven London clergyman who signed the address of 18 January 1649 against the trial and execution of the King. His 1652 edition of the works of Marcus Aurelius was described by Henry Hallam as the "earliest edition of any classical writer published in England with original annotations".[4]

In 1619 Gataker first published his tractate *Of the Nature and Use of Lots* (London: Edward Griffin, 1619; 2nd ed., here cited, published in 1627).[5] The topic was apparently put on the agenda of the day by the great lottery of 1612, described in Stow's *Annales* as follows:

> The King's maiestie in speciall favor for the present plantation of English Colonies in *Virginia*, granted a liberall Lottery, in which was contained five thousand pound in prizes certayne, besides rewardes of casualtie, and began to be drawne in a new built house at the West end of *Paul's*, the 29th of June 1612. But, of which Lottery, for want off filling uppe the number of lots, there were then taken out and throwne away three score thousand blanckes, without abating of any one prize; and by the twentith of July all was drawne and finished. This Lottery was so plainely carryed, and honestly performed, that it gave full satisfaction to all persons. *Thomas Sharpliffe*, a Taylor of London, had the chiefe prize, *viz.* foure thousand Crownes in fayre plate, which was sent to his house in very stately manner: during the whole tyme of the drawing of this lottery there were alwaies present diuers worshipfull Knights and Esquiers, accompanied by sundry graue discreet Cittizens.[6]

Gataker's book does not however focus exclusively upon lots and lotteries, but takes gaming in general into its purview, a lot being construed very generally an event whose outcome is due to chance.

Gataker's book was a work of substantial erudition running to over 300 pages. It comprised an elaborate survey of historical examples of the use of lots in the Old and New Testament — for example in the selection of a successor to the apostle Judas, *Acts*, 1: 23–26); in the assignment of priesthoods and public offices in Greece; in the allocation of benefits in Hebrew, Greek, Roman, and other legal practice; in customs governing the division of booty and the spoils of war; and the like. (All four gospels state that the Roman soldiers used lots to divide the garments of Jesus.[7]) Gataker defined a "lot" as an "event merely casual purposely applied to the deciding of some doubt" (p. 9), "casual events" being "such as might fall out in like sort diversely, and are not determined by any art, foresight, forecast, counsell, or skill of those that either act in them, or make use of them" (p. 14). He quotes with approval the dictum that "chance is founded, and dependeth upon Man's ignorance (*fortuna in ignorantia nostra fundatur*)" (p. 37).

Gataker follows St. Thomas in dividing the use of lots into the *divisory* (*sors diversoria*) for effecting distributions or allocations of goods or evils, the *consultory* (*sors consultatoria*) for settling on courses of action or determining matters of fact,[8] and the *divinatory* (*sors divinatoria*) for seeking the will of God or the decrees of Fate as to what will happen in the future. And be approved of their divisory use alone. Granting (with *Proverbs*, 16:24) that God knows the outcome of lots, Gataker rejected and criticized the view that "a Lot discovereth to men God's hidden will" (p. 25), arguing that "Lots are not to be used in [a] question of Fact past and gone . . . for what is no ordinarie Lot is able to decide; but where some question is who has the right to a thing; but where some question is who has the right to a thing; in which case, notwithstanding the Lot is not used to determine who in truth hath right to it, but who for peace and quietnesse sake shall enjoy it" (p. 148). Accordingly, Gataker insisted that, "concerning the matter or businesses wherein Lots may lawfully be used, the rule of Caution in general is this, that Lots are to be used in things indifferent only" (p. 125), for:

> . . . many good things there are that may at sometime be done, where off a man may make choise whether of them hee will doe, being not necessarily tied unto, or enjoyned any one of them: As for a student having divers bookes about him in his study, it is indifferent to choose one, this or that, refusing the rest, for present employment, there being no special occasion to urge the use of one more than another: Or for a man that carrieth a pair of Knives about him, it is indifferent to draw and use either when occasion requireth (as Plutarch says, *de Stoic. contradict.*). (P. 128.)

In the wake of his book, Gataker's promising clerical career was put in jeopardy by accusations of favoring games of chance through his defense of the use of lots.[9] This charge was certainly unjust. For Gataker's thesis was merely the employment of random-selection devices as a means of resolving matters of choice in cases where some preferential selection is desirable "for peace and quietnesse sake". Often, after all, lots are used in a perfectly harmless way — for example in the assignment of starting positions in swimming or rowing races (p. 119). Moreover, as the Bible notes, "The lot causes contentions to cease, and parteth between the mighty".[10] (A modern example of the sort of thing Gataker had in mind is afforded by the circumstance that when Hawaii was admitted as the fiftieth state of the Union, and two new senators were elected conjointly, random devices were used by the Senate to decide which of the two new Hawaiian senators would have seniority, with the decision made by a coin-toss, and which would serve the longer term, with the decision made by card-drawing.)[11]

In Chapter X, "Of Extraordinary or Divinatory Lots", Gataker expatiated negatively on the use of lots "either for the discovery of some hidden matter past, or present or for the presaging and fortelling of some future event". And the ensuing chapter argues at length that this sort of thing is superstitions and unlawful. Other uses of lots can in principle be appropriate, however.

Gataker's position in fact closely parallels that of Cicero's *De Divinatione*. Cicero there approved the use of divination, and especially of augury, as a practice that had certain social and political benefits in fostering cooperative action and communal solidarity, and not at all because it had any prophetic utility — any predictive efficacy.[12] He distinguished between an inappropriate (superstititious) use of signs, anguries, and portents for making informative *predictions* about the future, and an appropriate use of them for making communal *decisions* in controversial matters in a way that public peace can be preserved since the authorities are not burdened with a choice between the pro-faction and the anti-faction since the matter is left "in the lap of the gods". For the sake of fairness and impartiality, people want to assure the human desires and predilections are expelled from various sorts of choice situations. And the use of lots is an effective way to attain this objective.

It was in this sort of spirit that Gataker sought to maintain the principle that there is nothing heretical or immoral in the use of lots *per se*. In the realm of human affairs, we do sometimes want to assure that things happen accidentally and "by chance" rather than "by design". And the use of lots is simply a way of insuring that certain issues eventuate on this basis.

However, the last third of his treatise, Gataker ventured upon more dangerous ground. He suggests that games of chance can (and should) be harmless playthings. Gaming has its "lawful use" — for diversion good recreations — as well as its "unlawful abuse" (p. 194):

> But it is one thing to play at dice or cards and another thing to be a Dicer or a Carder; as it is one thing *to drink wine*; and another thing *to be a wine drinker* or, as we use (sic) to say *a wine bibber*. (P. 229.)

Provided people do not over indulge in such diversions to the point of neglecting their business, and do not gamble for money, no harm is done. "Play [is to] be used as play; for pleasure, not profit; for game, not for gain" (p. 251). It was doubtless this part of his discussion that led Gataker into difficulty with his pious critics.

The very circumstance that Gataker had to defend himself against the charge of favoring games of chance testifies to a rising concern with gambling, which religious people have generally regarded as not just a manifestation of private imprudence and time-wasting, but as a mode of immorality and even impiety, insofar as it abandons the use of God-given reason to arrive at a decision of affairs through the mediation of chance or fate.[13] Indeed there is something impious about thinking that there are any "casual" or "chance" occurrences. It is only from our human point of view that "casual events" exist at all; an omniscient God keeps track not only of the flight of the sparrows but of the roll of the dice as well.[14]

3. BALTHASAR DE GRACIÁN Y MORALES (1601–1658)

Gracián was a Spanish moralist who was educated in Toledo and entered the Jesuit order as a novice at the age of 18. He published his books under the name of Lorenzo Gracián. His *Pocket Oracle* (*Oráculo manual y arte de prudencia*), first published in 1647, was a series of three hundred pithy precepts, each accompanied with a brief commentary, setting out the guidelines of prudent action. The book enjoyed a great popularity, was echoed by La Rochefoucault, and admired and translated into German by Schopenhauer.[15]

Gracián's book depicted the human situation on analogy with card games and formulated much practical advice on this basis. His position in this regard ran as follows: In this life, fate mixes the cards as she lists, without consulting our wishes in the matter ("*Baraja como y cuando quiere la suerte*", sect. 196). And we have no choice but to play the hand she deals to us. But the wise man bides his time and places his bets when conditions are favorable ("*Pero el sagaz atienda al barajar de la suerte*", sect 163). He tests the waters, as it were, before getting in too deep, and if matters look inauspicious withdraws to play again another day. ("*Conocer el día aciago, que los hay. Nada saldrá bien, y aunque se varié el juego, pero no la mala suerte. A dos lances, convendrá conocerle y retirarse, advirtiendo si está de día o no lo está*", Sect 139.) The sagacious gambler never counts on luck's lasting and prepares for adversity amidst good fortune. ("*Prevenirse en la fortuna próspera para la adversa . . . Bueno es conservar para el mal tiempo, que es la adversidad cara y falta de todo*", Sect 113.)

There are *rules* for coping with risks and the sagacious person can facilitate good fortune. ("*Reglas hay de ventura, que no toda es acasos para el sabio; puede ser ayudado de la industria*", Sect. 21.) Of these rules the most important is to play well whatever hand fate may have dealt. ("*La mejor treta del juego es saberse descartar; más importa la menor carta del triunfo que corre, que la mayor del que pasó*", Sect. 31.) Another cardinal rule is to know when to quit: the knowledgeable gambler never "pushes his luck". (*Saberse dejar ganando con la fortuna. Es de tahúres de reputación . . . . Continuada felicidad fué siempre sospechosa: más segura es la interpolada y que tenga algo de agridulce aun para la fruición: Cuanto más atropellándose las dichas, corren mayor riesgo de deslizar y dar al traste con todo . . . . Cánsase la fortuna de llevar a uno a cuestas tan a la larga*", Sect 38.) Thus one crucial rule is not to deem oneself as destined for domination. To think oneself to be the ace of trumps is a fatal flaw. ("*No ser malilla. Achaque es de todo lo excelente que su mucho uso viene a ser abuso*", Sect. 85)

In this way Gracián analogized the conduct of life to card-play and reinterpreted the guidelines of good card sense as principles of life. Life and playing cards are both games of chance, as it were, and the precepts for effective operation in both contexts are fundamentally akin.

Gracián's perspective struck a resonant note among his countrymen. Gambling has long been a prominent fact of Spanish life. (The *Lotería Nacional*, established by Carlos III in 1763, is the oldest of surviving national lotteries.) Official estimates indicate that moneys spent in gambling currently amount to some 15 percent of family income, making Spain a world leader in this regard.[16] Spaniards widely view gambling not as a human weakness or vice, but as a plausible opportunity for improving one's condition.

The general attitude and outlook of Gracián's book accordingly came to play an important role in Spanish philosophy. And there is further good reason to see this as unsurprising. For while the Western philosophical mainstream has striven to imbue our understanding of the world with the intelligible order of rational system, the Spanish philosophers of the anti-scholastic tradition have generally viewed the world as an uncertain, unpredictable and unreliable setting for human life. Their position runs roughly as follows: Nature and man himself conspire in creating a difficult and largely intractable environment for us. Reality, or at any rate that part of it which constitutes the setting for human life, is chaotic, incoherent, pervaded by disorder. Life is precarious. In all of our doings and dealings we cannot count on things going "according to plan!" Planning, prudence, foresight, and the like can doubtless help to smooth life's path, but they are far from sufficient to assure a satisfactory outcome to our efforts. Chance, accident, and luck — fortune, in short — plays a preponderant and ineliminable role in human affairs. In all of our doings and undertakings we humans give hostages to fortune. The outcome of our efforts does not lie in our control: fortune (chance, contingency, luck) almost invariably plays a decisive part.

This fortunism did not, however, in general carry the Spaniards to an un-Christian extreme. They were not drawn to the endorsement of inaction, lethargy, and a supine resignation to the inevitable. In their view, active enterprise is called for because our actions set the stage for luck — someone who buys no ticket cannot win the lottery. To a degree — although a very limited degree — people are the authors of their own fortune. Man proposes, fortune disposes.

The changeable and unpredictable nature of the human situation means that flexibility and adaptability are major human virtues. People have to be many-sided, able to adapt to changing circumstances and play very different sorts of roles. Like a good actor, a capable person must be prepared to play very different parts. (Spanish literature offers the influential model of *el picaro* as a chameleon, a person who manages to attune himself to the requirements of the moment and is able to be all things to all people.)

On this sort of basis, versatility and adaptability were seen by the Spanish anti-scholastics from the time of Gracián onwards as crucial aspects of prudence. People whose life is too orderly — who place too great a reliance on the regularity of an established system — thereby risk disaster. The sagacious man is one who has the prudence to develop this flexibility to accommodate himself to difficult and changing conditions. He strives to be able to say, with the Abbé Sieyès, theorist of the time of the French revolution, who when asked about his activities during the heyday of Robespierre and the terror responded, "*J'ai vécu*".

In this way, Spanish anti-scholastics took the prominence of fortune in man's affairs to be token the limits of human power, setting the stage for a fundamentally pessimistic appreciation of the power of human reason for guidance in this sublunary sphere. We find thus attitude prominent not only in the great figures of the Golden Age of Spanish literature who were more or less contemporaries of Gracián (in particular Quevedo and Calderón), but also much later — in Unamuno's insistence that human reason is inadequate and unsatisfactory as a guide to life, and in Ortega y Gasset's rejection of the utility of scientific reason as directrix of human affairs. The idea that life is too chancy and fortuitous thing to be manageable by rational means runs as a recurrent Leitmotif through the history of Spanish thought. Gracián's recommendation of the gambler's perspective fell on fertile ground.

4. BLAISE PASCAL (1623—1662)

Though he died at the early age of 39, Pascal's many-sided genuis flowered early, and, despite the brief span of his life, he managed to make an astonishing array of contributions in mathematics, physics, philosophy, and theology. During his early years he was a member of fashionable society and a fixture of Parisian salons, but after a profound religious experience in 1654, he joined the Jansenite community in Port Royal and lived a reclusive and scholarly life there, dedicating his pen to theological exposition and controversy. His *Thoughts* (*Pensées*) was a collection of brief notes and drafts that he jotted down during the years 1657—62 in preparation for writing a projected *Apology for the Christian Religion.* Published after his death, Pascal's sketches proved to be a work of enduring value and popularity.[17]

Pascal saw the role of chance in human affairs as pervasive:

> Everyone ponders how to make the best of his condition, but as for the choice of condition and country, chance [*sort*, destiny] gives them to us. It is a pitiable thing to see so many Turcs, heretics and infidels, follow the ways of their fathers solely because they have been taught that this is best. And it is this that determines each to his condition of locksmith, soldier, etc. (*Pensées*, ed. Brunschvicg, No. 98.)

Change and caprice pervade human affairs. "Truth on this side of the Pyrenees, error on the other" (*ibid.* No. 294). In living this life we constantly "take our chances". And this holds with regard to the next life as well.

In one brief passage of two leaves that form part of this assemblage, Pascal presented his famous Wager in favor of the religious life. It was addressed to his former fashionable friends, such as the clever but somewhat shady Chevalier de Méré, a typical specimen of the worldly *libertin*: "brilliant talker, fearless freethinker, and inveterate gambler".[18] The core of the Wager argumentation ran as follows:

> When there is an equal risk of winning and of losing, if you had only two lives to win, you might still wager; but if there were three lives to win, you would still have to play (since you are under the necessity of playing), and being thus obliged to play, you would be imprudent not to risk your life to win three in a game where there is an equal chance of winning and of losing. But there is an eternity of life and happiness. That being so, if there were an infinity to chances to which only one was in your favor, you would still do right to stake one to win two, and you would act unwisely in refusing to play one life against three, in a game where you had only one chance out of an infinite number, if there were an infinity of an infinitely happy life to win. But here there is an infinity of infinitely happy life to win, one chance of winning against a finite number of chances of losing, and what you stake is finite. That removes all doubt as to choice, wherever the infinite is to be won, and there is not an infinity of chances of loss against the chance of winning, there are no two ways about it; you must risk all.[19]

This is not the place to enter into the complexities of Pascal's analysis. Suffice it to observe that, in effect, the Wager discussion says: "When gambling, you people act on the sensible principle of evaluating wagers by blending the chances of an outcome with the gain to be realized. Be consistent and do the same in matters of religion. You will then have to agree that no matter how small you deem the chances of God's existence, the infinite reward that will come to the faithful, should He exist, serves to render the gamble of religious commitment worthwhile". Pascal too saw chance as a pervasive rogue factor in human affairs: "By chance you are the son of a duke, and your very existence is due to an infinity of chances. Your birth is due to a marriage, or rather to a series off marriages of those have gone before you. But these marriages were often the result of a chance meeting, of words uttered at random, of a hundred unforeseen and unintended occurrences".[20] Our very lives are a gamble. The Wager is an invitation to think about the big issue of life in this world and the next in the manner of a gambler.

Recognizing the extent to which gambling had become fashionable among his worldly compatriots, Pascal thus seized on an ingenious scheme to capitalize on this circumstance for apologetic purposes. The Wager argument is, in effect, a clever stratagem for exploiting the attitudes and modes of thought current among the gamblers to turn worldly people to the ways of religion and the life of faith. The Wager was designed to accomplish its work by using the very latest intellectual technology in deploying the machinery of mathematical expectations that represents Pascal's enduring contribution to the development of probability theory. For a new mode of thought had entered upon an old arena of discussion; "St Augustine had seen that that one labors under uncertainty at sea, in battle, etc. But he did not see the law of expectation (*'la règle des partis,'* the 'rule of partition') which shows how one ought to proceed" (*Pensées*, ed. Brunschvicg, No. 234).

His work on games of chance made Pascal one of the founding fathers of the mathematical theory of probability, for Pascal and his co-workers managed the enormous achievement of imposing some degree of mathematical intelligibility on inherently fortuitous occurrences. His Wager — which turns on using the machinery of mathematical expectations to assess the acceptability of gambles — was an ingenious use of a mathematical resource for the purposes of religious apologetics. Pascal marks the point of transition when the philosophy of chance gave rise to the mathematics of chance. With the rise of probability theory, Pascal and his mathematical congeners discovered something that even the cleverest intellects of classical antiquity would have viewed as a virtual contradiction in terms — the existence of *laws of chance*.[21]

Our three theorists thus present a picture of rapid escalation. Gataker was concerned merely to defend the religious and moral legitimacy of using instruments of gaming to resolve a limited range of practical issues. In Gracián we find the idea of using the groundrules of gambling as guiding principles for the conduct of our everyday lives at large. With Pascal we have the application of the principles of rational gambling as a basis for decision in matters of religion governing our view of God and the life to come.

5. GOTTFRIED WILHELM VON LEIBNIZ (1646–1717)

In the next generation, too, philosophers continued to devote attention to matters relating to games of chance, with Leibniz as the most prominent example here.

It was among Leibniz's primary concerns to deny any room for chance on the stage of the world's events and processes. His basic ideological commitment was exactly that of Einstein: God does not play dice with the universe. A "Principle of Sufficient Reason" was accordingly one of the corner-stones of Leibniz's philosophy, and he insisted time and again on "the great . . . principle that *nothing takes place without a sufficient reason*, in other words, that nothing occurs for which it would not be possible for one who has enough knowledge of things to give a reason adequate to determine why the thing is as it is and not otherwise".[22]

Leibniz saw the ultimate reason for all of the world's arrangement, to lie in the will of God, and here, in the domain of God's will and choice, there is no room for any element of chance:

> If the will of God had not as its rule the principle of the best, it would tend towards evil, which would be worst of all; or else it would be indifferent somehow to good and to evil, and guided by chance. But a will that would always drift along at random would scarely by any better for the government of the universe than the fortuitous concourse of corpuscles, without the existence of divinity. And even though God should abandon himself to chance only in some cases, and in a certain way (as he would if he did not always tend entirely towards the best, and if he were capable of preferring a lesser good to a greater good, that is, an evil to a good, since that which prevents a greater good is an evil) he would be no less imperfect than the object of his choice. Then he would not deserve absolute trust; he would act without reason in such a case, and the government of the universe would be like certain games equally divided between reason and luck.[23]

After all, God always proceeds rationally, and rational choice must (by definition) proceed in line with reason. Indeed this replace he holds for any genuine act of will:

> In things absolutely indifferent there is no (foundation for) choice and consequently no election or will, since choice must be founded on some reason or principle. A mere will without any motive is a fiction, not only contrary to God's perfection but also chimerical and contradictory, inconsistent with the definition of the will.[24]

To be sure, we humans can sometimes reasonably effect selections by random mechanisms such as lots, but only because of our imperfect knowledge allows ignorance of outcomes:

> But, as I have declared more than once, I do not admit an indifference of equipoise,

and I do not think that one ever chooses when one is absolutely indifferent. Such a choice would be, as it were, mere chance, without determining reason, whether apparent or hidden. But such a chance, such an absolute and actual fortuity is a chimera which never occurs in nature. All wise men are agreed that chance is only an apparent thing, like fortune: only ignorance of causes gives rise to it.[25]

Of course such a selection does not qualify as a genuine *choice*: with us, as with God, rational choice must be based on reasons.

Leibniz was greatly interested in games of chance and gave much encouragement — and some active help — to the development of the theory of probability.[26] He saw the development of the formal theory of games of chance as useful to the human intellect and peculiarity suited to its powers (specifically as affording a rational means for reflecting their deficiencies). But putting *epistemological* considerations aside, Leibniz's *metaphysical* concern was precisely to deny any room for considerations of chance or probability in the ontology of the real. For Leibniz, metaphysics must approach the issue from God's eye point of view, and at *this* level there is simply no place for chance on probability, seeing that a benevolent God would not and does not "play games" with his creation. The opponents of Gataker worked to exclude gambling from England; Leibniz and his followers worked more ambitiously to exclude it from nature.

6. THE ETHOS OF THE ERA

How is one to account for the philosophers' concern during the 1610–1650 period with gambling and games of chance? The answer lies in the prominence of gaming in contemporary European society, a phenomenon that is itself largely the consequence of war.

The turbulent era of the Civil War in England and the Thirty Years' War on the continent saw an epidemic of gambling spread across the face of Europe. It began among soldiers and sailors faced with the need to kill time during long periods of enforced inactivity away from the more varied and constructive activities of civilian life at home. An anecdote from the Thirty Years' War related by the Spanish soldier of fortune Alonso de Contreras (1582 — ca. 1641) illustrates this phenomenon:

After the capture by us Spaniards of a moorish ship in the Mediterranean had produced much booty, the commander of the victorious galleon, knowing well the predilections of his men, strictly forbad gambling, so that each soldier would presevere his share and return to Malta a wealthy man. To enforce this order, he had all dice and cards thrown overboard. But the urge to gambling among his men was not to be so easily denied. They drew a chalk circle on the deck and each concurrently placed a louse near its center, the winner to be he whose louse first crossed the line and left the circle. Great sums were wagered on this basis. "When the commander learned of this, he let his men have their way, recognizing that this vice of gambling cannot be suppressed among soldiers".[27]

The civil and military authorities of the period struggled in vain to contain gambling among soldiers. All attempts to prohibit or restrict it ultimately came to naught.[28]

The prolonged criss-crossing of the continent by the armies and military hirelings of the major powers during the Thirty Years' War led to a diffusion of gambling throughout Europe. The mania for gambling soon spread from the soldiers to civil society in general — polite and otherwise. The situation is illustrated by the following episode:

In 1656, under the ministry of Mazarin, letters-patent were granted to an Italian, named Fonti, to establish a lottery in Paris, whose profits were to construct a stone bridge between the galleries of the Louvre and the faubourg St. Germain, the wooden bridge which formerly connected them having been destroyed by fire. This lottery, however, seems to have turned out a fiasco. The first royal lottery drawn in France was an improvised one, started in a moment of loyal enthusiasm at the marriage of Louis XIV. Parliament sanctioned this lottery, but refused to grant further patents. The people, having once acquired the taste for this kind of gamble, finding public lotteries denied to them, had recourse to private and foreign ones and tickets might be bought for five sous. Despite numerous attempts to suppress the private home lotteries, they continued to flourish.[29]

Initially the gambling mania pervaded France, Spain, and the Netherlands,[30] but with the Restoration it crossed the Channel to England as well, moving with the return of the exiled court from the Paris of the Chevalier de Méré to the London of the Restoration. The Gaming Act of 1665 (16 Car. II. c.7), directed "against deceitful disorderly and excessive gaming" was the first piece of English legislation to regulate gambling debts.[31] (Persons winning more than 100 pounds on credit were not only prohibited from recovering the excess, but subjected to penalties for winning it.)

The most graphic description we have of gaming in this period is due (as usual) to Samuel Pepys, and it is well worth quoting at length:

[Went] to see the manner of the gaming at the Groome porter's [the controller of gaming at court] . . . where . . . they begin to play at about 8 at night — where to see how differently one man took his losing from another, one cursing and swearing, and another only muttering and grumbling to himself, a third without any appearing discontent at all — to see how the dice will run good luck in one hand for half an hour together — and another have no good luck at all. To see how easily here, where they play nothing but guinnys, 100 pounds is won or lost. To see two or three gentlemen come in there drunk, and putting their stock of gold together — one 22 pieces, the second 4, and the third 5 pieces; and these to play one with another, and forget how much each of them brought, but he that brought the 22 did think that he brought no more than the rest. To see the different humours of gamesters to change their luck when it is bad — how ceremonious they are as to call for new dice — to shift their places — to alter their manner of throwing; and that with great industry, as if there was anything in it. To see how some old gamesters, that have no money now to spend as formerly, do come and sit and look on; as among others, Sir Lewes Dives, who was here and hath been a great gamester in his time. To hear their cursing and damning to no purpose; as one man, being to throw a seven if he could and failing to do it after a great many throws, cried he would be damned if ever he flung seven more while he lived, his despair of throwing it being so great, while others did it as their luck served, almost every throw. To see how persons of the best quality do here sit down and play with people of any, though meaner; and to see how people in ordinary clothes shall come hither and play away 100, or 2 or 300 guinnys, without any kind of difficulty. And lastly, to see the formality of the Groome-porter, who is their judge of all disputes in play and all quarrels that may arise therein; and how his under-officers are there to observe true play at each table and to give new dice, is a consideration I never could have thought had been in the world, had I not now seen it. And mighty glad I am that I did see it; and it may be will find another evening, before Christmas be over, to see it again; when I may stay later, for their heat of play begins not till about 11 or 12 a-clock; which did give me another pretty observation, of a man that did win mighty fast when I was there: I think he won 100 pounds at single pieces in a little time; while all the rest envied him his good fortune, he cursed it, saying "A pox on it that it should come so earely upon me! For this fortune two hours hence would be worth something to me; but then, God damn me, I shall have no such luck." This kind of profane, mad entertainment they give themselfs. And so I having enough for once, refusing to venture, though Brisband pressed me hard and tempted me with saying that no man was ever known to lose the first time, the devil being too cunning to discourage a gamester; and he offered me also to lend me ten pieces to venture, but I did refuse and so went away.[32]

In the reign of Charles II, gambling was all the rage at court, and this fashionableness helped to spread it widely through town and country. Pepys was shocked at finding that even Sunday was not respected at court. On 17th February, 1666—7, he writes: "This evening going to the Queen's side to see the ladies, I did find the Queen, the Duchess of York, and another or two, at cards, with the room full of great ladies and men; which I was amazed at to see on a Sunday, having not believed it, but contrarily, flatly denied the same a little while since to my cosen Roger". On 14th February, 1667—8, he observed: "I was told to-night that my Lady Castlemaine is so great a gamester as to have won L15,000 in one night, and lost L25,000 in another night at play, and has played L1000 and L1500 at a cast". The niece of Cardinal Mazarin, who came over to England, and was a favorite of Charles II, was an avid gambler. She won 1400 guineas from Nell Gwyn at basset in one night, and over L8000 from the Duchess of Portsmouth "in doing which she exerted her utmost cunning, and had the greatest satisfaction, because they were her rivals in the royal favour".[33]

In theory, the probability calculus could have arisen in a very different way, one that had nothing to do with games of chance — from the rationalization of insurance, for example, or from the need for mathematical expectations in legal contexts of inheritance.[34] But the actual impetus came from games of chance.

In the seventeenth century, then, gaming and gambling pervaded the soldiery and thereupon because diffused through society at large. Having first drawn the notice of the upholders of discipline and good order in the military services, games of chance soon attracted the attention of philosophers and moralists, and then, finally, that of the mathematicians. In the order of historical causes and effects, the mathematical calculus of probability roots in the gambling mania of the soldiers of the Thirty Years' War.

An instructive symbolism is at work in this conjunction of events. Indulgence in gaming and gambling paradigmatically illustrate the irrational side of human nature, since placing one's hard-gained resources at risk on the throw of a die or the draw of a card may be viscerally thrilling but deliberately hands one's fate over to circumstances wholly beyond one's control. The great faith of 17th Century philosophers in the power of reason to illumine and improve the condition of man — present throughout such otherwise varied thinkers as Hobbes, Descartes, Spinoza and Leibniz — is symbolically validated in the creation of a "calculus of chance" that manifests the capacity of reason to establish a base in the very heartland its most powerful enemy, the sphere of chance and random contingency. Its philosophical contemporaries saw the development of a calculus of probability as a powerfully encouraging sign of the capacity of human reason to master the vagaries of uncontrollable circumstance. Alas, three centuries of its development have not altogether validated this optimistic vision.

NOTES

1 Ian Hacking, *The Emergence of Probability* (Cambridge, 1975), p. 1.

[2] The ancients distinguished between *fortuna* and *fatum*, the latter of which, necessity, acts according to fixed laws, while the former, luck, operates by accident and chance.
[3] On Gataker see the article in the DNB, Vol. VII, pp. 939—940. See also *Encyclopaedia Britannica*, 11th ed., (London, 1910—11), Vol. XI, p. 527. The posthumous *Adversaria miscellanea* (London, 1659) edited by his son Charles was prefixed by his autobiography (in Latin).
[4] *Encyclopaedia Britannica, loc. cit.*
[5] In writing this tract, Gataker seems to have been motivated by opposition to the treatise of James Balmford (b. 1556), *A Short and Plain Dialogue concerning the unlawfulness of Playing at Cards, or Tables, or any other Game consisting in Chance* (London, 1593). And this writer was moved to an eventual reply to Gataker in his tract, *A Modest Reply to certain Answers which Mr. Gataker . . . in his treatise . . . gave to Arguments in A Dialogue concerning the unlawfulness of Playing at Cards, or Tables, or any other Game consisting in Chance* (London, 1623).
[6] Quoted in John Ashton, *The History of Gambling in England* (London, 1898; reprinted in Montclair, N.J. in 1969), pp. 224—225. Licenses for lotteries were granted for improving the supply of water to London in 1629 and 1631. On the Continent, lotteries had already been popular for a century — especially in Netherlands. See Simon Schama, *The Embarrassment of Riches* (New York, 1987), pp. 306—311, or for a fuller account G. A. Fokker, *Geschidenis der Loterijen en Nederland* (Amsterdam, 1862).
[7] *Matthew*, 27:35; *Mark*, 15:24; *John*, 19:23—24.
[8] This "sortilege" had long been condemed by Catholic theologians as a relic of paganism but was difficult to eradicate. It long continued among Protestants — Wesleyans in particular. John Wesley himself sought guidance in this ways as to whether or not he should marry, as is recorded in his journal for March 4, 1737:

Having both of us [Mr. Delamotte and himself] sought God by deep consideration, fasting and prayer, in the afternoon we conferred together but could not come to any decision. We both apprehended Mr. Ingham's objection to be the strongest, the doubt whether she was what she appeared. But this doubt was too hard for us to solve. At length we agreed to appeal to the Searcher of Hearts. I accordingly made three lots. In one was writ, "Marry": in the second "Think not of it the year". After we had prayed to God to "give a perfect lot," Mr. Delamotte drew the third, in which were the words, "Think of it no more". Instead of the agony I had reason to expect I was enabled to say cheerfully "Thy will be done". We cast lots again and know whether I ought to converse with her any more, and the direction I received from God was "Only in the presence of Mr. Delamotte". (Quoted from F. M. David, *Games, Gods and Gambling* [New York, 1962], p. 14.)

[9] See the references cited in note 3 above.
[10] *Proverbs*, 18:18.
[11] See the *New York Times* front-page report of August 29, 1957.
[12] *"haruspicina, quam ego rei publicae causa communisque religionis colendam censeo"* (*De divinatione*, II, xii, 28). And again "*retinetur autem et ad opionem vulgi et ad magnas utilitates rei publicae mos, religio, disciplina, ius, augurium collegi auctoritas*" (*ibid.*, II, xxxiii, 70).
[13] Compare Chapter Six, "The Metaphysics of Gambling," in Franz Rosenthal, *Gambling in Islam* (Leiden, 1975).

[14] "The lot is cast into the lap; but the whole disposing therefore is of the Lord" (*Proverbs*, 16:33).
[15] For a brief account of Gracián see the article by Neil McInnes in *The Encyclopedia of Philosophy*, vol. 3 (New York, 1967), pp. 375—76. For a fuller treatment see Alan Bell, *Balthasar Gracián* (Oxford, 1921).
[16] *The Economist*, August 29, 1987; p. 49. At the present writing, the largest-ever lottery was one held in Spain, with a prize in excess of $100 million.
[17] For Pascal's role in the origination of probability theory see pp. 57—72 of Ian Hacking, *The Emergence of Probability, op. cit.*
[18] H. F. Stewart, "Blaise Pascal", *Proceedings of the British Academy*, vol. 28 (1942), pp. 196—215 (see p. 204).
[19] Quoted from John Warrington's translation of the Pensées (London and New York: Everyman's Library, 1960); sect. 343 (Lafuma) = sect. 233 (Brunschvicg). For further details regarding the Wager argument see also Nicholas Rescher, *Pascal's Wager* (Notre Dame, 1985).
[20] "*Trois discours sur la condition des grands*", premier discours. The ruminations of this opuscule on the role of chance in human affairs are reminiscent of Gracián.
[21] Throughout the Western philosophical tradition, chance (*tuchē, casus, Zufall*) has been *defined* as an accidental concurrence of independent eventuations which, as such, is inherently unpredictable and exempt from any mode of lawful regularity.
[22] "Principles of Nature and of Grace", Sect. 7; tr. in L. E. Loemker, *G. W. Leibniz: Philosophical Papers and Letters*, 2nd. ed. (Dordrect, 1969), p. 639.
[23] G. W. Leibniz, *Theodicée*, Appendix ("Abridgment of the Controversy Reduced to Formal Arguments"), *ad fin.*; tr. by E. M. Huggard, *Theodicy* (New Haven, 1952), pp. 387—388.
[24] Loemker, *op. cit.*, p. 687.
[25] *Theodicée*, sect. 303 (tr. Huggard, p. 310).
[26] As J. M. Keynes noted (*A Treatise on Probability* [London, 1921], p. 311n.), the concept of mathematical expectation was first clearly articulated in Leibniz's 1678 essay *De incerti aestimatione* (Couturat, *Opuscules et fragments inédits de Leibniz* (Paris, 1903), pp. 569—571. But here Leibniz followed the precedent of Pascal. But here the measurement of probability as a ratio of "equally possible" cases (i.e., the "Laplacean" *principle of indifference*) was his invention. (See this same essay.) For a discussion of Leibniz's work on probability, see also Hacking, *The Emergence of Probability, op. cit.*, ch's 14—15 and *passim*.
[27] Adapted from Marcelin Defourneaux, *Spanien im Goldenen Zeitalter* (Stuttgart, 1986), pp. 239—240. For Contreras see A. Morel-Fatio, "Soldats espagnols du 17e siècle: Alonso de Contreras, *Bulletin hispanique*, vol. 3 (1911), pp. 135—58. His biography was first published in the *Boletin de la Réal Academía de la Historía*, vol. 37 (1900), pp. 129—270.
[28] Compare M. Defourneaux, *op. cit.*, p. 240. Regulations and orders of all sorts pervade the military scene of the day — to little practical effects as their very proliferation attests.
[29] Adapted from Charles William Heckthorn, *The Gambling World* (London, 1898), p. 95.
[30] In the Netherlands, gambling was intimately connected with the rise of capitalism. Speaking of the Amsterdam bourse, built in 1608, a modern scholar reports that: "Like

so much else in Amsterdam, both its architecture and its commercial practices were transplanted from Antwerp. The Flemish city in its day had been famous for its addiction to gambling, and in this too, its Dutch stepdaughter followed suit. Wagers were made on every conceivable opportunity, from the outcome of a siege to the sex of an impending baby. They were made on the street, in taverns, at home, barges. . . . The line between casual betting and organized trading in stock was often blurred". (Simon Schama, *The Embarrassment of Riches* (New York, 1987), p. 347.) The gambling mania of the earlier part of the century set the stage for the great bursts of speculative fever that were to follow: the tulip mania, and the Mississippi and South Sea bubbles.

[31] "Gaming and Wagering", *Encyclopaedia Britannica*, 11th ed., Vol. XI (Cambridge, 1910), pp. 446—450.

[32] *The Diary of Samuel Pepys*, ed. by R. Latham and W. Matthews, Vol. IX (Berkeley and Los Angeles: University of California Press, 1976), entry for 1 January 1668.

[33] Quoted from C. W. Heckthorn, *op. cit.*, pp. 62—63.

[34] For example, medieval Arabic scholars nibbled around the fringes of the issue in their treatment of the problem of legacies. Cf. Solomon Gandz, "The Algebra of Inheritance", *Osiris*, vol. 5 (1938), pp. 319—391. And medieval Jewish scholars addressed the question of whether meat found in the streets with x butcher shops only y of which were Kosher could be treated as such. Many such problems are canvassed in Nachum L. Rabinovitch, *Probabilities and Statistical Inference in Ancient and Medieval Jewish Literature* (Toronto, 1973), whose treatment of the matter rightly indicates that yet another, very different route to a calculus of probability was in principle available.

GERD BUCHDAHL

REDUCTIVE REALISM AND THE PROBLEM OF AFFECTION IN KANT

I

The problem alluded to in the title of this essay is one that has plagued Kant exegesis since the appearance of the first *Critique*. As we shall see, an attempt to come to grips with this problem involves a fresh presentation of the over-all structure of Kant's philosophical approach, of Kant's basic imagery; involving in particular a novel appraisal of the significance of two concepts that lie at the core of the *Critique Pure Reason*, and which have always been felt to stand in need of further explication: the concept of the thing-in-itself and that of the transcendental object.

We can see at once how even at first sight the problem posed by Kant's use of the notion of affection connects with these two concepts. Right at the start of the section on the Transcendental Aesthetic Kant writes:

> Intuition is only possible ... in so far as the mind is affected in a certain way. The capacity (receptivity) for receiving representations through the mode in which we are affected by objects, is entitled sensibility .... The action of an object upon the faculty of representation, so far as we are affected by it, is sensation. (A19—20/B33—34)[1]

And there are many other such passages. Thus, in the exposition of the concept of space, we read that "the receptivity of the subject, its capacity to be affected by objects, must necessarily precede all intuitions of these objects" (A26/B42). Or again, at A44/B61, Kant writes that our "representation" of a thing is not that of "an object in itself" but is "merely the appearance of something, and the mode in which we are affected by that something". A whole exegetical tradition has evolved from passages such as these and others like it. Roughly, Kant seems to be saying that the only things of which we can have knowledge are appearances which, as appearances, presuppose the affection of sensibility by something else. And it looks, moreover, that the 'objects' which — at least, in the passages just cited — are said to act on sensibility cannot themselves be appearances, since the latter are after

J. R. Brown and J. Mittelstrass (eds.), An Intimate Relation, 221—241.

all supposed to be the *result* of affection, and cannot thus at the same time be their *cause*; although it should already be noted here that this may not be the only interpretation to be put onto such passages. At any rate, what are these 'objects' which here seem to be posited as source of affection? The most frequent answer given to this question has been that the objects referred to must be the 'things-in-themselves'. And there are passages that seem to underwrite such an interpretation explicitly. Thus, in the course of his discussion of the Second Analogy Kant remarks:

> How things may be in themselves, irrespective of the representations through which they affect us, is entirely outside our sphere of knowledge. (A190/B235)

And in the *Prolegomena* "appearances" are explicitly stated to be "representations" effected by things in themselves "when they affect our senses".[2] Evidently there seems to be a clear implication here that it is things-in-themselves that affect us or sensibility; although, as again we shall see below, alternative interpretations of such passages are possible. At any rate, the distinction between things as they appear and as they are in themselves, underlying the former, is explicitly affirmed in Kant's Preface to the Second Edition of CPR, where he inveighs against "the absurd conclusion" that there could be an "appearance without something that here appears" (Bxxvii); a contention which again lends support to the idea that 'experience' involves being affected by things in themselves.

An alternative suggestion has been that it is the 'transcendental object' which is the cause of the affection. And at first sight, Kant seems again himself to be the source of such a view. Thus at A372, where he in fact criticises the idea that things in themselves "produce the appearances", he remarks that "we can indeed admit that something, which may be (in the transcendental sense) outside us, is the cause of our outer intuitions . . ."; where this 'something' a little later is labelled "the transcendental object". We shall below try to explicate this last contention. For the moment we note that the expression 'outside us' has — as Kant also immediately notes — a double sense; firstly, as "signifying what as thing in itself exists apart from us"; secondly, as "what belongs solely to outer appearance" (A373).

Passages like these have moreover suggested that we meet in Kant with two kinds of 'affection', corresponding to the two senses of 'outside us' just noted: firstly, an affection by the transcendental object or thing in itself; secondly, an affection by the thing or things qua appearance. In the literature this has come to be knwon as 'the theory of double affection', made prominent above all by the work of Erich Adickes.[3] In the summary of this argument given by R. P. Wolff[4], itself following the account of T. D. Weldon[5], we find that

> First the independent things-in-themselves affect the transcendental self; as a result of this affection, and in conformity with its inherent forms of intuition and conception, the transcendental self synthesizes the world of empirical objects, among which is the empirical self; in this spatio-temporal world certain physical objects interact with the body of the empirical self and thereby affect the self sensibly . . . by interpreting its perceptions as representations. (Wolff 1963, p. 170)

Now whilst the second form of affection does not (at least in the present context) pose any problems, the difficulties surrounding the first, affection by things-in-themselves, are immense, having been found more or less intractable since the time of Kant. At any rate, to mention just one relatively recent criticism, by Strawson[6], the latter argues that the "transposition of the terminology of objects 'affecting' the constitution of subjects takes that terminology altogether out of the range of its intelligible employment", if for no other reason that it is supposed to operate outside the context of space and time, which on Kant's view are after all themselves the *result* of (the primary) affection. So Strawson concludes that for this reason "we can no longer understand the doctrine, for we no longer know what 'affecting' means, or what we are to understand by 'ourselves'". And one might add that the whole idea of things-in-themselves standing in a *causal* relation to appearances is inconsistent with Kantian doctrine according to which causal action is valid only in so far as it holds *between appearances*.

There is however an even more weighty objection to this account, involving as it does the positing of quasi-*hypothetical* objects as the causes of appearances — an objection moreover which, it should be noted, is already to be found in the A372-passage of CPR alluded to before; indicating that the above interpretation can hardly represent Kant's views consistently. As he puts it, such a position raises "problems in regard to which the human mind is quite at a loss how to proceed". And he continues:

> For if we regard outer appearances as representations produced in us by their objects, and if these objects be things existing in themselves outside us, it is indeed impossible to

see how we can come to know the existence of the objects otherwise than by inference from the effect to the cause; and this being so, it must always remain doubtful whether the cause in question be in us or outside us. (A372)

Interestingly, this passage continues however by contrasting "things existing in themselves outside us" with "the transcendental object", as "the cause of our outer intuition" (*ibid.*); a position which — as already noted — Kant, at first sight surprisingly, appears to sanction, as something quite permissible; even though he there adds that this transcendental object "is equally unknown to inner and to outer intuition"; unknowable — we may add, and as will be explained in more detail below — in principle. But, so one might object, if 'things-in-themselves' are unknowable, why allow transcendental objects as the cause of intuition? (Clearly, the transcendental object is not always identical with the thing in itself! But more of that anon, when we shall try to unravel Kant's position with regard to the transcendental object.)

At any rate, all such objections are so obvious that it is likely that the structure of the Kantian argument has here been misconstrued, and that his whole philosophical imagery has to be recast in a different way. As a preliminary, it should be noted that the verbal formulation of 'objects acting on sensibility' is one which was subsequently rejected quite explicitly by Kant himself. In a letter to Beck, January 20, 1792, he writes:

Perhaps right at the outset you might avoid defining "sensibility" in terms of "receptivity", that is, [if this implies] the manner of representations in the subject insofar as he is affected by objects; perhaps you can locate it in that which, in a cognition, concerns merely the relation of the representation to the subject, so that the form of sensibility, in this relation to the object of intuition, makes knowable no more than the appearance of this object. That this subjective thing constitutes only the manner in which the subject is *affected by representations* [my italics], and consequently nothing more than the receptivity of the subject, is already implied by its being merely the determination of the subject.[7]

This looks like a locus classicus, with Kant stating that the 'objects' by which we are 'affected' have themselves only the status of representations, converting the 'affection-doctrine' into a purely phenomenological exercise characterising 'experience' — cf. the use of the term "subjective" in this passage; an insistence that we are always confined here to the 'experiential' or 'cognitive' context — as being in terms of 'representations'; thus coming close to the Humean doctrine of objects as mere 'sensations or ideas'. It should be noted, besides, that in many of the passages, such as those cited above, where Kant contrasts things in themselves with the representations which they effect in us, it is never clear whether the word 'they' refers to 'things in themselves' or to 'things', simpliciter; e.g. (perhaps) representations; leaving the status of these 'things' quite open. However, such a somewhat Humean interpretation cannot be our final conclusion either. For one thing, Kant never re-wrote the passages of the *Critique* where the older affection doctrine is invoked. Indeed, in the Second Edition, B72, he still speaks of "our mode of intuition" being "dependent upon the existence of the object" and being possible therefore "only if the subject's faculty of representation is affected by that object"; as contrasted with 'divine intuition' which "originates" the existence of the object. And similarly in the *Prolegomena* passage already referred to above.

Of course, all such passages may be read in a number of different ways; some identical with the Adickes view; others being interpreted in terms of the position sketched in the letter to Beck. Besides, there is even in the text of CPR itself a passage which already presages the position taken in the letter to Beck; for at A494/B522 Kant actually writes:

> The faculty of sensible intuition is strictly only a receptivity, a capacity of being affected in a certain manner with representations [mit Vorstellungen affiziert zu werden], ...

This already shows that terms like 'affizieren' and 'Affizierung' (to affect, affection) have themselves to be interpreted with extreme care. Thus it is perhaps worth noting that whilst the verb 'to affect' in English almost inevitably carries with it certain causal overtones, suggesting a reference to something 'external', *by* which one is affected, the German use is caught more properly in a locution like 'The thought of the death of my friend has touched me deeply [hat mich sehr affiziert]' — with a purely 'internal' or 'subjective' association.

There are certainly in Kant a number of other passages which suggest a far more 'internal' or 'subjective' grammar attaching to the term 'affizieren', than might seem to be the case at first sight. Thus B154 speaks of "inner sense" being "affected by" the "transcendental synthesis of imagination", as though we were here dealing with a form of 'self-affection'. Similarly on the next page: "Thus the understanding does not find in inner sense such a combination of the manifold, but *produces* it, in that it *affects* that sense". And still more explicitly on the following page, where the reference is to 'inner' and 'outer sense', we read:

> Since with respect to outer sense we cognize objects only in so far as we are externally affected [äußerlich affiziert], we must also admit, with respect to inner sense, that through it we intuit ourselves as we are affected internally by ourselves; i.e. as regards inner intuition our own subject is cognised only as appearance, but not in respect of what it is in itself.

Still more extreme, we have a case where Kant speaks of a "proposition" being "affected [affiziert] by the condition of time" (A152/B191).

In quoting these passages I am not so much concerned with the particular positions that we meet with here, but rather, with indicating something of what we may call 'the semantic imagery' that attaches to terms like 'affizieren, Affizierung'. As I shall argue, the intention that lies behind their use is primarily one of seeking to give what we may term a quasi-"subjective" interpretation (in the terminology of the letter to Beck) of experience as such, and thus in isolation from any extra-experiential factors such as the thing in itself; at least, in any 'realist' sense of that term.

At any rate, it will be clear that Kant's at first sight apparently confusing and contradictory positions with respect to 'affection' (as implied in the above quotations) make it necessary to give — hopefully in a somewhat novel way — a more general survey of the structure of his philosophy as regards the status and significance possessed by some of its central theoretical terms: not only of 'affection' but also of 'thing in itself', 'appearance', 'transcendental object', and so on.

II

We have noted how in some of the above passages Kant himself already gets away from the idea of 'affection' implying a relation of experience to some extra-experiential causal factor such as things in themselves. To reinforce this, we will now seek to counter more explicitly the image, often associated with Kant, of 'two worlds': the world of appearance, and the world of things in themselves; especially when the latter are viewed as located in a separate kind of existential domain, either beyond the world of appearance (like some extra-galactic nebula), or alternatively, underlying that world in the way in which the hidden and

Fig. 1. *Kant's transcendental structure.*

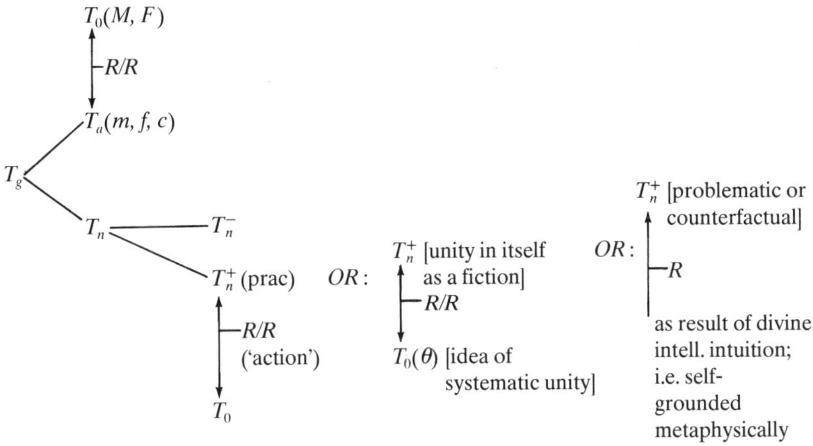

$T_s = T_n^+$ in its various interpretations; T_n^- ; T_0

Glossary

T_g	=	thing in general [Ding überhaupt]
T_0	=	transcendental object [$T_0(M, F)$ = 'matter' and 'form']
$T_0(\theta)$	=	transcendental object of the idea of systematic unity
T_a	=	thing or object as appearance [$T_a(m, f, c)$ = matter, spatio-temporal forms, categories]
T_s	=	thing in itself
T_n	=	thing as noumenon; T_n^- in 'negative sense'; T_n^+ = in positive sense T_n^+ (prac) = noumenon in moral context
R/R	=	realization and reduction, respectively
t_i	=	transcendental in 'immanent' sense
t_0	=	transcendental in 'idling' sense [categories not employed]
t_m	=	transcendental in 'dogmatic' or 'real', metaphysical sense

unobservable domain of atomic structure underlies the obvserved phenomena – an image frequently employed, especially by early 19th century readers of Kant. I want in a very definite way to replace this image by that of *one world*; or speaking more formally, of 'one object in general' (symbolically: T_g); with the latter being viewed by Kant under a number of different interpretations: as 'appearance' (T_a), in the context of 'experience', or, as we may term it, of some 'cognitive occasion', involving 'sensation' and 'thought'; alternatively, viewed as an 'object of thought' alone, or 'noumenon'; as defined by Kant in the chapter "Of the Ground of the Distinction of all Objects in general into Phenomena and Noumena" (A235/B294ff.) In thus concentrating on T_g, and the cognitive apparatus used to define its various interpretations, one of our objectives is to avoid from the start the imagery of a number of objects and subjects, possibly standing in various relations (such as causal relations) to one another.

In the diagram here appended (Figure 1) I have attempted to indicate in outline, though as far as possible to a fairly full extent, the details of the transcendental structure that in Kant's presentation evolves from the variety of interpretations which T_g receives from his hands. Only when the jungle of complexities and the rich detail of Kant's arguments are viewed in such a structural outline, is it possible to grasp the significance of the various terms involved; and only then, as we shall see, is it possible to understand the place of the notion of 'affection' in the Kantian usage.

As will be clear from our diagram, we start with the division of 'things in general' into 'phenomena' and 'noumena', central to Kant's whole account; in particular, we have indicated the various versions in which occurs the notion of the "noumenon in a positive sense" (B307–8); i.e., in a 'practical' or moral context (A538–9/B566–7); alternatively, within Kant's account of scientific theorising (A664/B692 – A675/B703); finally, as a counterfactual notion, when viewed as the result of intellectual intuition (e.g. at B72; A256/B312, etc.). And it will be fairly clear already, that the various versions of the thing in itself (T_s) when interpreted in its 'positive sense' (T_n^+) make it a most unlikely candidate for functioning as a 'source' or 'cause' of 'affection'.

However, perhaps the transcendental object is a more promising source, keeping in mind particularly that Kant himself, in some of the passages quoted already, seems to refer to it in this way; especially at A494/B522, where he speaks of "the transcendental object" as "the intelligible cause of appearances in general"; or again, at A538/B566, where "appearances" are said to have their "ground" in "the transcendental object". Besides, as our diagram indicates, T_0 seems to lie at the basis of most of the positive interpretations of T_g that we find in Kant: both of T_a and of the various uses of T_n^+. Let us begin, then, by first giving a few details concerning the relationship between T_0 and these 'interpretations'.

Consider the relationship between $T_0(M, F)$ and $T_a(m, f, c)$ – the significance of these terms will become clear as we proceed. I have labelled this relation to be one of 'realization' and 'reduction', respectively.[8] A locus classicus for Kant's use of the term 'realization' we meet with at A146–7/B186–7, where he says that "sensibility realizes" the understanding, and in particular, the categories (so far considered as purely logical notions), thereby yielding the "schemata" which correspond to the object qua 'appearance'. Prior to realization, the object has the status of something "merely transcendental" (A247/B304); symbolised by our T_0; involving the use of the term 'transcendental' which in our diagram I have labelled to; what we may term its 'idling sense'. Prior to realization, as Kant explains at A248/B305, the categories "have only transcendental [t_o] meaning" but no 'transcendental [t_i] employment'. In other words, it is the 'engaging' of the transcendental apparatus, in particular: of sensibility, which yields the 'transcendental' status (in what I have labelled the 't_i sense') of the categories, leading to what Kant elsewhere describes as their "immanent employment" (A327/B383); corresponding (in the diagram) to the realizational movement from $T_0(F)$ – where the "object in general" is only "thought by he categories" in their purely logical sense; cf. A254/B309 – to $T_a(c)$; with $T_a(c)$ denoting the 'appearances', characterised by the schematised set of categories.

Let us turn to some of the details of the realizational movement from $T_0(M)$ to $T_a(m, f)$; i.e. from the 'reduced' to the 'realized' component of the object, by placing $T_0(M)$ – denoting as will be explained presently, the "transcendental matter" – in an experiential context, corresponding to $T_a(m)$, or rather: $T_a(m, f)$. The latter stands for what Kant describes as "the matter" and "form of the appearance" (e.g. at A20/B34); where, as he there adds, the "matter in the appearance . . . corresponds to sensation", whilst its 'form' signifies the spatial and temporal "relations", viewed purely as "modes of sensibility". Moreover, just as the spatio-temporal forms are a function of the realization of $T_0(M)$ as $T_a(m, f)$, so the 'matter' itself depends for its qualitative characteristics on "the subjective constitution of the manner of sensibility" (B44); examples of the 'material' component (i.e. sensation or intuition) given by Kant being "impenetrability, hardness, colour" (A20/B35), as well as "sounds, and heat" (B44).

The movement in the reverse direction, e.g. from T_a to T_0, I have called 'reduction'; using this Husserlian term in the way in which it already occurs in one of Kant's *Reflexionen*, where he writes that in order to arrive at the noumenon we must "suspend everything sensory" so as to "free reality as far as possible via reduction from that which belongs to it qua phenomenon".[9]

Generalising all this, our diagram indicates a number of such realizational or reductive movements, respectively, the details of which we need not discuss here, confining ourselves instead to just two or three cases. One or the briefest and most succinct references in Kant, illustrating the movement from T_0 to T_a, occurs at A251, where he writes:

The categories represent no special object, given to the understanding alone, but only serve to determine the transcendental object (the concept of something in general) through that which is given in sensibility, in order thereby to cognise appearances empirically under concepts of objects.

This particular text illustrates our notion of 'realization' with such perfect clarity, it needs not further comments.

An instance of the movement from T_0 to T_n^+ (prac) we meet with in the *Critique of Practical Reason*, where Kant says that whilst in the context of 'speculative reason' there is no possibility for 'causality of freedom' to fill the "vacant place, *viz.* the intelligible" (clearly equivalent to our T_0 (F)), it is possible to "realize this thought" ["diesen Gedanken . . . realisieren"] in the context of practical reason;[10] with the "subject", acting in accordance with the causality of freedom, having the "character of a thing in itself" (as Kant formulates this in CPR, at A539/B567). Indeed, this latter passage supplies some further details concerning the realizational process with respect to the case of ethics in a form which graphically illustrates our account of the general structure of Kant's transcendental approach. He writes:

For since they [sc. appearances] are not things in themselves, they must be grounded in [zu Grunde liegen] a transcendental object which characterises [bestimmt] them as mere representations; and consequently there is nothing to prevent us from ascribing to this transcendental object, besides the property [causality] through which it appears, a causality which is not appearance . . . [sc. causality of freedom]. (A538/B566f.)

As an aside, it should be noted that the reference to 'things in themselves' as *not* being grounded in T_0 evidently implies the use of the term 'thing in itself' in the 'problematic', or really: counterfactual sense, as indicated in our diagram; contrasted with the use of the same term in the 'practical' context (T_n^+ (prac)), where, by contrast, T_s *is* 'grounded' in T_0, as per A539/B567. I mention these details, in order to suggest how the structure indicated in our diagram can help to clarify some of Kant's more puzzling statements here that have frequently hardly been understood at all.

But to proceed: take a second instance of realization, that of the idea of systematic unity (e.g. of a scientific system like that of Newton), which on one occasion Kant proposes to treat, for heuristic reasons — the details need not detain us here — as a "real object", subsisting "in itself".[11] At A677/B705 Kant writes that in order to "approximate to the highest possible degree of empirical unity, I shall not only be entitled, but shall also be constrained, to realize this idea [diese Idee zu realisieren], i.e. to posit for it a real object", although "only as a something which I do not at all know in itself"; the emphasis here being on 'know'; i.e. the 'object' does have the logical character of an 'in itself', even though we have no knowledge of this. And so Kant contends that this object can only be "thought on the analogy with a real substance" (675/B703); it has the status only of an "as if", of a fiction (e.g. A686/B714, and elsewhere).

Finally, the case of what I have termed the 'problematic or counterfactual realization', which functions as a sort of logical myth for the 'genuine' case of a thing as 'thing in itself', viewed as a quasi-selfsubsisting entity, not 'grounded', as we have already noted, in the transcendental object, unlike T_a, or T_n^+ (prac). And to remind ourselves again, we have seen that at A372 Kant has already rejected this case of the 'thing in itself' as a possible candidate for 'affection'; explicitly replacing it by a reference to the transcendental object as "cause [sc. ground] of our outer intuitions". The main point to register for the moment is that Kant's concept of the world must evidently be seen as operating entirely within the confines of the realization-reduction framework alone. From which also follows that it is within the boundaries of this framework alone that any interpretation of 'affection', and of 'cause of affection', will have to be placed.

For the moment, however, with these instances of realization-reduction in mind, let us consider in slightly more detail the significance of $T_a(m, f, c)$ and T_0 (M, F), by discussing this in terms of the 'reduction' of the former to the latter, so as to get a clearer idea of Kant's construal of the transcendental object. What remains of the object qua appearance under reduction? We have symbolised this as $T_0(M, F)$, with

$T_0(F)$ denoting the categories with merely "transcendental meaning" (B305); i.e. in the to sense of this term. What of the component $T_0(M)$? We have already noted that according to Kant $T_a(m)$ — the "matter in the appearance" — "corresponds to sensation". What then corresponds to 'sensation' under reduction, leading to T_0 (M)? (Remembering that under reduction all 'qualitative' aspects (the so-called 'secondary qualities') disappear likewise (cf. B44, quoted above)). Kant gives a reply in a passage which — at least in this connection — has seldom received the attention it deserves. In the chapter on the Schematism he writes:

> Reality, in the pure concept of the understanding [i.e. at the level of reduction], is that which [at the realizational stage] corresponds to a sensation in general; that therefore, whose concept points to a being (in time) [i.e. again at the level of realization] . . . [By contrast] that in the objects [qua appearance] which corresponds to sensation, is [at the level of reduction] the transcendental matter of all objects as things in themselves (Sachheit, reality). [Obviously our $T_0(M)$] (A143/B182)

Evidently this passage registers a correspondence between $T_a(m)$ and $T_0(M)$, provided we interpret — as we obviously must — Kant's reference to 'things in themselves' as being to T_0. The difficult term here is the German word 'Sachheit'; Kemp Smith translates it somewhat misleadingly as 'thinghood'. It denotes something like 'the (mere) concept of something being the case', without our being able to express either the qualitative or spatio-temporal aspects in which this 'something' occurs or 'is the case'. At the realizational level we characterise an object as occurring in such and such qualitative and spatio-temporal modes; and we say *that* it occurs in these modes. 'Sachheit' denotes what we may describe as 'the thatness' (at the reductive level) of the 'that' at the realizational stage; a 'thatness' which Kant describes as something falling under the categorial notion of 'reality', as a "pure concept of the understanding". However, what is absolutely essential here is to appreciate that under reduction the 'that' of 'being the case' is not converted into 'nothing'; rather, as Kant formulates this, there is 'reality', albeit entertained as something with purely conceptual import — "in the pure concept of the understanding [im reinen Verstandesbegriffe] (A143/B182), as the 'Sachheit' of those 'Sachen' (i.e. those states of affairs) which are verbalised as obtaining under realization, in the experiential context, i.e. where the concept of 'reality' acquires a transcendental function, in the t_i sense of this term.

The central image which is now beginning to emerge is clearly this:

the objects of Kant's world must always be interpreted as operating within the transcendental framework, involving the reductive and realizational 'stations', respectively; i.e. as operating between T_0 and T_a. It follows that T_a is always correlative with T_0. In this connection there is an interesting statement of such a position to be found in the course of the argument of the Second Analogy. Kant writes:

> Now as soon as I tighten [steigere] my concepts of an object up to a transcendental meaning, [an object such as] the house is no longer a thing in itself but only an appearance, i.e. a representation whose transcendental object is unknown [unbekannt] . . . (A191/B236).

The correlativeness of T_a and T_0 is here explicit. Notice that T_0 is said to be — both in this passage and elsewhere — unknown [unbekannt]; Kant in many other places speaks of T_0 as an 'x' (e.g. at A104; A109; A250, etc.); in German mathematical jargon called 'die Unbekannte'; in English: x, the unknown. In other words, T_0 has as-it-were the status of a mathematical variable, still to be given a 'value' (in our terminology: via realization). Now formally, for Kant the function of the concept of an object is to add connectedness or "unity" to the "manifold of our representations" or intuitions (cf. e.g. A105). As he puts it at A109:

> Appearances are not things in themselves; they are only representations, which in turn have their object — an object which cannot itself be intuited by us, and which may, therefore, be named the non-empirical, that is, transcendental object $= x$.

In what follows Kant at once explains that this transcendental object which "confers unity" upon our representations is equivalent to, and thus corresponds to, the "necessary unity of consciousness", or the "unity of apperception"; an equivalence or 'correspondence' which is also asserted explicitly at A250, where Kant, speaking of "the transcendental object", says that the latter is the "correlate of the unity of apperception"; a classical locus for illustrating the kind of 'correspondence' that we can now appreciate as universally holding between the 'reductive' and 'realizational' levels; here: between $T_0(F)$ and the unity of apperception.

The central point in all this is that the transcendental object, far from being something 'additional' to objects qua appearances, let alone functioning like some mysterious extra object as 'external cause' of the appearances, basically stands to the latter in a relation of 'correspondence'. To take the instance just given, it is as though Kant was saying that $T_0(F)$ in an experiential context 'boils down to' the unity of apperception, the latter being its 'true' explication. However, and even more importantly for our own purpose here, viz., of gaining insight into the status of 'affection' in Kant's position, this 'correspondence' covers not only the 'formal' side of the argument, i.e. not only the understanding whose categories are related via the unity of consciousness, but extends also to the 'materal' side of experience and its objects, dependent on sensibility; a correspondence which we have already met with in the passage from A143/B182, mentioned above. Thus, at A494/B522, Kant makes the revealing remark that whilst "the non-sensible cause of the representations is completely unknown to us" — reiterating the position we have already found him to take at A273 with respect to 'things in themselves (in the sense of our T_n^+) — he goes on to say that, we may, however,

entitle the purely intelligible cause [sc. ground] of appearances in general the transcendental object, but merely in order to have something corresponding to sensibility [viewed] as receptivity.

So, just as before $T_0(F)$ was said to correspond to the 'spontaneity' of the unity of apperception in relation to the understanding, now — as at A143/B182, except with the additional reference to sensibility — $T_0(M)$ is said to 'correspond' to the receptivity of sensibility, and its associated 'sensation', the 'matter in the appearance'.

III

What has become clear from the above is that the Kantian world is entirely 'enclosed' within its transcendental framework: the confines of the understanding and sensibility, of thought and intuition, at one or the other of our two 'corresponding' levels, reduction and realization, respectively. Thus, take the notion of 'Sachheit', already met with above, denoting the 'material' aspect of T_0, i.e. ($T_0(M)$). Now this notion occurs also in the chapter on the Ideal of Reason, where it is, however, first placed in a yet more general — what we might call 'pre-critical' — context; where Kant distinguishes between merely "logical affirmation" on the one hand and what he there calls "transcendental affirmation" (A574/B602); which may be "entitled", he goes on to say, "reality (Sachheit)", expressing "the realities which contain the data, and, so to speak, the material or transcendental content, for the possibility and complete determination of all things." However, as the argument then develops, it becomes clear that 'transcendental', in this context, is meant to denote the pre-critical, quasi-'metaphysical' version of 'transcendental' (in our diagram symbolised as t_m). Now, as Kant explains in what follows, according to his own philosophical position this "reality" or "matter" has to be re-interpreted as occurring solely "within [the field of] appearance", or of "single and all-embracing experience", and must thus be interpreted as something "given [which] . . . corresponds to sensation". (A581—2/B609—10) We thus see that $T_0(M)$ is so to speak the 'critical echo' of the pre-critical version of the 'real' or 'Sachheit'; a "reality" which only in this last-named version has the character of "a thing in itself" (A576/B604); whilst at the 'critical' stage $T_0(M)$ in its turn 'corresponds' to 'the given', i.e. to "sensation" (A581/B609); such that all the material of Kant's world is thus transferred to the level of $T_0(M)$ and $T_a(m)$, respectively. What is once again evident from all this is that the t_m-version of 'Sachheit', interpreted as the reality of a thing in itself, is on Kant's view quite clearly not available as a source of affection. In the light of all this, how is one to interpret the notion of affection? I think it is becoming clear that this is a reference to an 'ultimate' — in a sense 'uncaused' — element of experience, characterising its essential nature; operating within what Kant in the letter to Beck called the domain of "subjectivity"; naturally in the 'transcendental' (t_o-t_i) sense of that term. For as we have expressed it: viewing the world as 'object', there is just one object, and one object only, under different descriptions; and this, at one of its reductive-realizational levels or other, can hardly be the cause of itself. This is confirmed also by the formulation to which we drew attention earlier, and which occurs in the course of the passage from which we have just quoted, viz. A494/B522, where Kant defines "the faculty of sensible intuition" (i.e. sensibility) as "a receptivity, a capacity of being affected in a certain manner *with* [my italics] representations". Indeed, there is a passage in the 'Phenomena-Noumena' chapter where the notion of 'affection' is explicitly dissociated from any object; with Kant there alluding to "mere intuition [through] which nothing is thought", and then adding that "the fact that this affection of sensibility is in me, does not amount to a relation of such representation to any object" (A253/B309).

Finally, there is a passage where the relevance of an 'object' to affection is left more or less open and ambiguous. At B207 Kant remarks:

> Appearances contain in addition to intuition the materials for some object in general [Object überhaupt] . . . , that is to say, the real of sensation as merely subjective representation, concerning which [latter] one can only be conscious that the subject is affected, and which one associates with [or 'refers to': bezieht auf] an object in general [Objekt überhaupt].

Note that Kant leaves the logical nature of the 'object' referred to in this context at its most 'general'; i.e. the reference being merely to T_g. He certainly does not *say* that such an object functions as 'cause' of affection; at the very least, he leaves this open, with his reference to 'object in general' [Objekt überhaupt]. We may therefore take him here as reiterating simply the position outlined at A105, already discussed above, viz. that the manifold of representations has always to be referred to an object, in the sense of T_0, with the latter 'corresponding to' the unity of the representational manifold, i.e. the unity of apperception, in order to yield the 'empirical object' qua appearance, and thus not functioning in any way as 'cause' of the representations. It is this constant interplay between affection, sensation, object in general, transcendental object, empirical object, which not surprisingly has led to such confusing interpretations of Kant's intentions.

IV

And yet one may still be wanting to ask: what then is the *origin* of these representations *with* which sensibility is affected? In particular, since Kant invariably characterises sensibility as a "receptivity", one is almost inevitably tempted to ask whence sensibility receives its impressions of sensation, to use the Humean term. Nevertheless, this question still betrays the same old misunderstanding of Kant's position as met with so far. We are after all not in any way concerned with the physical or psychological origin of our experiences; to wit, with the external objects qua appearances that cause in us certain experiences, e.g. sensations of light, heat, etc., due, say, to the impact of photons; something which belongs purely to the uncontentious field occupied by the world of empirical subjects and objects, for Kant characterised as belonging to the domain of 'appearance', perhaps in line with the picture of Adickes' second form of 'affection', taking place between 'subject' and 'object', both viewed as 'appearances'.

Instead, as should be clear by now, we are at the start of our analysis — at its 'reductive' stage — confined to the consideration of a single abstract object-in-general; representing what only under realization comes to be converted into a plurality of objects *and subjects*; i.e. when placed within the context of the transcendental apparatus: sensibility, understanding, imagination, syntheses, etc.; from which then devolve the various pluralities of empirical subjects and objects and their mutual interactions, including causal interaction.

At any rate, the general result of all these various considerations must be that the Kantian expressions, 'receptivity of sensibility', the domain of 'the given', 'affection by objects', have to be seen as transcendental characterisations of what I have called the realizational process. For a start, we have to take it that the terminology of 'the receptivity of sensibility' is primarily intended to contrast with 'the spontaneity of the understanding'. It follows that the question as to from whence sensibility 'receives its material' is misplaced, if this is understood as a request for some additional object as the source of what is thus received. Instead, the locution: 'receptivity of sensibility', must be understood as a phenomenological characterisation of cognitive occasions, in the context of which sensibility simply *finds itself* with such and such sensations, as an *ultimate matter of fact*; where it should be remembered that the German for 'sensation' is 'Empfindung' — literally: 'a finding'; or also: 'something found within us'.

It is of the utmost importance to stress the logical nature of the 'ultimacy' of such 'finding'. This can be made clear by reverting to the notion of 'correspondence' which we have noted as central to the Kantian account. Summarising this, we have seen that at the pre-critical (t_m) level, things are characterised by Kant as simply expressing a "Reality (Sachheit)" (A574/B602); which latter, so we saw, at the reductive level (to) corresponds to the "Sachheit, reality" of the object qua $T_0(M)$ (A143/B182). Then, under realization, i.e. moving to the 'corresponding' experiential version, the Sachheit ('thatness') comes to expresses itself as the 'that' of some given sensation; where, as noted before, the 'what' and 'how' of the sensation is of course a function of "the subjective constitution of the manner of sensibility" (B44). It is thus evident that 'Sachheit' is here a reference to the fact that the 'that' of the qualities which sensibility generates is already 'determined'. But — and this is central — it is not determined *by* some additional 'outside source'; rather, the reference to 'Sachheit' is simply another way of *saying* that it is an essential characteristic of the realizational level, i.e.

of experience, that it *finds* itself with such and such sensations, as an ultimate matter of fact; simply *corresponding* to the ultimacy of things at the other two corresponding levels (t_m and t_0).

What is thus essential to grasp, with respect to the logical grammar of the language of 'affection', is that $T_0(M)$ should not be thought of as 'determining' $T_a(m)$, let alone 'producing' it, since both refer to one and the same thing, at the reductive and realizational stages, respectively. In other words, 'Sachheit' ($T_0(M)$) simply *expresses the fact* that realization involves the 'finding' of some contingent, 'given', element (the "matter of appearance"); i.e. $T_0(M)$ *says* that sensibility just *is* contingently determined as such, 'receptive' with, or "affected with" respect to this or that sensation or "representaion" (as we have found Kant to put it); affection being, at the realizational level, the equivalent of the Sachheit at the level of reduction.

It may be of interest to note that this reading echoes the approach taken by Fichte already during Kant's lifetime, except that we have sought to replace the Fichtean somewhat misleading emphasis on the "self", the *Ich*, by the more neutral language of the process of reduction and realization. Fichte describes experience as an "activity" [Tätigkeit], grounded in the self, which itself, as he puts it, "posits a limit" on itself, described as the "not-self". However, he charges Kant with the view that this 'limiting process' [Begrenzung] "must have a ground, [evidently in the sense of 'cause'] and this ground is thus the thing in itself". Had Kant grasped the true nature of this limiting process, Fichte then argues, he (Kant) would have appreciated "that he finds himself just as much, only without knowing it, in the same circle in which we [Fichte] find ourselves with our [account of] cognition."[12] May-be, Kant's letter of 1792 to Beck already reflected an awareness that the potential 'misunderstanding' of his position needed correction.

To formulate Kant's teaching adequately, and summarily, we might say that on his view sensibility is defined by or with respect to the Sachheit of the 'Sachen' (matters of fact) but that this is only another way of saying that the 'material' side of things is something 'given', as an ultimate matter of fact; sensibility being 'limited' with respect to the 'that' — where even this notion of the 'limit' is only a picture. For sensibility is not 'limited' *by* anything else, not even by 'itself' (to echo Fichte's idea). The notion of the 'limit' is here merely a picture for the fact that from among the infinitely possible material data that could come up, only such and such specific ones are given. The 'determining' or 'affecting' of sensibility is only a logical (sc. transcendental) locution, not a physical or factual one for the ultimacy of the 'givenness' of the sensory material.

At any rate, is it now possible to understand Kant's wording, when he speaks of sensibility as "the capacity (receptivity) for receiving representations through the mode in which we are affected by objects" (A19/B33)? Notice that Kant speaks here of 'representations' that are being 'received'. But we can see now that this is merely a quasi-phenomenological description of 'experience' itself, to be read in the way we have above interpreted the passage at B207. It is not a reference to some 'external influence', but instead should be read in the light of Kant's wording at A494/B522, mentioned before, where we saw sensibility being defined as a capacity of being affected *with representations*. But what of the 'objects' *by which* we are said here to be 'affected'? In the light of the foregoing, this can at most be a reference only to the object qua $T_0(M)$, or T_g, and thus be a misleading way of saying — as explained in what has preceded — that sensibility is determined with respect to, i.e. finds itself with, the 'that' of some given sensation, in turn reflecting the 'thatness' (Sachheit) of the object at the reductive level, which expresses the contingency of the 'finding' of the sensation of which experience finds itself appraised.

Of course, Kant is not always explicit about the type of 'object' that is involved in his appraisal of the experiential situation. Thus, in the *Prolegomena* he argues that

> Idealism consists in the assertion that there are none other than thinking beings; the other things which we believe we perceive in intuition are only representations in the thinking beings to which in fact no object outside the latter corresponds. I say on the contrary: things are given to us as objects of our senses situated outside us, but of what they may be in themselves we know nothing; we only know their appearances, i.e. the representations that they effect in us when they affect our senses.[13]

I think there is no need to analyse the cogency of such a passage once again; it obviously needs no more than a translation into the kind of actual position which we have attributed to Kant in what has preceded. However, it is of interest to note that Hume himself did not in fact hold that 'there are none other than thinking beings' and their 'representations'. In the *Treatise* he remarks on the contrary that

> As to those *impressions*, which arise from the *senses*, their ultimate cause is, in my

opinion, perfectly inexplicable by human reason, and 'twill always be impossible to decide with certainty, whether they arise immediately from the object, or are produc'd by the creative power of the mind, or are deriv'd from the author of our being.[14]

I quote this passage in order to indicate the sort of 'twilight' that surrounds any theory concerning the 'cause' of impressions or sensations in this literature. It has often been held that although Hume here seems to pay lipservice to the idea that sensations are causally conditioned, this was only a rhetorical device which in fact is meant to boil down to the kind of position Kant in the above passage attributes to the 'idealists'; i.e. possibly the view that it is 'senseless' to ask for the cause of sensations. What Kant has contributed to this problem is to offer a sophisticated apparatus in terms of which to define the kind of 'object' to which Hume refers in the above passage as the possible 'cause of impressions'. And as we have suggested, the way in which Kant has answered Hume's question goes deep into the very structure of the Kantian philosophical position, and cannot be grasped without an understanding of that structure. If the account given above is correct, we can say that what both Hume and Fichte lack is Kant's two-level approach, which we have defined as the reduction-realization levels of his account of reality. And I think, moreover, that only with such an account is it possible to tread one's way through the complexities of the Kantian story.

NOTES

[1] I have used the translation by Norman Kemp Smith. (1953). *Immanuel Kant's Critique of Pure Reason*, London: Macmillan and Co., abbr. CPR, except where the translation seemed unreliable, when I have provided my own. As usual, A and B refer to the first the second editions of CPR.

[2] Immanuel Kant, *Prolegomena to any Future Metaphysics*, transl. Peter G. Lucas. (1953), Manchester University Press, Para. 13, Note II, p. 45.

[3] Erich Adickes. (1929). *Kants Lehre von der Doppelten Affektion Unseres Ich als Schlussel zu Seiner Erkenntnistheorie*, Tubingen: Mohr.

[4] Robert Paul Wolff. (1963). *Kant's Theory of Mental Activity*, Cambridge, Mass.: Harvard University Press, p. 170.

[5] T. D. Weldon. (1958). *Kant's Critique of Pure Reason* (2nd ed.), Oxford: Clarendon Press, pp. 252—256. Similarly, J. N. Findlay. (1981), in his *Kant and the Transcendental Object*, Oxford: Clarendon Press, speaks of "the transcendent sources of experience", responsible for the 'appearances'; cf. pp. 2, 15, 96, 185.

[6] P. F. Strawson. (1966). *The Bounds of Sense*: *An Essay on Kant's Critique of Pure Reason*, London: Methuen, pp. 40—41.

[7] Arnulf Zweig (ed.). (1967). *Kant Philosophical Correspondence 1759—99*, Chicago: Univ. of Chicago Press, pp. 183—84.

[8] For some further details see my "Reduction-Realization: A Key to the Structure of Kant's Thought", in J. N. Mohanty and Robert W. Shahan (eds.), (1982), *Essays on Kant's Critique of Pure Reason*, Norman: University of Oklahoma Press, Also my "Realism and Realization in a Kantian Light", in E. Schaper and W. Vossenkuhl (eds.), (forthcoming) *Reading Kant*, Oxford: Blackwell. pp. 39—98.

[9] Ak. edn., vol. 18, pp. 554—56 (Refl. No. 6286).

[10] Ak. edn., vol. 5, p. 49.

[11] For Kant's account of Newtonian systematics, see my "Kant's 'Special Metaphysics' and *The Metaphysical Foundations of Natural Science*", in Robert E. Butts (ed.), (1986), *Kant's Philosophy of Physical Science*, Dordrecht: Reidel, pp. 127—162.

[12] Johann Gottlieb Fichte (n.d.) *Grundriss des Eigentümlichen der Wissenschaftslehre in Rücksicht auf das theoretische Vermögen* (originally published in 1795), Leipzig: Felix Meiner Verlag, pp. 36—37.

[13] *Prolegomena*, see note 2, p. 45. Interestingly, on p. 47 Kant again says that "sensible knowledge does not represent things as they are but only the way in which they affect our senses, and thus that appearances merely, not the things themselves, are given by it to the understanding for reflection". This can again be taken in two ways: what affects sensibility is (i) things; (ii) things [in] themselves.

[14] David Hume, *A Treatise on Human Nature*, Selby-Bigge edn., (1946), Oxford: Clarendon Press, p. 84.

JAAKKO HINTIKKA

THE PARADOX OF TRANSCENDENTAL KNOWLEDGE

1. KANT'S THEORY OF MATHEMATICS AND HIS ARISTOTELIAN MISTAKE

This paper is a second half of a longer argument.[1] In its predecessor,[2] I argued that Kant saw the gist of the mathematical method in what essentially amounts to instantiation rules, i.e., what he himself characterized as arguing in terms of particular representatives of general concepts.[3] But when can such an anticipatory introduction of representatives of general concepts yield synthetic knowledge *a priori*? Kant's transcendental viewpoint commits him to answering: Only in so far as we have ourselves put the relations and properties we are arguing about into objects. Then our mathematical knowledge does not pertain to things, but only to the structure of our processes of coming to know them.

But what are the processes by means of which we come to know particulars? Kant's Aristotelian fallacy consists in his following the philosopher and answering: perceptions. Accordingly, mathematical knowledge is due to the forms of our sense-perception, and reflects them.

Instead, I argued, the most general description of the relevant processes is as processes of seeking and finding.[4] Hence the true reconstructed Kantian theory of mathematical reasoning, including what we would call quantificational or first-order reasoning, is my game-theoretical semantics.[5] For in it, what first-order logic deals with are the games of seeking and finding by means we explore the world. The knowledge we gain by means of such logical (Kant would have called it mathematical)[6] reasoning is in the last analysis about the structure of our seeking and finding games, about what could or could not happen in them.

2. KANT'S MISTAKE: THE TRANSCENDENCE OF THINGS IN THEMSELVES

One sense in which it is very difficult to correct Kant's mistake (his

J. R. Brown and J. Mittelstrass (eds.), An Intimate Relation, 243—257.

identification of the way we gain knowledge of the existence of particulars in general with sense-perception) is that its consequences are virtually impossible to disentangle from the rest of Kant's philosophy. Here I can only call the reader's attention to some of the most obvious consequences. It was precisely Kant's Aristotelian mistake that turned that inevitable dark side of his transcendental position, things in themselves, from epistemological restraints on our knowledge into metaphysical reifications. For it was the assumption concerning the role of perception in our knowledge-seeking that implied that the epistemological unreachability by our knowledge-seeking activities, which things in themselves by definition enjoy, amounted to *transcendence with respect to sense-perception*, that is, noumenal existence. In other words, the self-same Aristotelian assumption is the ultimate reason why Kant identified the limits of the legitimate use of the categories of understanding to possible sense-experience, and forced him to seek the ground of the applicability of the categories in the apperceptive assimilation of sense-experiences to the texture of our knowledge. All this has to be re-evaluated if Kant's philosophy is to be purged of his crucial mistake. Needless to say, I cannot attempt this monumental task here.

Attempts have been made repeatedly both by Kant and by his followers to do better justice to those fundamental ideas of his which are independent of the mistake. For instance, one can start from empirical objects in Kant and consider things in themselves as the outer limit of what these phenomenal objects could become if we considered them in themselves, i.e., abstracted from them all traces of our knowledge-seeking activities. But as long as one assigns to sense-perception the role it in fact plays according to Kant in the acquisition of our knowledge, it is hard to see that one cannot help ending up with the noumenal objects all over again. However, this theme cannot be developed fully here, even though it is extremely important for our understanding and evaluation of Kant's philosophy. In particular, I don't want to deny that there were forces in Kant's own thinking which pushed him towards a position much closer to mine than his official one.

It is not hard to find specific manifestations of such forces. Within Kant's own system, there is for instance a reluctant and indirect admission of the conceptual fact that the logic of our knowledge of particulars (intuitions) is the logic of existence and universality, that is, twentieth-century philosophers' "first-order logic". For him, there are certain principles of understanding which correspond to the categories of quantity, which in turn correspond to the different quantities of a judgement: universality, particularity and singularity. These features of propositions (judgments) are precisely what first-order logic deals with. Now — *prima facie* quite surprisingly — the principles of understanding which correspond to these categories are according to Kant the *axioms* of *intuition*. (See A 161—166 = B 200—207.) However, in the light of what we have found, this identification is really not very surprising. It is merely Kant's oblique recognition of the fact that the logic of existence and universality is the "logic" of the axioms of intuition, i.e., of the axioms for particular representations.

Moreover, it is hard to avoid the impression that, for the purpose of the first axiom of intuition, the connection between intuitions and sense-perception postulated by Kant is otiose. The axiom is formulated by Kant as saying that "all intuitions are extensive magnitudes". Unpacked, it says that all particulars are subject to geometrical and kinematic conditions. For this purported conclusion, it is certainly easier to argue in terms of our concepts of existence and universality than in terms of the idea of intuitions as always being given to us in sense-perception. It is certainly much more difficult to think of forms of existence of particulars which are not spatial and temporal than to imagine possible forms of perception of which the same is true. This point is especially persuasive if some form of game-theoretical semantics is adopted. For how can one possibly look for a search for particulars except in space and time? Even though we can perhaps in the last analysis make sense of nonspatial and nontemporal seekings and findings, the paradigm case can nevertheless be maintained to be the spatio-temporal one.

Whatever further arguments one may want to put forward here, these observations strongly suggest that the viewpoint from which we are looking at his theory of space, time, mathematics, and particular existence is one toward which Kant himself was pushed to some extent by his own conceptual framework.

Admittedly, Kant tries to give a deeper foundation to the alleged connection between particulars and sensibility by means of his examination of the threefold synthesis through which objects of experience are according to him constituted. But this attempt is a belated one. The plausibility of Kant's account of synthesis is no greater than that of

Kant's initial Aristotelian mistake. It may be *prima facie* more plausible to think of the particular objects of our experience as being constituted in perception and apperception than as constituted to be potential objects of seeking and finding. On reflection, it is seen that this plausibility is merely another form of the same Aristotelian fallacy. The true account of individuation and identification will center crucially on the re-identification of objects in space and time, i.e., on those very properties of theirs which make them potential objects of seeking and finding.[7] In other ways, too, can it be seen that objects are in our actual conceptual system individuated primarily to be re-identifiable objects of search and recognition.

3. THE UNKNOWABILITY OF THINGS IN THEMSELVES AND THE UNKNOWABILITY OF OUR COGNITIVE PROCESSES

Behind Kant's use of the notion of the transcendental (and even more so behind some of his successors' use of similar ideas) there lies a problem situation which I have not seen discussed even though it clearly manifests itself in Kant and also in the subsequent history of Kantian problems. It is what the title of this paper serves to highlight. It is connected with the first and foremost function of the concept of the transcendental for the purpose of marking the limits of what we can know. This function we registered above. It is what is marked by the limiting concept of things in themselves. It inevitably gives rise to the paradoxical fact that *things in themselves are unknowable because and in so far as our knowledge-seeking processes are unknowable.*

The basis of this paradoxical-looking idea is straightforward enough. Things are allegedly unknowable in themselves because we can know them only by means of certain processes which "color", i.e., affect, the resulting knowledge. But if we knew these processes well enough to understand precisely how it is that they influence their eventual product, i.e., our knowledge, we could as it were subtract these influences from our knowledge, and the rest would tell what things are in themselves, in the sense of in so far as they have not been affected by our knowledge-seeking processes. Hence, in order for things in themselves to be unknowable, our knowledge-seeking processes must likewise be unknowable in certain respects.

An illustration — or perhaps rather an analogy — which may be useful here is to compare our knowledge-seeking activities, plus the conceptual system they use, with an elaborate measuring or registering apparatus. Its feelers touch the real objects, but we don't "see" them; we can reach them only by means of the apparatus. Our knowledge of reality is what we can tell of it on the basis of the registrations of the "machine" at our end. These are not due completely to the objects at the receiving end of the apparatus, but are influenced by the mode of operation of the instrument itself. Some registrations may even be caused completely by the structure of the apparatus.

It might seem that there is no way of eliminating this influence of our measuring technique on its output, as long as we are committed to using it and only it in our search for knowledge. This is parallel with the vaunted unknowability of things considered in themselves, i.e., independently of our cognitive processes. But this impression is mistaken. We can do so at least as a matter of degree by learning more about the principles of operation of the measuring apparatus. By so doing we can extract more information from its readings, by being able to tell which tremors of its needles merely reflect the resonances of our own machine and which ones can be traced all the way to the objects touched by the feelers of the apparatus on its hidden side. For instance, we might be able to disregard altogether some merely apparent registrations. By so doing, we could as it were subtract the influence of our own tools from our *prima facie* knowledge of things. Now it is precisely this influence that allegedly makes it impossible for us to know things as they are in themselves. Hence our insights into our own knowledge-seeking processes and into the paraphernalia they use will enable us to gain knowledge of things in themselves at least in the sense of bringing it about that our *prima facie* knowledge reflects as a result of these insights things in themselves more faithfully than it did before.

This observation puts a new complexion not only to Kant's pivotal concept of the transcendental but also to the uses he puts it to, which comprise much of his central philosophical thought.

Before examining these consequences of our observation, however, a couple of possible objections have to be met. First, it may seem that my use of the measuring-apparatus analogy is a typical twentieth-century anachronism. (It is actually used by Wittgenstein in his *Tractatus Logico-Philosophicus* 2.1511—2.15121.) It is therefore not irrelevant to note (as Charles Harvey first reminded me) that the same analogy was used before me by Hegel,[8] who speaks in so many words about our cognition "as the *instrument* to get hold of the Absolute" and considers the idea that the unknowability of the Absolute (i.e., things in themselves) "could be remedied through an acquaintance with the way in which the *instrument* works" (A. V. Miller translation), just as I have done. (Emphasis is added in the last two quotes.)

I cannot examine here the uses to which Hegel put the instrument analogy. Contrary to what I earlier believed (and even said in the earlier version of this paper), Juha Manninen has convincingly argued that Hegel in effect sided with me against Kant.[9]

4. THE PARADOX OF SEMANTICAL KNOWLEDGE

After having disposed of these objections, we can return to the consequences of the paradox of transcendental argumentation. Some of these consequences can be seen more clearly from a parallel case, which has in fact played a remarkable role in recent philosophy of language.[10] There the unknowability of reality as it is independently of our knowledge-seeking processes becomes the indescribability of reality independently of the articulating and conceptualizing processes whose products are embodied in our language. In brief, the unknowability of things in themselves is transformed into a linguistic relativity — or perhaps rather a paradigmatic relativity *à la* Thomas Kuhn. Indeed, the difference between our knowledge-seeking activities and the language-games which give the words of our language their meanings is not always very great. For instance, the language-games of seeking and finding, which we encountered earlier and which give the quantifier words of our language their meanings, can be thought of as activities of verification and falsification, i.e., as knowledge-seeking activities.

What is even more interesting, the paradox of transcendental knowledge is reflected on the linguistic side by the thesis of the ineffability of semantics. For semantics is precisely the study of those relationships which connect our language with the world and which are analogous with the activities which connect our knowledge with our objects. As I have shown elsewhere,[11] the assumption of the ineffability of semantics has played and extremely important role in contemporary philosophy of language even though this role has not been acknowledged in the literature in its full ineffability of semantics, and this belief has affected their respective theories of language and logic in important respects. A considerable part of the interest in the paradox of transcendentality is due to its semantical twin.

It is interesting to note that one of the most outspoken defenders of the ineffability of semantics, Ludwig Wittgenstein, in so many words identified his thesis with Kant's. In *Vermischte Bemerkungen*, p. 27 he writes:[12]

> The limit of language shows itself in the impossibility of describing the fact that corresponds to a sentence . . . without repeating that very sentence.
>
> What are we dealing with here is the Kantian solution to the problem of philosophy.

In general, Wittgenstein's concept of the limits of language is a close analogue to Kant's idea of the limits of human knowledge, which he builds into his idea of things in themselves.

5. THE INEXHAUSTIBILITY OF THINGS IN THEMSELVES vs. THEIR UNKNOWABILITY

At first, it might seem that the paradox of transcendental knowledge will defeat the whole enterprise of transcendental philosophy. The knowledge it aims at was seen to pertain to our knowledge seeking activities and to the conceptual system they employ, and its goal was seen to establish the limits of these activities, i.e., the limits of the correct employment of this system. But the paradox says that this task is possible only if transcendental knowledge is impossible. Hence, the program of *Transzendentalphilosophie* seems to be self-defeating. The linguistic counterpart to this conclusion, the impossibility of semantics as a serious theoretical enterprise, has in fact been embraced by some of the foremost philosophers of language, even in the teeth of their interest in semantics and of their having elaborate ideas about semantics.

This conclusion is more pessimistic, however, than that which my argument warrants. Even if the kind of transcendental knowledge is possible which enables us to increase our knowledge of things in themselves, there is no reason to believe that we can eliminate *all* those unwitting contributions of our own to our *prima facie* knowledge which rule out direct knowledge of things in themselves. This inexhaustibility conversely provides the idea of transcendental knowledge with a legitimate niche. We don't have to deny the possibility of such knowledge in order to do justice to the contribution of our own knowledge seeking activities to the total fabric of our knowledge and even to the uneliminability of those contributions. However, we must realize that the uneliminability is not due to any intransgressible boundary, but means only inexhaustibility. It is a fallacy to think that we are separated from things as they really are by a fixed impenetrable iron curtain.

This is one of the places where the consequences of Kant's fallacy of assigning to sense-perception a crucial role in all information gathering induced further mistakes in his philosophy. For it was this fallacy that led Kant to think of the boundaries of possible experience as such inescapable limits of our knowledge in general. This in turn gave rise to a strong form of the paradox of transcendental knowledge in Kant's philosophy. As we may express this idea in terms of my measuring apparatus analogy, in order to know that there is an absolute limit to what the instrument can do is to possess knowledge of the very kind which presumably will enable us to transgress that limit. To change the metaphor, in order to draw a boundary of what we can know, we must know what there is on the other side of the boundary. If we cannot do the latter, we cannot do the former, either.

It is interesting to see that the same twist I am trying to give to the concept of the transcendent — *viz.* to replace its unknowability by its inexhaustibility — was given to it earlier by Edmund Husserl. For him the transcendence of objects means that we cannot even know them adequately, that is, completely.[13] Husserl notes quite correctly that on this concept what is transcendent is still in principle possible to reach in experience.[14]

It is no wonder, in view of this problem (paradox of transcendental knowledge), that Kant was not clear about the contrast, transcendental *vs.* transcendent. If you try to draw absolute "transcendental" limits to legitimate knowledge, you will *ipso facto* find yourself overstepping them and straying to the transcendent fields where Kantian angels fear to tread. It seems to me, however, that in the happier formulations of his own idea of things in themselves Kant did in fact treat them as an unreachable asymptotic limit rather than a fixed boundary; but I cannot argue adequately for this exegetical point here.

In the philosophy of language, this interpretation amounts to the claim that semantics is inexhaustible rather than ineffable. This view has not been studied in the philosophy of language in a relation to its Kantian ancestry, however, even though it is in many ways a more realistic view than the extreme Wittgensteinian and Quinean distrust of explicit semantics (model theory). It also provides an answer to the stock objection first raised by Bertrand Russell, in his introduction to Wittgenstein's *Tractatus*[15], to the semantical counterpart to Kant's transcendental position, *viz.* to the thesis of the ineffability of semantics. This objection consists in allegedly circumventing the problem by proposing to formulate the semantics of our given language (object language) in another one (metalanguage). The answer is that the object language cannot be the whole of our language. Indeed, we can draft to the service of this answer the well-known results, originating with Tarski, which show that the metalanguage must be stronger than the object language in order for the semantics (truth definition) of the latter to be formulated in the former.[16]

6. THE INSEPARABILITY OF CONCEPTUAL AND OBJECTUAL KNOWLEDGE

My interpretation of things in themselves as unreachable goals rather than as being hidden behind an impenetrable curtain (and the associated interpretation of transcendental knowledge) does not water down the resulting philosophical theory, either. On the contrary, the paradox of transcendent knowledge is closely related with a most remarkable conclusion. Once again, it can be seen from my instrument analogy. How are we to classify the knowledge which enables us to disregard irrelevant *prima facie* information about the reality "measured" by our apparatus? Is this knowledge *about* our knowledge-seeking activities or *about* objects? At first, there seems to be no problem. Such knowledge is — didn't I say so? — about our notional measuring instrument. But it also enables us to increase our knowledge of the reality which the feelers of our instrument touch. It is in this respect on a par with ordinary knowledge about the reality, and it can in principle be quite as useful, practical, and otherwise down-to-earth as regular factual knowledge about the usual objects of knowledge. Now, which one is it "really" about, our own activities and concepts, or about reality? The right answer is that there is no single right answer here. The right way of looking at the situation is to realize a most important consequence of the transcendental viewpoint which I have illustrated by means of the measuring apparatus analogy. There is an area where *knowledge about our own conceptual system* (and of the activities on which it is based) *is inseparable from knowledge about the world.* This is one of the most important conclusions from the transcendental viewpoint, stripped of its mistakes.

From a suitable vantage point, this inseparability of conceptual and objectual knowledge (as I shall call it) is of course virtually a reformulation of the paradox of transcendental knowledge discussed above in section 7. For why is it that things in themselves can be knowable only if our knowledge-seeking processes are unknowable? The reason is that knowledge about the latter can yield knowledge about the former. But such knowledge must have not only a conceptual but also an objectual component if it is to threaten the vaunted unknowability of things in themselves.

Moreover, the inexhaustibility of things in themselves discussed above is very closely related to a special form of the inextricability of conceptual and objectual knowledge from each other. Even if we think of the borderline knowledge (knowledge that *prima facie* can be taken to be either about our concepts and conceptual activities or about objects) as being conceptual in nature, even then the inexhaustibility amounts to a somewhat different kind of inseparability. It means now that the newly redefined conceptual knowledge (conceptual knowledge in the sense that includes the ambiguous knowledge just mentioned) cannot be completely disentangled from unambiguously objectual (factual) knowledge. In the following I will consider both kinds of inseparability. Even though the inseparability of conceptual and objectual knowledge has not received the attention it deserves, it has not remained completely unknown to philosophers. We can perhaps see an unwitting reflection of it in Kant's difficulty in keeping transcendental knowledge and transcendent knowledge apart from each other. The conclusion we have reached says precisely that you cannot separate the two completely. As the case is so often in philosophy, terminological confusion is here a manifestation of an underlying problem.

More self-consciously, the inseparability of "the given" from a priori assumptions of our own making was one of the central tenets of that remarkably Kantian work of early twentieth-century philosophy, C. I. Lewis's *Mind and the World Order*.[17] Witness, for instance, Lewis's statement on p. 58 to the effect that "what is given (given in part) is the real object. But the whatness of this object involves its categorical interpretation; the real object, as known, is a construction put upon this experience of it, and includes much which is not ... given in the presentation".

Even more importantly, our insight into the inseparability of conceptual and objectual knowledge puts into dramatically new light the puzzle about Kant's notion of transcendental knowledge which emerged in op. cit. (note 2 above), sec. 1. Now we can see that Kant's confused and perhaps even contradictory statements as to what transcendental knowledge is about can be given a partial justification. He may be essentially right in sometimes calling transcendental knowledge conceptual (about our own mode of knowledge) and sometimes treating it as objectual, for it turns out that no sharp line between the two can be drawn anyway. This is an interesting example of how even the apparent confusions of a great philosopher can receive a deeper interest through a close examination of the underlying conceptual situation.

7. LOGIC AS A TRANSCENDENTAL DISCIPLINE

The inseparability of conceptual and objectual knowledge has a neat manifestation in the precise languages of logicians.[18] One way in which first-order languages can be thought of as operating is as follows: Each sentence *S* presents to us a number of pairwise exclusive and collectively exhaustive alternatives concerning the world and says that one of them is in fact realized. Other alternatives are thereby exluded. The more such alternatives are thus excluded by one's knowledge, the more one knows in a most down-to-earth sense of being able to dispense with all preparations for the excluded alternatives.

It turns out, however, that some of the alternatives admitted by *S* may be merely apparent ones, not possibly realizable in the actual world. This inconsistency is due to their structure. However, it may not be *prima facie* apparent, and as long as I don't realize that an apparent alternative cannot be actualized, I have to heed it (even on the level of practical preparations) in the same way as the "real" ones.[19] Hence the elimination of a merely apparent alternative increases one's knowledge about objects in a very real sense. Yet it can be thought of as merely conceptual knowledge. Indeed all our logical knowledge amounts to the elimination of some such merely apparent possibilities. Hence it all belongs to the twilight region in which a sharp distinction between conceptual and objectual knowledge cannot initially be drawn, and cannot ever be drawn once and for all. Hence logic offers us a vivid example of the kind of knowledge which a Kant *redivivus* would be forced to classify as transcendental knowledge.

Notice how extremely close we are here to the "measuring instrument" analogy employed above in Section 3. For instance, dismissing a merely apparent alternative is just like eliminating one of the spurious readings of the measuring apparatus as being merely due to its own mode of operation. Indeed, we may think of the unavoidable presence of spurious alternatives in much of the information conveyed, received, and stored by means of first-order sentences as the inevitable result of relying on that knowledge-seeking instrument which is called first order logic (and which has to be interpreted so as to be based on the language-games of seeking and finding mentioned above in Section 1). The spurious alternatives are, as it were, the internal resonances of this particular instrument.

The measuring instrument analogy will take us even closer to certain recent developments in logic, however. The merely apparent possibilities, whose discovery and elimination we have discussed, can be interpreted as Rantala's *urn models* mentioned above in Section 4. As such, they are not mere unreal shadows, completely unrealizable illusions. They can in fact be realized, if it turns out that our knowledge-seeking activities ("language-games of seeking and finding") can affect reality, which can thus change in certain ways between successive moves in our "games of exploring the world". By eliminating certain "apparent" possibilities we are thus in effect ruling out certain ways in which our knowledge-seeking activities can affect the reality we are seeking knowledge about. This fact shows especially strikingly what a close connection there is between contemporary logic and Kant's idea of transcendental knowledge.

At the same time, we have here an excellent illustration of the inextricability of conceptual and objectual knowledge. Eliminating a merely apparent alternative (an urn model) can increase one's knowledge in a most realistic sense perhaps even relieving one of the burden of concrete preparations. At the same time the knowledge so gained is conceptual, and in the same sense "about" our own conceptual system (about our language-games of seeking and finding).

Moreover, these two kinds of knowledge are inseparable also in a more sweeping sense. It is the second sense of inseparability mentioned above in this section. The merely apparent and the real possibilities are effectively (recursively) inseparable. That means that there is no mechanical way of eliminating all merely apparent alternatives. However many of them we have been able to uncover and dismiss, there is typically a lingering uncertainty as to whether the remaining ones are all "real" ones. This is in fact the gist of the undecidability of first-order logic. This famous result thus turns out to have a striking transcendental meaning. It highlights a special case of the inseparability of conceptual and objectual knowledge. I don't think that it is an illegitimate exaggeration to aver that this result is a partial but nevertheless remarkable vindication of Kant's ideas about transcendental knowledge.

We can pinpoint its Kantian role even more closely. It is the inseparability of conceptual and objectual knowledge that causes the inevitable residue where these two types of knowledge are intertwined and whose eliminability is what Kant highlighted by means of his idea of things in themselves. Thus the effective inseparability result in the special case of first-order discourse, which Church's undecidability theorem establishes, vindicates things in themselves in this area. If it were correct to reify things in themselves in the way Kant did (as it is not), we could go even further and say that Church proved the existence of things in themselves for first-order languages.

After my examination of the paradox of transcendentality above it may not come as a surprise that this demythologized version of the Kantian idea of transcendentality does not imply any iron curtain between us and objects considered in themselves, i.e., considered otherwise than as potential objects of seeking and finding. Transcendent knowledge is not occult knowledge on this scheme; it is merely the unreachable ultimate limit of our transcendental knowledge. Objects in themselves are noumenal objects of thought but not perception; they are merely regular objects of knowledge, considered in abstraction from our knowledge-seeking activities — or at least some of them, if we prefer to relativize the concept of *Dinge an sich*. For instance, if the activities in question are the language-games on which first-order logic relies, things in themselves are but a logician's ordinary individuals, considered counter-logically in abstraction from the language-games of seeking and finding (in the sense that these games are assumed to have all been played *ad infinitum*).

The prospects of transcendental knowledge and transcendental argumentation are thus virtually unlimited in our own contemporary philosophical work. The transcendental element in contemporary thought is not restricted, either, to the types of discussion which are identified explicitly as descriptive metaphysics or transcendental philosophy. I have tried to illustrate here the remarkable fact that there are transcendental insights already obtained in contemporary philosophy of logic and philosophy of language. It is a sad comment on the growing departmentalization of philosophical studies if this transcendental element in some of the best logical and analytical philosophy of our times is not recognized as such. I would go so far as to surmise that the most clear-cut and instructive paradigm cases of transcendental argumentation are to be found in that direction. (Witness, e.g., the role game-theoretical semantics and the undecidability of first-order logic played in my discussion above.)

NOTES

[1] This argument is presented in my paper, "Das Paradox transzendentaler Erkenntnis", in W. Vossenkuhl and E. Schaper, (ed.), *Die Bedingungen der Möglichkeit*, Klett-Cotta Verlag, Stuttgart, 1984, pp. 123—149. The present paper is the English version of the second half of that paper.

[2] This predecessor is "Kant's Transcendental Turn and His Theory of Mathematics", *Topoi* **3** (1984), pp. 99—108. It represents an English version of the first half of the paper mentioned in note 1.

[3] See here my earlier work on Kant. Most of it is collected either in *Logic, Language-Games, and Information*, Clarendon Press, Oxford, 1973, and *Knowledge and the Known*, D. Reidel, Dordrecht, 1974 but see also "On Kant's Notion of Intuition (*Anschauung*)", in T. Penelhum and J. J. MacIntosh (eds.), *Kant's First Critique*, Wadsworth, Belmont, CA, 1969, pp. 38—53, "Kantian Intuitions", *Inquiry* **15** (1972), pp. 341—345, and "Kant's Theory of Mathematics Revisited", *Philosophical Topics* **12**, 2 (1981), pp. 201—215.

[4] *Op. cit.*, note 2 above and "Semantical Games and Transcendental Arguments", in E. M. Barth and J. L. Martens (eds.), *Argumentation: Approaches to Theory Formation*, Benjamins, Amsterdam, 1982, pp. 77—91.

[5] For this theory, see Esa Saarinen (ed.), *Game-Theoretical Semantics*, D. Reidel, Dordrecht, 1979; Jaakko Hintikka, *The Game of Language*, D. Reidel, Dordrecht, 1983; Jaakko Hintikka and Jack Kulas, *Anaphora and Definite Descriptions: Two Applications of Game-Theoretical Semantics*, D. Reidel, Dordrecht, 1985.

[6] The best argument to this effort is to show that the Kantian concept of "construction", in which he sees the gist of the mathematical method, has to be identified with the instantiation procedures which are the essence of the logic of quantifiers. For Kant, see A 713 = B 741. For evidence for my identification of Kantian constructions and logicians' instantiations, see the work referred to in note 3 above.

[7] On the subject of individuation and identification, see Jaakko Hintikka and Merrill B. Hintikka, "Towards a General Theory of Individuation and Identification," in W. Leinfellner *et al.* (eds.), *Language and Ontology: Proceedings of the Sixth International Wittgenstein Symposium*, Hölder-Pichler-Tempsky, Vienna, 1982, pp. 137—151.

[8] See *Phänomenologie des Geistes*, Introduction to Part I, sec. 73, pp. 3—5 of the 1807 edition; pp. 62—63 of the Hoffmeister edition, Felix Meiner, Hamburg, 1952.

[9] See Juha Manninen, "Tietokyky mittauslaitteena" (in Finnish), in his book *Dialektiikan ydin*, Pohjoinen, Oulu, 1987.

[10] See Jaakko Hintikka, "Wittgenstein's Semantical Kantianism", in Edgar Morscher and Rudolf Stranzinger (eds.), *Proceedings of the Fifth International Wittgenstein Symposium*, Hölder-Pichler-Tempsky, Vienna, 1981, pp. 375—390.
[11] Note 10 above, and cf. also my paper, "Semantics: A Revolt Against Frege," in G. Flöistad (ed.), *Contemporary Philosophy: A New Survey*, vol. 1, Martinus Nijhoff, The Hague, 1981, pp. 57—82.
[12] Ludwig Wittgenstein, *Vermischte Bemerkungen*, Suhrkamp, Frankfurt am Main, 1977. I have departed here from the translation of Peter Winch (see Ludwig Wittgenstein, *Culture and Value*, The University of Chicago Press, 1980, p. 10): "This has to do with the Kantian solution of the problem of philosophy". This is far weaker than Wittgenstein's point, which is not just that we are here *dealing with Kant's* solution, but that we *have* the Kantian solution right here.
[13] See his *Ideen zu einer reinen Phänomenologie und phänomenologischen Philosophie*, Halle, 1913, Sec. 144, p. 298 of the original, and cf. Sec. 149, p. 311 of the original.
[14] *Op. cit.* Sec. 45, p. 84 of the original, and Sec. 47, pp. 88—89.
[15] Bertrand Russell, "Introduction", in Ludwig Wittgenstein, *Tractatus Logico-Philosophicus*, Kegan Paul, London, 1922.
[16] Cf. Alfred Tarski, "Einige methodologische Untersuchungen uber die Definierbarkeit der Begriffe", *Erkenntnis* **5** (1935—36), pp. 80—100; English translation in Alfred Tarski, *Logic, Semantics, Metamathematics*, Clarendon Press, Oxford, 1956, pp. 296—319.
[17] Charles Scribner's Sons, New York, 1929.
[18] Cf. here Jaakko Hintikka, "Information, Deduction, and the '*a priori*'", in *Logic, Language-Games, and Information* (note 3 above).
[19] Cf. Jaakko Hintikka, "Impossible Possible Worlds Vindicated", in Saarinen (ed.), note 5 above.

GEREON WOLTERS

MESMER IN A MOUNTAIN BAR: ANTHROPOLOGICAL DIFFERENCE, BUTTS, AND MESMERISM

Here the learned do not bargain for paper treasures,
One measures not the roads to Rome and Athens
Reason one does not tie to scholastic precepts,
And no one teaches the sun to move in its orbits
(. . .)
And here nature gave the teaching of how to live right
To man in his heart and not in his brain.
(Composed 1728)[1]
Albrecht von Haller: "The Alps"

I would really rather be climbing a mountain, tending bar, or driving a taxicab than writing a philosophical book. (Robert E. Butts, 1968)[2]

I. INTRODUCTION I

I am not sure whether Bob Butts at the present time would still prefer to be a taxi driver rather than a philosopher. That twenty years after this confession he still prefers bars and mountains to philosophical seminars and conferences, is something I believe I can present as a confirmed fact on the basis of extensive, pleasant experiences shared at such places. Among Butt's favorite mountains are certainly the Swiss Alps, which Albrecht von Haller (1708—1777) praised in his great poem, "The Alps". I hope I am not being too partial to the flatter foothills when I say that in the last decade our friend of the mountains has grown very fond, too, of this part of the country.[3] Especially that magic triangle between *Constance* with its lake, *Wildhaus* in Toggenburg with the simple wooden house where Huldrych Zwingli (1484—1531) was born and grew up, and *Zürich* with the *Kronenhalle*, where James Joyce used to keep his beer from getting flat.

This magical German-Swiss triangle is Mesmer country. Franz Anton Mesmer was born in 1734 in the small village of Iznang, on that part of Lake Constance called the "Lower Lake". He attended school

J. R. Brown and J. Mittelstrass (eds.), An Intimate Relation, 259—282.

(Gymnasium) in Constance, where he returned in 1812 after a very eventful life in the European urban centers of Vienna and Paris and after a long exile in the neighboring Swiss canton of Thurgau. In 1814 the by now 80 year old Mesmer moved to Meersburg on the northern shore of the lake, at that time the Summer residence of the bishops of Constance. One year later Mesmer's stormy life found its final resting place in the cemetary at Meersburg.

II. INTRODUCTION II

Like theologians philosophers too have a tendency to stylize themselves in profound and mystifying ways. Especially in Germany, the home of dark profoundity, one scarcely finds a philosopher who does not enjoy creating the impression that he is concerned with nothing other than the fate of reason or morals, the end of Western metaphysics with its '*ontological* difference',[4] the future of all of mankind, and whatever else there is of importance and sublimity. Almost everyone who has the opportunity wants to be known as a guru, even if it is only as one who leeds through the shallow waters of, say, the second-order predicate calculus with identity. Some are certainly astonished when a philosopher confesses that, instead of writing profound and epoch-making books, he "would really rather be climbing a mountain, tending bar, or driving a taxicab".[5] And yet such preferences can, like Haller's "Alpen", reveal a philosophical attitude. By no means do they need to be an obstacle to writing excellent philosophical books. Butts's *magnum opus* is itself a prime example.[6] This book stands out among the wealth of literature on Kant for several reasons. I would like to mention three of these: first, a philosophico-historical reason, second a reason connected with the method of writing the history of philosophy, and finally a purely philosophical reason.

First of all, Butts has succeeded in explaining Kant's pre-critical and critical writings in terms of a single motivation: Kant's enlightened mission to "help to secure and maintain individual and social mental health" (p. 4). In Butts's reconstruction of Kant spriritual health is endangered by the belief in a *possible* theoretical approach to the supernatural. People who believe in such theoretical access to the supernatural do not usually sit in insane asylums. On the contrary[7], they stand, according to Kant, before the learned pulpits of scholastic metaphysics or publically praise their superstitious metaphysics of the marketplace. Kant's prophylactic *medicina mentis* ("the philosopher's diatetic of the mind" (p. 295)) has two main ingredients: (a) the spatio-temporal conception of everything that should be considered sensory intuition and (b) 'grammatical'[8] forms, which steer the conceptualization of perceptual data.[9] The decisive point of this conception is that sensory intuitions have something very important in common with alleged experiences of the supersensible: they are *private*. Only the grammatical forms for processing data "mould private sensations into publically available objects of possible experience" (p. 7). It is the public sphere or intersubjectivity of spatio-temporally ordered sensory intuitions produced by these transsubjective, grammatical (Kant: 'transcendental') forms of knowledge that separate possible experience from secular enthusiasm (*Schwärmerei*).

Leibniz divided the world into two completely different realms: the realm of supernatural and spiritual objects and that of sensually perceivable things or natural bodies. The realm of nature as the domain of necessity constantly stands under the reign of mechanical forms of explanation. The realm of the spirit is also the realm of freedom. In it have their home, if I am permitted to use this earthly and spatial metaphor, the forms of natural things, our souls, and also God. Mechanical forms of explanation naturally are not predominant in the realm of the spirit. This realm stands instead under the rule of the principles of noncontradiction and sufficient reason. But describes this methodological dualism in Leibniz with the beautiful metaphor '*Double Government Methodology*' (DGM). In short, according to Leibniz, there are two fundamentally different areas of experience which are both accessible to human knowledge. All we have to do is choose the proper method in each case.

Thus we come to Kant and his relationship to Leibniz's DGM. Here one can do nothing better than quote Butts himself (p. 11).

> Although Kant surely accepts the Double Government Methodology of Leibniz, he turns it on its head. Mechanical explanation that for Leibniz was only required because as defective human knowers we have to be able to deal with physical objects, becomes for Kant the preferred method of knowing because its principles are seen to constitute nature for the only knowers we can comprehend — ourselves. Metaphysical explanation that for Leibniz captures the basically real features of the universe by resolute attention of pure thought, becomes for Kant a set of demands of reason in its effort to regulate, to methodize, human inquiry. Drop Kant's demand for data, for publically available objects of inquiry, and the Leibnizian form of DGM is restored, along with the threat of a new outbreak of the disease of the soul that takes frenzy, not data, as the proof of genuine knowledge.

This should establish Kant's philosophical antidote for all types of

Schwärmerei as well as the value of Butt's book on Kant for the history of philosophy. A *Second* merit of this book lies in its contribution to *method in the history of philosophy*. In it Butts exemplifies the usually insufficiently considered thought that philosophy (among real philosophers) is not to be separated from the life of those who produce it.[10]

The *third* and final merit of this book lies in the way Butts implicitly and indirectly presents — if I am not mistaken — a systematical philosophical thesis that goes beyond methodological considerations about the history of philosophy, a thesis concerning a possible task and a possible goal of philosophy: *philosophy is the dietetic of the mind.* Apart from all that ostentatious business about Gurus, Butts shows that philosophical reflection can be a preventive measure against a whole range of those frightening mental afflictions which, sometimes quite out of the blue, seem to threaten our mental health[11]: "They all result from attempts to objectify things thought to exist beyond the space and time of the ordinary human experience" (p. 298).

In my opinion the three points just mentioned here contribute to the special attraction of *Double Government Methodology*. It seems to me that all three have something in common that I would like to call *anthropological difference*. By 'anthropological difference' I intend a formal parallel to Heidegger's 'ontological difference'. Heidegger accuses Western metaphysics of having up to now taken an interest only in what lies in the foreground, in being things; not, however, in what is ostensibly real, the being, that is presupposed by all 'being' things and for all knowledge about them. Whatever else the status of being and being things may be, 'anthropological difference' means something more modest and closer at hand. With this concept I want to point out how little so many philosophical efforts have to do with everyday life. I want to name that commonly accepted axiom according to which philosophy and life would have as little to do with each other as mathematics and life or physics and life; the axiom is that philosophy is 'pure' thought or at least concerns 'pure' thought. Butts shows us that even the philosopher of 'pure reason' cannot be properly understood solely as a pure thinker. Kant saw this difference. Indeed, one can say that Kantian philosophy — and along with it philosophy in Butt's sense — *is* philosophy of reflected anthropological difference.

An important application of philosophy as dietetics of the mind was in Kant's time the so-called animal magnetism that Mesmer believed he had discovered.[12] Completely in accordance with his philosophical dietetics Kant takes Mesmerism to court:

> Against this nonsense there is nothing other to do than to let the animal magnetizer magnetize and disorganize, so long as it is pleasant to him and other gullible people. It should be suggested to the police that morality should not be approximated too closely, and to carry on oneself, following the sole path of natural research, through experience and observation, which make recognizable the characteristics of objects of external sense.[13]

The Kantian philosophical dietetics of the mind has still another side, besides distinguishing between possible experience and *Schwärmerei* with respect to data: the *distinction between science and pseudoscience*.[14] It was precisely this problem that in connection with Mesmerism strongly moved the spirits of established scientists and which I will consider more thoroughly in the following sections.[15] In the process, I will first sketch (in Section III) Mesmer's conceptions and their resonance. Then (in Section IV) I will take up the problem of demarcation between science and pseudoscience. With respect to Mesmerism, this problem found a viable solution in a report of the Paris Academy of Sciences in 1784. This report is based on principles of successful scientific practice. Finally, in Section V, I take a stand on the rejection of the very idea of scientific rationality and of the demarcation between science and pseudoscience.

III. MESMER'S DOCTRINE

Mesmer's views can be grouped into three distinct areas: (a) a *comprehensive theory* covering a variety of astronomical, physical, and organic phenomena (TAM); the specialization of this theory to a *medical theory* (nosology; henceforth abbreviated NAM); and finally a practice of healing based in turn on his nosological theory of animal magnetism (and which I will naturally call HAM).

When Mesmer himself views TAM as 'theory', he finds himself in agreement with the common use in everyday language where almost everything can be called theory that does not deal with mere description. Mesmer was a successful practitioner, but not a great scholar. His theory TAM remains strangely unclear, although he elaborated it in a series of papers and pamphlets. This is all the more remarkable since he himself viewed TAM as a *mechanistic theory of fluids* of the type so highly valued in 18th century physics as a means of explaining phenomena like electricity, heat, and mineral magnetism. Theories of fluids conceive of these as well as other phenomena as results of the movements of most subtle, invisible, and imponderable matters that were called *fluida*. Their movements are to be described by means of the mechanics of fluid bodies.[16] Rudolf Tischner, perhaps the most knowledgable person on Mesmer's work, holds the very cogent view that TAM is in principle merely a *mechanistic* version of the old magical magnetic theory of medicine. This theory had appeared prior to Mesmer only in the guise of vitalism.[17] In his *Mémoire sur la découverte du magnétisme animal* (Paris 1779), Mesmer summarizes TAM in 27 short 'propositions'. The most important are: (i) There exists a multitude of mechanical interactions between heavenly bodies, the earth, and living organisms (proposition 1 = p 1). A lawlike expression of these interactions remains unknown. (ii) The medium of these interactions, to which gravitation also belongs, is a universally dispersed subtle fluid that I would like to call *fam* (fluid of animal magnetism). (iii) All events in Mesmer's cosmos occur in places where *fam* is unevenly distributed. In the case of the earth the moon has the greatest influence on the uneven distribution of *fam*. Its occurrence has a periodicity comparable with the ebb and flow of the tides. (iv) All characteristics of animate as well as inanimate bodies on earth are influenced by the effects of the bodies surrounding them as well as by heavenly bodies. *Fam* is the means by which these effects are mediated. (v) Animal organisms (including man, who is an animal) reveal a special receptivity for effects mediated by *fam*, although the nature of the receptivity may differ from one organism to the next. The interaction between bodies and animal organisms mediated by *fam* affects the nerves. These effects are polar and analogous to common magnetism. This is why the recptivity for the fluid *fam* is called 'animal *magnetism*' (pp. 9, 10). (vi) Not only animals possess this receptivity; that is, animal magnetism. Plants and inanimate objects such as water and minerals also possess animal magnetism. At this point, I should make a terminological remark. Mesmer uses the word 'animal magnetism' in two different senses. Up to now, it has been used to refer to the general receptivity of natural bodies for the universal fluid *fam*, making it a characteristic of natural bodies. To this original meaning of 'animal magnetism', Mesmer adds a second and more important one (p. 20)[18].

According to this conception, 'animal magnetism' refers not only to the characteristic of natural bodies to be receptive to the fluid but also to the special form that *fam*, distributed throughout the universe, assumes in animal organisms. For the sake of brevity, I will refer to the general fluid fam, when it occurs in animal organisms, as 'A. M.' instead of 'animal magnetism'. A. M. is the focal point of both NAM and HAM. (vii) A. M. has the following physical properties: (a) like electricity, A. M. can be accumulated, stored, and transported (pp. 12, 17); (b) A. M. can act at a distance without the necessity of a conductor (p. 14); (c) A. M. can be transmitted and amplified by sound (p. 16); and (d) A. M. can be reflected and amplified by mirrors.

Mesmer's nosological-medical theory, NAM, can be characterized as a varaint of *humoral pathology* commonly found in classical antiquity, particularly in Galenus.[19] For Mesmer, health consists in an age-dependent, harmonic relationship between *movement*, on the one hand, and *solidification*, particularly muscular solidification, on the other.[20] According to Mesmer movement is sustained by an invisible '*fire of life*' received by human beings at birth. He defines *illness* as muscular inactivity resulting from a solidification that exceeds the solidification typical of aging. The disturbed muscular activity is responsible in turn for obstructions in the circulation of the bodily humors. The symptoms of illness result from these obstructions. The cause of illnesses, the abnormal solidification of the muscles, comes from a deficient 'fire of life'. The complete extinction of the flame means, finally, the *death* of the organism.

Mesmer's *practice of healing* (HAM), which is based on TAM and NAM, should not be difficult to guess by now. In structural terms, it is closely related to the *iatromechanical* therapeutic conceptions prevalent at the time.[21] The fire of life is nothing other than A. M.. The art of medicine consists in the physician concentrating A. M. in his own body and transferring it to that of his patient. The transmission affects the patient's nerves, sets the muscles in motion and infuses the patient with the fire of life. This eventually results in the orderly circulation of the bodily fluids and the reinstitution of health. So much for Mesmer's conceptions of TAM, NAM, and HAM.

Mesmer always and vehemently viewed these three conceptions as a *contribution to the serious, scientific study of physics and medicine in his time*. This means that he places his efforts — following Kant completely — under the rule of the mechanical section of DGM. Of course, he may have feared right from the beginning that his assessment would not be accepted everywhere. Mesmer sent his first publication on animal magnetism[22] (which he wrote in 1775 shortly after its discovery), a small work of 12 octavo pages in the form of a letter, to all the European academies of science with the request that they evaluate it.[23] Only Berlin responded. One of the members of the Berlin Academy, Johann George Sulzer, however, had to inform the interested public that, after a short account of Mesmer's booklet before the academy, he did not find "this society disposed to become involved in a closer examination and judgement of it. One judged throughout that both what Dr. Mesmer says about his magnetic cures, as well as what he says in particular about his experiments concerning the communication of the magnetic force to very different types of bodies and its collection in bottles, has been subjected to quite significant doubts; and that the latter contradicts present reliable experiences of magnetic force to such an extent that one has no sufficient reason, on the basis of Mr. Mesmer's too indefinite report, to take his supposed discoveries under serious consideration".[24] Only Sulzer's discrete suggestion that one would "probably do his royal imperial [Austrian] majesty's plenipotentiary at this [i.e. the Prussian] court, his highness Baron van Swieten, a favor" brought about the formation of a small commission.[25] The Berlin commission, in the space of less than one printed page, justified why the Academy did not find it "necessary to become involved in a closer investigation and judgement of this matter which rests on such uncertain and indefinite foundations."

The Bavarian Academy also came to a decision. This was favorable to Mesmer and animal magnetism. But in Munich one did not subject himself to the rigours of making an argument. The academy designated its approval of animal magnetism by making its apparent discoverer an academy member.[26] This impressive recognition probably rests on the fact that, in the meantime the Catholic church which controlled the Munich academy, was highly indebted to Mesmer. With the help of Mesmer's expert opinion, it succeeded in neutralizing a famous-infamous village priest and exorcist from Vorarlberg (Austria) by the name of Gassner. Even more than this successful exorcizing of the devil by Beelzebub, the fact that the president of the academy considered one of Mesmer's magnetic cures to have caused relief of pains that had been plaguing him, spoke on behalf of Mesmer.

Not without reason do I mention these apparently extraneous factors that seem to determine the assessment of Mesmer's theory in its early stages. Since Thomas Kuhn's book on scientific revolutions, quite a few people are of the opinion that such *external* factors having little to do with the actual object of scientific investigation are precisely what guides the acceptance or rejection of theories and the demarcation between science and pseudoscience. Today, of course, it is less common for external factors speaking for or against theories to present themselves in the successful fight against exorcists or the curing of the various ailments of leading academics. And yet it is a hard blow to established belief in the rationality of science if, according to a commonly held conception, factors *external to* science and no longer *internal* arguments *in* science should determine the direction of the development of science and its difference from pseudoscience.

It is clear that empirical data cannot determine the choice between alternative theories. The reason for this is the 'underdetermination of theory by data'. As far as the demarcation between science and pseudoscience is concerned, empirical data cannot help either. Here the problem is not simply to identify 'real' data in the spirit of Kant's dietetic of the mind. Rather it is the problem of correctly explaining these data that is at stake. Here again, as at the level of producing data, we encounter two radically different ways of explaining data: on the one hand, alleged experience of the supersensible corresponds to pseudo-scientific explanation; on the other, spatio-temporal intuited data (that in addition have been shaped by transcendental categories) correspond to reliable scientific explanation. At this level of explanation it seems hopeless to strive for a criterion or criteria of the sort used by Kant at the simpler level of producing data.[27]

Mesmer's wish in 1775 that the established scientific bodies of his day seriously investigate the scientific nature of animal magnetism was only fulfilled ten years later, in 1784 in Paris (that is, taking a side-glance at what was happening elsewhere at the same time, three years after the publication of Kant's *Critique of Pure Reason*). But his request for investigation was fulfilled in a way that differed from what Mesmer had expected and wished for. For in 1784 TAM, NAM, and HAM were no longer the ideas of an unkown Viennese physician fighting for quasi-official recognition among the scientific community. On the contrary, Mesmerism in 1784 was a highly controversial *cause célèbre* that had *tout Paris*, right into the Chambers of Marie Antoinette, holding its breath.

But in 1784, Big Brother, too, was already closely watching the turbulent scene. As Robert Darnton observes in his excellent book dealing with the role of Mesmerism on the eve of the French Revolution, the government was suspicious of Mesmerism for two principle reasons.[28] *First*, Mesmer had been joined by a number of would-be philosophers and scientists who were not taken seriously by official bodies such as the Academy or the university. These individuals, failures in the eyes of the scientific 'establishment', viewed Mesmer as one of their own, despite his renown and, even if disputable, his fame. For Mesmer's ideas, too, had been jeered and laughed at by large parts of the scientific establisment. This frustrated group of would-be philosophers and scientists (among them Jean-Paul Marat, although he was not an advocate of Mesmerism) viewed the scientific establishment as part of the established political regime. Only the massive assistance of the government could keep so repressive and deceitful a system as established science in power. Thus, the resentment of those who felt they had been mistreated by the scientific community was finally destined to turn against the government, which, in time it did. *Secondly*, because of its allegedly detrimental influence on morals and religion, Mesmerism was considered to be a "matter for the police". This was the view expressed by Jean-Pierre Lenoir, lieutenant-general of the Paris police at the time.[29] Mesmerism finally became so dubious in the eyes of the ruling powers that they thought they must act. Of course, simply banning Mesmerism, together with Mesmer's expulsion, could not have been seriously considered in Paris, which was the major center of European Enlightenment. Things really could be, so the rulers may have thought, the way Mesmer and his followers said they were; namely that the triad of TAM, NAM, and HAM represented a matter of serious scientific interest. If that were really the case, officially banning Mesmerism would mean suppressing scientific truth by decree; and surely no one in the land of Diderot, d'Alembert, La Mettrie and Voltaire would want to be guilty of such an outrage.

IV. THE PARIS REPORT

So what happened is what would happen today in such a situation. On March 12, 1784, Louis XVI, king of France, called upon "the physicians [chosen by the medical faculty of the University of Paris] de Borie, Sallie, d'Arcet, Guillotin to investigate the cures which Mr.

Deslon [a physician and follower of Mesmer] was applying and to submit a report to him about this. At the request of the four physicians five members of the Royal Academy of Science, Messieurs Franklin, Bailly, le Roi, de Bory, and Lavoisier were also named by his majesty."[30] The task of the commission consisted in two points concerning the question of whether Mesmer's teachings were part of the realm dominated by mechanical methodology: (1) to examine the existence of animal fluids within the limits of the theory TAM, and (2) to examine the therapeutic value of HAM.

The above-mentioned gentlemen began to work quite industrously and, divided between medical faculty and Academy, had already prepared separate, comprehensive and extensive reports by August 1784. The physician's report, however, is comparatively weak and a little diffuse, whereas the report of the Academy members, written for the most part by Lavoisier[31], was very comprehensive and analyzed and argued with precision. Despite all the precision in the argumentation, however, it is clear that in 1784 we find ourselves in what claims to be an enlightened, but not a democratic age. For belonging to the commission of physicians was an additional member, of whom no word is mentioned in the report, not even in the list of the members of the commission. And yet in the Paris of 1784 the publication of an officially suppressed dissenting vote was still possible: almost at the same time as the report was printed, the distinguished Botanist and physician, Antoine-Laurent de Jussieu published his dissenting opinion which did not agree with the rejection of Mesmerism by both commissions.[32]

The commission's reports on Mesmerism offer us virtually unique historical material for the examination of theses concerning the demarcation between science and pseudoscience. A commission consisting in part of highly qualified people attempts to examine a new theory. How will they ground their judgement? Will it rely on internal criteria of scientific rationality that are also in and of themselves valid? Or will it base its decision on ostensibly rational criteria that are in reality extraneous, external criteria such as political opportunism? In fact this last possibility suggests itself in Mesmer's case. For the members of the commission must have known that the heads of government were suspicious of Mesmerism. They had to know that the report demanded of them was intended as scientific support for the government's impending ban on Mesmerism. It did not matter much that at the same time Mesmerism also had a number of adherents at court and among the nobility.

It is Robert Darnton who in fact suggested that the members of the commission argued against Mesmerism in an ostensibly scientific way whereas in reality, political factors played a decisive role in causing them to anticipate obediently the wishes of those in power. And this was exactly Mesmer's opinion too.[33] According to Darnton the arguments presented by the Academy members would be applicable to a theory of one of its members, Lavoisier, as well as to Mesmer's animal magnetism. For Lavoisier's so-called caloric theory of heat also relied on a subtle fluid of the type that was not accepted in the case of Mesmer's TAM. According to the caloric theory, all material bodies are permeated by the smallest invisible pores. The more caloric fluid is in these pores, the warmer the body. Caloric fluid is extremely subtle, virtually weightless and indestructible. Its existence and its characteristics cannot be observed directly, because it is invisible. The existence as well as the properties of the caloric fluid have to be deduced from the effects ascribed to it.

Now stories in which generals are cowards, cardinals godless, members of the salvation army alcoholics, bankers fraudulent, and dentists have a bad set of teeth, will on the whole, generate more interest than those in which the generals are heros, the cardinals saints, the members of the salvation army teetotalers, the bankers trustworthy, and dentists flash their beautiful white teeth. It appeals to us more when we see that others are hewn out of yet more crooked wood than ourselves. In this sense an account of how scientists, ostensibly the guardians and supporters of rationality, turn out to be unreasonable hardheads, frauds seeking fame, cunning intriguers or nimble opportunists can count on sympathy of a broad segment of the public. This seems to me to be the source of a good part of the fascination that external explanations for the developement of theories encounters everywhere. This is, of course, not to deny that in the history of science, and probably also in contemporary scientific practice, extraneous, external criteria have often enough influenced or even temporarily determined the acceptance or rejection of theories and the distinction between science and pseudoscience. In the present case as well, I assume that political opportunism as well as — with the physicians — simple envy of their successful colleague, influencced the commission's ruling. These motives may have actually been dominant. In any event the members of the commission viewed themselves as a policing body which was to serve the state by regulating reason. Bailly explained when the report was presented before the Academy: "If, however , such an error [like mesmerism] emerges from the realm of science, and spreads among the common masses, dividing opinions and making rebellious when it offers the ill a fraudulent remedy and prevents them from seeking other remedies [. . .], a good government will find it useful to eradicate it. What a marvelous use does not one make of one's authority when one uses it to spread light! The commissioners have taken great pains to fulfill the intentions of the administration and to make credit to the honor of having been chosen."[34]

Despite the fact that the Academy's rejection of Mesmerism seems to rest heavily on external grounds, I would like to advance the view that *the rejection of Mesmerism was sufficiently justified by internal motives of scientific rationality.* This can be shown by refuting Darnton's view that the commissioners noticed the splinter in Mesmer's eye while completely ignoring the log blocking Lavoisier's vision. What I intend to show is that, whereas TAM contradicted fundamental principles of scientific practice, Lavoisier's caloric theory, at least at this time, cannot be accused of similar defects.

I am talking here of internal theoretical *motives*, not of *criteria* of scientific rationality guiding the rejection of Mesmerism as pseudoscientific. In doing so, I return to a point mentioned already in section III: it seems hopeless to look for a demarcation criterion between science and pseudoscience. Science is an extremely varied and heterogeneous enterprise. It mirrors, in this respect, the varieties of life. Scientific rationality can as little be covered by one or two criteria as can life. Given this situation, it seems more sensible to check scientific and pseudoscientific practices in order to find reasons for demarcating science from pseudoscience.[35] According to this view scientific rationality is not a property of scientific statements, but a characteristic of the acts and procedures that produce these statements. Thus, *flawed practices* are the basis of pseudoscience. In speaking of flawed practices, we have left the field of criteria with its yes/no decisions. Here *judgement* is needed. Judgement cannot be learned like ABC; it has to be acquired while practicing it. The best practice for acquiring judgement in matters of scientific rationality is active participation in scientific research. But we know that practicing scientists do not reflect very much about these matters.[36] In the Paris of 1784 things were different, for first rate philosopher-scientists were at work. They, in fact, answered the king's question on the existence of the animal fluid *fam* by investigating the practices that led Mesmer and his disciples to claim to have demonstrated it. These mesmeristic practices differed considerably from those performed by Lavoisier and others with respect to the caloric theory of heat. There are, according to commission's review, *three aspects of flawed practice* that make Mesmerism a pseudoscience. These practices would, I believe, *mutatis mutandis* destroy every-day life as well as science: (1) disregard for the law of causality, (2) disregard for the difference between facts and hypotheses,[37] (3) disregard for what I would like to call the 'principle of experience'. That principle says that one is not allowed to contend what one likes, without being prepared to let one's assertions pass the test of the experience of reality. Mesmerism failed in all of these points.

To prove that, I would like to touch briefly on the caloric theory of heat as Lavoisier formulated it, i.e. the theory that the temperature of bodies is determined by the amount of caloric fluid contained in them. My first question is this: Did Lavoisier insist on the *existence* of the caloric fluid?

In two papers he wrote with Laplace on the caloric theory of heat, we find, in the first place the fundamental distinction between *observable facts* and *hypotheses*. The two authors discuss two competing contemporary hypotheses explaining the phenomenon of heat[38]: (1) the caloric theory of heat, and (2) the so-called mechanical theory of heat. According to the mechanical theory, heat "is nothing other than the product of imperceptible movements among the molecules of matter".[39] *However*, for Lavoisier and Laplace, the observable data on the nature of heat are not sufficient to justify favoring one theory over the other: "We do not want to decide between the two [. . .] hypotheses. Some observations tend to favor the latter, for example, the fact that heat is created by rubbing two solid bodies together. Other observations, however, can be explained more easily according to the first theory [i.e. the caloric theory]."[40] Lavoisier and Laplace place more confidence in the explanatory power of the caloric theory, using it as a basis for their research.[41] Perhaps they even believe — privately as it were — in the existence of the caloric fluid. And yet they leave no doubt that their imponderable caloric fluid is only an aid in explaining empirical phenomena, but not an observable object or fact. The fact that caloric theory can be used successfully to explain phenomena is, for Lavoisier and Laplace, insufficient evidence for the assumption that the caloric fluid exists. Additional deliberations are necessary here in order to prove the existence of a hypothetical substance. Although they do not comment on the nature of such deliberations, they do state very clearly that the evidence available is insufficient for deciding between mechanical and caloric theories and even more so for assuming the existence of the caloric fluid.

Mesmer is a different story altogether. From the very beginning, he is completely and (unfortunately) unshakably convinced of the existence of his fluid *fam*. Certain effects that occurred when steel magnets were placed on diseased parts of the body gave him the idea that *fam* was at work there. Within a short period of time he constructed on the basis of this idea, a comprehensive, though scarcely coherent system that relied on the mechanistic terminology of established science, though not on its methodology and customary practices. Most importantly, he was (1) not familiar with the difference between hypotheses and facts, (2) did not want to accept, at least for his own experimental practice, the usual standards for conducting successful experiments (including the law of causation, according to which identical causes have to have identical effects), and (3) was unwilling to accept the principle of experience according to which theories have to be tested by experience. It is precisely because of these three deadly sins that the members of the academy criticized TAM. I believe they were amply justified in doing so.[42]

In a series of experiments the commissioners determined that, first of all, the effects of the animal fluid A.M. among the persons examined, depended on whether they possessed magnetic sensibility. Of fifteen people being tested only five demonstrated this sensibility. Among the magnetically sensitive persons, allegedly mostly women, the magnetic effect only appeared when these individuals *knew* they were being magnetized. The effects ranged all the way from sensations of warmth to convulsive states and spontaneous excretions; effects that were considered to be signs of a 'crisis'. If the magnetized persons did not know they were being magnetized, then the effects predicted by the theory did not appear, or they appeared in the wrong place, or in a similarly defective way. In this situation, keeping conditions constant, the ceteris-paribus principle, becomes particularly important. The members of the commission were also very careful to observe this. The inevitable conclusion is that here an allegedly existing physical agent should, under constant conditions, exert an effect at one time, but not at another. That, however, violates the law of causality.

What, then, caused the obvious presence of these magnetic states or processes M if they occur independently of the actions A required in TAM? The Academy commission answered this question much the way John Stuart Mill did later with the "method of agreement", which he condensed into the following rule: "If two or more instances of the phenomenon under investigation have only one circumstance in common, the circumstance in which alone all instances agree will be the cause of the given phenomenon".[43] If we apply this to the case of Mesmerism, it means that the knowledge of the operations of the magnetizers is the only event occurring in all experimental situations with magnetic effects, and that it is thus the decisive cause of these effects. According to the words of the commissioners, it was thus 'imagination', based on a knowledge of magnetic operations and effects, that was ultimately responsible for the appearance of mesmeristic states among the persons observed.

Thus (and here it is the *distinction between fact and hypothesis* that is concerned) it is not necessary, in order to explain the phenomenon of Mesmerism, that one assumes the existence of a physical substance, the fluid *fam* or A. M. respectively, supposed to cause this. On the contrary, the investigations of the Academy members did not reveal the least indication of the existence of *fam*, though this seemed to make absolutely no impression on Mesmer. To his dying day, he clung tenaciously to his belief in the existence of a physical agent called 'animal magnetism'.

This perseverance was further fed by his supreme lack of consideration for the *principle of experience*, which left him indifferent to the results of empirical tests of his theories. Once he decided that he was in possession of the truth, he had nothing more to learn from the test of experience. Mesmer's disregard of the law of causality and the principle of experience corresponds to an apparently indestructible trust in the validity of the old fallacy 'post hoc ergo propter hoc'.

Mesmer's own conviction, and probably also his ability to convince others, depended on patients, believing, after a magnetic cure, that they were healed — whether justifiedly or not is of no importance here. On the other hand, the members of the commission pointed out that these supposed therapeutic effects of magnetic treatment could be the result of imagination. The self-healing powers of nature should also not be left out of consideration. And finally, even aspects of the magnetic treatment, such as strongly pressing or rubbing parts of the body, could create effects that Mesmer ascribed to A. M. This argument, too, did not impress Mesmer. Mesmer's distorted understanding of causality led him to view the apparent proof of the success of HAM as support for TAM and as proof of the existence of *fam* and A. M., for that matter.

Moreover, it is not the case that the members of the commission cast doubt on the existence of mesmeristic phenomena as such. On the contrary, in his exposé for the Academy, Bailly stated explicitly that magnetic phenomena are "facts for a still new science, the science of the influence of the moral on the physical".[44] We see here in Bailly's remark the anti-Leibnizian program of taking psychic phenomena out of the realm of metaphysical government into that of mechanics. The science projected by Bailly, which simply examines in other words the interaction between mental and physical states has since itself been established in many variants. Full knowledge of the "influence of the moral on the physical" would consist, however, in solving the so-called mind-body problem, which we are very far from doing and perhaps will never achieve.

Mesmer's supporters now made it clear, entirely in the spirit of their hero, that the — in their eyes — corrupt and stupid academy report was counterbalanced by Jussieu's minority vote.[45] This assessment is not justified by Jussieu's report. Jussieu by no means supported radically flawed practices in science. Neither did he doubt the distinction between fact and conjecture, nor believe that one can do without respect for the law of causality. Finally it is also clear that he did not hold that scientific theories need not care about experience. Jussieu's report contradicts that of the majority only in expressing doubts that the experiments conducted by the majority are themselves sufficient to justify a negative vote on Mesmerism. Moreover, Mesmer may have used an unsuitable theory to explain magnetic phenomena. Jussieu knew of a better one that was worth investigating. His own theory, however, is indebted to a type of theory that in his day had already been surpassed by scientific progress. Similar to alchemical theories, it is a so-called theory of principles according to which all natural events can be traced back to the 'principles' of matter and movement. The principle of movement reveals itself in apparently different ways, for example, in magnetism, in electricity, and in heat. Heat, more than anything else, is responsible for the mesmeristic phenomena, since it is conveyed through the contact taking place during the magnetic treatment. The therapeutic effects of HAM are essentially those of the time-honored method of contact medicine which was simply renewed in this case. Jussieu, moreover, gives no indication in investigating these alternative theories that he does *not* intend to accept the above-mentioned criteria for distinguishing between science and pseudoscience.

As for the second question put to the commission by the king, namely, the question about the therapeutic value of Mesmerism, the argument of the commission is less convincing. Among its objections to the therapeutic value of Mesmerism is, first, that the resulting violent convulsions are not compatible with supporting the healing powers of nature by 'soothing' means. In view of the treatments advocated by academic medicine at the time, this is a hypocritical argument. In addition, the commission feared that convulsions could become habitual, finally occurring without magnetic stimulus. They would reach epidemic proportions and become inheritable. In fact, the Review of the physicians even claimed that the convulsions were carcinogenic. Just as convincing as I find the argumentation of the members of the commission concerning the theoretical status of TAM, NAM, and HAM, their arguments concerning the use or abuse of magnetic treatment are just as unconvincing. In my view, the only convincing argument for or against the therapeutic use of HAM, or any therapy for that matter, is whether it eases the suffering of the patient without keeping him from what might be a better therapy. Whether such a therapy also possesses a corresponding theory is of secondary importance.

IV. ANYTHING GOES?

Paul Feyerabend is frequently said to have voiced the opinion that all attempts at establishing criteria of scientific rationality, and perhaps even all attempts at identifying characteristics of successful scientific practice, are irrelevant and to be rejected in their entirety. Scientific progress has never been furthered by criteria of rationality, but rather came into existence for the most part by neglecting or even infringing upon such criteria. On the whole, it is clear "that the idea of a fixed method, or of a fixed theory of rationality rests on too naive a view of man and his social surroundings".[46] If this is referring to eternally fixed, universal *theories* of rationality and demarcation between science and pseudoscience that have failed historically many times over (cf. fn. 14), then one can only agree with the Feyerabendian conception. If, however it means that all *practical principles* for good science are to be declared obsolete, then I would no longer agree. In such a conception I see rather a certain unenlightened, one could even say, postmodern naivité that is not completely harmless. Where Feyerabend is right is in pointing to numerous examples from the history of science which appear to show that important scientists were not bothered when their new hypotheses apparently contradicted established facts. But it is important to recognize that in so doing, these scientists did not seek to nullify the principle of experience, but only to call for a new interpretation of certain facts or to point out a wrong way of looking at things. The *application* of principles of good science like those used by the members of the Paris commission may be controversial in concrete cases. But the principles themselves in all successful theories in the history of modern science have not been. One can rightfully say that the undeniable success of modern science is due in the last analysis to such principles.

In recent times, the attacks mounted against such principles seem to have increased. Think of the growing strength of anti-evolutionary creationism in the U.S., or the dismal mélange which blends modern physics with far eastern conceptions of wholeness that is supposed to constitute the theoretical nucleus of a 'New Age'. Against these pseudoscientific errors, I would point to the firm anchoring of scientific practice in an enlightened way of life, i.e. a way of life grounded in independent thinking and in self-responsibility. Thus, finally, the distinction between science and pseudoscience is fixed in a rational life-practice. From that it follows that there is no Archimedian point from which one might distinguish between science and pseudoscience. Pseudoscience springs from a flawed way of life and cannot, therefore, be refuted by arguments.

Rational life-practice as a basis for scientific practice and thus of science not only borders on irrationality, in a "downward" direction as it were, but also on the centers of political and religious influence in an "upward" direction. One of the members of the commission, Benjamin Franklin, knew already by the end of August 1784 that the persistent reliance on principles of rational scientific practice in the commission's report was also felt here and there as a threat. In one of Franklin's letters he writes: "The Report is publish'd and makes a great deal of Talk. Everybody agrees that it is well written: but many wonder at the

Force of Imagination describ'd in it, as occasioning Convulsions &c. and some fear that Consequences may be drawn from it by Infidels to weaken our Faith in some of the miracles of the New Testament."[47]

Whoever resists the arrogance of a privileged access to exclusive truths about nature which are not subject to the law of causality, whoever requires the sober empirical control of theoretical flights of imagination, in short, whoever demands generally acceptable reasons for statements, holds onto a conception of man as a being who "should have enough courage to use his own mind". Kant, who said this, wrote in his letter to Borowski (see section II, fn. 13): "A comprehensive rejection here [of Mesmerism] is against the dignity of reason and does not accomplish anything; contemptuous silence is more appropriate for such a type of insanity." Kant's actions, however, stand in opposition to these words. As Butts has shown, Kant did *not* keep silent. His whole philosophical and theoretical work aims in a fundamental way at overcoming the anthropological difference between scientific knowledge and an irrational way of living. Kant himself, however, considered this goal to be unattainable or at least only partially attainable. He closes his letter to Borowski with the apparently resigned comment that "such events [as Mesmerism] last only a short time in the moral world in order to make room for other follies".

The goal of a perfect rationalization of life, however, is not just unattainable. It does not even seem to me to be desirable. In a world where anthropological difference has been removed, i.e. in a perfectly rationalized world Mesmer would never drive a taxicab to a mountain bar, in such a world we would not like to live any more.[48]

(Translation Steven Gillies)

NOTES

[1] [. . .] Zwar die Gelehrtheit feilscht hier nicht papierne Schätze,
Man mißt die Straßen nicht zu Rom und zu Athen,
Man bindet die Vernunft an keine Schulgesetze,
und niemand lehrt die Sonn in ihren Kreisen gehn.
[. . .]
Und hier hat die Natur die Lehre, recht zu leben,
dem Menschen in das Herz und nicht ins Hirn gegeben.

[2] Slightly modified from the editor's introduction to Butts (ed.), (1968). The fact that I am writing this on December 15th, 1987 is a mesmerizing coincidence since Butts wrote his forword on December 15th, 1967.

[3] Perhaps the bars are better there. — But, *de gustibus non est disputandum*: D. H. Lawrence (1956, p. 123), for example, wrote in September 1913: "When I was in Constance the weather was misty and enervating and depressing, it was no pleasure to travel on the big, flat desolate lake" — Christiane Schildknecht, to whom I owe this quotation, finds even misty days at the lake neither depressing nor enervating.
[4] Heidegger (1965, p. 15) coined the phrase 'ontological difference': "Ontic and ontological truth concern to different matters: being things in their being and the being of being things. They belong together essentially because of their relation to the difference between being and being things (ontological difference)".
[5] See fn. 2.
[6] Butts (1984). In the following quotes from this book cited in the paper will be referred to only by page number.
[7] Cynical observers of universities will perhaps find the normal meaning of the term 'insane asylum' too narrow.
[8] In the sense in which Wittgenstein used this term.
[9] Kant speaks (a) of the "forms of pure intuition of space and time" and (b) of the "concepts of pure understanding" or "categories".
[10] Fichte (1845, p. 434) expressed this thought with respect to the reasons for choosing between 'idealism' and 'dogmatism' (= realism) as follows: "What philosophy one chooses depends [. . .] on what kind of person one is: for a philosophical system is not a lifeless household item that one could get rid of or acquire as one likes; a philosophical system is animated by the soul of the man who has it".
[11] Here are a few rhapsodic passages that I have collected: "The apothecary and the surgeon seek to help ailing bodies by empirical and mechanical means; the philosopher directs people to appreciate the powers of reason over feeling" (p. 312). "The power of the mind, of reason, is the power of self-regimentation; it seeks to prevent mental distress by automation of healthy response patterns" (313). "Philosophy produces a feeling of strength that can compensate for bodily frailties by providing a rational estimate of life's worth" (315).
[12] I hope that the reader still has together in his mind the different parts of the somewhat dadaistic title of this paper.
[13] Kant (1972, p. 446) to Ludwig Borowski, written between March 6th and March 22nd, 1790.
[14] According to Laudan (1983) previous attempts at providing a general criterion for distinguishing between science and pseudoscience were not convincing.
[15] I draw for support on my article Wolters (1988).
[16] In view of the claim that TAM is part of mechanistic science, it is surprising that a quantitative or numerical account is nowhere to be found in Mesmer's writings. Also missing are complete diagrams, of which there is otherwise no shortage in 18th century physics books.
[17] Tischner (1928, p. 71). Thus, for example, the theory of the 17th century Scottish physician William Maxwell is nearly identical with Mesmer's TAM, if one replaces Maxwell's vitalistic terminology by Mesmer's mechanistic one of subtle fluids. F. A. Pattie (1956) mentions another point of similarity.
[18] Cf. Mesmer (1800, pp. 31).
[19] Cf. Rothschuh (1978, pp. 185).
[20] Cf. Schott (1982, pp. 205) with reference to Mesmer (1814, p. 166).
[21] Cf. Rothschuh (1978, p. 224); Schott (1982, p. 241).

[22] Mesmer (1775).
[23] Cf. Tischner/Bittel (1941, p. 40).
[24] Sulzer (1775, p. 190).
[25] The father of the ambassador, Gerard van Swieten, was personal physician to the Empress Maria-Theresia of Austria and Professor at the University of Vienna. For this reason he probably had the best contacts to Mesmer's Viennese enemies, and it was probably also the case that van Swieten wanted to support these people with an expected negative report from Berlin.
[26] Cf. Bittel (1940, p. 27).
[27] Cf. fn. 14.
[28] Darnton (1968, pp. 83).
[29] Darnton (1968, pp. 86).
[30] Bericht (1785, p. 15). — Only those two French members of the Academy commission who played a decisive role in preparing the final report experienced the French Revolution five years later. These were the astronomer and later revolutionary mayor of Paris, Jean-Sylvain Bailly, and the very successful (not just as a tax collector) Antoine-Laurent de Lavoisier, considered to be the father of modern chemistry. It is not without tragic irony that both these individuals were executed by the machine named after their colleague on the commission, Guillotin.
[31] Cf. Duveen/Klickstein/Fulton (1954, p. 253): "The first and more extensive *Rapport* is in the style of Lavoisier; a slightly differing version which exists is actually in his handwriting (Oeuvres III, pp. 513—527)."
[32] Jussieu (1826).
[33] Mesmer (1800, pp. 8).
[34] Bailly's exposé before the Academy is in Bericht (1785, pp. 8). It is striking that Bailly, like Kant in his letter to Borowski (see section III, fn. 13), also wants to call in the state to assist in the fight against Mesmerism; with the difference, however, that Kant wants to do this only when morality is threatened. Bailly, on the other hand, seems to have been an eighteenth century version of McCarthy.
[35] This useful suggestion has been made by Lugg (1987, p. 228).
[36] That gives us philosophers a chance.
[37] Here the level of correct data processing in Kant's sense reappears.
[38] Lavoisier/Laplace (1862, pp. 285).
[39] Lavoisier/Laplace (1862, p. 287).
[40] Lavoisier/Laplace (1862, pp. 286).
[41] Lavoisier/Laplace (1862, p. 288).
[42] I would not like to contend, however, that these three characteristics cover all aspects of good scientific practice.
[43] Mill (1973, p. 390).
[44] Bericht (1785, p. 17).
[45] Cf. e.g. Tischner/Bittel (1941, p. 291).
[46] Feyerabend (1979, p. 27).
[47] Letter to William Temple Franklin, August 25, 1784, in Franklin (1970, p. 268).
[48] I am grateful to Mic Detlefsen for revising the text.

BIBLIOGRAPHY

Bericht. (1785). *Bericht der von dem Könige von Frankreich ernannten Bevollmächtigten zur Untersuchung des thierischen Magnetismus*, Altenburg.

Bittel, K. (1940). *Der berühmte Hr. Doct. Mesmer vom Bodensee*, 2nd ed., Friedrichshafen.

Butts, R. E. (ed.). (1966). *William Whewell's Theory of Scientific Method*, Pittsburgh.

Butts, R. E. (1984). *Kant and the Double Government Methodology. Supersensibility and Method in Kant's Philosophy of Science*, Dordrecht/Boston/Lancaster.

Darnton, R. (1968). *Mesmerism and the End of Enlightenment in France*, Cambridge, Mass.

Duveen, D. I., Klickstein, H. S., and Fulton, J. F. (1954). *A Bibliography of the Works of Antoine Laurent Lavoisier (1752—1794)*, London.

Feyerabend, P. K. (1979). *Against Method. Outline of an Anarchistic Theory of Knowledge*, London.

Fichte, J. G. (1845). *Erste Einleitung in die Wissenschaftslehre* [1797], in *Sämmtliche Werke*, Vol. I., Berlin. Repr. under the title *Fichtes Werke*, Berlin, 1971.

Franklin, B. (1970). *The Writings of Benjamin Franklin*, vol. IX (1783—1788), New York.

Heidegger, M. (1965). *Vom Wesen des Grundes*, Frankfurt, 5th ed. (1st ed. 1929).

de Jussieu, A.-L. (1826). 'Rapport de l'un des commissaires chargés par le Roi de l'examen du magnétisme animal', in A. Bertrand (ed.), *Du magnétisme animal en France et des jugement qu'en ont portés les sociétés savantes*, Paris.

Kant, I. (1972). *Briefwechsel*, 2nd ed., Hamburg.

Laudan, L. (1983). "The Demise of the Demarcation Problem", in L. Laudan/R. S. Cohen (eds.), *Physics, Philosophy and Psychoanalysis. Essays in Honor of Adolf Grünbaum*, Dordrecht/Boston/London, pp. 111—127.

Lavoisier, A.-L. (1862). *Oeuvres*, Vol. II, Paris 1862.

Lawrence, D. H. (1956). *Twilight in Italy*, London.

Lugg, A. (1987). "Bunkum, Flim-Flam and Quackery: Pseudoscience as a Philosophical Problem", *Dialectica* **41**, 221—230.

Mesmer, F. A. (1775). *Schreiben über die Magnetcur an einen auswärtigen Arzt*, Wien.

Mesmer, F. A. (1800). *Über meine Entdeckungen*, Jena.

Mesmer, F. A. (1814). *Mesmerismus oder System der Wechselwirkungen. Theorie und Anwendung des Thierischen Magnetismus als die allgemeine Heilkunde zur Erhaltung des Menschen*, K. C. Wolfart (ed.), Berlin (Reprint Amsterdam, 1966).

Mill, J. S. (1973). *A System of Logic* [. . .], in: *Collected Works*, Vol. VII, Toronto/London.

Pattie, F. A. (1956). "Mesmer's Medical Dissertation and its Debt to Mead's 'De imperio solis ac lunae', *Journal of the Histoiry of Medicine* **11**, 275—287.

Rothschuh, E. (1978). *Konzepte der Medizin*, Stuttgart.

Schott, H. (1982). "Die Mitteilung des Lebensfeuers. Zum therapeutischen Konzept von Franz Anton Mesmer (1734—1815)", *Medizinhistorisches Journal* **17**, 195—214.

Sulzer, J. G. (1775). "Gutachten einiger Mitglieder des Königl. Academie der

Wißensch. über den Inhalt des Schreibens über die Magnetcur von Herrn D. Mesmer in Wien etc.", *Allgemeine Deutsche Bibliothek* **26.1**, 190–192.

Tischner, R. (1928). *Franz Anton Mesmer*, München.

Tischner, R. and Bittel, K. (1941). *Mesmer und sein Problem. Magnetismus — Suggestion — Hypnose*, Stuttgart.

Wolters, G. (1988). "Mesmer und sein Problem: wissenschaftliche Rationalität", in G. Wolters (ed.), *Wissenschaft, Scharlatanerie, Poesie: Franz Anton Mesmer und der Mesmerismus*, Konstanz.

ANDREW LUGG

HISTORY, DISCOVERY AND INDUCTION: WHEWELL ON KEPLER ON THE ORBIT OF MARS*

William Whewell's stature among philosophers has slipped considerably since the late 1960s and early 1970s, when he was widely portrayed as providing an alternative to positivist philosophy of science. Partly as a result of shifting philosophical fashion and partly because the shortcomings and idiosyncracies of the historicist approach have become clearer, philosophical interest in Whewell now tends to focus on his discussion of particular issues such as the role of consilience in theory choice rather than on his more general conception of scientific inquiry. Yet in redressing the balance we are in danger of losing sight of the gains that Whewell undoubtedly made. There remains much to be said for his contention that philosophy of science should be rooted in a close examination of actual scientific practice. And many of his specific insights concerning scientific discovery have still to be fully assimilated by philosophers of science.

As is well known Whewell takes Kepler's discovery of the elliptical orbit of Mars to be an exceptionally clear example of scientific induction. While recognizing that Kepler's "works are . . . extremely curious and amusing", Whewell contends that they "are a very instructive exhibition of the mental process of discovery". These works, he argues, "exhibit to us the usual process (somewhat caricatured) of inventive minds: they rather exemplify the *rule* of genius than (as has generally been hitherto taught) the *exception*" (1857/1967, I, p. 318). In particular he regards Kepler's discovery as deserving special attention since it illustrates especially well the crucial element in all induction, namely the introduction of a new conception. For him Kepler is remarkable if only because he clearly "apprehended that colligation of facts which is the main business of the practical discoverer" (1860/1971, p. 121).

Whewell accepts the standard conception of "induction . . . as the process by which we collect a *General Proposition* from a number of *Particular Cases*" (1847/1967, II, p. 48). But he rejects the assumption that "the general proposition results from a mere juxta-position of the cases" on the grounds that "there is a New Element added to the combination by the very act of thought by which they are combined".

J. R. Brown and J. Mittelstrass (eds.), An Intimate Relation, 283—298.

"The peculiar import of the term *Induction*", he insists, is that "some Conception is introduced, some Idea is applied, as the means of binding together the facts, and thus producing the truth" (pp. 49—50). However "*Induction* . . . was originally or anciently employed", the plain fact of the matter is that "in every inference by Induction, there is some Conception *superinduced* upon the Facts" (p. 50).

In the special case of Kepler's discovery we should think of "the Invention of the Conception [of the ellipse as] the great step in the *discovery*", Kepler's main achievement having been to "superinduce" this geometrical conception on Tycho's data (p. 51). To appreciate the character of the discovery, we need to recognize that Kepler was the first to have "bound together the particular observations of several places of Mars by the notion, or . . . the *conception*, of an *ellipse*, which was supplied by his own mind" (1860/1971, p. 253). This was "an essential element in his Induction": before his investigations the facts were "detached, separate, lawless"; afterwards they were "connected, simple, regular" (p. 254). As in every discovery a "Conception . . . which did not exist in any of the observed facts" had been introduced, "a Principle of Connextion" had been supplied, a "String" on which to hang "the Pearls" had been provided (1847/1967, II, p. 48).

Whewell thus stands foursquare against Mill's suggestion that the sole induction that Kepler performed was to infer from observed positions of the planet to unobserved ones. Whewell agrees that Kepler inferred that the planet "would continue to revolve in [the] same ellipse" and that "during the time which intervened between two observations [the planet's positions] must have coincided with the intermediate points of the curve" (see Mill 1843/1973, p. 293 and Whewell 1860/1971, p. 248). But he is unequivocally opposed to Mill's further suggestion that these inferences constituted "the only real induction concerned in this case". In his view the inferences isolated by Mill contributed little if anything to Kepler's fundamental achievement. They required no special expertise and would have been made routinely by any competent astronomer at the time.[1]

According to Whewell, Mill and like-minded philosophers overlook the very element in scientific discovery that requires special talent, the invention of appropriate theoretical conceptions. It is, he argues, a mistake to think that the facts determine their own interpretation and even more so to suppose that scientists obtain their conclusions by generalizing from unconceptualized facts. In cases of induction such as

Kepler's the relevant inference is to the conclusion that *A*s should be classified as *B*s, not merely to the conclusion that all *A*s are *B*s (given that some of them are). When Kepler found that Tycho's data could be colligated by means of an ellipse, he had — says Whewell — in effect already made the various inductions that Mill singles out for special discussion. As he puts the point, the Keplerian colligation included all that Mill mentions since "a continuous line, a periodical motion, are implied in the term *orbit*" (1860/1971, p. 248).

In arguing that Kepler "superinduced" a conception on Tycho's data, Whewell does not deny that the ellipse law describes the phenomena. On the contrary, he explicitly holds that "the orbit of Mars is a Fact — a true Description of the path" (1860/1971, p. 250). In his view the "description of ellipses" was fundamental to the Keplerian inference (1847/1967, II, p. 51) and the theory that Kepler finally adopted was "a mere representation of the motions and distances as they were observed" (1857/1967, I, p. 331). Indeed Whewell is even willing to go along with Mill's much criticized suggestion that a person with "adequate visual organs and a suitable position" would be able to see the ellipse in the phenomena (see Mill 1843/1973, p. 297 and Whewell 1860/1971, p. 249).[2]

Where Whewell and Mill differ is with regard to the propriety of calling the ellipse law an induction given that it merely describes, represents, sums up the phenomena. The butt of Whewell's criticism is Mill's conclusion that Kepler "found the expression only, not the inference" (see Mill 1843/1973, p. 294). For Whewell finding the expression — i.e. the description — was itself a matter of inference. "There is", he maintains, "no validity discoverable in the distinction which Mr. Mill attempts to draw between 'descriptions' like Kepler's law of elliptical orbits, and other examples of induction" (1860/1971, p. 252). "Description [of the sort involved in Kepler's discovery] is a kind of Induction, and must be spoken of as Induction, if we are to speak of Induction as the process by which science is formed" (p. 249).

To understand what Whewell is driving at here, it is important to remember that he takes uninterpreted observations to be nondescript. "In whatever manner facts may be presented to the notice of a discoverer", he tells us, "they can never become materials of exact knowledge, except [in the event that] they find his mind already provided with precise and suitable conceptions by which they may be analyzed and connected" (1847/1967, II, p. 23). "The fact of the elliptical orbit was not the sum of the observations *merely*; it was the sum of the observations, *seen under a new point of view*" (1860/1971, pp. 256—257). In other words Mill erred when he argued that "conceptions do not develop themselves from within, but are impressed from without" since these "must have both origins, or they cannot make knowledge" (p. 258). Even someone with "adequate visual organs and a suitable position" would — according to Whewell — still be obliged to contribute something from his or her own mind in order to see the planet's orbit as an ellipse.[3]

It is undoubtedly a difficult question "whether or not abstract orbital paths are found in astronomical data" (Butts 1968, p. 28). But some of the mystery can perhaps be dispelled by noting that the positions of planets define ellipses in much the same way as sets of stars define constellations. While Whewell is not always clear about these matters, what he presumably means is that an ellipse can properly be said to be in (and not in) the phenomena in exactly the same sense that a certain figure can properly be said to be in (and not in) the night sky. When we say that a planet has an elliptical orbit or that a group of stars forms a constellation, we do not project new facts into the world but see old ones in a new way. In both cases what we say about the world depends on the concepts that we employ but it is the world itself that determines whether and to what degree what we say is true.[4]

Be this as it may, Whewell can hardly be faulted for failing to recognize the ingenuity involved in scientific discoveries like Kepler's. When reading Whewell one never feels — as one does reading some philosophers — that such discoveries are relatively straightforward affairs, that they are more a matter of good fortune than of intellectual brilliance. It is one of his constant themes that discovery involves "inventive talent" (1847/1967, II, p. 41), that "at each step of the progress of science, are needed invention, sagacity, genius" (1847/1967, I, viii). "A facility in devising hypotheses", he insists, "is so far from being a fault in the intellectual character of a discoverer, that it is, in truth, a faculty indispensible to his task" (1847/1967, II, p. 54). It is not enough to recognize that "the discovery of new truth requires . . . minds careful and scrupulous in examining what is suggested"; we must also acknowledge that "it requires, no less, such as are quick and fertile in suggesting" (pp. 55—56).

Unsurprisingly then Whewell looks askance at the idea of "a logic of induction" construed as a set of general principles for devising new theories. In his view "an Art of Discovery is not possible" and "we may hope in vain, as Bacon hoped, for an organ which shall enable all men to construct scientific truths, as a pair of compasses enables all men to construct exact circles" (1847/1967, I, p. viii). For besides taking it to be indisputable that "no maxims can be given which inevitably lead to discovery", Whewell also maintains that "in every inductive inference, an act of invention is requisite" and that "Induction mounts by a leap which is out of the reach of method" (1847/1967, II, pp. 20, 51 and 92). Indeed Whewell's ultimate position would seem to be that there are no methods for deriving laws from data apart from special methods appropriate to special circumstances (compare Book XIII on the "Methods Employed in the Formation of Science").[5]

Should we therefore conclude that Whewell is — as is commonly assumed — a proponent of the hypothetico-deductive method?[6] Given his emphasis on hypotheses as "guesses" (p. 41), it is tempting to assume that he is. However it is also true that much of what he says runs counter to the deductivist conception. Contrary to Venn, "Whewell's account of the Inductive process [does not resolve] itself into making guesses, and then justifying these guesses by subsequent deduction" (1907/1972, p. 356). Nor is it plausibly regarded — as Ducasse would have it — as "a comprehensive and systematic theory of induction . . . in terms of the so-called Newton method of Hypothesis — Deduction — Verification" (1966, p. 217). In actual fact Whewellian induction is less easily characterized and very much more subtle.

Whewell does not think of scientists as merely showing that particular hypotheses fit the available data, nor does he use the expression "inductive inference" solely to indicate that such a fit has been achieved. In his view scientists do more than deduce observational consequences and check to see whether or not they agree with the facts. Rather they should be seen as proving that hypotheses hold, as establishing that they are true, even as demonstrating their necessity. In this regard at least Whewell is closer to the inductivist who takes inductive inference to be a technique for deriving general conclusions from premises about particular individuals. He does not have anything unusual in mind when he speaks of induction as "the genuine source of all our real general knowledge" (1847/1967, II, p. 47). He is not a deductivist in disguise.[7]

In the final analysis Whewell remains faithful to the Baconian view that "true knowledge is . . . obtained from Facts by Induction" (1847/1967, I, p. vi). To appreciate his position we need to remember that he holds that whenever a new law of nature is discovered "a number of facts . . . are brought into a point of view in which order and connexion becomes their essential character" and "it is seen that each fact is but a different manifestation of the same principle; that each particular is that which it is, in virtue of the same general truth" (1834/1984, p. 315). For all his emphasis on conjecture and supposition, Whewell still takes the "Formula for the Colligation of Facts by Induction" to be "'The Several Facts are exactly expressed as one fact if, *and only if*, we adopt the Conception and the Assertion' of the inductive inference" (1847/1967, II, p. 90).

In the particular case of Kepler's discovery, the relevant "induction" was not to the conclusion that the elliptical-orbit hypothesis is consistent with Tycho's data but to the law itself, Kepler's conclusion having been "the true doctrine, that the planet's path is an ellipse" (p. 42). When Tycho's data had been colligated under the conception of an ellipse, "Kepler's *laws* of the elliptical motion of the planets were established [and] immediately became the *facts* on which the mathematicians had to found their mechanical theories" (1847/1967, I, p. 48). Indeed Whewell even endorses Kepler's claim — which he made in response to a challenge posed by Ramus — to have constructed "an astronomy without hypotheses" (1857/1967, I, p. 331). In Whewell's opinion "this was not saying too much".

These observations notwithstanding, it may still seem that Whewell should be regarded as a deductivist if only because he fails to provide an account of inductive inference. While it is true that he holds that hypotheses are proved, it is also true — as has often been argued — that he has little if anything to say concerning what counts as a proof of a hypothesis. In this regard at least Mill was right: "Dr. Whewell's theory of the logic of science [passes] over altogether the question of Proof" (1843/1973, p. 304). As Buchdahl has observed "there is in Whewell no provision for forms of argument which involve the drawing of conclusions, or the detaching of such 'conclusions' from the evidence [and what is] more extreme even, there is no attempt to assess in any way quantitatively or quasi-quantitatively the inductive probability of any hypothesis" (1971, p. 350; see also Butts 1973, p. 56).

Nor is it helpful to recall that Whewell introduces his "inductive tables" under the rubric of "the logic of induction". True, he avers that "the analysis of doctrines inductively obtained, into their constituent facts, and the arrangement of them in such a form that the conlusiveness of the induction may be distinctly seen, may be termed the *Logic of Induction*" (1847/1967, II, p. 82). But the arrangement of doctrines that he envisions does not by any stretch of the imagination supply "the means of ascertaining the truth of our inductive inferences, so far as the form in which our reasoning may be stated can afford such a criterion" (p. 83). Moreover it should not be forgotten that when challenged on the question of whether inductive tables actually constitute a "logic", Whewell allowed that they did not and retorted that this was "*so much the worse for logic*" (see Todhunter 1876, II, p. 417).

Nonetheless it would be wrong to conclude that Whewell has nothing of importance to say about discovery and to treat his position as tenable only insofar as it accords with strict deductivism. While Whewell does frequently seem to be attempting to reconcile the irreconcilable, his remarks are undoubtedly informed by a definite vision of how scientific discoveries are made. Understanding them in the way he intended — as complementary to historical investigations — it is not difficult to see why he considered the issue of the "logic of induction" to be unimportant even as he insisted that scientists prove their results. Indeed there is no need to look further than his analysis of the "Inductive Epoch of Kepler" in *The History of the Inductive Sciences* to appreciate what he had in mind.

In the discussion of the Keplerian discovery in the *History*, which is considerably more detailed than the discussions of *The Philosophy of the Inductive Sciences* and "Mr. Mill's Logic", Whewell outlines some of the more important steps leading to Kepler's final conclusion. Interestingly he does not argue that Kepler made his discovery by superimposing a conception on Tycho's observations but instead focuses on the manner in which Kepler deployed these observations in the course of developing his new view of planetary motion. Here Whewell's main object is to relate the discovery to the wider framework in which Kepler pursued his astronomical investigations and to clarify the various technical considerations that prompted him to revise his inital assumptions and ultimately to accept the elliptical-orbit law itself.[8]

Whewell starts by noting that "the occasion of the discovery of [Kepler's] laws was the attempt to reconcile the theory of Mars to the theory of eccentrics and epicycles [and] the event of it was the complete overthrow of that theory, and the establishment, in its stead, of the Elliptical Theory" (1857/1967, I, pp. 324—325). Then he goes on to detail the "repeated struggles" that led Kepler to conclude that "the path of a planet is [not] a perfect circle", to adopt "the supposition of the oval", to resort to the use of an ellipse to simplify calculation, and finally to appreciate that "he might take another ellipsis, exactly intermediate between the former one and the circle, and that this must give the path and motion of the planet" (pp. 327—329). In a nutshell then Whewell takes Kepler to have "combined the observed geocentric places with successive modifications of the theory of epicycles, till at last he was led, by one step after another, to change the epicyclical into the elliptical theory" (p. 325).

Here what Whewell is describing is not a single "act of colligation" nor a single inductive inference but rather a long and highly complex investigation. Although he undoubtedly holds that Kepler's final colligation was of immense importance, it is the argument leading up to it that receives the bulk of his attention. In his view it is not the specific interrelationships of the Keplerian law and Tychonic data that require investigation but rather the character of the inquiry that led Kepler from the data to the law. To appreciate what Kepler achieved we must attend to the process whereby he arrived at his result as well as to the nature of the result itself. In this and similar discoveries what is needed is not a logical analysis (in terms of the abstract propositional content of the laws and data in question) but a historically-oriented analysis (in terms of scientists' actual procedures).[9]

Whewell does indeed hold that Kepler "made nineteen hypotheses with regard to the motion of Mars, and calculated the results of each" (1847/1967, II, p. 42). But he does not take him to have engaged in nineteen separate inquiries. Rather the opposite: according to Whewell the various stages in Kepler's long struggle were linked together and the whole investigation comprised a single "train of researches" (1857/1967, I, p. 326). As he interprets the historical situation Kepler's formulation of the ellipse law would have been arbitrary and ill-motivated in the absence of the results (both positive and negative) of the astronomer's early inquiries. Instead of thinking of Kepler as having found the ellipse law after a number of unfortunate false starts, we should think of him as having found it because he had thoroughly explored other hypotheses and figured out why these fell short. Here as elsewhere "the discoverer [had] constantly to work his way onwards by means of hypotheses, false and true" (1847/1967, II, p. 59).[10]

In this respect the difference between Whewell's view and the empiricist view often attributed to Mill could hardly be greater. While allowing that the ellipse hypothesis provided a compendious summary of Tycho's observational data, Whewell rejects the suggestion that Kepler summarized them by adding them together. For him it is a fundamental error to assume that Kepler obtained the elliptical-orbit law this way if for no other reason than that Tycho's data had to be interpreted, seen from a point of view, organized in a coherent fashion, reduced to an appropriate form. The calculations that Kepler made were of a quite different order from those envisioned by partisans of the strict empiricist point of view. The ellipse law was not the result of a straightforward process of summation but rather the culmination of a long and complex theoretical investigation.[11]

Bearing these remarks in mind it is not difficult to see why Whewell took the criticism that he had not supplied principles of inductive validity to be beside the point. He would not have seen the need for such principles for the simple reason that he regarded scientific discovery as a process of careful thought and detailed analysis. The clear assumption underlying his observations is that scientists reason their way to their conclusions, that their specific acts of colligation take the form they do because they are informed by a deep understanding of the various theoretical issues involved. It is not for nothing that Whewell compares the discovery of laws with the deciphering of codes (see for instance his 1834/1984, p. 315). Nor is it by chance that he constantly refers to the indispensibility of "sagacity" for the development of science and fruitful colligation (see for instance his 1847/1967, II, p. 40 and p. 55).[12]

This is particularly clear in Whewell's discussion of the ellipse law, which is largely devoted to detailing Kepler's reasons for revising his earlier views and for embracing new ones. "At every step", he observes, "he [i.e. Kepler] endeavoured to support his new suppositions by what he called, in his fanciful phraseology, 'sending into the field a reserve of new physical reasoning on the rout and dispersion of the veterans': that is, by connecting his astronomical hypotheses with new imaginations, when the old became untenable" (1857/1967, I, p. 325). Moreover it should not be overlooked that Whewell maintains that Kepler's discovery of "the right figure was a matter requiring research, invention, resource" (1860/1971, p. 254) and that he draws attention to the trouble that Kepler took to record "the notions by which he had been led to invent or to entertain [the various suppositions that he had made]" (1857/1967, I, p. 326).[13]

Whewell means what he says when he speaks of scientists as learning from their mistakes. Unlike philosophers such as Karl Popper for whom errors merely provide information about what is not the case, Whewell recognizes that scientists can also glean information from them concerning what is the case, that they frequently learn something positive from their mistakes. In particular he reminds us that early astronomers were able to advance their investigations considerably by determining where the errors of inadequate hypotheses lay. While "the *Doctrine of epicycles* . . . was erroneous", he argues, "it was . . . of immense value to the progress of astronomical science; for it enabled men to express and reason upon many important truths which they discovered respecting the motion of the stars, up to the time of Kepler" (1847/1967, II, pp. 60—61). Indeed Whewell even suggests that "all who discover truths, must have reasoned upon many errors" (p. 56).

Nor should the importance that Whewell accords to discussion and disagreement in science be forgotten. For him scientific progress depends in large measure on the clash of ideas and on the ability of scientists to negotiate their way among conflicting demands. "Discussions of Ideas" can, he tells us, be crucial for scientific advance in that they constitute "the Method (if they may be called a *method*) by which the Explication of Conceptions is carried to the requisite point" (1847/1967, II, p. 376).[14] In his view it is a simple but profound fact that "controversies make up a large portion of the history of each science; a portion quite as important as the study of facts; and a portion, at every stage of science, quite as essential to the progress of truth" (1860/1971, p. 255). Furthermore along these lines it should be remembered that the emergence of truth from the clash of ideas is one of the central themes of Whewell's famous account of "The Transformation of Hypotheses in the History of Science" (see pp. 492—503).

It is true that Whewell confounds the contexts of discovery and justification, that he was — as Butts succinctly puts it — "both historically and spiritually a preReichenbachian" (1973, p. 56). But this does not mean that he was committed to an untenable form of psychologism. Quite the reverse: his failure to separate the two contexts should put us on our guard concerning the import of Reichenbach's distinction. If nothing else, Whewell's close analysis of Kepler's discovery provides us with an incentive to look twice at the common assumption that there is a logical gulf between induction construed as a process of discovery and induction construed as a mode of proof (see also my 1985, section

VI). One might even argue that Whewell establishes — albeit inadvertently — that there is a sense in which discovery and justification go hand in hand and that there is a viable alternative to the kind of rigorous anti-psychologism that has long been the vogue.

What I am suggesting then is that Whewell's approach to the issue of discovery differs both from the approach of the traditional inductivist and from that of the modern-day deductivist. From Whewell's standpoint each of the two sides is partially right and partially wrong. On the one hand he rejects the inductivist's assumption about the existence of canons of proof even as he insists that scientists reason to their conclusions. On the other hand he takes exception to the deductivist's contention that reasoning in science is always from (rather than to) hypotheses even as he embraces the thesis that there is no general "Art of discovery". By concentrating on the process or "method" of reasoning to conclusions, he is able to straddle the fence and to account both for the element of creativity, invention and genius in science and for the fact that the discovery of new hypotheses is — to use a phrase of Hanson's — often a "reasonable affair" (1958, p. 71).

One reason why philosophers tend to overlook views such as Whewell's is that they assume that scientists must argue either from data to laws or from laws to data and they forget that scientific conclusions are normally obtained from a vast range of premises. Whewell certainly did take Kepler to have reasoned from Tycho's data but he did not for one moment assume these to have been his only premises, his view being that the final Keplerian colligation rested on a wide range of theoretical, conceptual, mathematical and methodological assumptions. Whewell's picture of Kepler was — as he himself put it — one of an individual proceeding "from observations by means of hypotheses" (1847/1967, II, p. 222). To understand Kepler's achievement we need to keep firmly in mind the point — stressed by Whewell — that he uncovered the planet's true orbit by revising the body of background information that he began with (using Tycho's data and various physical and mathematical principles).

It is also in this light that we should understand the "inductive tables", on which Whewell sets so much store. He does indeed refer to these tables loosely as constituting a logic of discovery but he also expressly states that they represent "the elements and order of [scientists'] inductive steps, [not] the whole signification of the process in each case" (1847/1967, II, p. 77). To read these tables properly, we must consider them in conjunction with the detailed historical accounts of the *History*. For what they in fact schematize is not a set of principles of "inductive validity" but rather Whewell's earlier historical analyses of the steps of specific discoveries (compare Butts 1973, p. 77). In Whewell's eyes the entries comprise summary statements of the starting and finishing points of especially significant scientific investigations and should be thought of as being mediated — as Whewell himself notes in connection with Kepler's discovery — by a variety of further physical assumptions (see the inductive table for astronomy in Whewell 1847/1967, II, between pp. 118 and 119).

Furthermore we can now better appreciate Whewell's insistence that Kepler "established" the laws of the elliptical motion of the planets. The critics are right to point out that the ellipse law cannot be established as an "impregnable fact" merely by noting that it is more accurate and simpler than the epicyclic hypothesis (compare Wilson 1974, p. 246). But this is fully compatible with Whewell's view that Kepler had proved the ellipse law and even with his contention that he had constructed "an astronomy without hypotheses". For it was no part of Whewell's account that Kepler never made use of hypotheses, still less that the ellipse law would never require modification. What he was mainly concerned to argue was that Kepler could justifiably claim to have provided a proof of the law given Tycho's data and a host of auxiliary assumptions. For Whewell, Kepler was justified in thinking of the ellipse law as nonhypothetical because he had shown how it can be derived, because he had established it without relying on assumptions "merely postulated and not proved" (p. 252).[15]

In this paper I have focused on those parts of Whewell's discussion in which he attends most closely to actual scientific practice. This is in line with his expressed aim of developing a philosophy based on a "systematic and regular" study of science and its history rather than "casually and arbitrarily" (1847/1967, I, p. 12). But it is also true that some of what Whewell says appears to have been shaped more by *a priori* philosophical considerations than by a study of science. Thus it is far from obvious that the various Kantian and Platonic elements that permeate Whewell's thinking were — or can be — derived from his historical analyses. And one might certainly query the status of his views (discussed under the rubric of "the fundamental antithesis of philosophy") concerning the inseparability of thoughts and things, theories and facts, ideas and sensations. Doubtless it would be foolhardy to suppose that the various concessions to his philosophical preconceptions that Whewell made in the course of his deliberations were always minor or beside the point.

Nonetheless it seems clear that the theme that I have extracted from Whewell's discussion is central to his thinking about scientific inquiry. Even granting that he compromised his initial vision of a philosophy of science subservient to science and its history, it can hardly be denied that his thinking continued to be informed by his analyses of particular historical cases. The important thing to bear in mind is that Whewell's general philosophical remarks about scientific discovery cannot be detached from his remarks about discoveries such as Kepler's without making them seem ill-motivated, confused, even incoherent. When considering his conception of discovery (and induction) it is crucial that the discussion of the *History* be accorded at least as much attention as the discussion of the *Philosophy* itself.

University of Ottawa

NOTES

* I am pleased to have the opportunity to dedicate this paper to Robert Butts on the occasion of his sixtieth birthday. While preparing it I have been often reminded of how much Butts has done to alert us to the importance (and limitations) of Whewell's philosophy of science, Hopefully some of what I have learned from studying his work is reflected in what follows.

1 Compare Venn (1907/1972, p. 355): "Whewell, in fact, almost ignores the generalizing element in our inductions; not of course that he would deny its existence, but rather because he takes for granted that it would be sure to follow as a matter of course".

2 Note also that it is beside the point that the Martian orbit cannot be seen to be elliptical since it is not exactly an ellipse and even if it were, it would be too close to being circular to be detected as an ellipse (compare Venn 1907/1972, p. 354 and Ducasse 1966, pp. 214–215). The question at issue is not whether the Martian orbit describes an ellipse exactly but whether it describes one within specifiable limits and whether it could conceivably be seen as so doing.

3 Thus I take Whewell's treatment of induction to foreshadow Norwood Russell Hanson's account of the "logic of discovery" (see especially Hanson 1958, chapter 4). Whewell's view that Kepler saw Tycho's data from a "new point of view" is much the same as Hanson's view that he discovered a pattern in these data by "pull[ing] together [the data] into a geometrically intelligible pattern" (p. 83). Interestingly Hanson takes Whewell's account of Kepler's discovery to be quite different from his own even though he recognizes that Peirce — whose views he endorses — regarded discovery as "begin[ning] always with [a Whewellian] colligation" (quoted by Hanson, p. 88).

[4] Compare Ducasse's discussion of Whewell's comparison of colligations with cryptograms (1966, p. 202). Ducasse is surely right to note that a new conception is no more an additional fact than is the key to a cryptogram, but presumably the relevant comparison is between the conception and the message embedded in the cryptogram (rather than the key). Also it should be borne in mind that Whewell holds that it is no more possible to discern an ellipse in Tycho's data without an appropriate conception than it is possible to find Kepler's law in his book without a knowledge of Latin (1860/1971, p. 257).

[5] Significantly Whewell devotes the bulk of his discussion of the "General Rules for the Construction of the Conception" to a historical account of Dulong and Petit's investigation of the law for cooling bodies (see 1847/1967, II, pp. 389—395). In fact he not only fails to supply "general rules", he freely admits that "we cannot give rules which will be of much service" (p. 395).

[6] Compare Buchdahl 1971, p. 345, Butts 1968, p. 17, Ducasse 1966, p. 217 and Venn 1907/1972, p. 356. It is perhaps also worth recalling here that the view championed by Hanson (see note 3) is often held to reduce to a form of deductivism and that Hanson himself explicitly states that Whewell's account of Kepler's discovery is "little better than the modern hypothetico-deductive account" (1958, p. 84).

[7] In formulating this point I have benefited from Buchdahl's discussion of the deductivist and inductivist approaches in his 1971, section I (even though I disagree with his interpretation of Whewell as a deductivist). Also I might note here that Butts observes that the standard view of Whewell as a hypothetico-deductive theorist overlooks distinctive features of his methodology (see his 1973, pp. 66ff) and that Ducasse is on record as holding — in apparent opposition to his characterization of Whewell as a proponent of the Newtonian method — that Whewell regards scientists as "*proving* true or false each hypothesis thought of, in its turn" (1966, p. 215).

[8] Whewell's account might be criticized — e.g. on the grounds that it fails to do justice to Kepler's physical arguments (see Wilson 1974, p. 247) — but this does not detract from his discussion of the philosophical issues involved. More complete and more accurate historical accounts can be found in Aiton 1969, Koyré 1973 and Wilson 1974.

[9] Compare Buchdahl's contention that Whewell's "emphasis is all on 'process'" (1971, p. 350), Butts's observation that "Whewell viewed science as an historically developing process" (1973, p. 57) and Strong's point that Whewell employs "the term 'induction' to cover the *modus operandi* of discovery" (1956, p. 231). (Also compare Stoll 1929, chapter IV, which is devoted to "The Process of Discovery".) Furthermore in connection with the present point it is of interest that both Peirce and Hanson treat Kepler's discovery as a process even as they attempt to reduce it to a single "retroduction" (see Peirce 1960, p. 31 and Hanson 1958, p. 76).

[10] It is, I believe, in this context that Whewell's discussion of consilience and other "tests of hypotheses" needs to be considered. According to Whewell these are important aids for "working onwards" in that they constitute "some of the most general ... processes by which, in certain cases, the discovery of the laws of nature may be materially assisted" (1847/1967, II, pp. 59—60).

[11] See also Ducasse (1966, p. 214): "Whewell would acknowledge that the idea of ellipticity as applied to the orbit of Mars does summarize our observations in the sense that it states them all at once, but not in the sense that it was obtained from them by a

mere process of summation". In passing I should also mention that Mill's position is considerably more sophisticated than usually supposed. For he not only allows that Kepler tried a number of hypotheses before obtaining the correct one, he also explicitly states that his guesses were "skilful" and not merely "lucky" since he was "abounding in knowledge and disciplined in intellectual combinations" (1973, p. 297).

[12] The point here is similar to one to which Ducasse alludes, namely that Whewellian colligations are like diagnoses (see his 1966, p. 215). Unfortunately Ducasse does not pursue this suggestion but adopts the implausible view that to diagnose is to state "that what one perceives is such as it would be if a certain conjecture were true". In addition it should be kept in mind — as Niinuluoto has stressed — that "Whewell's theory of induction is . . . related to the idea of analysis as an 'upward movement'" (1977, p. 292). (Compare in particular Whewell's discussions in his 1847/1967, II, pp. 382—383 and pp. 389—395.)

[13] Here again Whewell foreshadows Peirce and Hanson, for whom Kepler "never modified his theory capriciously, but always with a sound and rational motive for just the modification he makes" (Peirce 1960, p. 31; see also Hanson 1958, p. 84). (In addition compare Whewell's discussion of Dulong and Petit's investigations referred to in note 5. Here too Whewell's emphasis is on the process of reasoning that led to the discovery.) Moreover I should mention that my argument runs counter to Wilson's claim that Peirce was the first to recognize that Kepler's discovery depended on a "succession of reasonings, testings, choices, and exclusions" (see his 1974, p. 249).

[14] In her discussion of Whewell, Stoll observes that it is strange that he discusses the explication of conceptions before the colligation of facts (1929, pp. 68—69). However if the present view is correct, it made good sense to Whewell to have discussed the explication of conceptions first if only because such explication may figure prominently in the development of correct colligations.

[15] Hence I disagree with Wilson's contention that Whewell misunderstands Kepler's response to the Ramean challenge (1974, p. 247). On my account Whewell's reading is similar to the one that Wilson attributes to Kepler himself (see pp. 251—254).

BIBLIOGRAPHY

Aiton, E. J. (1969). "Kepler's Second Law of Planetary Motion", *Isis*, **60**, pp. 75—90.

Buchdahl, G. (1971). "Inductivist Versus Deductivist Approaches in the Philosphy of Science as Illustrated by Some Controversies Between Whewell and Mill", *Monist*, **55**, pp 343—367.

Butts, R. E. (ed.). (1968). *William Whewell's Theory of Scientific Method*, Pittsburgh: University of Pittsburgh Press.

Butts, R. E. (1973). "Whewell's Logic of Induction", In R. N. Giere and R. S. Westfall (eds.), *Foundations of Scientific Method: The Nineteenth Century*, Bloomington: Indiana University Press, pp. 53—85.

Ducfasse C. J. (1966). "William Whewell's Philosophy of Scientific Discovery", In E. Madden (ed.), *Theories of Scientific Method*, Seattle: University of Washington Press.

Hanson, N. R. (1958). *Patterns of Discovery*, Cambridge: Cambridge University Press.

Koyré, A. (1973). *The Astronomical Revolution*, Ithaca: Cornell University Press.

Lugg, A. M. (1985). "The Process of Discovery", *Philosophy of Science*, **52**, pp. 207–220.

Mill, J. S. (1843/1973). *A System of Logic*, Toronto: University of Toronto Press.

Niiniluoto, I. (1977). "Notes on Popper as Follower of Whewell and Pierce", *Ajatus*, **37**, pp. 272–327.

Peirce, C. S. (1960). *Collected Papers*, C. Hartshorne and P. Weiss (eds.), Cambridge: Harvard University Press, Volume I.

Stoll, M. R. (1929). *Whewell's Philosophy of Induction*, Lancaster: Lancaster Press.

Strong, E. W. (1955). "William Whewell and John Stuart Mill: Their Controversy about Scientific Knowledge", *Journal of the History of Ideas*, **16**, pp. 209–231.

Todhunter, I. (1876). *William Whewell, D. D. An account of His Writings*, London: Macmillan.

Venn, J. (1907/1972). *The Principles of Empirical or Inductive Logic*, Second Edition, London: Macmillan. Reprinted: Burt Franklin, New York.

von Wright, G. H. (1957). *The Logical Problem of Induction*, Second Edition, Oxford: Basil Blackwell.

Whewell, W. (1834/1984). *Astronomy and General Physics Considered with Reference to Natural Theology*, London: William Pickering. Reprinted in part in Y. Elkana (ed.), *W. Whewell, Selected Writings on the History of Science*, University of Chicago Press, Chicago.

Whewell,W. (1857/1967). *The History of the Inductive Sciences*, Third Edition, London: John W. Parker. Reprinted: Frank Cass, London.

Whewell, W. (1860/1971). *On the Philosophy of Discovery*, London: John W. Parker. Reprinted: Burt Franklin, New York.

Whewell, W. (1847/1967). *The Philosophy of the Inductive Sciences*, Second Edition, London: John W. Parker. Reprinted: Johnson, New York.

Wilson, C. (1974). "Newton and some Philosophers on Kepler's 'Laws'", *Journal of the History of Ideas*, **35**, pp. 231–258.

LARRY LAUDAN

FOR METHOD: OR, AGAINST FEYERABEND

During the last quarter century, the theory of scientific methodology has come in for more than its share of drubbing from a variety of sources. Polanyi, Quine, Hesse, Kuhn, Wittgenstein, and a host of others have urged the abandonment of the methodological enterprise. Most of the arguments against methodology boil down to one of two sorts. They typically allege either (a) that the rules of scientific investigation radically *underdetermine* theory choice (and are thereby presumed to be impotent), or else (b) that the explicit rules of methodology are so vague and ambiguous that they forbid nothing[1]. I have tried to show elsewhere that the arguments lying behind these particular allegations will not support the conclusions drawn from them.[2] But I have yet to come to terms in print with the writings of the most forceful and persistent critic of methodology in our time, Paul Feyerabend.

Feyerabend's critique of methodology involves a number of ingenious and original arguments, quite distinct from those commonly found in the literature. Because that is so, because Feyerabend's views have had enormous impact (especially in the social sciences), and because — their ingenuity aside —they are a series of non sequiturs, it seems time to confront them directly.[3]

SCIENCE WITHOUT RULES (METHODOLOGICAL ANARCHISM)

Feyerabend's *general* views on methodology are so well known as scarcely to need extensive summary. A self-styled methodological and epistemological 'anarchist', he seeks to liberate science and science-based cultures from the heavy yoke imposed by assuming that scientific rationality has a determinate and specific character. "Anything goes", the ubiquitous Feyerabendian slogan, stands in sharp contrast to the non-permissivist orientation of traditional methodology and philosophy of science. Less well known, or at least rarely discussed, are the specific threads of argument which lead him to the repudiation of methodology. Since his general conclusions rest upon those interwoven strands, I want to disentangle a few of the more striking and central ones in order to scrutinize them with some care.

J. R. Brown and J. Mittelstrass (eds.), An Intimate Relation, 299—317.

Central to Feyerabend's numerous and influential attacks on scientific rationality and methodology is this common thread: the insistence that scientific methodology is sterile and bankrupt because there is no rule or principle of theory assessment which has not (or which, if implemented, would not have) impeded the course of scientific advance. As he sees it, the great breakthroughs in the history of science (he is especially keen on citing the work of Galileo, Newton and Einstein) have been the products of bold pioneers, prepared to ride roughshod over every sacred canon of careful and rigorous reasoning. Moreover, insists Feyerabend, scientists have frequently utilized methods which fly directly in the face of all our 'rational' guidelines. Thus, scientists sometimes proceed 'counter-inductively', deliberately ignoring contrary evidence and the testimony of their senses. Oftentimes, scientists refuse to give up hypotheses which appear to have been decisively refuted. Frequently, they lie, cheat, suppress information, propagandize and resort to all manner of other trickery to persuade others to come around to their point of view. More to the point, such apparently deceitful behaviour is, on Feyerabend's view, absolutely crucial to the advance of science. The over-arching conclusion which he draws from all these specific generalizations is that there are no viable rules of scientific rationality. Scientists, at least the *best* scientists, make up their own rules as they go along, and try to persuade others to play the game of science using their non-standard rules. They may or may not succeed at this persuasive and rhetorical exercise. But succeed or fail, there is — so Feyerabend intones — no scope for an enterprise called 'scientific methodology'. There can be no methodology for science simply because there are no general or over-arching rules about how to compare and assess scientific theories. Circumstances alter cases so radically that scientific reason can be found only in the specific and the particular. In science, anarchism reigns; the only relevant rule is "anything goes". Scientists, he insists, have succeeded in building science only "because they did not permit themselves to be bound by 'laws of reason', 'standards of rationality', or 'immutable laws of nature'",[4]

Apparently keen to emulate what he takes to be scientific behavior, Feyerabend himself frequently resorts to propagandizing, distorting the evidence, and rhetorical hand-waving to make his case plausible. These techniques make it difficult to know when to take the man and his arguments seriously and when he is deliberately trying to pull a fast one. Yet because many of his readers take him very seriously indeed (probably far more than he intended), we shall spend some time here dissecting Feyerabend's more fundamental arguments — at least those which have ramifications for the general status of methodology. In effect, his numerous harangues against methodology involve variants of two straightforward arguments. The first of these, which I shall call the 'historical argument', involves pointing to situations in which science 'progressed' or 'advanced' (terms which Feyerabend, as self-professed epistemological anarchist, uses but is not supposed to believe in) allegedly because scientists behaved in ways which modern philosophers of science would regard as irrational. They broke the rules of scientific rationality and thereby furthered the aims of science.[5] How silly, he seems to say, to continue to look for the rules of scientific method when we see scientists breaking those rules, with impunity and to good effect, all the time.

Feyerabend's second argument goes well beyond saying that scientists have broken the rules; it propounds the thesis that scientific progress could have occurred *only if* scientists had broken the rules of scientific rationality. Science works best, perhaps science works at all, only when scientists flout the canons of scientific reasoning concocted by overly rationalistic philosophers. Despite its global character, this normative thesis of Feyerabend's rests on one specific argument, which I shall call the 'discovery of anomalies argument'. In brief, that argument says that weaknesses in a prevailing theory can generally be discovered *only* by the *proliferation* of rivals to it. Since (he plausibly imagines) many of those rivals will have much less going for them — at least at first — than the reigning theory does, all conventional rules of rationality will tip the epistemic balance in favor of the dominant theory. Because dominant and long-lived, the latter will appear to have a substantial body of evidence in its favor. Its fledgling rivals, by contrast, will normally be neither very well developed nor very well supported by the data. Hence these rivals can be explored and elaborated only if scientists are prepared to go against the rules of rational assessment which philosophers recommend. Yet (he holds), without these rivals, we could never discover where our reigning theories break down. As Feyerabend sees it, the rationalist about science is caught in a classic 'catch-22' situation: he can discover the flaws in the dominant theory only by the development of rivals; yet those rivals can never get off the ground because all our appraisal measures are 'stacked' in favor of well-developed theories! If these embryonic rivals were not to be developed, science would fail to progress; for, if he is right, the only way in which we can discover many of the refuting instances of the dominant theory is by exploring these ill-developed alternatives. I shall deal with these two arguments in turn.

The "Exception Disproves the Rule" Strategy

In a series of classic essays written between 1962 and 1975, Feyerabend adduced several historical case studies to show that science is a great deal richer and more flexible methodologically than its philosophical caricature might lead one to believe. Specifically, he purports to have shown that scientists, especially great and creative scientists, do not accept that theory which is best supported by evidence. On the contrary, they will often develop and articulate theories which, at least in their initial stages, are less well confirmed than older and more familiar rivals. Equally, scientists will sometimes flout the rule that one's worldview should be *consistent*, proceeding quite deliberately to elaborate a theory for one domain which is fundamentally inconsistent with well-confirmed theories in other domains. To make matters worse, scientists will frequently proceed 'counter-inductively', indifferent to the evidence before them. Equally, scientists will sometimes ignore falsifying evidence which runs counter to their favorite theories, resorting to all sorts of *ad hoc* strategems and face-saving devices to explain away apparent anomalies.

None of these claims would be very troubling, not even to an arch-rationalist, if they concerned themselves with how a minority of aberrant scientists behaved. After all, not even the most devout believer in scientific rationality holds that all scientists are rational all the time! The polemical force of Feyerabend's historical examples is that the very chaps breaking the rules of scientific rationality are precisely those responsible for setting modern science on its present course. It is not scientific charlatans he is describing; rather (they may — if he is right — be charlatans, but) the figures he is writing about have always been considered as the folk heroes of our scientific culture. It is the likes of Kepler, Copernicus, Einstein and Galileo who turn out to be the chronic 'cheaters' in the scientific game, for they are the ones who proceed counter-inductively, who ignore falsifying evidence, and who resort to all sorts of chicanery to win a hearing for their views.

Feyerabend's message comes down to this: as a matter of fact, we did not get modern science because a group of very clever thinkers once followed rules of reasoning and theory assessment which moderns like Carnap and Popper have explicitly described. Quite the reverse, says Feyerabend, the theories of the likes of Galileo, Copernicus and Kepler emerged when they did, and perhaps could only have emerged *at all*, because certain very clever people were prepared to shape their beliefs in ways which appear by current philosophical lights (and by the standards of their own time) as irrational. But he is concerned with drawing more general morals than the one which says that this or that methodology fails to describe how modern science developed. The larger point of his historical gambit is to establish that the 'rationalist enterprise' of seeking to spell out the rules of scientific methodology is generally ill-conceived.

How stands Feyerabend's argument? Not very well, despite the fact that some readers — perhaps because of a prior disposition to believe that methodology was bunk — have seen in his historical researches the final undoing of the rationalist tradition.

A host of difficulties beset Feyerabend's analysis. To begin with, it is not clear that he has got his historical facts straight. Numerous scholars, many with an expertise in 17th-century languages and texts which he cannot equal, have challenged his treatment of several of the key episodes in his story.[6] (It is arguable, for instance, that his treatment of Galileo is more heavily influenced by reading Brecht than by reading Galileo!) Although debate has raged about the historical reliability of Feyerabend's examples and although I am myself persuaded by his historical critics that there are some serious flaws in the 'history' he writes, I will, for the sake of argument, assume that everything he says about these masters of the scientific revolution is basically correct.

I make this concession both because this is not the place to engage in long-winded historical excursions and because I want to show that, even if all of Feyerabend's historical claims were correct, most of the conclusions which he draws from the history simply do not follow. An analysis of the reasons why they do not follow will expose several key Feyerabendian confusions about the nature of methodological rules. So, the central issue before us is whether, *if* we take his history as veridical, it warrants the blanket indictment of the mothodological enterprise which he draws from it.

Before we proceed further, we should remind ourselves about the manner in which methodological rules can be appraised. As I have argued at length elsewhere,[7] a sound methodological rule is one which optimizes the prospects for realizing a certain set of cognitive aims. Hence, we can legitimately criticize a rule by pointing to evidence that (a) it chronically fails to promote a certain goal, or else that (b) it promotes that goal less effectively than other rules could do. On the face of it, Feyerabend's strategy with the historical gambit is an exemplification of these techniques for he criticizes methodological rules of modern philosophy of science by showing that they were repeatedly violated by several prominent scientists. To be more precise, he shows that certain scientists promoted the goals of science even though they violated the rules under scrutiny.

The question is this: what, if anything, follows from the fact that the goals of science were avowedly promoted by these prominent instances of 'rule-breaking ' behavior? The answer is: relatively little. When we claim that a certain rule is methodologically sound, we are not commited to saying that the ends of science can be promoted *only* by following the rule in question; nor are we saying that the ends of science will *always* be furthered by following said rule. Rather, when we endorse a rule, we are asserting our belief that following that rule is more likely to realize one's goals than violating it will. What makes a rule acceptable as a rule is our belief that it represents the *best strategy* we can imagine for reaching a certain desired end; but it need not be, and commonly will not be, either a necessary or a sufficient condition for reaching that end. In other words, the advocate of a methodological rule as a means to a particular cognitive end need not be dismayed by the discovery that the end can sometimes be achieved by violating the rule. As Newton-Smith has observed,

> it cannot count against a particular inductive rule that it has on some occasion led us to hold false beliefs. To have [good] reason to abandon such a rule we would have to have reason to think that it has led us wrong more often than right.[8]

But if one cannot discredit a rule by showing that it is unnecessary for achieving a certain end, then what would ever be rational grounds for abandoning a rule? As I have shown elsewhere, there are several answers to this question.[9] One might, for instance, find evidence that acting contrary to a particular rule has brought about a desired goal in a larger proportion of cases than when the rule has been followed. Under such circumstances, we would be bound to say that the rule is apparently *not* the best way to achieve the end in question. Alternatively, one might be able to show that we had good grounds to believe that a particular rule could *never* be followed, or that if followed, it would never bring about the desired consequences. In any of these sitauations, the rule would be discredited. But it is noteworthy that Feyerabend's historical researches exhibit *none* of these results. The fact (assuming it to be a fact) that a few successful scientists such as Copernicus, Galileo and Kepler broke certain familiar rules of scientific rationality and nonetheless 'advanced' science in the process, does not establish that the rules they broke should be regarded as inadequate or inappropriate. It may still be reasonable to regard the rules as well-established bi-conditionals, linking optimal means to desired ends, even if their violation sometimes 'works'. If Feyerabend seriously intended to discredit methodology then he should have shown, but nowhere does, that *most* of the instances of successful science have been the result of scientists violating what we regard as the methodological norms of science. Until that is shown, it is no more than a rhetorical gimmick to suggest that occasionally successful rule-breaking on the part of a handful of scientists is legitimate grounds for repudiating certain widely-advocated rules of scientific inquiry.[10] (In structure, his position is rather like arguing that, since some people get spontaneous remission of cancer without taking any treatment, it follows that it is unreasonable to recommend that cancer patients should seek medical advice.)[11]

But Feyerabend's inferential leaps from the historical record are a good deal grander than I have thus far suggested. For his ultimate strategy is to argue that — because he has discredited certain modern-day methodologies of science (especially those associated with Carnap and Popper) — we are thus forced to conclude that the whole enterprise of delineating the rules of scientific methodology is ill-conceived. This is a monumental non-sequitur. It is rather as if one argued that, since most scientific theories eventually fail, it follows that we should stop doing theoretical science altogether. Let us, again for the sake of argument, assume that Feyerabend's arguments against certain prominent methodological rules (especially those associated with his prime target, Karl Popper) are telling criticisms of those particular rules. Nothing whatever follows from that about the inadequacy of attempts by *other* methodologists to characterize scientific activity as a rule-governed activity. The fact (if it be a fact) that the methodology of Popper is unreasonable will obviously not sustain Feyerabend's general charge that all methodologies are bound to fail. To move from the alleged failure of two of three methodological rules to the presumption that all methodologies are hopeless is to engage in just that sort of naive inductivism about which he is otherwise so abusive. To claim, as Feyerabend does in effect, that if Popper's methodology does not work, then "anything goes" reveals more about the narrowness of Feyerabend's philosophical ancestry at the London School of Economics than it does about the legitimate conclusions to be drawn from the collapse of Popperian methodology. In so far as his anarchism is grounded on this part of the historical argument, it has nothing to recommend it.

Feyerabend himself has belatedly come to realize as much. "A case study", he has recently conceded, "is not supposed to eliminate all rules, but only some . . .".[12] But what he gives with one hand he takes away with the other. For he immediately goes on to conjecture that "for every rule and every standard a case can be constructed that goes against the rule of standard".[13] Well, he might be right. But what is transparently clear is that neither Feyerabend nor anyone else has shown that all the *extant* rules of scientific methodology are inadequate, let alone that all possible rules are discredited. Under pressure from critics, he has conceded that this claim about the inadequacy of all rules is a 'bold conjecture',[14] but having said that he goes on to assert that he has shown that conjecture to be 'plausible'.[15] I trust he will forgive those of us whose plausibility threshold stops short of his own. So far as I can see, Feyerabend's anaysis gives us no reason whatever to follow him in his hyperbolic globalisation of his results. Indeed, he has not yet established that all or most of the *extant* rules of scientific methodology are inadequate, let alone that all possible rules are so.

The 'Discovery of Anomalies' Argument

But there is more to Feyerabend's repudiation of methodology than the assertion that existing methodologies have been persistently violated by prominent scientists; he claims as well to have a fully general argument to the effect that a careful understanding of the nature of theory testing shows the necessity of repudiating virtually all of the general principles which have been proposed for the assessment of scientific theories. His technique here is to focus on two or three widely-held methodological rules and to show that they are epistemically suspect — apparently hoping that the reader will share his own inductivist penchant for extrapolating from the failure of a couple of familiar methodological rules to the hopelessness of the whole methodological enterprise.

More specifically, Feyerabend wants to show that a particular methodological principle, one which he believes to be universally advocated by philosophers of science, is inconsistent with one of the central cognitive aims of science, namely, the discovery of empirical refutations of our theories. It is to the consideration of that principle which we now turn. Feyerabend attacks what he calls the 'consistency condition', a rule which says (in effect) that we should not accept theories inconsistent with our best-established theories. His discussion is designed to discredit the consistency condition by showing that the development of mutually inconsistent, rival theories is a prerequisite for scientific progress. His critique of the consistency condition, in turn, is the primary lynch-pin of his general assault on the methodology; for, he reasons, if a rule as ubiquitous and as plausible as the consistency condition is badly flawed, then surely methodology as such is highly suspect.

In his classic "Problems of Empiricism [I]," Feyerabend defined the consistency condition as follows:

> only such [new] theories are then admissible in a given domain which either *contain* the theories already used in this domain, or which are at least consistent with them inside the domain.[16]

As he rightly points out, such a requirement would tend to "encourage a *theoretical monism* and discourage a *theoretical pluralism*".[17] His strategy, in discrediting the consistency condition, is to show that it rests on, or presupposes, what he calls "*the autonomy principle*".[18] That latter principle asserts

> that the facts which belong to the empirical content of some theory are available whether or not one considers alternatives to *this theory*.[19]

He claims that the autonomy principle is almost universally taken for granted: "it is clearly implied in almost all [philosophical] investigations dealing with questions of confirmation and tests".[20]

The strategy underlying Feyerabend's argument is (a) to show that the autonomy principle is flawed, so that (b) the consistency condition — which allegedly presupposes autonomy — must be given up. That, in turn, is taken as warrant for asserting that (c) methodology is bunk. Of course, one might seek to block this notorious slide at several points.

On might, with Newton-Smith, argue that philosophers of science do not in fact advocate the consistency condition, at least not in the form Feyerabend gives it.[21] One might reasonably challenge the claim that the repudiation of the consistency condition (even assuming that scientists and philosophers accepted it) provides a reasonable warrant for the generalization that all methodological rules are as flawed as that principle.

These have been the favored responses to Feyerabend's challenge. I have no quarrel with them, but I think we can cut deeper into his approach by directly confronting his arguments against the autonomy principle. As we shall see, his critique of the autonomy principle is a series of non-sequiturs. Because they fail, we can block the slippery slope from "no autonomy" to "anything goes" right at its starting point. Feyerabend offers two distinct (although related) arguments against the autonomy principle. The first appeared in his "Problems of Empiricism [I]" (1965), the second received fullest expression in *Against Method* (1978). Both discussions are open to a set of powerful objections. So far as I am aware, objections of this sort were first developed by John Worrall.[22] Those objections, and some important corollaries which follow from them are worth elaborating in some detail, for they put us in a position to show that each version of Feyerabend's argument fails to sustain the anti-methodological morals he draws from it.

(a) The core principle of Feyerabend's attack on the consistency condition is the thesis that there are 'facts' which refute or falsify a theory, *T*, and that those 'facts' can be discovered to be refutations of *T* *only if* one develops rivals to (i.e., contraries of) *T*. As he puts it:

> the relevance and the refuting character of many decisive facts can be established only with the help of other theories that, although factually adequate, are not in agreement with the view to be tested.[23]

This passage is making two separate claims. It asserts, first, that the (epistemic) *relevance* of certain facts to the appraisal of certain theories can be recognized or 'established' only via the development of contraries to the theory being evaluated. I shall call this the doctrine of *relevance proliferation*. Secondly, the passage asserts that the *refutational* character of certain facts which falsify a theory can be shown only via the exploration of rival theories. This is the doctrine of *refutational proliferation*. Although Feyerabend is committed to both doctrines, he never gives an independent argument for relevance proliferation. Rather, he 'piggybacks' the thesis of relevance proliferation onto his arguments for refutational proliferation (which he does lay out in detail). Such tactics make sense because refutational proliferation obviously entails relevance proliferation.[24] Unfortunately, however, this tactic backfires because Feyerabend's arguments for refutational proliferation are not compelling. Hence, even the relatively mild claim of the importance of theory proliferation for the recognition of the epistemic relevance of certain facts is left dangling.

Since virtually all of Feyerabend's specific arguments concern refutational proliferation, I will limit my discussion to that issue. We must note at the outset what a strong claim is being made in the passage just quoted. He is not merely making the point that a theory is generally too incomplete to lead us to the discovery of all the 'facts' about the world. Nor is he making the familiar Duhemian point that, in order to test a theory, we will typically require a number of collateral theories (e.g., about measuring instruments, boundary conditions and the like). Nor is he merely saying that a rival theory, T^c, to the theory under test, T^b, will enable us to find refuting instances for T^b. He is claiming, rather, that certain refuting instances of T^b can be 'identified' *only* by virtue of exploring the contraries of T^b. I shall call this claim *the thesis of theory proliferation in its strong form* ('TP', for short). My aim in this section is to show that the arguments for TP fall wide of the mark.

Before I examine those arguments, however, we should note an important ambiguity in my (and Feyerabend's) formulation of TP. It concerns the sense of the term 'recognized' in my definition (and its counterpart term, 'established', in the Feyerabend passage quoted immediately above). The ambiguity arises because it is unclear whether he is making a semantic or an epistemic claim, or both. Is he, to be precise, making the *semantic* claim that we can ascertain that certain observation statements are logically inconsistent with T^b *only* by considering contraries of T^b? Or is he making the *epistemic* claim that we can establish the 'facticity', the evidential warrant, of those observation statements only by considering contraries to T^b? In other words, does the exploration of T^c's contraries enable us to tell that certain observation statements fall within T^b's empirical content or do those contraries rather enable us to know that certain entailments of T^b (determinable as entailments of T^b in the absence of rivals) are false? In some of his wilder moments, Feyerabend suggests the former. He says, for instance, that if one excludes contraries of the theory under test, one thereby "decreases the empirical content [sic] of the theories that are permitted to remain".[25] On its face, this claim is quite extraordinary. Since the empirical content of a theory is defined as the class of its potential falsifiers, and since the potential falsifiers are negations of the empirical consequences of the theory, Feyerabend is committed to saying that we can acertain what a statement (*viz.*, a theory) *entails* only by considering its contraries. One may have reservations about the Tarskian semantics, or about its elaboration by Popper into a semantics of theories. But if one is going to utilize that analysis to the point of talking (as Feyerabend does) about 'the empirical content of theories', then one cannot possibly imagine that we require knowledge of a theory's contraries in order to know what the theory entails. Indeed, this approach requires us to ride so roughshod over what is commonly meant by 'entailment', 'content' and 'potential falsifiers', that one is tempted charitably to assume that this semantic version of TP was just a slip of the pen. And several of Feyerabend's remarks (though by no means all) do lend themselves to the more plausible epistemic construal of TP. However, as he does not avail himself of this distinction, and because his arguments and examples are ambiguous as between these two different versions of TP, I shall recommend only that we bear the distinction in mind while trying to sort out his arguments. Having before us a 'clearer' — if still unhappily ambiguous — version of TP to work with, we can finally explore Feyerabend's case for TP.

In "Problems of Empiricism [I]," the central relevant passage is this:

> assume that a theory T has a consequence C and that the actual state of affairs in the world is correctly described by C', where C and C' are experimentally indistinguishable. Assume furthermore that C', but not C, triggers or causes, a macroscopic process M that can be observed very easily and is perhaps well known. In this case there exist observations, *viz.*, the observations of M, which are sufficient for refuting T, although there is no possibility whatever to find this out on the basis of T and of observations alone. What is needed in order to discover the limitations of T . . . is another theory, T', which implies C', connects C' with M, can be independently confirmed . . . Such a theory will have to be inconsistent with T, and will have to be introduced not because T has been found to be in need of revision, but in order to discover whether T *is* in need of revision.[26]

What Feyerabend claims to show in this passage is that we need rival theories to the one being tested in order to discover (at least some of) the refuting instances for that theory. Unlike Kuhn, who maintained that a prevailing paradigm would eventually accumulate a sufficiently large number of counter-instances that it would collapse of its own accord even without a viable rival in sight,[27] Feyerabend insists that many of the counter-instances to a theory can be found *only* by developing rivals to it. Specifically, he asserts in this passage that there are certain observations, *M*, which are 'sufficient for refuting *T*', although we could never know that they refute *T* except by considering a certain rival thory, *T′*. (*T* and *T′* are rivals in the strong sense; because each entails the negation of the other, both cannot be true.) Feyerabend asserts, moreover,

> that this situation is typical for the relation between fairly general theories, or points of view and 'the facts'. (*ibid.*)

The key question is whether, in the case before us, he has presented us with a 'refutation' of *T* — specifically *M* — which could be recognized as a refutation of *T only* by developing *T′*. Schematically, what he tells us is that:

(i) *T′* entails *C′*
(ii) *T* entails *C*
(iii) *T′* entails not-*T*
(iv) *C′* entails not-*C*
(v) *T′* entails that *C′* causes *M′*
(vi) *M* is observed.

To begin with, we should ask whether, as Feyerabend claims, *M* is 'sufficient for refuting *T*'. Well, there are many things we might want to say about the role of *M*. We might for instance, note that *M* confirms *T′* (since *T′* entails *M*). In a pinch, we might even say that *M* confirms *C′* (by virtue of Hempel's so-called 'special consequence condition'). But whatever else is going on here, *M* does not 'refute' or 'falsify' *T*, for the truth of *T* is compatible with *all* the conditions (i to vi inclusive) which Feyerabend spells out. The introduction of *T′* is just a red herring so far as the 'refutation' of *T* is concerned. Because *T* has not been shown to entail the negation of *M*, *M* cannot legitimately be regarded as a refutation for *T*. It thus seems clear that Feyerabend's 1965 argument for TP is not especially compelling.

(b) Perhaps sensing that his argument of 1965 was not watertight (but evidently unwilling to abandon its conclusion), he drops from his discussion of autonomy in *Against Method* the lengthy passage I quoted above (most of the *rest* of his discussion of autonomy is carried over intact from his 1965 paper). Instead he is content in *Against Method* to make his argument for TP on the strength of a single historical example (which had already figured in his 1965 paper). His example is this: we now regard the Brownian particle as a perpetual motion machine of the second kind: "its existence refutes the phenomenological second law [of thermodynamics]".[29] But, says Feyerabend, we can demonstrate *that* the Brownian particle *refutes* the second law *only* by accepting a theory of heat (i.e., the kinetic theory) which contradicts the second law. As he points out, Einstein utilized the kinetic theory in his calculation of the statistical properties of Brownian motion. Svedberg and Perrin then staged a crucial experiment which was widely construed as showing that the Brownian particle was a *perpetuum mobile* and thus a refutation of the second law.[30] But we have to ask (a) whether the Perrin experiments (which I will call M) *refute* the second law (T) and (b) whether postulating a contrary to T was the only way to ascertain M? Feyerabend's TP requires an affirmative answer to *both* questions. But his analysis gives us little grounds for confidence.

Consider, first, whether M refutes T. If, as Feyerabend claims, (i) kinetic theory is inconsistent with phenomenological thermodynamics (PT) and (ii) kinetic theory is essential for deriving M, then no proponent of PT is going to be logically compelled to change his mind about T by experimental results *whose very interpretation depends on accepting a theory which denies PT*. If physicists have come to regard M as disconfirmatory of the second law of thermodynamics, that is because they have *independent* grounds for preferring the kinetic theory, grounds which lead them to acquiesce in the interpretation of Perrin's experiments provided by the kinetic theory. But if one rejects the kinetic theory (and what serious proponent of the second law would not, knowing that it entails a denial of the second law), then one is hardly compelled to regard Perrin's experiments as refuting T.

Consider, secondly, whether (as TP requires) M — even assuming it refutes PT — could have been discovered to pose problems for PT absent the kinetic theory. Nowhere does Feyerabend establish this necessity. He does say that, without the kinetic theory and its statistical resources, we could have found results like M only by directly measuring heat transfer and particle motion with an accuracy which goes "beyond experimental possibilities".[31] He concludes that a 'direct' refutation of the second law . . . [by] Brownian motion is impossible".[32] Even granting that (which 'impossibility' is itself crucially theory-dependent), it does *not* follow that no conceivable theory, compatible with PT, could have provided the statistical and interpretive resources needed to link PT and M in the manner in which kinetic theory links them. He is here simply sliding from a contingent feature of the historical situation to a very strong claim about logical possibilities.[33]

To sum up, Feyerabend's critique of methodology fails on all counts. The fact that prominent scientists have sometimes succeeded by failing to follow what we regard as sound rules of scientific inference does not challenge the legitimacy of those rules as general strategic guidelines. Moreover, his one detailed effort to mount a general epistemic argument against methodology (i.e., his argument against the consistency condition) falls flat on its face, because it will not sustain the conclusions which he draws from it.

DIGRESSION AND CONCLUSION.

By way of concluding this paper, I think it might be useful to step back briefly from the fine-grained texture of Feyerabend's critique of methodology in order to examine one or two of the general implications of his methodological anarchism. As we have seen, he is committed to the view that there are never rational grounds for believing that any method for reaching one's ends is any better than another, or rather that there are never objective grounds for making the judgment that one method is better than another. If one thinks of methods as means to ends (where the ends in question are the aims of science or of a particular scientist), then Feyerabend's denial of methodology is equivalent to the denial that *any* means for reaching certain ends is better than *any* others.

Brief reflection will show what an extraordinarily perverse perspective this represents. The world we live in presumably has some definite constitution or other; even Feyerabend, with his residual epistemic realism, is committed to that much. But to say that the world has a certain make-up is just to say that it behaves in accordance with certain regularities. The task of the methodologist, on my view, is to make (and appraise) conjectures about the regularities governing inquiry. To assert with Feyerabend that all methods work equally well is to hold that *there are no regularities about inquiry*, that there are no facts of the matter about how to put questions to nature. Since he generally holds that the world has a determinate structure, it is not a little peculiar that he would hold that there are no regularities to be discovered about such matters as designing experiments, testing theories, and discovering theories. For my part, I see no justification whatever for taking the program of constructing theories about inquiry (for that is what methodology is) any less seriously than we take programs for constructing any other sorts of theories about natural processes.

If there is any enduring legacy from Feyerabend's treatment of methodological issues, it centers around the question how new theories can possibly come to be rationally propounded. One might call it the *innovation problem.* As Feyerabend correctly observes, new theories — almost in the nature of the case — will generally be less well articulated and less well tested than their older rivals.[34] Thus, if customary requirements of empirical support are strenuously insisted on, it will *never* be reasonable to accept such theories in their early stages. Yet (and this is what the innovation problem comes to) unless some scientists commit themselves to elaborating, modifying and strengthening such new theories as are produced,[35] then science would never change.

Feyerabend believes that the only solution to the innovation problem is to suspend all methodological judgments, thereby permitting scientists who are minded to explore new alternatives to do so without appearing to run counter to scientific rationality. Anarchy alone, as he sees it, will solve the innovation problem. Of course, anarchy solves the problem only by producing a more intractable one: when anything goes, everything is gone — including any grounds for picking out some theories as more acceptable than others. Fortunately, there are less drastic solutions to the innovation problem. Specifically, I have proposed distinguishing sharply between the rules of appraisal governing *acceptance* and the much weaker and more permissive rules or constrains which should govern '*pursuit*' or '*employment*'.[36]

More generally, I hold that it is crucial to recognize that there is a whole spectrum of cognitive stances which scientists can adopt towards their theories (suggested by phrases like 'entertain', 'consider', 'utilize as a working hypothesis', etc.). 'Accept' and 'reject' are but two (rather remote) points on this spectrum; yet they virtually alone have received serious attention from epistemologists and philosophers. Feyerabend is quite right that, insofar as scientific methodology has been *identified* among the logical empiricists with providing rules of acceptance, then methodology has no resources for solving the innovation problem. But a little reflection will make clear that one need not (and, as the innovation problem shows, one ought not) limit methodology to the analysis of acceptance or belief. Once the conception of methodology is broadened to include the analysis of the semantics and epistemology of these other cognitive attitudes, then Feyerabend's plea for anarchism becomes wholly groundless.

University of Hawaii

NOTES

[1] This second criticism is especially prominent in Kuhn's critique of methodology.

[2] For a lengthy discussion of the first point, see my "De-Mystifying Underdetermination", in W. Savage (ed.), *Minn. Stud. In Phil. of Science*, forthcoming . . . For a briefer critique of the argument about the ambiguity of rules, see my "Kuhn's Critique of Methodology", in J. Pitt, (ed.), *Change and Progress in Modern Science* (Reidel: Dordrecht, 1985), pp. 283—289.

[3] Concerning Feyerabend's influence, largely pernicious, in the social sciences, consider the fact that Derek Phillips wrote a widely-acclaimed book in the 1970s, *Abandoning Method* (San Francisco, 1973), which argues — as one might guess from the title — for the wholesale repudiation of methodology. It is hardly necessary to add that Feyerabend's influence on Phillips' is substantial.

[4] *Against Method* (Verso: London, 1978), pp. 190—191.

[5] There is a certain ambiguity in asserting that historical figures violated the rules of method, as between violating 'our' rules of method or violating the canons of research which prevailed in their own time. Feyerabend makes it clear that he believes major researchers chronically do *both*. (See, for instance, Paul Feyerabend, *Philosophical Papers*, Cambridge, 1981, vol. II, p. 13.)

[6] For a small sampling, see for instance, P. Machamer, "Feyerabend and Galileo", *Stud. Hist. Phil. Sci.* **4** (1973), 1ff.; J. McEvoy, "A 'Revolutionary' Philosophy of Science", *Phil. Sci.* **42** (1975), 49ff.

[7] See my "Progress or Rationality?" *American Philosophical Quarterly*, **24** (1987), 19—31.

[8] W. Newton-Smith, *The Rationality of Science* (London, 1981), pp. 133—134. I would quibble with Newton-Smith's suggestion that rules can only be rationally abandoned when we have good reason to believe that they would lead us to be wrong more often than right. One can readily imagine a situation in which a methodological rule, although producing 'right' results more often than 'wrong' ones, was nevertheless to be rejected; specifically, if we had a rival rule whose track record was (or which we had good reason to believe to be) significantly better than the rule under consideration. But the general sense of Newton-Smith's point (*viz.*, that the fact that a methodological rule sometimes leads us to make mistakes is by itself no grounds for abandoning it) is surely right-headed.

[9] See chapter three of my *Science and Values* (Berkeley, 1983) and "Progress or Rationality?" *American Philosophical Quarterly, loc. cit.* note 7.

[10] Of course, showing that most of the episodes in the history of science violate certain methodological canons is a tall order, in terms of the research and analysis which would

be involved. One can fully understand why Feyerabend has not completed that ambitious enterprise. But, in that absence of that task or something comparable, his global conclusions simply cannot be sustained.

[11] Given Feyerabend's advocacy of acupuncture and other forms of non-standard 'medicine', he would probably find this parenthetical aside something other than the *ad hominem reductio* it was intended to be!

[12] *Science in a Free Society* (NLB: London, 1978), p. 212.

[13] *Ibid.*, p. 212.

[14] *Ibid.*

[15] *Ibid.*

[16] P. Feyerabend, "Problems of Empiricism", in R. Colodny (ed.), *Beyond the Edge of Certainty* (Englewood Cliffs, 1965), p. 164. Italics in original. (Hereafter: PE.)

[17] *PE*, p. 164. Italics in original.

[18] *PE*, p. 174.

[19] *PE*, pp. 174—75. Italics in original.

[20] *PE*, p. 175.

[21] Newton-Smith, with some justification, holds that Feyerabend's version of the consistency condition "has not been advocated by any influential scientist or philosopher in this century." (*op. cit.*, p. 131)

[22] See, for instance, John Worrall, "Is the Empirical Context of a Theory Dependent on Its Rivals?" *Acta Philosophica Finnica*, 1982, pp. 298—310.

[23] *PE*, p. 176.

[24] Assuming that if a statement refutes a theory it is *eo ipso* relevant to the assessment of that theory.

[25] *PE*, p. 176.

[26] *PE*, p. 176.

[27] This is the force of Kuhn's dictum that 'normal science is its own undoing'.

[28] *PE*, p. 176.

[29] P. Feyerabend, *Against Method,* (London, 1978), p. 39. (Hereafter: *AM*.)

[30] In particular, the Brownian particle constitutes no sort of difficulty at all for classical thermodynamics *unless* one also assumes that the medium through which the Brownian particles move is in thermal equilibrium (and that the system is closed in the relevant sense). Since those assumptions are clearly themselves theoretical and distinguishable from classical thermodynamics, no observed motions of the Brownian particle can directly or incontrovertibly refute classical thermodynamics.

[31] *AM*, p. 40.

[32] *AM*, p. 40.

[33] I think that the correct epistemic account of the Brownian motion case proceeds very differently from Feyerabend's version of it. What Einstein showed was that Kinetic theory possessed the explanatory or problem-solving resources to *account for a phenomenon* which its rival, thermodynamics, had not been able to make any sense of. Brownian motion thus challenged thermodynamics *not* because the former was a 'refutation' of the latter but because it raised serious doubts about the latter's heuristic potential. Because a rival to thermodynamics had shown its extendibility to the Brownian domain, and because thermodynamics, had not been shown to be similarly extensible, it was quite reasonable for scientists in the early 20th century to revise their

assessments of the relative worth of the two theories. But, of course, that is a far cry from the claim that Brownian notion 'refuted' thermodynamics. To use earlier language of mine, Brownian motion became a '*non-refuting anomaly*' for classical thermodynamics. (For the development of this notion, see my *Progress and Its Problems*, pp. 26ff, and my "Anomalous Anomalies," *Philosophy of Science*, **48** (1981), 618—19)

[34] This will be especially so if (as I believe) Feyerabend is right in maintaining that later theories do not entail or contain their predecessors. Precisely because they do not, those later theories cannot 'piggyback' on the empirical suport enjoyed by their predecessors, but must establish their credentials directly.

[35] And to developing the collateral theories which those new assumptions require in order to make 'contact' with the world.

[36] I have drawn such a distinction in *Progress and Its Problems*. For a different sort of approach to the innovation problem, see Rachel Laudan and Larry Laudan, "Two Problems about Methodology" *Philosophy of Science*, forthcoming.

JÜRGEN MITTELSTRASS*

WORLD PICTURES: THE WORLD OF THE HISTORY AND PHILOSOPHY OF SCIENCE

1. THE HISTORY OF SCIENCE AND THE MODERN WORLD

Science is not an external aspect of rational cultures. On the long road from their highly praised discovery of the scientific mind by the Greeks, these cultures first learned to define their rationality primarily in terms of the concept of the scientific, and later also in terms of the technical. Rational cultures in this sense are technical cultures supported by science. As such they constitute the modern world.

This world is therefore a product of a rational history; at any rate a history whose motivating force is the will to be rational — not only in its triumphs, but also in its failures and its shortcomings. It is, moreover, the historical product of the *epistemic* and *technical* nature of man — his searching and finding, knowledge and doubt, planning and building, thinking and doing. Epistemic nature here means that the formation of knowledge is not external to man. Rather, it constitutes the medium in which man *orients* himself — in everyday and scientific situations. Orientations are here understood as including not only knowledgable ways of dealing with objects and their relations, but also their interpretation and explanatory appropriation. In scientific knowledge this appropriation acquires its true rational form. What Aristotle formulates in the proposition that all men by nature strive for knowledge[1] is, if understood in this way, the grounding of rationality in the epistemic nature of man.

His technical nature corresponds to it on another level, the level of productive activity. When man creates tools for himself, and relies on tools for his way of life, he adds, as it were, artificial organs to the other organs through which he interacts with the world. Technology not only supplies what is needed for the 'completion' of man's nature; it also develops its own laws, acquires its own rationality, and creates a new nature. The historical product of this constructive power, along with the epistemic power previously mentioned, is the modern world. *Homo sapiens* and *homo faber* — that is not Socrates on the one hand and Leonardo da Vinci on the other, but the peculiar double aspect in

J. R. Brown and J. Mittelstrass (eds.), An Intimate Relation, 319—341.

which the rationality of man and his world develops historically. Expressed in a different way, this means that in a culture which is in its existence a product of scientific and technical rationalities, not just any of man's abilities or 'talents' are realized but precisely those that make him a *rational subject* and his world a *rational world*. The modern world, i.e. the reality of scientifically supported, technical cultures as they are represented by modern industrial societies, is not a contingent, but rather a logical result of man's developed nature. This also means a way of legitimizing the modern world on the basis of man's epistemic and technical nature.

But if the modern world is the product of a rational history, based on man's epistemic and technical nature, then a significant portion of the way rational man understands himself and his world is also included in this history. What I mean here is not only the general insight that what we (including the world in which we live) are, we are historically, but also the more specialized, methodologically applied insight that systematic (and empirical) analyses of a particular present state of affairs must be supplemented with genetic analyses.[2] Systematic understanding comprehends its object only incompletely when the nature of this object is historical; just as historical understanding gives an incomplete account of objects of a systematic nature. Both are the case in dealing with the modern world. That this world has a history makes it an object of historical understanding. In fact that this history is rational and that its rationality is founded in man's epistemic and technical nature, makes it the object of systematic understanding. Both forms of understanding are necessary here; the one or the other alone is insufficient for genuine understanding.

This poses high hurdles for those disciplines whose professional concern is the investigation of the history of the modern world. Among these is the *history of science*. It is primarily the function of the history of science, on the description given above, to comprehend the course of epistemic and technical rationality insofar as it is given form in science. On the whole, however, the history of science is poorly prepared for this task. This is already apparent from the atmosphere of the museum which hangs over it. This is a result of an unspoken agreement between the philosophy of science and the history of science; as if the philosophy of science were charged with investigating contemporary problems in science, and the history of science only those of the past, whose solution has not found a place in the textbook knowledge science provides about the modern world. On this view, philosophy of science, while not always finding favor among the sciences, is seen as contributing to scientific progress, with the history of science satisfying the perfunctory historical curiosity.

Such a division of labor between the philosophy and the history of science, which makes the one an element in the progress of knowledge and the other an element in the appreciation of knowledge, is surely inadequate for comprehending a scientific reality. (The same applies to the so-called science of science, which belongs neither to the history nor to the philosophy of science.)[3] This is true in part because of the teleological unity of historical and systematic understanding, but also for another reason. For many orientations in the history and philosophy of science it is still valid to say that the present state of science determines the facts that belong to its prehistory. History of science in this sense is always the prehistory of any textbook knowledge. This conception, however, is false. Not only because it stands in opposition to current models of scientific development inspired by the philosophy of science, but also because it implicitly includes the inadequate assumption that scientific knowledge does not require a genetic analysis. What science knows, to state this view in another way, it knows only systematically (and empirically), not historically. The history of institutions and the history of research would be external to scientific knowledge. But they are not in fact. This is already illustrated by the fact that scientific *planning* and the *historiography* of science, thus the future of science and its past, are never determined solely by the state of textbook knowledge at the time, but always also by aims and developments that connect it with a general, and therefore also a social, practice. For this reason the history of science is not an isolated segment of a rational history, but a history of the development of the rational subject and his world, viewed through the eyes of the scientist, whose work is itself part of this development.

With respect to the functions and capacities of a non-archival history of science, I shall mostly discuss, in what follows, the concept of the world typical of science; or, perhaps better, the view of the world that is formed in scientific work, both in research and in theory construction. That in rational cultures it is not (as in pre-rational cultures) myth but rather the world that science makes, does not mean that one of the functions of mythical cultures, namely the production of world pictures, is no longer in operation. Science, too, has the power to generate world pictures. It not only makes the world in which rational cultures live, but also provides it with a picture in which it appears as what it ostensibly is 'in itself', as 'nature', as 'creation', as 'evolution', regardless of how we transfer the conception of a 'world without us', including its assumed cultural virginity, to 'our world', the world in which we live. For an understanding of science it is thus necessary to understand its power to constitute the world and to generate world pictures. The question is what the world of (and in) science looks like, and what it looks like in the history of science.

2. THE WORLD OF SCIENCE

The *Aristotle world*, the *hermetic world*, the *Newton world*, and the *Einstein world* are all examples of worlds of science which, in their philosophical interpretations of nature, recommend themselves as the correct view of the world. These examples are not only representatives of the history of science, but also constitute preferred demonstration-material for philosophical assessments and construction of the developments of science. Here it is not a question of what the authors of these 'world pictures' meant in detail, but a question of the world they created with their scientific conceptions and theories — for the scientific as well as the non-scientific mind.

The *Aristotle world* is a world of natural things that consist of matter and form and have within themselves a source of motion. Motions caused by such a 'natural' source are 'teleological' motions, i.e. they make a thing into what, according to its own nature, it really is (motion as the realization of the form of a being), or they lead it, in the form of a 'natural' local motion, to its 'natural' place. A theory of natural positions, incorporated in a theory of elements, corresponds in this sense to a theory of simple (natural) bodies (bodies that have a source of motion in themselves) and simple motion (the motion of simple bodies). A perfect order would result in an arrangement (in terms of lightness or heaviness) of the four elements in concentric layers around the center of the world. The inclusion of a fifth element, aether, leads to a conception of finite space, with a theory of (natural) position (as the positional relationships between bodies) taking the place of a theory of space. This is a logical development insofar as the foundation of the Aristotelian world picture is not a theory of nature (in the sense of the realm of nature in its entirety), but a theory of natural things.

Thus, in the cosmological dimension, the Aristotle world consists of eleven spheres grouped around the central body, earth. Each such sphere is constituted by two concentric spherical surfaces: the three inner spheres housing the elements water, air, and fire (in a state of perfect order), and the eight outer spheres housing the moon, Mercury, Venus, the sun, Mars, Jupiter, and Saturn as planets, and the system of fixed stars (with a daily rotation about the axis of the heavens). The *geocentrism* of the Aristotle world is thus a result of the theory of elements or the theory of natural positions. That a heavy body falls to the earth is a result of the center of the cosmos being the natural position for this body, i.e. the motion of heavy bodies is not toward the earth (this is only *per accidens*), but toward the center of the cosmos (*per essentiam*). A causal relationship, moreover, exists between the *supralunary* and the *sublunary* worlds. The diurnal movement of the heavens is accordingly the cause for the eternity of earthly processes, insofar as these involve the linear movements of elements. The counteracting movements of the sun, the moon, and the planets in the zodiac are thus the origin of such different processes as generation and degeneration (substantival change), growth and diminution (quantitative change), and augmentation and abatement (qualitative change or alteration). Substantial changes, in contrast to all other changes, occur instantaneously, as the direct transition from one state to another, i.e. without motion in a narrow sense. This is because motion, according to Aristotle's conception, is always motion between opposites, an aspect that does not apply to substantial changes or substances.

In oppostion to the atomistic conception of the constant movement of atoms, the Aristotle world is characterized by the notion that every movement requires a mover. Thus, not only the change of motion, but also the uniform motion of a body requires a causal force. This force must either reside in the moving body itself (in the form of a motivating 'soul' or as a natural movement) or exist in direct contact with it; action at a distance is not permitted. The place of atoms in atomistic conceptions is filled by so-called *minima naturalia*, i.e. the smallest particles of matter that place a natural limit on its divisibility without altering its substantival form. Correspondingly, all matter has quantitative *minima* that possess the characteristics of macrobodies made from it. These *minima* also possess a characteristic size, though their geometric form is not predetermined. In chemical processes, *minima* in immediate proximity to each other constitute a '*qualitas media*', which is the basis for the '*forma mixti*' of matter which possesses a particular substantival form. (According to atomistic conceptions, all that changes in chemical processes is the configuration of the smallest particles, which lack qualitative characteristics and whose geometric form is constant).

The Aristotle world is thus characterized by a high degree of *experiential evidence*. The scientific propositions describing this world are confirmed by the experience acquired in everyday life, or are derived through generalizations made on the basis of experience. Examples of this are (1) the Aristotelian law of gravitation, according to which the velocity of a falling body is proportional to its weight and inversely proportional to the density of the medium, (2) the above mentioned Aristotelian 'law of inertia', which states that all things moved have a mover, and (3) the Aristotelian theory of elements with its familiar concepts derived from the experience of daily life, for example, 'above', 'below', 'natural', and 'unnatural' (as in the case of violent movements that run counter to natural movements). The Aristotle world, moreover, is always in the process of becoming a natural order, embedded in the inner teleology of this world or the teleological nature of all things. This natural order never appears as a perfect state, but it is constantly present in the form of an astronomically ordered, supralunary world. In other words, disorder as well as the tendency to order is the normal state of the (sublunary) world. It is the world of experience and hence — despite physics and natural philosophy which seek to interpret it — a very human world.

As opposed to the Aristotle world, the *hermetic world* — by which I mean the world of alchemy, astrology, and parts of natural philosophy in the Renaissance — is a world of mysterious interactions. Occult powers and living substances take the place of the simple bodies characteristic of the Aristotle world. Nature consists of different combinations of primary substances that originated in undifferentiated primordial matter. At the same time, these combinations are conceived of as developmental processes that man can accelerate or retard, though always with methods that 'imitate nature', for example, by 'refining' metals and other substances (transmutation). Inorganic processes are viewed analogously to organic processes. Explanations of the world take the shape of allegorical interpretations: coming into being and passing away as birth and death, separation and unity as the polarity of the sexes (the *conjunctio* as sexual union or the hermaphrodite as the overcoming of sexual differences).[4]

This conception finds its cosmological expression in the correspondence between macrocosm and microcosm which interprets the world in antiquity and in the hermetic tradition as a great organism mirrored in the microcosm, particularly in man: "what is below is like what is above; what is above is like what is below: both reveal the miracle of the one".[5] The influence of the macrocosm on the microcosm corresponds to the everpresent assumption in magical thought that it is possible to effect a change in the macrocosm through changes in the microcosm. This conception, as the 'sympathetic' relationship between all of the parts of the world, is still at work within the context of natural philosophy in the Romantic period: man as a microcosm "in which the universe looks at itself".[6]

For the Flemish scientist and philosopher Franciscus Mercurius van Helmont (1614—1699), whose natural philosophy can serve here as an example for a hermetic world, every body possesses a certain *archeus*, i.e. a personalized natural power (in the case of inorganic bodies also known as '*vulcanus*'). This power is concentrated in the seeds and determines growth and development as well as the 'signature' through which a body reveals itself in its inner nature. In addition to the *archeus*, there are also the physical and spiritual *minima* out of which the physical and the spiritual world is constructed. Central monads represent a 'central spirit', which comprehends the external world as the very likeness of God. The physical world functions as the last stage of several worlds that, according to neoplatonic and kabbalistic conceptions, 'emanated' from God. The earth is conceived of as a living being, namely, a nourishing mother. More empirically minded, yet no less involved in a hermetic world picture, van Helmont's father, the physician and chemist Johannes Baptista van Helmont, also stressed the deeper unity of all three realms of nature (minerals, plants, animals) in the seeds created by God, specific to all bodies, and including an *archeus*. The world is a *symbolic continuum*.

The English physician and natural philosopher, Robert Fludd (1574—1637), prepared an allegorical sketch representing such a continuum between microcosm and macrocosm:[7] a naked female figure, representating nature as the world-soul, stands on the elements earth and water, and is connected directly with God through the *catena aurea* over her right hand. The woman sets the sublunary world in motion and, through the influence of the planetary spheres, the sublunary world as well. The solar features of her breasts and the lunary features of her abdomen symbolize the creative and preservative power of nature. From her left hand the chain of being leads to a man sitting on the planet earth, the 'ape of nature' who represents the arts and sciences. A funnellike section running from the earth out to the sublunary world reveals the realm covered by knowledge of natural philosophy.

In this hermetic world everything becomes a riddle or a key to solving its secrets. The familiarity of the Aristotle world gives way to a demonic world that is only accessible through ritual and mystical forms of knowledge. The scientist becomes in this way the mediator between two worlds, a life-world and a hermetic world, and at the same time the real 'addressee' of his own hermetic knowledge. The alchemical *separatio* reproduces itself as the *separatio* of the material and mystical body (the 'diamond body') in the scientist. It constitutes the actual *magisterium*, i.e. the 'great work', the self-development or spiritualization of man. Thus the hermetic world stands not only in opposition to the familiarity of the Aristotle world, but also in opposition to the 'mechanistic world' that in the modern age begins to supplant both the Aristotle world, as well as Aristotelian physics.

The foundation for this mechanistic world picture is the *Newton world*. In this world it is only (gravitational) mass that moves in absolute time, through absolute space. Matter and space are the real elements of this world. The smallest particles of matter, hence, the actual atoms, combine to build complex formations or second degree particles. Several of these combine in turn to become third degree particles and so forth. The inner structure of matter is thus characterized by a complex hierarchy of particle formations. These formations are not massive corpuscles, but contain empty space. As the order of the particle hierarchy expands, the amount of empty space in them increases while the extent of solid matter decreases correspondingly. Matter in the world is thus only seemingly solid. In fact, the world is a vacuum for the most part. The actual amount of solid matter in the universe could fit into a nutshell (atomistic nutshell theory).

Characteristic of the Newton world, moreover, is the assumption that a fundamental dualism exists between passive matter and active immaterial principles. According to this notion, which can be traced back to Cambridge Platonism and hence to hermetic conceptions of the world, matter can be the origin only of mechanical effects, that is, effects mediated by pressure and impulse. Matter itself does not exert force, but only withstands the effects of forces (through its own inertia). Gravitational pull, in particular, is not a trait of matter. Gravitation has more the status of an active principle and finds its origin in a non-material aether that exerts an effect on matter. Matter, 'inanimate and brute', is not able to guarantee even halfway stable processes of development through its essential characteristics. Since in this world a general principle for the conservation of energy is not valid, mechanical interactions lead to a steady loss of movement which cannot be fully compensated for by the active principles that bring forth new movement. All the regularly functioning causes (material or immaterial) taken together would not be able to impede the movement of the world toward disorder and chaos. The stability of the world, i.e. compensation for the energy loss, is a matter only for God or an occasional divine intervention in this world.

The nutshell theory of matter on which this world is based corresponds, as regards its concept of space, to a container or arena theory. The space of the Newton world is not formed by spatial relations of material bodies (concept of relational space), but exists 'in and of itself' as an ontological entity on the same level as matter. Space is independent of matter. In proving the existence of inertial forces, Newton attempts to endow the related concept of *absolute space*, i.e. the conception of a stationary system of coordinates that differs from the mere relative state of rest between bodies, with experimental content. He himself tried to show that the centrifugal forces generated by rotational movement cannot be traced back to relative rotations of whatever type, i.e. they have to be conceived of as 'true' rotations, as rotations against an absolutely stationary space. This absolute space is analogous to the '*sensorium Dei*', i.e. the omnipresence of God put into law (following here the Cambridge Platonists as well). Just as the mind of man can receive sense impressions through its presence in his brain, so God perceives the processes in the universe through his presence in absolute space.

The 'mechanism' of the Newton world, expressed in a mechanics of gravitational movement, in Newtonianism not only determines how inorganic nature is understood but also proliferates itself in the organic, psychic, and social cosmos. In the theological aspects it still retains, this mechanism documents their fundamental dispensability. The criticism of the effects of occult powers (qualities) in an hermetic world also applies to Newton's theological legitimations. The Newton world, the quintessential 'mechanization of the world picture', becomes a 'world of machines' — with God as a 'retired engineer'.[8]

In contrast to the concept of absolute space in the Newton world, a concept of relational space is dominant in the *Einstein world*. Here space is constituted only by matter, with energy also being matter. In order to do justice to the special effects of rotation discovered by Newton, Einstein refers to 'Mach's principle', which considers the forces of inertia not as the result of true rotation (rotation against absolute space) but as the effect of rotation relative to distant masses (that is, the center of gravity in the universe). Einstein's general theory of relativity attempts to give Mach's programmatic idea a physical dimension in order to establish the validity of a theory of relational space in terms of epistemology as well as physics.

Particularly relevant philosophically is the idea of a *geometricization of nature*. In the general theory of relativity, gravitation (with certain restrictions) is no longer conceived of as a force that diverts bodies from their natural trajectory, but as an entity that is inseparably bound up with the structure of space and time. If one examines the trajectory of a body in the field of gravitation from an adequate standpoint, one would recognize that this body actually follows the most linear trajectory. Later Einstein also tried to apply this idea to electromagnetic forces in order to achieve a unified theory of gravitation and electrodynamics. The central idea was that all interactions between particles can be traced back to space-time structure. This means that a particle has an effect on space-time, which in turn has an effect on a second particle and in this way mediates the interaction between the two.

Matter itself, however, is not included in the unification. Particles are singularities of the fields, i.e. particles are not themselves solutions to the field equations. Where particles are, the field equations are not valid. Both ponderable matter as well as fields can be viewed as matter in a broader sense. Both possess physical reality in the same sense. Since the field of gravitation in particular is of this type and at the same time is identified through the general theory of relativity with the space-time structure, matter is everywhere where space is. The Einstein world is not virtually empty like the Newton world, but full like the Cartesian world. Like the Newton world, however, it is subject to deterministic considerations in the form of a unified theory of interactions. Thus, in the Einstein world there are no essentially accidental elements; everything is predetermined from the beginning and takes place necessarily. God does not throw dice.

It is characteristic of the special as well as the general theory of relativity that the essential geometric quantity is a *four*-dimensional metric interval. This can be divided into a spatial and a temporal component, and yet this division is dependent on the system of coordinates used. Einstein draws the conclusion from this situation that the 'transient now' (the idea of a shifting present) possesses no objective meaning. He draws the same conclusion from the symmetry of equations in mechanics and quantum mechanics against time reversal. In all elementary processes there is no difference between past and future. Such a difference is a mere illusion. In reality there is no development, no actual change. All that is real is a static, four-dimensional state of being. In this sense the Einstein world is neither Aristotelian nor hermetic nor Newtonian, but Parmenidean.

3. THE WORLD OF THE HISTORY OF SCIENCE

The examples cited here should demonstrate the power of science to constitute worlds and generate world pictures. There can be no doubt that they fulfill this role. But doesn't the brief discussion of the Aristotle, hermetic, Newton, and Einstein worlds demonstrate more than was intended, with their different answers to the question of what the world of (in) science and its development are like? In the final analysis, these examples reveal not only that science makes worlds, but that there is also a certain *relativity of scientific worlds*. What reveals itself in the history of science, i.e. in an analysis of the history of science, is not a world on which the scientific mind constantly builds, but a multitude of worlds that have little to do with each other. Each of these worlds, with respect to the scientific (or philosophical) view of things on which they are based, has its own plausibility, and each is somehow consistent — only not as work on a common world. Which world is the right one, and in which one do we want to live? Has scientific progress already answered this question or does the relativity of scientific worlds lead to a deeper, more unsettling relativity, a *relativity of science*?

Actually, the possibility of this type of relativity, the expression of which would then also induce a relativity of scientific worlds, is a topic of modern philosophy of science. This has been discussed particularly under the heading of *incommensurability*, where it concerns the relationship between two paradigmatic theories in transition from one to the other (e.g. the shift from Aristotelian to Newtonian theory). For the advocates of the incommensurability thesis there are no argumentative bridges between two different scientific paradigms because there is no common language; hence no translatability on a terminological level and no common standards of argumentation. Like methodological norms and philosophical background theories, these are determined paradigmatically; that is, they are unique, theoretical 'singularities'. Only scientific 'revolutions' go beyond them — by creating other paradigmatically constituted contexts. Different scientific worlds would be expressions of different scientific paradigms, which are not connected with each other in terms of scientific progress, since the criteria for scientific progress are not transparadigmatic criteria.

There are, however, many indications that the incommensurability thesis in this form is exaggerated. This is true, for example, both for the area of *intertheoretic relations*, which is connected with explanation, justification, and reduction, as well as for the precise formulation of a theory of meaning implied by the thesis of incommensurable languages. (One of several questions this raises is that concerning whether a semantic embedding of concepts is valid not only for theoretical, but also for empirical concepts.) But there are other alternatives for assessing a paradigm shift or establishing heuristic criteria for the development of a new theory from an old theory. These alternatives do not begin with the strong assumption that complete discontinuities exist between different paradigms or between different theories determined by paradigms, and yet they are able to deal with the same exemplary material that the incommensurability theoreticians also rely on.[9] The last word has not been spoken on this question.

The incommensurability thesis, which, taken alone, very likely makes too strong a claim, is nevertheless part of a particular orientation, namely *historism* in the philosophy of science.[10] It is this historism which represents precisely the relativity in the development of science and scientific progress that the relativity of scientific worlds also seems to reveal. What is involved here is Kuhn's theory of scientific development. Within the limits of this theory, questions of validity can only be answered with reference to factual developments, and assessing historical patterns in the development of scientific knowledge is dependent on criteria that are themselves part of these developments. In such a conception truth and falsity, or grounding, are no longer topics for the philosophy of science. Every theory and every scientific tradition has its own rationality: a transparadigmatic rationality, if we take the incommensurability thesis literally, does not exist. Only historical developments determine our views on scientific progress and the validity of scientific orientations.

It is apparent that in such a conception of scientific development, the relativity of scientific worlds, as the above mentioned examples reveal, acquires a type of theoretical legitimacy; one that allows an individual to decide for himself about the world in which he wishes to live. For Feyerabend, this conception of scientific development acquires an additional philosophical 'foundation' in epistemological anarchism, i.e.

the dissolution of all methodological standards in favor of a maxim clearly expressing a contempt for rationality: 'Everything goes!' All that can be said against a decision for older worlds (for example, the hermetic), is that one would have to step outside the dominant historical development, whose world today would rather be the Einstein world. There is no argument from rationality against this. Such an argument would have to have a transparadigmatic power, which, according to the presuppositions of scientific historism, it cannot have. The history of science, within the limits of historism in the philosophy of science, "only mirrors a change of conceptions, not a logic of research",[11] which one would have to follow in the philosophy of science.

Now, to maintain an attitude of scepticism toward the relativities which are a product of historism in the philosophy of science, one does not have to believe that it (i.e. the history of science) is a French garden where the main axis leads directly to the splendid palace of modern knowledge. A more precise analysis of what I have here referred to as the relativity of scientific worlds already provides some illumination. With respect to the Aristotle, hermetic, Newton and Einstein worlds, relativity means the dependency of these worlds on scientific theories. These acquire, in the terminology of modern philosophy of science, a paradigmatic status — with all the attendant difficulties of theoretical comparison and reconstruction of rational developments from one theory to another. These theories represent the conditions relative to which it is possible to speak of the validity of these worlds or world pictures. Phrased in a manner freer of jargon, this means that world pictures follow the development of the sciences. Fundamental changes in scientific theory result in fundamental philosophical reinterpretations of the world. This *diachronic* relativity documents, as I have already stated, the power of science to constitute the world and to generate world pictures.

Diachronic relativity, however, is not the only relativity that comes into play here. Up to now it may have seemed as if scientific theories, which are followed diachronically by philosophical interpretations of the world, completely determine these interpretations, as if there were *the* Aristotle world of Aristotelian physics and *the* Newton world of Newtonian physics. It may have seemed as if this relativity exists only between the scientific paradigms and their worlds, and not within a scientific paradigm with respect to its constitution of worlds and the generation of world pictures. Such a view, however, is inaccurate. Scientific theories actually do not predetermine their philosophical interpretation. Just as an empirical fact can be explained by different theoretical statements (and also be theoretically determined), so a theory can be explained by different 'philosophies'. As a rule, different philosophical interpretations of the same theory are possible. This, however, means that *synchronic* relativity supplements diachronic relativity. It does not necessarily mean that problems of orientation, in so far as they are connected with scientific theories, are inevitably insoluable. In the context I have outlined here, synchronic relativity only means that science does not completely determine world pictures. It constitutes the stimulus for the development of world pictures and provides the conditions that facilitate decisions concerning its adequacy. What it does not do, however, is show that such world pictures are directly involved in theory formation. Science restricts the degrees of freedom in the development of world pictures, but it does not allow any definitive distinction of a particular world picture.

Formally speaking, what we are concerned with here now are *hermeneutic* relationships: the translation of philosophical questions into the language of the respective scientific (for the most part physical) theories. This translation can be given different ways, so that the same theory can provide different answers to the same philosophical question. Philosophical rationality is not the same as scientific rationality: 'the place of philosophy is not in the equations but between the equations'. Again, some examples may serve to illustrate.

(1) Whereas Newton conceived of gravitation not as an essential characteristic of matter, but as an active, immaterial principle, early Newtonianism already began to view gravitation as an essential property of matter. Lesage, on the other hand, attempted to derive the law of gravitation solely from particle collisions. Whereas gravitation thus represents a simple phenomenon for mainstream Newtonianism, for Newton himself, as well as for Lesage, it is a derived, mediated effect. According to Newton gravity is caused by the immaterial corpuscles of the aether; for Lesage, by the collisions of material particles. In other words, the Newtonian theory provides no clear answer concerning the natural-philosophical character of gravitation. One and the same theory provides different answers depending on the interpretation. (2) Similarly, the mechanistic tradition of the 19th century interpreted the electromagnetic field as a state of a mechanical aether, whereas Einstein

considered it to be an independent entity. In both cases, the same Maxwellian theory of electromagnetism is subject to differing interpretations. (3) In 20th century physics, different interpretations of quantum mechanics differ according to whether the indeterminacies in quantum mechanics are taken to be a product of irreducible chance, or merely the result of insufficient knowledge of the initial conditions involved. The latter is assumed to be the case, for example, in an interpretation in terms of variables. (4) With respect to the theory of relativity, it is still uncertain whether, as Einstein himself believed, a theory of relational space fits the general theory of relativity. One obtains different answers, depending on how one translates classical relationalism into the conceptual language of relativity theory. At least at the moment, it is still not possible to decide in favor of one or the other of these translations.

Whereas — and all of the examples clearly reveal this — a scientific theory that is the object of philosophical interpretation is definitely distinguished by criteria of validity, this is not true in the case of the interpretation itself. Interpretations are ambiguous. In their hermeneutic character, philosophical interpretations always reveal particular uncertainties which cannot be completely avoided even in a so-called rational reconstruction of the principles underlying a theory.[12] Concerning this aspect, the interpretation of the quantum theory does not significantly differ here from an interpretation of the Kantian theory of space and time.

This, in turn, does not mean that historism in the philosophy of science, with its incommensurability thesis and the *historical relativism* derived from it, is justified. Diachronic and synchronic relativity is involved in the distinction between scientific or methodological criteria of theory assessment on the one hand and philosophical interpretations of the theory on the other; but this does not imply an (historicist) rejection of such criteria that consistently include the concept of a transparadigmatic rationality. Diachronic and synchronic relativity in the form I have presented here, teaches us, once again, that a scientific theory does not as a rule determine its philosophical interpretation, particularly with respect to world pictures. That is why this interpretation cannot be relegated to science (i.e. to non-philosophical, scientific treatment). Against the suspicions of a thoroughgoing relativism (and its support through historism in the philosophy of science), diachronic and synchronic relativity does not teach us that scientific rationality itself is ambiguous.

4. HISTORY OF SCIENCE INSTEAD OF PHILOSOPHY OF SCIENCE?

Scientific rationality is expressed not so much by individual theories or theoretical traditions as by the *criteria of rationality* which guide the assessment of theories and make the choice between them unambiguous. In view of the historism in the philosophy of science, the question is whether these criteria of rationality are not in fact secondary, or in specific cases, even dispensable as *criteria of assessment* for scientific theories, given the so-called 'normativity' of facts (i.e here empirical control), and assuming, of course, that not all modes of conducting science are violated (so that shamans have the say). The scientific facts themselves speak against such a conjecture: "The Copernican theory was preferred to the Ptolemaic, although the former did not account for the facts more precisely than the latter. Bohr's atomic theory was retained despite the fact that it did not adequately account for the Stark effect, and not at all for the helium atom. Planck's second theory of radiation was never accepted because it could not account for the Franck-Hertz experiment".[13] This means that 'conformity to facts,' even in science, is not the last word. Or, to put this differently: "The question of how a theory should be assessed is simply not identical with the question of whether it is right or wrong. There are excellent wrong theories, just as there are probably correct theories that are poorly conceived. Newton's mechanics is wrong, but it nevertheless represents the paradigm of a good theory; the character psychology at the beginning of the century, despite its conformity to facts, is completely inadequate from a methodological point of view".[14]

Other criteria are thus necessary in addition to 'conformity to facts.' Rationality criteria provide precisely these additional criteria. Among these are not only lack of contradiction, lack of circularity, verbal clarity, intersubjectivity, and the reproducibility of scientific results, but also, for example, *optimal applicability* (theories should explain more than those problems that they were formulated to solve), *simplicity* (not in an absolute, but in the relative sense of theoretical clarity), and *fertility* (though this should be sought less in the explanatory power of theories than in their ability to deal with new problems).[15] Viewing such criteria as dependent on paradigms would mean depriving the concept of paradigm of all applicative meaning. The opposite of theory-rationality, in the sense given by the above mentioned criteria, would not be another rationality, but rather myth and irrationality. But that would also mean that a strict historism in the philosophy of science, which also refers to this type of criteria, is not practicable. Or, in other words, that even so-called incommensurable theories can be assessed not only 'internally' (by means provided by the theory) but also 'externally' (by means of the universal criteria of rationality acting as a standard). At the same time, this means that 'progress' is not just an 'internal' concept.

Science, moreover, has a *research form* and a *theory form*.[16] In its research form, science is concerned with discovering what is the case. In its theory form, it is concerned with representing theoretically what it has discovered by conducting research in this way. In its research form, science is an expression of object competency or *object rationality*. In its theory or representational form, it is an expression of meta-competency or the *rationality of justification*. Scientific rationality always includes this dual aspect of science, although now it is particularly the theoretical or representative aspect that makes the world being researched the world of science (in the sense of what was said about the role of world pictures in science). These world pictures, in other words, lie enclosed within the theory form of science, and less in its research form (insofar as this relies on its theory or representational form when it expresses itself theoretically). The same is true of the criteria of rationality. These belong both to the research form as well as to the theory or representational form of science. They exert an influence on research, e.g. in experimental contexts (criteria of reproducibility etc.), and determine the construction and the assessment of theories. Their explication itself has a theoretical or representational form. Even a 'logic of research', like Popper's fallibility model of scientific rationality, is, in this sense, a representation of research with respect to (fallibilistic) criteria of rationality.

Not only in this way is the previously stated diachronic and synchronic relativity of scientific worlds relativized by a reliance on a transparadigmatic theory or representational form of rationality criteria. Something has also been demonstrated that, in epistemological terms, is the essence of all 'knowledge' and all 'truth': the *appropriation of the object through its representation*. We only acquire knowledge of the world through its representations, and only against the background of the double aspect of scientific research and representation does it become clear that (scientific) explanations are also representations based on principles which cannot simply be inferred from what is represented. How we view the world is the way it looks to us. There is no simple reason of facts that could enable us to circumvent our own scientific representations. In terms of a *Leibniz world*, the correlate of knowledge is not an 'objective world' and the correlate of the world is not an 'objective knowledge.' Things are much more the way we see and represent them — through our daily experiences in the everyday-life world and through our theories. And just as things cannot take the place of experiences and theories, so experiences and theories cannot take the place of things. This is precisely the source of the 'true relativity' of all knowledge in science and in everyday life and of all scientific and everyday life worlds. This relativity does not lead to relativism, but — once again with reference to the appropriation of an object through its representation — to the *constructive nature* of every orientation, even the orientation of science.

After this short epistemological excursion into the philosophical world of constructivism let me return to the history of science. The modern philosophy of science has discovered the transitoriness of textbook knowledge. It *historicized* scientific knowledge and, in this way, made analyses in the history of science look uncommonly like the philosophy of science. Is the history of science the last word of the philosophy of science?

The history of science itself does not know the correct answer to this question. Its image shifts between mere history and ambitious theory. On the one hand, it has been satisfied (for a long time) with being the history and philology of science. On the other hand, it considers itself (in recent times) to be Minerva's owl, which begins its flight over the fields of science when the forms of scientific rationality have grown old. Only in this evening flight, does scientific rationality — thus the self-confident idea of the history of science — comprehend itself. Correspondingly, in the historism model of scientific development, scientific rationality appears essentially as *research on representations*; and thus as second level research, and not, as is still the case in the fallibilist model, as the representation of research (second level representation). Philosophy of science attempts to become the history of science.

In vain. The 'historiographical revolution', which Kuhn claims his historicist conceptions to be,[17] has confused rather than illuminated historians of science. It promised the history of science a new competency in theoretical matters, and a new self-consciousness; yet it has led to misunderstanding in the philosophy of science. Specifically, the misunderstanding that systematic orientations would all be historical orientations, and that history would serve not only as the arena for scientific battles, but also as their judge in questions of validity. But this is just what the history of science, even in its modern form, cannot be. Historical reflections are, of course, one side of understanding; and this holds in judging scientific developments. But they cannot at the same time take the place of the other, systematic reflections. Not history, but reason decides between truth and falsity — even when it is often only history that opens one's eyes.

The discovery of the history of science within the modern philosophy of science, moreover, took a heavy toll on the history of science that many historians of science seem not to have noticed; namely, the virtual loss of its *historical* character. An expression of this loss is the concept of the historical case study. With this concept, comes a theoretical claim that is unknown to the traditional history of science. This claim is an expression of a new epistemological concern and discredits at the same time the prior historical concern. The concept of case study, applied to the history of science, gives the impression that history is an arsenal of 'facts' which have to be given an 'explanation.' The *case* is always a case *of something*. In the notorious simplicity of logic, 'case' is a two-place predicate. This presupposes that something, the special case of which one analyzes, is already known as something (theoretically) general; in this case the 'structure' of the history of science. That, however, is an enormous claim that could only be hatched in the minds of philosophers of science (and philosophers in general), and not in the minds of historians of science. It does not fit in with these at all; nor with the history of science, when this is seen as *history* and not reason.

This does not mean that the history of science has no place within orientations of reason (here within the field of science). On the whole, what is still involved is the question of the *understanding of history*. It is not possible to speak of such an understanding either where there is only mere familiarity with history (the standpoint of the *archive*), or where the historical material is apparently adapted to particular theories of the history of science (the standpoint of *metahistory*). In both cases, history is not understood but narrated (or translated) into philosophical or logical structures. Histories and their models replace historical *experiences* reflected in the understanding of history. Thus, one can only speak of understanding history where both alternatives are avoided. This happens when historical reconstructions are based on constructive models of reasoned developments (and critical judgements of factual developments), without losing sight of the contingent character of historical developments.

This is not the place to discuss in depth what the 'methodology' of such an understanding of history should look like.[18] At this point it is enough to have pointed out the extent to which the history of science has been both undertaxed in terms of history and overtaxed in the philosophy of science, and to have formulated aims for the history of science that might make these aspects of enlightenment. This may also serve to hinder a misunderstanding that has begun to affect the relationship between science and the modern world. Here what I am thinking of is the strange conception of a *science without scientists*, or the conception that scientific developments essentially take place behind the scientist's back — in a way reminiscent of so-called technological change which is also not propelled by acting subjects in any real sense. In this view, philosophical or evolutionary conceptions of history do the work of individuals in science. The historism model in the philosophy of science is among these. Scientific being determines subjective being — just as in Marxist theory social conditions determine private conditions. Scientific rationality would be read out of the prevailing context of science, as it were.

The *non-empirical* character of reason, as well as the afore mentioned fact that the scientific world is a *Leibniz world* under the aspect of representation, reveals that this is a false, or at least a misleading view of science. Neither reason nor the world is simply given. Both are rather a result of the work man does, and, in the context of science and the modern world, particularly of the *scientific* work he does. Historical understanding does not serve just any purpose, including, again, the aims of mere history or the philosophy of science. It helps us to understand our own world, which, in rational cultures, is essentially a scientific world. The history of science and the philosophy of science should thus have a common task — an enlightening one. What is to be developed in the history and philosophy of science is a critical understanding of the modern world, and thus more than what philosophy, in the form of a primarily affirmative trust in scientific reason, has been saying for some time.

University of Konstanz

NOTES

* I would like to thank Dr. Martin Carrier (Konstanz) for his assistance on numerous questions in the history of science and for giving the manuscript the benefit of his critical attention.

[1] Met. A1.980a21.

[2] For the methodology for this distinction see F. Kambartel, "Wie ist praktische Philosophie konstruktiv möglich? Über einige Mißverständnisse eines methodischen Verständnisses praktischer Diskurse", in *Praktische Philosophie und konstruktive Wissenschaftstheorie*, F. Kambartel (ed.), Frankfurt 1974, pp. 22ff.

[3] For a more detailed account see J. Mittelstrass, "Theorie und Empirie der Wissenschaftsforschung", in *Grundlegung der historischen Wissenschaftsforschung*, C. Burrichter (ed.), Basel/Stuttgart 1979, pp. 71—106.

[4] Cf. Ch. Thiel, "Alchemie", in *Enzyklopädie Philosophie und Wissenschaftstheorie*, vol. I, J. Mittelstrass (ed.), Mannheim/Vienna/Zurich 1980, pp. 67—74.

[5] The first sentence of an apocryphal text attributed to Hermes Trismegistos. See Ch. Thiel, "Makrokosmos", in *Enzyklopädie Philosophie und Wissenschaftstheorie*, vol. II, J. Mittelstrass (ed.), Mannheim/Vienna/Zurich 1984, p. 749.

[6] J. J. Wagner, *System der Idealphilosophie*, Leipzig 1804, p. LIII.

[7] R. Fludd, *Utriusque cosmi majoris scilicet et minoris metaphysica, atque technica historia*, vols. I—II, Oppenheim/Frankfurt 1617.

[8] E. J. Dijksterhuis, *De Mechanisering van het Wereldbeeld*, Amsterdam 1950, p. 539.

[9] See, for example, H. R. Post on the 'inductive' model in "Correspondence, Invariance and Heuristics: In Praise of Conservative Induction", in *Studies in History and Philosophy of Science* 2 (1971), pp. 213—255. G. Wolters discusses this also with respect to overemphasized conceptions of discontinuity in "Topik der Forschung. Zur wissenschaftstheoretischen Funktion der Heuristik bei Ernst Mach" in *Technische Rationalität und rationale Heuristik*, C. Burrichter/R. Inhetveen/R. Kötter (eds.), Paderborn etc. 1986, pp. 123—154, here pp. 138f.

[10] On the use of this term in the philosophy of science see J. Mittelstrass, "Historismus in der neueren Wissenschaftstheorie", in *Die Bedeutung der Wissenschaftsgeschichte für die Wissenschaftstheorie* (Symposium of the Leibniz Society in Hanover, Nov. 29th and 30th, 1974) Wiesbaden 1977 (*Studia Leibnitiana*, Sonderheft 6), pp. 43—56. See also my article "Historische Analyse und konstruktive Begründung", in *Konstruktionen versus Positionen. Beiträge zur Diskussion um die Konstruktive Wissenschaftstheorie*, vols. I—II, K. Lorenz (ed.), Berlin/New York 1978, vol. II (*Allgemeine Wissenschaftstheorie*), pp. 256—277.

[11] M. Carrier, "Wissenschaftsgeschichte, rationale Rekonstruktion und die Begründung von Methodologien", *Zeitschrift für allgemeine Wissenschaftstheorie* **17** (1986), p. 203.

[12] On the concept of (rational) reconstruction and the difference between reconstruction and interpretation, see J. Mittelstrass, "Scientific Rationality and Its Reconstruction", in *Reason and Rationality in Natural Science. A Group of Essays*, N. Rescher (ed.), Lanham/New York/London 1985, pp. 83—102, here pp. 89ff.

[13] Carrier, *op. cit.*, p. 202.

[14] *Ibid.*

[15] See here also Carrier, *op. cit.*, p. 203.

[16] On this distinction see K. Lorenz, “The Concept of Science. Some Remarks on the Methodological Issue ‘Construction’ versus ‘Description’ in the Philosophy of Science”, in *Transcendental Arguments and Science. Essays in Epistemology*, P. Bieri/R. P. Horstmann/L. Krüger (eds.), Dordrecht/Boston/London 1979, pp. 177—190.

[17] Th. S. Kuhn, *The Structure of Scientific Revolutions*, 2nd ed., Chicago 1970, p. 3.

[18] For additional comments see J. Mittelstrass, “Rationale Rekonstruktion der Wissenschaftsgeschichte”, in *Wissenschaftstheorie und Wissenschaftsforschung*, P. Janich (ed.), Munich 1981, pp. 89—111 and 137—148.

JAMES ROBERT BROWN

LEARNING FROM THE PAST

Here is a big fact: Scientists are good at doing science. It seems a platitude, yet there are those, such as David Bloor, who would deny it.[1] On the other hand, most people, including most philosophers, would agree that scientists are good at doing science; nevertheless, these same philosophers don't seem to think this fact is in any way interesting or important. But it is. There are few facts in philosophy; let's not let this one slip through our fingers.

Conceding the big fact leads to two questions. Firstly, *why* are scientists good at doing science? A variety of answers are possible; the candidates include: Scientists have grasped the platonic form of good scientific method; scientists are 'hard wired' or genetically programmed for rationality in the same way (according to Chomsky) we are hard wired for grammar, a result of the evolutionary process; scientists, in trial and error fashion, have learned from experience what works. I have no idea which, if any, of these are true, nor will I make any attempt here to find out.

The second question is, what do scientists *do* when they do good science; that is, what is good scientific method? Once again there are a variety of answers: From Popper: conjectures and refutations; from Lakatos: the method of scientific research programmes; from Salmon: Bayesian inference; from Glymour: bootstrapping.

What I am interested in doing in this essay is finding out how to answer the second question. And here's why the big fact is no mere triviality: We can use it to find out what good scientific method is. This is in marked contrast to the way philosophers normally go about answering this question. Even though they gave wildly different answers to the question, 'What is good method?', Popper, the positivists, and most others have all agreed on the grounds for any answer: *a priori* reflection. The use of actual historical scientific practice has been mainly illustrative, never evidential; like a diagram in a geometry book, the example helps the imagination but provides no justification.

The other strategy for answering the second question is based on the big fact. Since scientists are good a doing science, so let's study them to

J. R. Brown and J. Mittelstrass (eds.), An Intimate Relation, 343–367.

find out what good method is. Of course, these two approaches needn't be exclusive. Indeed, I favour a mixed strategy; I think they are both necessary to get the whole story. But this is a side issue since my aim is to spell out and defend the notion of appealing to historical instances as evidence for methodological norms, whether the resulting procedure be thought the only way to determine methodology or as merely part of a mixed strategy.

Neither philosophers nor historians of science are as tuned in to this issue as they ought to be. A survey of some recent views (all of which I think wrong) may provide a useful introduction to some of the vexing questions as well as set the stage for the view I shall propose on this matter.

FALSE PROPHETS

Giere

Kuhn's *Structure of Scientific Revolutions*, when it appeared in 1962, was a blow to the then prevailing empiricism and a stimulus for a much more historically oriented outlook. Increasingly, the new sentiment has been that philosophers must 'do justice to the history of science'. (The works of Butts, Buchdahl, Feyerabend, Hacking, Hanson, Hesse, Lakatos, Laudan, and many others are representative.) But traditionalists had their champion in Ron Giere (1973) who steadfastly held that the relationship was 'not intimate', but rather a 'marriage of convenience'. Chief among his objections was a perceived problem of circularity which he took to be

> . . . especially acute for any historical approach to the criteria for the rational choice of theories. Suppose . . . that history provides *empirical* data for one's account of theory choice. In this case the account is itself an empirical conclusion, or, broadly speaking, a theory. But to choose a theory of theory choice on the basis of historical data one must already have some criteria for theory choice. (1973, p. 292)

In other words, trying to get norms out of history is question begging. Others (e.g., Laudan 1976) have claimed that it would be even viciously circular. Actually it is neither, though showing that it is not is a somewhat long and tedious procedure. (See Brown 1988 for details.) The situation is a rather complicated special case of the coherence theory of knowledge, and so long as that account of epistemology is acceptable in principle, then the use of historical cases as data for choosing among rival methodological theories must be acceptable, too.

The great virtue of Giere's objection, however, is that it focuses attention on the right issue: Just *how* is it that history works as data for choosing among rival norms? It is one thing to say philosophers ought to be historically informed, but as advice goes, that's on the level of telling physicists they ought to pay attention to experiments.

Lakatos

Imre Lakatos had some specific advice for scientists about experiments: Make novel predictions about them; if right, count yourself progressing; if wrong, make some adjustments which save the core of your theory (that is, modify the initial conditions or auxiliary assumptions); never take any single experiment too seriously, none is crucial. On the other hand, if you can not seem to make any successful predictions over a long haul, then your research program is degenerating and you should switch to another.

How do we know that this or any other methodology is right? Lakatos's answer to this question is the same as his answer for deciding among scientific theories, with one difference: Instead of accounting for empirical data as a scientific theory must do, a methodology should account for the 'value judgments of great scientists'. The process requires what he calls a 'rational reconstruction'. This is because, as he likes to paraphrase Kant, 'History without philosophy is blind, philosophy without history is empty'. (1971, p. 91) All of this becomes enormously complicated since

> In writing a historical case study, one should . . . adopt the following procedure: (1) one gives a rational reconstruction; (2) one tries to compare this rational reconstruction with actual history and to criticize both one's rational reconstruction for its lack of historicity and the actual history for lack of rationality. (1970, p. 138)

As for particular anomalous value judgments, we needn't be too concerned, says Lakatos, since he wishes to

> emphasize the far-reaching analogy between scientific and methodological research programmes and stress that just as empirical 'basic statements' can be overruled by theory, normative 'basic judgments' can be overruled by methodology. (1971, p. 180)

The idea is to reconstruct the history of science through the eyes of different normative philosophies of science with the aim of making the history of science come out as rational as possible. Needless to say, it has led to all sorts of objections. Holton says it makes his 'hair stand on end' (1974, p. 75), and Kuhn that this is 'not history at all but philosophy fabricating examples.' (1971, p. 143)

On the contrary, I consider this to be actually a very sophisticated account the matter; its sins are largely sins, not of substance, but of exposition. The view I give below is much influenced by Lakatos. However, there are some problems of content with Lakatos's view, as well. For instance, his reliance on intuitions is excessive; it seems to be the only kind of data available for testing rival methodologies, but surely there are other types of consideration as well. Moreover, the idea of a rational reconstruction is in great need of being clarified. And finally, it appears that the 'winning' methodology on Lakatos's account would make *every* episode in the history of science rational — something which is surely absurd.[2]

Laudan

In his important and influential *Progress and Its Problems* (1977), Larry Laudan sees a number of ways that history and philosophy interact. When it comes to using history as evidence for or against methodoligical theories he takes an intuitionist line.

> ... [T]here is, I shall claim, a subclass of cases of theory-acceptance and theory-rejection about which most scientifically educated persons have strong (and similar) normative intuitions. This class would probably include within it many (perhaps even all) of the following: (1) it was rational to accept Newtonian mechanics and reject Aristotelian mechanics by, say, 1800; (2) it was rational for physicians to reject homeopathy and to accept the tradition of pharmacological medicine by, say, 1900; (3) it was rational by 1890 to reject the view that heat was a fluid; (4) it was irrational after 1920 to believe that the chemical atom had no parts; ...
>
> ... What I shall maintain is that there is a widely held set of normative judgments similar to the ones above. This set constitutes what I shall call *our preferred pre-analytic intuitions about scientific rationality* (or 'PI' for short) ... [O]ur intuitions about such cases can function as decisive touchstones for appraising and evaluating different normative models of rationality ... *[T]he test of any putative model of rational choice is whether it can explicate the rationality assumed to be inherent in these developments ... [and] the degree of adequacy of any theory of scientific appraisal is proportional to how many of the PIs it can do justice to.* (1977, p. 160f)

This proposal seems to satisfy the yen we have to do justice to the history of science; those events we feel exemplify great science will turn out to be rational by this criterion. In spite of the attractiveness of Laudan's proposal, there are numerous problems with it.

First, there is a growing conviction that good methodology will be comparative; yet Laudan's PI criterion is absolute, a fixed standard for all to meet. Second, and this is related, Laudan's 'intuitions' are either absolutely fixed, in which case they work as a kind of foundation for methodological knowledge, a doctrine which is highly suspect, or else they are changeable, that is, they are historically conditioned, in which case they are quite possibly contaminated. Either way, Laudan's intuitions seem unworkable.

Third, there is only a small number of historical episodes which do any work; the great bulk of history is evidentially neutral. This is analogous to saying that the great bulk of observations we make have, in principle, no relevance to scientific theories. Laudan, in spite of his own best intentions, doesn't seem to take (most of) the history of science so seriously after all.[3]

Butts

Robert Butts has been a paradigmatic example of someone who does philosophically informed history and historically informed philosophy. The interaction of history and philosophy has been illustrated by his work on numerous occasions; in his (1980) his view of the relation is spelled out. The account of the interaction he gives there, though only a sketch, is remarkably naturalistic. The major claim, which he calls 'the functional identity of philosophy and science', is this: '[T]here is no interesting and substantial difference between philosophy and science; at least no difference that can mark off the two enterprises as different in kind.' (1980, p. 256). The second ingredient (which is more or less implicit) is this: In order to do good science/philosophy familiarity with its history is essential. He then concludes that methodology (i.e., a particular methodological theory) is then the product of science/philosophy in just the same way that, say, quantum mechanics or the theory of evolution are products of science/philosophy. In every case empirical and conceptual factors in varying degrees play a role. Nevertheless, according to Butts, there is no principle difference between methodology and what we ordinarily think of as scientific theories; nor is there a principled difference in how history functions in either case.

The view is sophisticated, coherent, and quite naturalistic. (The 1980s have seen a rapidly growing number of naturalistic accounts of just about everything; this was one of the first of the recent crop of naturalistic accounts of methodology.) I doubt that it can be decisively refuted, but (even in an essay which is a tribute to the author of this view) it should not get away scott-free.

What we can do, I think, is grumble a bit about the premises. Butts's first claim is that philosophy and science are the same kind of enterprise. This is undoubtedly true when philosophy is taken to be metaphysics. But the case is a bit harder to make out when norms are at issue, as they are when methodology is being considered. As for the second premise, it, too, may beg the question. Why should we think history is important to science/philosophy? For a Bayesian, it certainly is, since *a priori* probabilities must be plugged into the formula to get an answer. These probabilities don't come *ex cathedra*, but rather from familiarity with the history of science, that is, from our somewhat vague knowledge of what sort of thing has and has not worked in the past. Of course, if we know Bayesianism is the right approach then we need not concern ourselves further with any of this; but, on the other hand, if we don't know what the right methodology is then we don't know whether history is relevant. There is a circularity involved. However, Butts calls this sort of thing a 'happy circularity', and, in a sense he's right; it is not vicious.

If the naturalistic view championed by Butts is to be defeated, it is only by comparison with a superior (I hope) rival, but that will have to wait until later.

Giere (II)

The 1980s have seen a tremendous growth of natualism. Ron Giere is a prominent example of someone who has done a complete about-face. Initially the enemy of history, he is now the whole-hearted friend. The normative methodologist must look to actual scientific practice for evidence. And what is the precise nature of the history-methodology relationship? According to Giere, 'the nature of the relationship . . . is clear. It is the standard relationship between theoretical models and empirical data'. (1984, p. 28)

At first glance this may seem to be the same proposal as Butts's. It is true that they are both naturalistic, but otherwise they are quite different. For Butts, philosophy/science confronts the world and modifies itself in its attempts to get things right. For Giere, however, it is science alone which theorizes about the world; philosophy comes after and theorizes about science. Giere's naturalism consists in claiming that the mode of theorizing (either philosophy about science or science about the world) is the same in each case.

Of course, we can have no quarrel with aspects of Giere's elegant proposal; but ultimately it won't do, and the reason is perfectly simple: Nature doesn't cheat, or make mistakes, or react to social forces; scientists do. A theory of, say, gravity cannot allow the existence of a single (real) counter-example, yet the correct theory of rationality can tolerate the existence of lots of irrational acts, just as the existence of murderers does not refute our (correct) moral belief that murder is wrong.

I doubt that any brand of naturalism will work; this one seems especially hopeless, in spite of its attractive simplicity.

Garber

Even though he is a sensitive historian himself, Daniel Garber seems to side with Henry Ford ('History is bunk'), at least when it comes to providing grounds for norms. According to Garber, '. . . the history of *actual* scientific change is, surprisingly enough, largely irrelevant to the *normative* theory of scientific change . . .'. (1986, p. 92). What this means, in consequence, is that it is entirely *possible* that all actual scientific practice, past and present, is irrational and 'unscientific'. But to accept this is to accept the absurd possibility that scientists might be bad at doing science.

Garber's point of departure was Laudan's account discussed above which involved normative intuitions of historical events. Like Laudan, he wants to do justice to these intuitions, but claims that the cases we reflect upon needn't be real historical cases. '[W]hat is important about the cases [that Laudan] cites is not their historicity, the fact that they *actually happened*, but the fact that they are a concrete embodiment of our intuitions about what is rational and what isn't.' (1986, p. 105)

By stressing *our* intuitions Garber makes it possible, and on occasion quite explicitly maintains, that the intuitions of past scientists have been different. If rationality is just rationality *by our lights*, then we cannot offer rational explanations of what happened in the past. We can only offer psychological and sociological accounts. But once we have done this we slip into a relativism that cannot distinguish between Newton's achievements and those of a contemporary of Newton who thought the universe is a poached egg.

Can we really treat actual and fictitious examples equivalently, as Garber would have us do? He claims that what is important is the intuitions themselves; whether they are real or imaginary is irrelevant. However, there is a world of difference. Very likely, there are influences on our intuitions which are not under our control, so there is a problem of relative contamination. Moreover, just think of the analogous situation in physics. Suppose a theory could do wonderful justice to all thought experiments, but failed miserably when faced with real ones. Would we give it much credence?

Laudan (II)

Laudan has recently abandoned his PI criterion, so heavily dependent upon intuitions, in favour of a naturalistic view. A stimulus for this change of heart was Garber (1986). If we are to use intuitions at all, then, Laudan allows, it matters not whether they are intuitions of actual cases or of fictitious ones. Garber is right on this point, as Laudan now sees it; either way it is *our* intuitions which are being used for testing purposes. So, in consequence of this, Laudan (1986, 1987) says we must throw out any reliance on intuitions entirely. The earlier PI criterion is replaced (in Laudan 1984, 1986, 1987) with a kind of naturalism.

Before getting on to the new meta-methodology, let me point out some problems with the way Laudan has abandoned his earlier account. Firstly, if the problem with intuitions is that they are *ours*, then there is still Lakatos's version of intuitionism to consider. Lakatos, as we saw above, does not rely on *our* intuitions, but on the intuitions of the scientific elite. These are spread throughout history and not just the product of here and now.

Second, Laudan over reacts. This is a common sin in using history to specify a meta-criterion; people try to give both necessary and sufficient conditions for choosing a methodology when they should be trying only to give necessary ones. When we specify how a scientific theory should be tested by experimental evidence we do not thereby give both necessary and sufficient conditions for accepting the theory. There can be other (i.e., conceptual) considerations as well. The empirical and the conceptual are both necessary but neither is sufficient (though together they might be).

Similarly, a methodology should do justice to history, and to give a criterion is to give necessary and sufficient conditions for it to do justice to that history. This much is necessary, but it is not sufficient for the choice of the best overall methodology. Taking account of conceptual issues here is also required. A meta-methodology does two things: (1) It does justice to history, (2) It does justice to conceptual matters.

Laudan's PI criterion, his new naturalism, and Lakatos's meta-methodology based on the intuitions of the scientific elite are all attempts to give necessary and sufficient conditions for (1). They then, quite mistakenly, further claim to have provided a *complete* meta-methodology.

Garber and other more traditional *a priori* methodologists focus in on (2). They too are mistaken, but they often realise what they are doing and usually provide some sort of argument for dismissing (1) (such as, that to countenance (1) would be to commit the naturalistic fallacy). Those like Lakatos and Laudan who mistakenly ignore (2) do so, I think, unwittingly.

Consequently, Laudan's response to Garber should not have been to attack Garber's avowal of (2), but rather to attack his disavowal of (1). In fact, that Laudan *must* embrace (2) follows almost trivially from his new Naturalism. On his new view the way we test methodologies is the same as the way we test ordinary scientific theories. When it comes to testing the latter, Laudan stresses the crucial importance of conceptual issues, so it follows that they must figure in a meta-methodology as well; we must do justice to conceptual considerations just as much as to history.

A critical exposition of Laudan's detailed views on the issue cannot be gone into here. I must be content with merely undermining some of the motivation.[4]

THE THEORY-LADENNESS OF HISTORIOGRAPHY

I shall now take up the topic of the nature of 'observation', especially as it pertains to historical investigation. The notion is unclear and problematic; yet it is important, because the relation between observing, on the one hand, and the great body of our beliefs, desires, and expectations on the other, is far from straightforward.

The question is this: Can something akin to *neutral observations* be made of historical events or are such perceptions conditioned, influenced, and even prejudiced by prior beliefs, hopes, anticipations? In short: Are historical observations theory-laden?

What does a historian look at anyway? Archive materials, for one thing: books, letters, and notes. Implements and artifacts for another: Galileo's telescope, for example. The historian will also, on occasion, try to reproduce certain experiments and results: Newton's rings, Fresnel's diffraction experiments, Faraday's magnetic induction work, and so on. In such examples as these, is it the case that historian's perceptions are theory-laden? Will rival historians have different perceptions of what went on? In order to answer this we first need to make some distinctions.

(1) Consider, to begin with, the case of two rival historians who are both looking at, say, a magnet, or Newton's rings, or a diffraction pattern. Do they see different things? Not if they have the same beliefs and expectations about the world. They may differ in their accounts of Gilbert on the magnet but they agree on what they take to be factual truth of the matter about magnets. The same goes for perceiving Newton's rings or a diffraction pattern. If both historians accept *contemporary* optics then they will both see the same thing. Using Hanson's famous example: If Kepler and Tycho had held the same planetary theory then they would have seen the same sun. Thus, contemporary historians may not see what historical personages saw, but each will see the same thing as other contemporary historians since, as a rule, they will share the same theories about the physical world, namely our best contemporary theories. Are such observations neutral? In one sense, no. They are theory-laden, at least if Kuhn, Hanson, Feyerabend, and others are right, and for present purposes let's assume they are. But in another sense these observations are neutral, or at least relatively neutral, because *they are the same for all historians.* So, rival historians need not be in conflict on this score.

(2) Historians often try to put themselves into the historical figure's frame of mind. They try to 'get into Newton's head' and see things the way Newton saw them. If two historians — we imagine them to be rivals — were both successful in getting to see things the way Newton saw them, then, consequently, each historian would be seeing things the same way the other sees them. When they put on their 17th-century Newtonian mechanics thinking caps they see the world alike, and, moreover, they see it just as Newton did. Of course such perceptions are theory-laden. And the theory is a different one from contemporary theories, so the historians see the world differently with their Newtonian thinking caps on from the way they see it with their contemporary science thinking caps on. But the important thing is this: any historian who sees things the way Newton saw them will see things the same way as any other historian who sees things the way Newton saw them. Once again we have neutrality of observation; not absolute neutrality, but neutrality in the sense of inter-subjective agreement about what is being seen.

(3) There are still other possibilities. A notebook entry for instance,

might be smudged. One historian expecting the author to have written x may see x, whereas another historian with different expectations might see y. But this will be a relatively rare event. The debates historians have among themselves hardly ever turn on this sort of thing. Doubtless, it does sometimes occur and we should note its existence; however, it is not very significant. If we are to take seriously the thesis that historical observations are theory-laden, it must be in some other sense of the word. A historian's *sensory* perceptions will be of little concern to us since the perceptions of rival historians will be the same.

Let us try a different tack. Historiographies of science are explanatory, but the way they explain is sometimes in one respect quite different from the way an explanation functions in the physical sciences. Explanations for human actions often appeal to reasons and rationality considerations. So, for instance, we might ask why Galileo accepted Copernicanism when he did. An explanation might go like this: Galileo had considerable telescopic evidence for Copernicanism and one should always accept a theory with such support. In other words, we explain the event by pointing out the rationality of the action. The evidence is the cause of the belief. We show that, in the circumstances, the action performed was the rational thing to do. (And if we cannot give a rational explanation then we try for some other, usually of a psycho-social nature). Considerations of rationality must appeal to some methodology or other. This is the point where norms and values come into play.

There is a trivial sense in which values get injected into the history of science. A historian chooses, perhaps on the basis of taste or perhaps because of some social interest he or she is promoting, whether to write a book on Copernicus or on Darwin. This is no different from the same trivial injection of values into the natural sciences. A scientist chooses, on the basis of interest and temperament to pursue astrophysics rather than microbiology. This is not sufficient for getting values into the very content of history. A much more interesting injection of values occurs when the historian sorts through all of the available material and picks out what is 'important'. To do this the historian must have some idea as to what it is that makes for *good* science. This usually involves such things as choosing to include the results of experiments and deciding to note what background assumptions were held by the scientist in question. What the scientist had for breakfast or where he or she took summer vacations is usually omitted.

We cannot explain actions in terms of their rationality without presupposing some account of rationality or other. That is, we must drag in some normative philosophy of science. And with it come those special concepts and categories which are often unique to particular philosophies of science. Consider, for example, concepts like 'crucial experiment', 'falsified theory', 'confirmation function', 'research programme', 'positive heuristic', and 'paradigm'. Clearly, we do not observe as a neutral given, the *crucialness* of a particular experiment. And a research programme is a sequence of theories, an ordered set; but a set is an abstract entity which can no more be seen than humankind or the number seven.

Yet some historians claim they can see crucial experiments while others are blind to them; some see evidence and method where others see only irrelevant data and madness. There seems but one inference to make: These historical observations are theory-laden. Seeing crucial experiments, or seeing research programmes requires seeing events through the eyes of some theory, a normative theory, a theory about how science *ought* to be practiced.

We believe there are electrons not because we can see them directly, but because we accept the particular atomic theory which says they exist. Analogously, we believe there are crucial experiments (if we do so believe) because we accept the particular normative theory which says they can occur and which tells us how to recognize them when they do.

History is theory-laden in the senses I mentioned at the outset of this section; but these ways of being theory-laden are of little significance to us. The most important sense, for our purposes here, is that a historiography must employ concepts from some normative theory. A thorough history of science is explanatory, and the correct explanation for many cognitive decisions is in terms of their rationality. The only way to give such an explanation is by invoking methodological concepts. This is the significant sense in which history is theory-laden; it is laden with normative theory. When rival historians see different things it is usually because they are seeing the past through the eyes of different normative philosophies of science. Of course this sort of perception is not sense-perception, but it is a kind of perception, nonetheless, and it *is* theory-laden. These considerations set the stage for my own view of the history-methodology relation which now follows.

LEARNING FROM THE PAST

I claim that the following criterion for justifying a normative methodology based on actual history is the best characterisation of the evidential relation; that is, here is how to learn from the past:

(*R*) That methodology is best which (all other things being equal) makes its theoretical reconstructions and normative reconstructions coincide for the greatest number of episodes in the history of science, while cohering with relevant independent sociological theories.

Just as theories of scientific method provide criteria by which rival scientific theories are to be evaluated, so *R* provides a meta-criterion, by which rival methodologies are to be evaluated.[5] As I stressed above when discussing Laudan's naturalism, a methodology should do two things: it should do justice to history, and it should also do justice to conceptual concerns. *R* is only trying to satisfy the first of these; everything else is covered by the 'all other things being equal' clause. I don't mean to downplay conceptual or *a priori* issues in normative methodology; it is just that I am concerned here with how the 'empirical' aspect of things.

Some explanations of criterion *R* are in order here. The term 'rational reconstruction' has been used by several philosophers, especially Lakatos, in a number of distinct and equivocal ways; so I shall avoid it in favour of two new terms, 'theoretical reconstruction' and 'normative reconstruction'. In the previous section, I noted that historical observations, and hence historical descriptions, are theory-laden. It is methodological theory that they are laden with. A *theoretical reconstruction* is supposed to capture this notion. It is a description of a particular historical episode using the concepts of some methodology. For example, if we describe an historical episode using Popperian methodological concepts, then the account will be in terms of 'crucial experiments', 'basic statements', 'falsified theories' and so on. On the other hand, if we were to describe the same episode using Lakatos's so-called 'methodology of scientific research programmes' then the history would be written employing such terms as 'research programmes', 'heuristics', 'hard core', 'progressive', 'degenerating', etc.

These terms, 'crucial experiment', 'heuristic', and so on, are theoretical terms. They come from a methodological theories. Any written historiography of science must employ such normative concepts if it is to be explanatory at all. It is in principle impossible to write a history of science without utilising such notions, that is, without using concepts which owe their being to some normative philosophy of science. (At least, if it is a rationalist history of science; a sociological history will invoke different theoretical entities, such as 'professional interest' or 'class interests' to explain events.) When a history does describe some episode using the conceptual apparatus of some methodology it is to be called a 'theoretical reconstruction'.

The reason why an historiography must invoke methodological concepts if it is to be an explanatory account is very simple, but it should be repeated and stressed. In order to explain an event we cite the cause. In the case of a cognitive decision, this amounts to citing a reason — at least, if the decision was a rational one. Thus, the reason why someone switched theories could, for example, be the occurrence of a crucial experiment. So, to explain the event is to cite the cause, which is to cite a *reason*, which is to invoke a methodological concept, e.g. a crucial experiment.

Whereas a theoretical reconstruction is an attempted description of actual history, a *normative reconstruction* is quite another thing. Like a theoretical reconstruction it employes the conceptual apparatus of some methodology or other; but a normative reconstruction does not attempt to say how history *actually* went. Rather, it declares how history *ought* to have gone *according to that methodology*. The initial starting point for the theoretical and the normative reconstructions will be the same, but from that point forward they may or may not diverge.

An example will illustrate these two notions. We might have the following (very simplified) theoretical reconstruction using, say, Popperian methodology: Jones *boldly conjectured* a theory *T* and submitted it to a *crucial test*. *T* was *falsified*. Jones continued to maintain *T* and looked for ways to confirm it.

On the other hand, using the same Popperian methodology, the normative reconstruction of the historical episode would run: Jones *boldly conjectured* a theory *T*. He submitted it to a *crucial test*. *T* was *falsified*. Jones *rejected* T.

The same historical episode might be theoretically reconstructed as follows, this time using, say, Lakatos's methodology of scientific research programmes: Jones conjectured a thoery *T* which was part of a *research programme RP*. *T* made a number of empirical predictions some of which succeeded and others of which failed. Jones did not abandon *RP*, but made slight modifications, which amounted to changes in the *protective belt*, leaving the *hard core* of *RP* intact.

Now we switch to the normative reconstruction of this episode still using Lakatos's methodology: the account, let us suppose, runs exactly the same as the theoretical reconstruction. That is, Jones conjectured a theory *T* which was part of a *research programme RP*. Some of *T*'s predictions panned out while others didn't. Jones did not abandon *RP* but made modifications in it which amounted to a new theory leaving the *hard core* of *RP* intact and modifying only the *protective belt*.

What I have called a theoretical reconstruction is an attempt, albeit highly theory-laden, to describe how history *actually* happened. And what I have called a normative reconstruction is an account of how history *ought* to have gone *according to that methodology*. My rule *R* says that it is a plus for a methodology if the theoretical and the normative reconstructions coincide in any particular episode. In the Jones case, Popperian methodology does not lead to a coincidence between the theoretical and the normative reconstruction, but Lakatosian methodology does. Thus, the Jones episode is evidence *for* Lakatosian methodology and evidence *against* Popperian methodology.

If all of this seems overly elaborate, a simple example from physics may help it appear familiar. Suppose we see a great mish-mash of streaks in a bubble chamber photograph and interpret it as the decay of a neutron into an anti-proton, a positron (or anti-electron), and a neutrino: $n \rightarrow \bar{p} + \bar{e} + v_e$. This is a highly theory-laden observation. It corresponds to what I have called a theoretical reconstruction of historical events. A particular theory of elementary particles may say that if a sequence starts with the decay of a neutron it *must* continue as follows: $n \rightarrow p + e + \bar{v}_e$. This is a prediction; it is derived from the theory and it corresponds to my normative reconstructions. (The theory rules out processes such as the one described first, since they violate conservation of baryon number.)

R says we should think more highly of a methodology when theoretical and normative descriptions coincide, just as we think more highly of a theory of elementary particles when the description of actual events (theory-laden though that description is) and the prediction of those same events coincide.

A great many historically-oriented, rationalist-minded philosophers of science have been at pains to present careful case studies. But what role do these case studies play? Clearly, they are intended to do more than merely illustrate some new *a priori* philosophy of science; they are supposed to be some sort of evidence. But how? In the past it has not been clear, but we have an answer now. *R* very straightforwardly says just what it is that the case study does; it says just how the case study serves its evidentially supporting role.

Many will be tempted to lambaste *R*. They will see it as implying that the best methodology is the one which makes *everything* rational, and we do not have to be cynics about the history of science to believe that that is an absurd consequence. But why do we doubt that the entire history of science is rational? Surely, it is because we have background beliefs about typical human behaviour. We know that on occasion people make mistakes, cheat to promote their own interests, and often enough are just plain silly. These are perfectly reasonable beliefs, not the result of some jaded misanthropy. They are covered, however, by the final clause in *R*. Let me elaborate.

Recall that when I discussed Lakatos, one of the objections raised against him was that on his account the best methodology might be the one which made everything rational (at least everything the scientific elite thought was rational). This will not occur on my account. The proviso in *R* prohibits this sort of absurdity. The coherence demanded by *R* tells us to maximise rationality as far as we can while still cohering with the best psycho-social theories (which in effect tell us to what degree scientists can be expected to be irrational.) There is no question about *R* making every episode in the histroy of science rational. A methodology which did would violate *R*.

At this juncture I should give examples of the sorts of theory, especially sociological theory, that a methodology must cohere with. I'll turn to one of the most prominent sociologists of science, Robert Merton, for illustrative examples. Whether his theories are true or false does not matter here since my aim at this point is merely to illustrate *R*. According to Merton (1973), the *scientific ethos* is characterised by the following properties: (1) Universalism: The evidence is open to all; there are no privileged observers. (2) Communism: Knowledge is collectively arrived at and owned by all. (3) Disinterestedness: We approach nature without prior wishes that it be one way or another. (4) Organised scepticism: Nothing is immune from doubt. According to

Merton, these are the values which most scientists explicitly and implicitly hold. Moreover, these are the values which the scientific community as a whole promotes.

There is a conflict, says Merton, between the reward system of science and these values which characterise the scientific ethos. The reward system of science is based almost entirely on peer recognition, specifically, recognition for originality. Scientists are obliged to advance knowledge and the only way to do that is by being original. The pay-off is in the form of having the contribution acknowledged. (Often this takes the form of eponymy: for example, Newton's Laws, Planck's constant, Mendelian genetics, the Freudian slip, Maxwell's equations, etc.) Merton thinks that it is not so much that egotistical people enter science as a way to become recognised; rather, the institution makes them desire it. The institution of science makes originality and the recognition that goes with it an *imperative.* This is the only way, says Merton, that we can explain why such ordinarily shy and self-effacing people as, say, Henry Cavendish get caught up in disputes over priority of discovery. The desire for recognition does not stem from an innate egoism, but must instead stem from institutional sources. Thus, according to Merton, we have an inevitable conflict between the reward system of science and the scientific ethos.

There are a number of different responses which scientists make to these conflicting demands. One is to do brilliant, original work which is recognised by the community. This, of course, is not a path open to everyone. Other responses include fraud, hoax, plagiarism, the construction of anagrams, and, most worrisome of all, the prudential cooking of data. The reason this last is so worrisome is that it may be hard to separate it from honest but theory-laden observation.[6] There is, of course, a difference between theory-laden and ambition-laden observations, but it isn't a difference that is easy to detect.

Frauds, hoaxes, data cooking, and so on are forms of deviant behaviour which result in part from the conflicting values of the scientific community. Merton proposes the following general law:

> Any social institution which gives emphasis to aspirations for all, which cannot be realised by many, exerts a pressure for deviant behavior . . . . (1973, p. 321)

Merton has also detected what he calls the Matthew Effect. (1973, p. 457ff) The name comes from St. Matthew: 'For unto every one that hath shall be given, and he shall have abundance; but from him that hath not shall be taken away even that which he hath.' Put bluntly: the rich get richer and the poor get poorer.

The Matthew Effect can manifest itself in many ways in science. For instance, credit in the case of simultaneous discovery, goes disproportionately to the more famous of the co-discoverers. And a new contribution to the scientific literature will be much more visible if it is made by a scientist who is already famous. When it comes to the allocation of resources, Merton notes that in 1962, 38% of U.S. federal support for science went to only 10 institutions; the situation has not changed significantly since then.

There is a recent nice illustration ofthe Matthew Effect in Ceci and Peters (1980). These two sociologists selected a number of articles which had been already accepted for publication in various scientific journals, but had not yet appeared. They changed the titles, the authors' names and institutional affiliations so that fame and prestige were considerably reduced. The articles were resubmitted to the same journals. Three-quarters of them were rejected on the second submission by the journals which had previously accepted them. They concluded that being a famous author or being affiliated with a prominent institution has something of a 'halo effect'.

There is another interesting phenomenon noted often by social psychologists (e.g. Barnes (1985, p. 79)) concerning peer pressure and authority. When everyone else deliberately exaggerated the length of an object, the test subject tended to agree with the judgments of the peer group.

Sexism is science is another recognised sociological phenomenon. As reported by James Watson in *The Double Helix* (1968), he and Francis Crick largely ignored the X-ray diffraction work of Rosalind Franklin *because she was a woman.*

There are also psychological constraints on acceptable normative theories. For instance, only self-conscious logicians and statisticians could actually make *explicit* use of Bayes's theorem in decision making. No one prior to Bayes's discovery could possibly have thought (that is, computed) in accordance with the theorem. We can say, almost *a priori*, that it is not the methodology that was actually used by Galileo or Newton. It is not enough that a methodology reconstruct history so that the episodes in question come out rational; it must also be the case that the methodology employed in the theoretical and normative reconstructions is a methodology which, psychologically speaking, *could* have been actually used. We are trying to find the one which actually was used, and Bayesianism, I suspect, is far too complicated for anyone to have actually calculated in accord with, at least prior to the mid-18th century. (Remember that one of the things we want a methodology for is to explain why people believed what they did, so even if a Bayesian analysis of an episode makes it rational, still that could not have been the *cause* of the decision.)

Psychological constraints, sexism, the Matthew Effect, and the pressures for deviant behaviour brought about by the reward system of science are typical sociological phenomena at work in the scientific community. These are the sorts of things that a methodology must cohere with according to *R*. Consequently, a reconstructing methodology will likely diverge in its theoretical and normative reconstructions when it considers the behaviour of Watson and Crick with respect to their appraisal of the work of Rosalind Franklin. Should the *correct* methodology diverge in its theoretical and normative reconstructions of this episode then we can say that it truly was irrational of Crick and Watson to behave as they did.

One thing that must be stressed is this: We can establish such things, as sexism in science and the existence of a Matthew effect *without* begging any normative questions. I do not say *in advance* that ignoring Franklin's work because she was a woman was irrational, only that it did happen. And I do not say in advance that the reconstructing methodology must diverge in its theoretical and normative reconstructions on this point, but only that it must cohere with the best psychological and sociological theories available. It may well turn out — though I certainly doubt it — that sexism is rational. But then note that by *R*, if it was rational to ignore Franklin because she was a woman, then it was *ir*rational to pay attention to the work of Marie Curie.

There are a number of things that can be said in favour of the criterion, *R*. I shall now outline a few of them.

WHY RULE R?

First, *R* captures the spirit of the sentiment expressed at the outset, namely, that the history of science must in some significant sense be considered rational. *R*, by trying to maximise rationality in the history of science, captures this spirit and so it has at least *prima facie* plausibility.

Second, *R* overcomes each of the problems of Laudan's PI account.

Recall some of the difficulties with his criterion. They were, for example: (1) The PI criterion is non-comparative. (2) It is foundationalist; and not only is this doctrine in widespread disrepute but it is in conflict with Laudan's own general methodology just as (1) is. (3) It relies on intuitions which can be understood either foundationally or as common beliefs. Neither of these interpretations of 'pre-analytic intuitions' yield a satisfactory criterion. And finally, (4) Most of history plays no role in the testing process.

Third, *R* is formally similar to the principle of charity. This is a methodological principle of linguistics which says that when translating the sentences of a radically different cultural group into our own, we should, as far as possible, translate their sentences into true sentences of our language. The principle was first enunciated in Wilson (1959); it is now advocated and employed by many, including Quine, who writes:

> The maxim of translation underlying all this is that assertions startingly false on the face of them are likely to turn on hidden differences of language. This maxim is strong enough in all of us to swerve us even from the homophonic method that is so fundamental to the very acquisition and use of one's mother tongue.
>
> The common sense behind the maxim is that one's interlocutor's silliness, beyond a certain point, is less likely than bad translation . . . . (1960, p. 59)

Donald Davidson also uses the principle:

> What matters is this: if all we know is what sentences a speaker holds true, and we cannot assume that his language is our own, then we cannot take even a first step towards interpretation without knowing or assuming a great deal about the speaker's beliefs. Since knowledge of beliefs comes only with the ability to interpret words, the only possibility at the start is to assume general agreement on beliefs. We get a first approximation to a finished theory by assigning to sentences of a speaker conditions of truth that actually obtain (in our opinion) just when the speaker holds those sentences true. (1973, p. 18f)

As a rule of thumb the principle of charity has a good deal of plausibility to it. It does have some shortcomings, but what I want here is to play on the initial plausibility of the principle and its similarity to *R*: As charity tries to maximise truth, so *R* tries to maximise rationality. On the other hand, there is an important difference. Charity says to interpret *others* as believing what *we* believe. This is how it maximizes truth. *R*, to put it simply, says to make the most scientists come out rational *whoever* they may be. In consequence, *we* may have to radically overhaul what we take rationality to be. *R*, unlike Charity, is not an imperialistic principle.[7]

Let me elaborate on this a bit. Charity tells us to maximize the truth of the beliefs of others. This principle has been rightly criticized on a number of scores. In its stead, a 'principle of humanity' has been suggested (Grandy 1973) which is being looked upon with increasing favour (e.g., by Lukes 1982). Humanity says that we should make the beliefs of others as intelligible to us as we can. This does not mean making them true as far as possible, but rather, making it clear to us that, in the context, the beliefs of others are quite sensible.

At first glance, it looks as if my *R* is just the principle of humanity by another name. Assuming a belief rational is not the same as assuming it true, but does it amount to the same thing to assume a belief rational as to assume it intelligible? If the answer is yes, then the two principles are the same; however, the correct answer is a flat no. In trying to maximize rationality in the history of science, it could turn out that making the beliefs of all maximally rational we end up by judging *our present* beliefs highly irrational. If making the beliefs of others intelligible necessarily means assimilating them to our present outlook, then I do not advocate making them intelligible at all. *R* and the principle of humanity will likely yield similar guidelines in most situations; but, in principle at least, they are rivals.

There is an old view of explanation which had it that to explain is to reduce to the familiar. Such a view is totally untenable in the light of our best explanations of the physical world. Quarks, neutron stars, and DNA are as unfamiliar as anything could be, though they are the stuff of the most wonderful accounts of nature going. The principles of charity and humanity both tell us to deal with foreign beliefs by making them familiar. *R* looks for something common in the beliefs of others and the beliefs of ourselves, but it needn't be anything familiar at all. The results of *R* could be as shocking as Freud's psychology or quantum mechanics. The results of employing charity or humanity could never be novel; *a fortiori*, they could never lead us to an *improvement* in *our* beliefs. *R*, on the other hand, is capable of making our science better.

Forth, *R* is supported by the arationality principle. The arationality principle is a methodological principle like the principle of charity. What it does is select a preferred explanation from among possible alternative explanations of human actions. The principle says that when given a choice between a rational explanation of an action and some other explanation, such as a psychological or sociological account of the same event, then the explanation we ought to choose is the rational one. That is, the right explanation is the one where the cause of the action was a good reason (Laudan 1977). Consider an example: Suppose we have two rival explanations for why Einstein abandoned Newtonian mechanics in 1905. One says that he did so because a rival theory, his own special relativity, better fit the available evidence. The other explanation says that he abandoned Newtonian mechanics because he was in rebellion against his elders, and so overthrew their most basic beliefs; it was a case of the conflict of generations, an inevitable fight between the son and the father (Feuer, 1974). The arationality principle tells us that of these two rival explanations of Einstein's action the one we should choose is the rational one, the one that says he did it for good cognitive reasons.

The arationality principle does not say that psycho-social factors never play a role. Often for a given action no rational explanation can be found. Then psycho-social (or other 'external') causes must be appealed to.

In their respective ways these two principles, *R* and the arationality principle, are trying to do the same thing; they are trying to maximise rationality. They are *not equivalent* principles; but they are strikingly similar and intimately related. It is because of this relation that I claim support for *R* from the arationality principle.

The difference between the two principles is this: *R* is a criterion for choosing among rival methodologies. The arationality principle assumes the nature of rationality is already given. It is the business of *R*, for example, to choose between Bayesianism and falsificationism (among others). Only *after* this choice is made can the arationality principle be employed. Suppose Bayesianism is the winner. Then given two rival explanations of why a scientist made a certain decision, one saying he or she did it because it was the rational (i.e., Bayesian) thing to do and the other explanation saying that he or she did it because of toilet training, then the arationality principle says we should choose the Bayesian explanation. So clearly, these two principles work at differenct levels; but there is a common motivation behind *R* and the arationality principle and that is why they are mutually reinforcing. Both try to maximise rationality at their respective levels. Both tend to regard the *rational* explanation as the best and preferred explanation of any action.

The logical relationship between the two principles seems to be this: What *R* does is to give the greatest possible scope for the potential application of the arationality principle. By maximising the coincidence between theoretical and normative reconstructions, *R* maximises rationality in the history of science. It chooses that methodology which makes more episodes rational than any other methodology. Moreover, any theoretical reconstruction which coincides with a normative reconstruction automatically constitutes an explanation of the episode in question. The action is explained as being the rational thing to do; but this, of course is just what the arationality principle calls for. On the other hand, if the theoretical and normative reconstructions do not coincide for some event, then no choice is possible between a rational and some other type of explanation. No rational explanation exists, according to the methodology in question, and so the arationality principle cannot be invoked in its favour. Thus, when *R* picks out the methodology which maximises rationality in the history of science, it inevitably satisfies the aim of the arationality principle. And it does this by giving the principle the largest possible scope for employment.

CONCLUDING REMARKS

Philosophers have long had a schizophrenic attitude toward history. They want to 'do justice to history', that is, they want the scientific greats to come off looking good by the philosophers' methodological lights. On the other hand, they think of themselves as doing something conceptual and *a priori*, quite independent of the facts of daily life in the lab.

It is possible, indeed, it is even imperative to have it both ways. *R* states a procedure for testing rival methodologies for their adequacy to history (including contemporary history, of course). There is no reason not to add *a priori* constraints to *R* to yield a complete rule. The correct normative methodology of science should do justice to history (*R* says how) and also satisfy various *a priori* conditions as well. This would make it analogous to the common claim that scientific theories have empirical virtues *and* non-empirical, metaphysical, pragmatic virtues, as well. (I mention this analogy only to illustrate; it is not meant to support my account since there is no reason to think methodology and meta-methodology need be the same.)

One of the more striking features of *R* is the large measure of sociological input required, even though it is extremely rationalistic in its intent. Some sociologists of knowledge, especially the strong programmers of the Edinburgh school, have complained that greedy rationalists leave only a few crumbs for them. As they see it, the arationality principle, for instance, gives sociologists only the irrational residue to explain. However, this is not the case on *R*. It is true that sociologists get to explain what is irrational, but that is not their only or even their chief task. Much more important is the circumscribing (in effect) of the realm of the rational in the first place. *R* does not just call upon sociologists *after* rationalists have done all they can; rather, sociologists have a fundamental role to play right from the start. This is quite a different role than either sociologists of knowledge or rationalists have in the past envisioned.[8]

University of Toronto
Toronto, Ontario
Canada

NOTES

[1] Bloor (1976) would deny that scientists, or any others for that matter, are good at acquiring a better understanding of how the world really works, though he and other sociologists of knowledge would cheerfully agree that scientists are good at promoting their own interests.

[2] For these and other criticisms of Lakatos in more detail see Brown 1989.

[3] Additional criticisms of Laudan can be found in Brown 1989.

[4] However, detailed criticisms can be found in Brown 1989.

[5] John Worrall (1976) has given an account which overlaps mine in many ways. See his excellent study for additional convincing reasons to adopt this general outlook.

[6] This is a confusion which often besets a recent study by Broad and Wade (1982).

[7] Martin Hollis (1982) says the principle of charity is a necessary 'bridgehead' for understanding different cultures. He is right; a bridgehead is needed; but it is our resulting common rationality, not a common truth.

[8] I have greatly benefited from numerous discussions with Robert Butts and Kathleen Okruhlik on the topic of this paper and I owe them many thanks.

BIBLIOGRAPHY

Barnes, B. (1985). *About Science*, Oxford: Blackwell.

Bloor, D. (1976). *Knowledge and Social Imagery*, London: Routledge and Kegan Paul.

Broad, W. and N. Wade. (1982). *Betrayers of the Truth*, New York: Simon and Schuster.

Brown, J. R. (ed.). (1984). *Scientific Rationality: The Sociological Turn*, Dordrecht: Reidel.

Brown, J. R. (1989). *The Rational and the Social*, London: Routledge.

Butts, R. E. (1980). "Methodology and the Functional Identity of Science and Philosophy", in Hintikka *et al.* (eds.). *Pisa Conference Proceedings*, vol. II, pp. 253–270.
Cesi and Peters, (1980). *Science*, (September 5, 1980).
Davidson, D. (1973). "The Very Idea of a Conceptual Scheme", *Proceedings of the American Philosophical Association, 1973–74.*
Feuer, L. (1974). *Einstein and the Generations of Science.* New York: Basic Books.
Garber, D. (1986). "Learning From the Past", *Synthese.*
Giere, R. (1973). "History and Philosophy of Science: An Intimate Connection or a Marriage of Convenience?", *British Journal for the Philosophy of Science.*
Giere, R. (1984). "Towards a Unified Theory of Science, in Cushing *et al* (eds.), *Science and Reality*, Notre Dame.
Grandy, R. (1973). "Reference, Meaning, and Belief", *Journal of Philosophy.*
Hollis, M. (1982). "The Social Destruction of Reality", in Hollis and Lukes (1982).
Hollis and Lukes (eds.). (1982). *Rationality and Relativism*, Oxford: Blackwell.
Holton, G. (1974). "On Being Caught Between Apollo and Dionysus", *Deadalus* (Summer).
Kuhn, T. (1962). *The Structure of Scientific Revolutions*, Chicago: University of Chicago Press.
Kuhn, T. (1971). "Notes on Lakatos", in Buck and Cohen (eds.), *PSA 1970*, Dordrecht: Reidel.
Lakatos, I. (1970). "Falsification and the Methodology of Scientific Research Programmes", in Lakatos and Musgrave (eds.), *Criticism and the Growth of Knowledge*, Cambridge: Cambridge University Press.
Lakatos, I. (1971). "History of Science and Its Rational Reconstructions", in Buck and Cohen (eds.), *PSA 1970*, Dordrecht: Reidel.
Laudan, L. (1977). *Progress and Its Problems*, Berkeley: University of California Press.
Laudan, L. (1984). *Science and Values*, Dordrecht: Reidel.
Laudan, L. (1986). "Some Problems Facing Intuitionist Meta-methodologies", *Synthese.*
Laudan, L. (1987). "Progress or Rationality? The Prospects for a Normative Naturalism", *American Philosophical Quarterly.*
Lukes, S. (1982). "Relativism in its Place", in Hollis and Lukes (1982).
Merton, R. (1957). *Social Theory and Social Structure*, (Revised and enlarged edition). New York: Free Press.
Merton, R. (1973). *The Sociology of Science*, Chicago: University of Chicago Press.
Quine, W. (1960). *Word and Object*, Cambridge: MIT Press.
Watson, J. (1968). *The Double Helix*, New York: Atheneum.
Wilson, N. (1959). "Substances without Substrata", *Review of Metaphysics.*
Worrall, J. (1976). "Thomas Young and the 'Refutation' of Newtonian Optics", in Howson (ed.). *Method and Appraisal in the Physical Sciences*, Cambridge: Cambridge University Press, 1976.

LORENZ KRÜGER

REDUCTION WITHOUT REDUCTIONISM?*

It is a time-honored piece of theoretical speculation that nature is a kosmos, i.e. an ordered, perhaps even purposeful whole. Kant may be taken as a typical exponent of this view under conditions of modern science: he defines nature in the material sense as the total set of potential contents of our experiences and, at the same time, attempts to show that this set is subject to pervasive laws which express nature in the formal sense of the term, i.e. the internal unitary principle (or set of principles) underlying the existence of the experiential world (Kant 1781/87, A 216/B 263; 1785, §§15—16; 1786, Preface).

On the other hand, there is the undeniable fact that experience comes to us through a multitude of different channels so that inevitably it is organized in a rich variety of different and at least partially independent systems of description and explanation, e.g. mechanics, optics, chemistry, psychological theories etc. Since the early beginnings of philosophical and scientific speculation reduction of one field, branch, or theory to another field, branch or theory has, therefore, been a theme of major importance, though possibly not always under this name. Questions I want take up in this paper are such as: To what extent does/can reduction of theories contribute to the programme of discovering nature in the formal sense as embodied in nature in the material sense? This is an ontological as well as epistemological problem: e.g. a nominalist or instrumentalist will claim that the search for formal nature is a ghost hunt, or should at best to be construed as the attempt at more theoretical economy. Furthermore, does the idea of unity of nature imply reductionism, i.e. something like the claim that, ideally and in the limit of research, all theories should be reducible to a comprehensive basic theory? This question indicates that the nature of the alleged formal unity of nature is itself a problem. Robert Butts has elucidated the character of this problem by showing how it naturally arises within the framework of the standard hypothetico-deductive view of science held by Logical Empiricism (1976, pp. 48—49). He also pointed to the Kantian background of the problem (p. 39).

Now, the term 'reduction' covers a number of different things and

J. R. Brown and J. Mittelstrass (eds.), An Intimate Relation, 369—390.

carries so many connotations that it seems advisable to begin this paper with a brief explication of the terms that appear in its title.

Let us assume that we are confronted with two qualitatively different domains of experimental phenomena or two different domains of certain entities and their changes, e.g. with the domain of heat and cold on the one hand and that of moving bodies on the other. For the purposes of this paper I shall understand by the reduction of one of these domains, call it A, to the other, B, any procedure or argument which shows that the A-phenomena/things/ processes are nothing more than B-phenomena/things/processes (possibly arranged or combined in a suitable way), e.g. that heat is nothing more than motion or nothing more than a complex arrangement of motions. It is obvious that reductions thus understood help to prepare the way towards a unified world view; and this is presumably one of the main reasons for our interest in them.

More specifically, in accordance with the analytical spirit of most of our intellectual efforts, we tend to believe that reductions should be possible whenever the things of a domain A are known to consist, or can be assumed to consist, of small parts with at least partially specific properties of their own. Apart from the analytical prejudice the interest in this subclass of reductions is enhanced once again by the possibility to limit, with their help, the huge diversity of qualities or phenomena which is so characteristic for the level of our immediate experience. At any rate, for whatever reasons, the widespread belief in the priority of smaller dimensions is certainly a rather typical feature of our time. It seems to me that it is the guideline of practically all attempts at reduction and the core of the philosophical view called reductionism.

For the purposes of this paper I want to characterize "local" reductionism by the following claims:

(i) The nature and changes of the entities of a given domain are solely and completely determined by the nature and the changes of a suitably chosen class of their parts (in the ordinary sense of this term) arranged in a certain way.

(ii) A complete account of the nature and the changes within this domain can be given in terms of the parts and their properties alone.

"Global" reductionism can then be characterized by the assertion:

(iii) Local reductionism is true for any given domain, with the possible exception of a basic domain whose entities do not consist of parts. This implies: Should such a basic domain exist, global reductionism would amount to the thesis that a complete account of *all* domains could, in principle, be given in terms of the basic domain.[1]

Assertion (i) is intended to catch the ontological intuition expressed by the phrase "nothing more than", whereas (ii) is added in order to pay due attention to Ernest Nagel's (1961, ch. 11, part III.2) and Carl Hempel's (1969) advice which says that one should rephrase reduction claims about reality in terms of the language which can meaningfully be used in describing this reality. In other words: the *ontological* problem of reduction is construed as that of an empirically meaningful (and hopefully adequate) unified theoretical *representation* of the world. Hence local reductionism is taken to be a claim which refers to a given or anticipated body of knowledge.[2] Global reductionism, on the other hand, is certainly to be understood as nothing more than a research programme, perhaps only as a regulative idea for research.

To argue for or against reductionism in this sense obviously amounts to proving or disproving the feasibility of reductions that connect different levels or orders of magnitude (and thereby also qualitatively different fields of phenomena). And from now on, I shall reserve the term 'reduction' to these interlevel connections.[3]

Now which interlevel reductions are feasible and which are not is, I take it, to be judged in the light of science. Our scientific knowledge is, indeed, given as a comprehensive and rather inhomogenous collection of various theories, theory fragments and facts on nearly all conceivable levels. Reduction claims have, therefore, since Nagel's pioneering work nearly invariably been construed as assertions about pairs of theories: one, T_A, that accounts for domain A and is to be reduced to the other, T_B, that accounts for domain *B*. In the spirit of the Hempel-Oppenheim model of explanation Nagel and most others after him have assumed that the reduction relation is essentially that of *derivability* of T_A from T_B, given, of course, additional premises and allowing for approximations in both theories and in the derivation. The additional premises include propositions connecting the primitive terms of T_A that do not occur in T_B with terms of T_B and possibly suitable boundary conditions in T_B, but no statement of T_A nor boundary conditions of T_A. A more detailed account was given by Schaffner (1977, sect. 2).

The advantages of the derivability view are obvious: it is logically clear; it says in what sense domain *A* is fully accounted for in terms of domain *B*. Moreover, this view is essential for the satisfaction of reductionism as here defined: nothing short of derivability guarantees that all reference to A-things or -processes can, in principle, be eliminated from our account of reality without making it thereby incomplete. Only if reduction is construed as derivation, the entities of the reduced theory do not appear as something new or "emergent" which exists beside and above the entities of the reducing theory; and in this sense they are then eliminated by the reduction, as was intended.

In order to avoid misunderstanding it is worth noting that this elimination does *not* imply the eliminability of the *theory* T_A. It may be the case that this theory cannot be dispensed with, even after the reduction has been completed. One reason may be that the reducing theory can only, or only satisfactorily, be confronted with the empirical evidence via the reduced theory T_A, or via a corrected version of T_A. This case of epistemological ineliminability is quite typical of micro-reductions which involve entities of T_B that are innaccessible to direct observation. In other cases it will at least practically be impossible to dispense with T_A. This pragmatic ineliminability leads naturally to a certain actual "conservatism" in science. Both kinds of ineliminability, especially the epistemological one, provide an increased motivation for reduction, since both T_A *and* T_B will continue to be used, but the two ontologies implied or suggested by T_A on the one hand and by T_B on the other hand will, as a rule, be significantly different, e.g. continuous matter endowed with primary and secondary qualities in the theory of heat versus atomism with only primary properties in particle mechanics. More generally speaking, derivability is desirable because it is a sufficient (though of course not necessary) warrant against inconsistencies, be they labeled 'ontological' or otherwise, between two theories which are about the same objects and simultaneously in use, but different in structure and content. This additional motivation for reduction is widely operative.[4]

The problem I want to discuss is now simply this: Is the account of reduction as derivation tenable? And the answer I wish to defend is 'NO'. Granting for a moment that this be the right answer, it would obviously entail (1) that the usual account of reduction in science would have to be altered and (2) that *reduction as a scientific procedure* would not be a sufficient tool to implement the philosophical idea of reductionism. And one might then wonder how much attractiveness or credibility reductionism will retain.

At this point I have to emphasize that in the following I do not claim to offer anything like a proof of the thesis that interlevel reduction in science never amounts to derivation. A proof of this kind would certainly be very difficult, if only because nearly no intertheoretic relation in science exactly resembles any other. In view of this situation most authors nowadays turn to case studies and try to develop more general conjectures on the basis of them.

I propose to follow their procedure also in this paper. My aim is, therefore, modest: I would already be happy if I could arouse the reader's suspicion against the derivability construal of reduction.

Since I am interested in the basic philosophical question of reductionism, I thought it might be good to return once more to a case that has not quite accidentally been chosen as a paradigm case for the whole species: the case of heat and motion or else that of the reduction of thermodynamics to (classical) particle mechanics. For, first, this is a non-trivial example of the ontological reduction program, in particular that of reducing secondary to primary properties, but still a fairly basic one, hence less plagued by ideological worries that crop up in reductions involving life, the human mind, or even society or history; and, secondly, it is certainly less complex and better known than any other interlevel case.[5] One may therefore surmise that it will mark something like an upper bound to the strength or the completeness one is likely to achieve in reduction in general. At any rate, one counter-example to the usual account of scientific reduction would already force us to review our concept of reduction and to question the validity of global reductionism.

All problems with a derivational account of this case seem to be connected with the introduction of the concept of probability which here, for the first time, not only figures in auxiliary theories (like error theory) or metatheories (like theories of induction), but in an *object* theory of physics (Gillispie 1963, Garber 1972, Schneider 1974/5, Krüger *et al.* 1987, part VI). This feature is likely to carry over to a number of further micro-reductions, adding thereby to the interest of the case study. Beyond that the case has additional interest in itself insofar as it contributes important aspects to a phase of the history of science which may be called "the probabilistic revolution". I shall therefore from now on concentrate on the role of probability in the reduction of thermodynamics to mechanics.

It is interesting to observe that the full awareness of the probabilistic nature of any mechanical account of heat did not diminish the confidence of the leading physicists in the possibility of reduction, indeed a derivational account of it. This is true of, among others, Maxwell who invented his demon as early as 1867 (Klein, 1970), Boltzmann (1871b, p. 295), and Einstein. Einstein in particular who is known for his adherence to determinism claimed that, in his treatment, "the second law appears as a necessary consequence of the mechanical world picture" (1902, p. 432). Nevertheless he described, in the very same paper, the task of proving his claim as the task of "deriving the theorems concerning thermal equilibrium and the second law solely on the basis of the mechanical equations and the calculus of probability" (1902, p. 417). The additional appeal to probability obviously did not bother him at all. This unquestioned coexistence of a deterministic world view with a liberal use of probabilistic concepts seems to be fairly typical for pre-quantum physics in general. And as long as quantum theoretical aspects are excluded, present-day opinions on thermodynamics are often not far from Einstein's.

The history of the case suggests four different reasons for believing that the derivational account of reduction is compatible with probabilistic reasoning. These may be stated as follows:

(A) Mathematical probability is directly translatable into mechanics; more exactly: the procedures used in classical probability theory for obtaining numerical values of probabilities can be directly applied to the mechanical subject matter.

(B) There is a purely mechanical account of probability; the so-called ergodic theory which is designed to derive statistical properties of deterministic mechanical motions.

(C) Suitable initial conditions for thermodynamics can be found in mechanical theory. Once found, they would make probabilistic arguments dispensable.

(D) Probability is merely epistemic, hence does not disturb the intended ontological picture which remains deterministic and purely mechanical.

I now want to turn to a critical discussion of these four possible grounds of rendering probability, as it were, harmless for reduction. For reasons of clarity I shall neglect the mutual relationships of points A through D; this is a simplification of the history as well as of the problem structure of the case, but one which will not substantially change the argument. Furthermore, for reasons of brevity I shall be unable to develop the material in its actual historical context (cp. Krüger 1981); I nevertheless hope to avoid the rightly lamented distortions produced by a philosophy which invents its own examples.

The approach on the basis of A appears in two independent variants: (A1) the translation into collision theory (most prominently in Maxwell 1860 and 1866 and in most of Boltzmann's work), and (A2) the translation into the combinatorics of phase space (Boltzmann 1877). The second version is intended to back up the first. Typical tools for computing actual probability values in the classical mathematical theory of probability are, as is well known, the following two assumptions: (i) that certain basic alternatives are equally possible or equiprobable, and (ii) that certain events are statistically independent from each other. Exactly these two assumptions reappear pervasively in the kinetic theory of gases which was the paradigm model for connecting thermodynamical and mechanical theories.

For instance, it is nearly invariably assumed that all places and all directions in space are indifferent, hence equally likely to be occupied by a given molecule. Assumptions of this type, innocuous as they may appear at first sight, already involve the further assumption, that no inhomogeneities of distribution in small dimensions are created by dynamical processes, e.g. by collisions between different molecules. It is very instructive but impossible within the scope of this paper to pursue the genesis of the relatively strong assumptions which were not even noticed as additional empirical hypotheses on top of the usual mechanics. The most notorious of those assumptions is the Stoßzahlansatz (first used in Maxwell 1866). Boltzmann originally believed that he had given a "strict proof", which was "free from all hypotheses" (1872, p. 345) for his famous H-theorem, i.e. for his purely mechanical equivalent of the thermodynamical law of increasing entropy. In other words: his first reduction of thermodynamics (which completed all earlier work on these lines from Daniel Bernoulli in 1738 through Maxwell and Boltzmann in the 1860s) was only possible, because the probabilistic assumptions were taken as quasi-mathematical or *a priori* true.

Unfortunately they could easily be shown to be false, simply by pointing out the existence of mechanical counter-examples. A class of such examples was first presented by Loschmidt in 1876. Boltzmann, in defense of his theory, replied that the counter-examples were so improbable that they could safely be neglected. This assertion of improbability once again relied on a *priori* assumptions of indifference between alternative possibilities of assigning states to the molecules in a gas (version A2 of approach A). In fact, Boltzmann used the model of an urn from which random drawings of possible mechanical magnitudes were made (1877). Moreover in *this* approach the mechanical theory was *not* used for what is supposedly its primary function, namely for describing the development of a system through time. The dynamical treatment of the motion is sidestepped by a further *a priori* reasoning whose persuasive force is entirely due to its tautological ring: "In most cases the initial state will be very improbable; the system passes from this through ever more probable states, reaching finally the most probable state, that of equilibrium" (1877, p. 165). But Loschmidt's objection had just shown that statements of this type could *not generally* be true. Thus reduction via (A) was finally abandoned as being a failure, also by Boltzmann.

Attempts at reduction on the grounds of (B) start from the following truly physical definition of probability: the probability of a certain state is construed as the fraction of time which a system under consideration spends in that state during a sufficiently long reference time. This conception (often expressed, but never actually used by Boltzmann) encounters a serious difficulty: the movements of microscopic particles cannot be followed, be it in experiment or in calculation, nor can *a fortiori*, the entire micro-history of a macroscopic system. This lack of information can only be compensated by a bold assumption called the "ergodic hypothesis". Maxwell developing an idea implicit in Boltzmann (1868, 1871a), was the first to offer an explicit formulation of the hypothesis to recognize its importance. His version reads: "The system, if left to itself in its actual state of motion, will, sooner or later, pass through every phase which is consistent with the equation of energy." (1878, p. 714). Under these conditions the averages over time can be replaced by other averages that are accessible to calculation.

But Maxwell's hypothesis was, as he pointed out at once, again not generally true. In order to save it, he assumed that *random* perturbations from outside would actually make it true (1878, pp. 714—715). It is obvious that this amounts to nothing more than pushing the problem one step further without adding to its solution. Thus, most physicists after Maxwell did not accept the ergodic hypothesis, although they recognized its potential usefulness. In particular, this is true of Boltzmann, the inventor of the hypothesis.[6]

It is true, though, that after a time of decline ergodic theory has attracted an increasing attention since the 1930's. Unfortunately it exceeds my capabilities and the limits of this paper to pursue this very complex line of research in any detail (cf. for a recent account e.g. von Plato 1987). But I think it is fair to say this much: In one semi-realistic case a reformulated version of the hypothesis has been proved by Sinai (1973), so that more results may reasonably be expected. Yet Sinai's very proof has enhanced previous doubts concerning the relevancy of ergodic mathematics to statistical physics, since it shows that ergodic theorem is not only valid for large numbers of particles, as it should be, but already for small numbers, eventually for two particles, i.e. systems which do *not* exhibit thermodynamic properties at all. From this we can infer that ergodicity can at best be *necessary* for the derivation of thermodynamics, but it cannot possibly be sufficient (Balescu 1975, Tisza 1977). Still more important misgivings arise from the fact that ergodic theory promises results only for the limit of *infinite* time intervals and does, therefore, not admit a correlation with physically relevant time developments that are characterized by specific finite times which depend on the nature of the actual system at hand. Indeed, an empirical interpretation of ergodicity-based statistical mechanics involves the assumption that results of actual measurements are averages of the corresponding magnitudes taken over a practically infinite time. Hence the theory can at best be applied to equilibrium states, but does not include irreversible processes within its scope (Balescu 1975, pp. 695 ff.).

In short: though the prospects of founding statistical behaviour of (classical) physical systems on underlying deterministic mechanics are not entirely negative, they are not very promising either. The attitude of the majority of physicists since Gibbs and Einstein is therefore hardly surprising and certainly not inadequate: they sought the true account of thermal phenomena elsewhere. The following way out suggests itself: if the *laws* of mechanical theory are insufficient for deriving the thermal properties of matter, one may strengthen the premises by adding assumptions about the *contingent initial conditions* that actually obtain in all cases we observe (Basis C for reduction). The advantage of this move is that it promises to avoid the whole issue of probability; we are not talking about averages anymore, but about picking out the right subclass of systems. The disadvantage, on the other hand, lies in the fact that this step marks the transition to a program of significantly weaker explanatory force: Not only the laws, but both laws *and* initial conditions are invoked as premises. It is precisely the strength and the beauty of the otherwise unconvincing ergodic theory that it promises to achieve the reduction whatever the initial conditions might be. Furthermore, the appeal to suitable factual statements does not provide any deeper insight as to why these conditions obtain. Their pervasive presence assimilates them to the laws of the theory, but they remain nevertheless entirely contingent with respect to the theory. Thus the appeal to this contingent fact has a certain air of ad-hocness, Boltzmann certainly was not enthusiastic about it, but under the impact of continued criticism (mainly by Zermelo in 1896) he came to the conviction that the unidirectional increase of entropy was based "uniquely and solely in the initial conditions" (Boltzmann 1896—98, §87). Since these conditions appear to be invariably of the desired kind in all initially independent systems observed by us, Boltzmann conceived a common cause: the improbable microstate of our cosmic neighborhood from which these systems branch off. Thanks to the work of Hans Reichenbach and Adolf Grünbaum we know that this idea can be developed into a satisfactory theory of the direction of time (Reichenbach 1956, Grünbaum 1973, ch. 8).

Does it equally well serve as a basis for successful reduction? If success is here taken to imply a derivation of the laws of the reduced theory T_A from reducing theory T_B, the answer is 'no', since the derivation shrinks to an ad-hoc postulate.[7] But if one recalls the ontological purpose of the derivational account, viz. the elimination of the upper level entities from a coherent and complete world picture, the answer may perhaps be 'yes'. As Kenneth Schaffner has argued against Polanyi (Schaffner 1969, p. 345 with note 1), the use of contingent conditions in itself is insufficient to prove emergence. A claim of emergence, he thinks, would require a proof to the effect that the initial conditions either are not characterizable in the language of T_B or else are the outcome conditions that are not so characterizable. In accordance with this metatheoretical view (which I accept) an argument against emergence would conversely require a proof to the effect that initial conditions can be characterized in T_B.

What does this modality 'can' mean in the present context? If we require the *epistemological* possibility of actually sorting out the desired initial conditions, we have lost the case, since we cannot deal with 10^{23} particles at a time. Let us therefore assume that only a *logical* or semantical possibility is intended, i.e. the mere availability of the appropriate language in T_B for describing the microstates in question. Granted that the dynamical properties of the molecules are known, this possibility is surely at our disposal. We can then characterize the huge set of potential candidates for the initial conditions we are postulating.

But further problems immediately arise: There is, as far as I am aware, no indication that there is a non-trivial or non-question-begging property in the language of mechanics that would be common to all microstates which behave normally, but absent from all those which do not. It is not even clear which set of cases we should try to characterize by such a property, if it could be found. It should, of course, pick out those states that develop towards thermal equilibrium and no others. But, as a theorem due to Poincaré tells us, all those states will in the long run develop away from equilibrium again. How long do we require them to stay there? Should we let this time depend on the size of the system? There is a host of conceptual problems here whose significance should not be underestimated. As things stand now, the appeal to initial conditions, I believe, boils down to an existence claim without any proof, indeed even without a clear conception of the proposition to be proved.

One may object at this point that the approach on the basis of initial conditions has been misrepresented: What is invoked in some treatments, e.g. Grünbaum's, is not a particular microspic condition of a macroscopic motion, or a class of such conditions, but rather particular *probability* distributions of initial conditions. And for these distributions an explicit expression in the language of mechanics *is* offered. The trouble is only that we are then again confronted with the problem of interpreting probability, so that we fall back on one of the alternative attempts at finding a basis for the reductionist construction: either in ergodic theory[8] or in the ensemble approach to be discussed next.

In order to avoid a misunderstanding it should be clearly said that the preceding discussion does *not* amount to the claim that there are no mechanical initial conditions which result in thermally normal behavior. For states of thermal equilibrium and natural processes that lead to them abound in our experience. The problem is that we do not characterize them solely in *mechanical* terms or select them according to their *mechanical* properties. In other words: we need an additional piece of theory to do this job for us; and this will be one form or other of thermodynamics. E.g. in order to ascertain that a body has a certain temperature, we have to presuppose that it is in equilibrium. There is no immediate observational test for this assumption (Tisza 1966, p. 119). The experimenter therefore checks it indirectly by examining whether the system continues to obey the laws of thermodynamics (Callen 1960, p.13).[9]

But one may ask the following question (as Adolf Grünbaum once did in a very illuminating discussion that helped much to clarify my thoughts): If mechanics is independently confirmed as the basic theory of motion and if the molecular constitution of matter is likewise confirmed, do not also the probabilistic assumptions concerning initial states of the motion derive confirmation from their success of explaining the second law of thermodynamics. Of course, they do. It is, however, important to stress that the assumptions are *essentially* probabilistic. They cannot be sharpened into a full mechanical specification of the single real system under investigation. It is not just *practically* impossible to back up any such hypothetical specification by refined empirical evidence that would differentiate the actual state from alternative states which are also compatible with the probabilistic description. It is, as a matter of *theoretical* principle, impossible to adopt any full specification, because it would fail to reproduce the features that are typical for the explanandum, i.e. thermally normal behavior. The description would simply be of the wrong kind in view of the phenomena to be explained.

Conversely, adequate descriptions that involve the language of micro-mechanics at all essentially refer to distributions of position or momenta[10] or to ensembles. Moreover, as we saw, the indispensable appeal to experience requires some form of thermodynamics. The most widely adopted version makes use of the notion of the ensemble, which finally leads us to consider the last attempt, D, at dealing with the additional probabilistic premises needed for the derivational reduction of thermodynamics to mechanics. In this approach probabilities are construed as describing our imperfect knowledge rather than any actual distribution of magnitudes in reality.

Gibbs was the first to voice this view clearly and to base his entire statistical mechanics on it. In the preface of his book "Elementary Principles in Statistical Mechanics" he says: "The laws of thermodynamics . . . express the laws of mechanics for such systems as they appear to beings who have not the fineness of perception to enable them to appreciate quantities of the order of magnitude of those which relate to single particles, and who cannot repeat their experiments often enough to obtain any but the most probable results." This is his justification of adopting the ensemble approach throughout, as the most adequate form of a genuinely statistical inquiry (Gibbs 1902, pp. vii–viii). In this context the ensemble of figures as a *conceptual* entity: viz.

as an assembly of similar systems only one of which is actual. In a decisive chapter of the book, that on irreversibility, he makes further explicit use of this epistemic point of view, firstly in order to derive the increase of entropy from a given state of low entropy, and secondly in order to justify the limitation of the result thus derived to the future.

The intrusion of epistemic terms into the connection between mechanics and thermodynamics may well appear reasonable in the light of epistemological considerations and is more often than not justified in these terms in modern textbooks (e.g. Tolman, 1938, §§1, 23, 25). But it frustrates reductionism as defined above. For, according to clause (ii) of our definition of reductionism, we require that a unified account of two different domains, A and B, be given on the basis of the theory T_B which deals with the entities B alone. In order to satisfy this demand we have to accept propositions connecting the entities and properties of A with entities and properties of B. It is now time to make a little more precise what the qualification 'unified' for the account to be given on the basis of T_B implies for those connecting propositions. I have pointed out at the beginning of this paper that idea of nature as being a unitary whole subject to laws is the main source of the search for reductions. The term "reduction" is, accordingly, understood as having an essentially ontological meaning. This is primarily expressed in clause (i) of the above definition of reductionism, but it affects the meaning of clause (ii) as well: As already remarked above in commenting upon (i) and (ii), the unified account, though comprising terms of T_A as well as of T_B, will have to be a representation of the *world*, here of domains A and B. Hence it must not draw upon explicitly epistemic conditions without which the simultaneous occurrence of terms from both T_A and T_B cannot be understood and justified. Moreover, it must not relate a macroscopic object considered as an actual thing localized in space and time with something on the microlevel, the ensemble, that is a theoretical construction carried out in a "space" of possible things only one of which can be actual.

In the example of heat (domain A) and motion (domain B) it is compatible with these requirements to complement T_B by propositions like: (a) The gas is the assembly of so many particles of such and such sorts; (b) the mass of the gas is identical with the sum of the masses of all particles mentioned in a; or (c) the pressure is nothing but the momentum transferred to the walls of the container per time; and so on. On the other hand, it violates the requirements to say things like: (d)

The gas is an ensemble of systems of micro-particles; (e) its actual pressure is a certain average of momentum transfer taken over the ensemble. Still less would it do to say: (f) The actual temperature is a property characterizing the ensemble but undefined for any of its members. True, statement (d) sounds somehow wrong; but it *should* be correctly assertible if the ensemble were to be a representation of the macro-system open to a realistic interpretation, and not only a (perfectly acceptable) way of dealing *somehow* with the macrosystem. Statements of types (e) and (f) are standard talk within the ensemble approach, but they don't admit of a non-epistemic interpretation, so that any account of *A* involving them falls short of being unified or of representing a unitary object in the sense intended. As Fowler (1936, p. 7) has aptly put it, the procedure in terms of ensembles is lacking the "sanity or physical reality" of Boltzmann's approach.

The missing sanity or reality can at least partly be restored by a less epistemic interpretation of the ensemble. In Gibbs ensembles figure as more or less convenient calculational devices (1902, p. 33), whose merits are to be judged by comparisons with thermodynamical equations (1902, ch. XIV). Einstein developed the same form of statistical mechanics independently from Gibbs, but he connects the ensemble immediately with a physical interpretation: it describes the distribution of energies for a collection of systems each of which is in contact with an infinitely large reservoir of a given temperature, a so-called heat bath (Einstein 1902, §5; 1903, §§3—4). It seems to me that this non-epistemic interpretion of the ensemble is, in fact, unavoidable. Under a purely epistemic interpretation it must be possible to express every physical magnitude as a weighted average of micromagnitudes where the weights are given by the distribution function characterizing the ensemble. But there are magnitudes which are *not* so expressible; and these are precisely the thermodynamical magnitudes like temperature and entropy. In other words: the ensemble has a *physical* significance: e.g. only certain ensembles can be used to represent the real state of thermal equilibrium (Balescu 1975, pp. 43—44).

Einstein's interpretation of the ensemble, then, is the only acceptable one. It secures the realistic character of the ensemble approach. But there is a price to be paid: notions like equilibrium and temperature (and thereby entropy) must be given physical meaning on a basis of more than just mechanics. These properties are then treated, as Gibbs once put it, as "independent variables" (Gibbs, 1902, pp. 179-180).

According to this proposal both mechanical *and* thermodynamical elements enter into a unified description of reality, as we require it for "reduction".

This completes the discussion of the four approaches to our test case of reduction. Let me briefly state the balance: Approach *A* is disqualified, because it leads to physical statements that are false, or at best approximately true under very restrictive circumstances. Approach *B*, in spite of its enormous mathematical sophistication, is still nearly without physical results; and the results obtained so far appear to be irrelevant for founding statistical mechanics on ordinary mechanics.

Approach *C* is not very attractive because of its weak explanatory power. A more important argument against *C* is based on the fact that *C* boils down to a pure existence claim whose content cannot be expressed without the use of extramechanical language and whose proof is missing.

Approach *D* finally is connected with the introduction of large sets or ensembles of macroscopically similar, but microscopically different non-interacting systems. Because of its unrestricted generality and its formal elegance it has become the standard treatment of statistical mechanics. The original and still frequent epistemic interpretation of the ensemble is at variance with the program of ontological reduction. But this interpretation is not essential for the ensemble approach; indeed, it does not even seem to be admissible. A realistic reinterpretation on the other hand involves both mechanical and thermodynamical terms in the very same theoretical context.

What this last feature implies for the programme of reduction will be discussed in a minute. Let us first ask a somewhat less demanding question: which approach has achieved the most coherent and most fruitful interrelation between mechanics and thermodynamics? "The ensemble approach in Einstein's realistic interpretation" is the answer that suggests itself from my presentation. The reader may perhaps be dissatisfied with the fact that the reasons for this preference are somewhat less than strictly compelling, which is certainly right. But this siutation admits of no easy improvement once the intimate connection between philosophical and scientific issues which I have presupposed has been granted. If it is granted, and if this is the conception of the philosophy of science one adopts, then, I think, one should not evade the problem of assessing the relative merits of different research programmes. In this sense I take it as established that the last method,

D, of dealing with our test case deserves the most serious consideration of the philosopher.

If we adopt it: what does it teach us about the procedure called reduction? And what import does it have for the cause of reductionism? The scientific argument in favour of the view that heat is nothing more than motion, in other words: the mechanical account of thermal phenomena, turns out not to be a derivation of one theory, T_A, from another, T_B, but rather the construction of a third interlevel theory, T_{AB}, that meets the theoretical boundary conditions set by both, T_A and T_B, though perhaps not without correcting one or the other or both. The interlevel theory achieves a reduction in a sense that differs from the sense which was originally intended: it does not eliminate the terms and the laws of one theory by "absorbing" (Nagel, 1961, p. 397) them into the other, but it creates a new unified description of both domains with the help of both theories involved.

This being said about reduction, what is the import for the philosophical view called reductionism? With respect to the above characterization it is immediately clear, that assertion (ii) of local reductionism is not true in our example: a unified account solely in terms of micro-entities and their properties has not been found; overall properties of large assemblies of them, like equilibrium and temperature, have not been eliminated from the theory. It is less clear whether or not we should withdraw from assertion (i) of local reductionism as well. The formulation is such that nothing of what has been said is incompatible with (i). On the contrary one might argue that (i) is eventually supported by any successful version of statistical mechanics as opposed, for instance, to a caloric theory of heat. On the other hand some thermal properties of macroscopic bulks of matter are "emergent" or "non-additive" with respect to particle mechanics in the technical sense of these terms defined by Nagel (Nagel, 1961, pp. 369—374, 394—397). I do not know how to weigh these two arguments, one pro, one contra ontological reducibility, with respect to each other.

Whatever one's intuitions are in this kind of situation, I think it is fair to recommend some caution with respect to philosophical claims of the reductionist kind, e.g. the location or even the relevance of a distinction between primary and secondary qualities or the priority of smaller dimensions in the order of nature. Metaphysical ideas like these may be correct, but they need not, for that reason, also form part of science. There is some reason to develop a revised conception of reduction in science and to separate scientific reduction from philosophical reductionism.

Although the single example I have reviewed in this paper can in no way serve as a proof of general statements about the unity of nature, I may, in a heuristic vein, perhaps be permitted to conclude by stating some of the background views that are suggested to me by the example. These may be summarized roughly as follows:

(1) If 'reduction' is to be understood as a term that describes an actual move of scientific practice, it should not (or at least not generally) be construed as the discovery or the construction of a derivability relation between two theories, but rather as the formation of an integrated representation of two (possibly more) separate domains of phenomena. This integrated or unified account will, usually, contain elements from the theories of both (all) domains involved; and it will, in general, not be possible to eliminate those elements that belong to the so-called "reduced" theory without rendering the account unworkable or at least incomplete.[11]

(2) It may, nevertheless, remain possible (perhaps important) to interpret the unified account in terms of the ontology of the "reducing" theory, especially if that theory deals with parts of the things the "reduced" theory is about. However, one should not conclude from this possibility that aggregates of parts could not have "emergent" properties, or properties which the parts don't have, indeed which they could not even meaningfully be said to possess.

(3) Nature in Kant's formal sense of the term, i.e. the internal lawful order of the experiential world, if it can be found at all, should not be conceived as consisting of a hierarchy of laws: basic ones referring to the smallest parts of matter and a series of derivative ones dealing with aggregates of increasing size or complexity. Rather, the theoretical unity may realistically be conceived as resulting from a network of theories referring to different orders of magnitudes and/or different kinds of phenomena, where the ties between the knots may, as it were, vary in strength and material, depending on where they are located in the network and what their function is with respect to the overall theoretical structure (cp. Krüger, 1980, part II).

(4) Our cognitive experience with the actual development of a multitude of partially interrelated theories about nature does not contradict a highly unified ontological model of reality; yet it does not support the hope that this model could, if only in principle or ideally,

be characterized in one scientific theory. The representation of the ontological order of the world is likely to reflect strongly the conditions under which we have to explore it, possibly including elements from the historical order of actual research. Reduction, therefore, should be reinterpreted realistically as one type of move in this actual development, and reductionism recognized as one of those dubious extrapolations or exaggerations which are so typical for philosophy.

NOTES

* This article is a revised version of a paper that was first read in the Lecture Series of the Center for Philosophy of Science at the University of Pittsburgh, and subsequently also at Princeton and Harvard. I am indebted to the Center for its support and hospitality during a most stimulating Research Fellowship in 1978/79. Moreover, I gratefully acknowledge the benefit I received from critical discussions; the participants will forgive me that I can no longer sort out what I learnt from whom. It is certain, though, that I was offered more than I was able to understand and retain.

[1] Thus W. Heisenberg argued that elementary particles form the basic domain, since they do not consist of still smaller particles but are rather transformed into each other. From this he inferred that "a satisfactory theory of elementary particles would, at the same time, have to be a theory of physics in general" (1967a; p. 236; cp. also 1967b, pp. 1—2, 137).

[2] The case of merely anticipated knowledge should not be excluded from the outset, since a reductionist claim may be perfectly legitimate in the role of a working hypothesis (Oppenheim/Putnam; 1958), though there is, of course, the permanent danger — in the field of reduction perhaps more than elsewhere — of mistaking the plausibility of research programmes for the established truth of wishful beliefs. Thus local reductionism may be just an hypothesis or it may be well confirmed.

[3] This terminological decision implies that many intertheoretic relations that are usually called "reductions" are here excluded, e.g. domain preserving reductions in the sense of Nickles (1973), i.e. accommodations of theories as limiting cases of more comprehensive theories concerning the same field of objects, similarly replacements of a theory by another theory that explains the same data, i.e. the case discussed by Kemeny and Oppenheim (1956).

[4] That, besides ontological interests, this more pragmatic motive plays an important role is indicated by the fact that an increasing number of reduction theorists agree that actual research strategies are hardly ever aimed at effective elimination of reduced or reducible theories. K. Schaffner has aptly baptized this feature "the peripherality of reductionism" (1974).

[5] As is well known, extensive scientific work has been devoted to this particular example of reduction from about 1850 up to the present. Careful historical studies of this development are also available; see especially Brush (1965, 1966, 1976). For philosophical analyses compare e.g. Nagel (1961, chap. 11); and more recently Brush (1977) and Tisza (1977).

[6] This fact does not quite harmonize with the widely accepted story, started by Boltzmann himself (1881), according to which he based the concept of probability on that of time fractions whereas Maxwell based it on the notion of the ensemble. Klein (1973) pursued Boltzmann's various and shifting notions of probability and showed that the time fraction idea was at no time the only in Boltzmann nor even unambiguous in its application. As a matter of fact, this idea acquires a legitimate status in the explication of probability only within the framework of the ergodic hypothesis, which was accepted by Maxwell but suspected by Boltzmann.

[7] There are, however, cosmological theories that may be used to explain the low entropy of the initial states and thereby remove the ad-hoc character of the reduction. All irreversibility might result from the irreversible expansion of the universe (see e.g. Gold/Schumacher 1967, esp. Introduction, Chps. II, IV, X.4; Gal-Or, 1972). But to grant this amounts to abandoning rather than advancing the reductionist program.

[8] This is Grünbaum's choice (1973, pp. 256 ff.): He invokes a condition of randomness to be satisfied by a space-ensemble of physical systems; the space-ensemble is assumed to be strictly analogous to the time-ensemble of the successive states of a single closed and ergodic system.

[9] This situation may also be expressed by calling concepts like 'equilibrium', 'temperature' or 'entropy' "theoretical" in the sense of J. D. Sneed (1971); the case has been argued by Moulines (1975).

[10] Precisely this feature permitted Boltzmann to derive an irreversible process governed by his H-theorem. Ilya Prigogine's current program extends this approach into a new kind of irreversible dynamics (see e.g. his 1962).

[11] The unified account, e.g. T_{AB} above, may be called an "interlevel theory" (Maull, 1977, see specially pp. 160 f.). Referring to analyses of biological theorizing Darden and Maull (1977) have introduced the more general notion of "interfield theories", i.e. such unified accounts that do not necessarily relate theories of aggregates on the one hand and of their parts on the other, but simply different domains that were originally studied separately and in independent ways ("fields"). The resulting conception of the unity of science closely resembles that of this paper, at least as far as the pragmatics of research is considered.

BIBLIOGRAPHY

Balescu, Radu. 1975. *Equilibrium and Nonequilibrum Statistical Mechanics*. New York: Wiley.

Boltzmann, Ludwig. 1868. "Studien über das Gleichgewicht der lebendigen Kraft zwischen bewegten materiellen Punkten", in Boltzmann 1968, vol. I, pp. 49—96.

Boltzmann, Ludwig. 1871a. "Einige allgemeine Sätze über das Wärmegleichgewicht", in Boltzmann 1968, vol. I, pp. 259—287.

Boltzmann, Ludwig. 1871b. "Analytischer Beweis des Zweiten Hauptsatzes der mechanischen Wärmetheorie aus den Sätzen über das Gleichgewicht der lebendigen Kraft", Reprinted in: Boltzmann 1968, vol. I, pp. 288—308.

Boltzmann, Ludwig. 1872. "Weitere Studien über das Wärmegleichgewicht unter Gasmolekülen", Quoted from: Boltzmann 1968, vol. I, pp. 316—402.

Boltzmann, Ludwig. 1877. "Über die Beziehung zwischen dem zweiten Hauptsatz der mechanischen Wärmetheorie und der Wahrscheinlichkeitsrechnung respektive den Sätzen über das Wärmegleichgewicht", in Boltzmann 1968, vol. II, pp. 164—223.

Boltzmann, Ludwig. 1881. "Referat über die Abhandlung von F. C. Maxwell 'Über Boltzmanns Theorem betreffend die mittlere Verteilung der lebendigen Kraft in einem System materieller Punkte'", in Boltzmann 1968, vol. II, pp. 582—595.

Boltzmann, Ludwig. 1896—98. *Vorlesungen über Gastheorie.* 2 volumes. Leipzig. Translation by S. G. Brush, Berkeley: University of California Press, 1964.

Boltzmann, Ludwig. 1968. *Wissenschaftliche Abhandlungen*, F. Hasenöhrl (ed.), Leipzig: Barth, 1909. Reprinted New York: Chelsea.

Brush, Stephen G. 1965—66. *Kinetic Theory — Selected Readings*, Oxford: Pergamon. 2 volumes.

Brush, Stephen G. 1976. *The Kind of Motion We Call Heat.* 2 volumes. Amsterdam: North Holland Publ. Co.

Brush, Stephen G. 1977. "Statistical Mechanics and the Philosophy of Science — Some Historical Notes", in Suppe, F. and P. D. Asquith (eds.), *PSA 1976*, vol. 2, East Lansing/Michigan, pp. 551—584.

Butts, Robert E. 1976. "The Hypothetico-Deductive Model of Scientific Theories: A Sympathetic Disclaimer", in Shea, William R. (ed.), *Basic Issues in the Philosophy of Science*, New York: Science History Publications, pp. 36—57.

Callen, H. B. 1960. *Thermodynamics.* New York: Wiley.

Darden, Lindley and Nancy Maull. 1977. "Interfield Theories", *Philosophy of Science* **44**, 43—64.

Einstein, Albert. 1902. "Kinetische Theorie des Wärmegleichgewichtes und des zweiten Hauptsatzes der Thermodynamik", *Annalen der Physik* **9**, 417—433.

Einstein, Albert. 1903. "Eine Theorie der Grundlagen der Thermodynamik", *Annalen der Physik* **11**, 170—187.

Fowler, R. H. 1936. *Statistical Mechanics*, 2. ed. Cambridge: University Press.

Gal-Or, Benjamin. 1972. "The Crisis about the Origin of Irreversibility and Time Anisotropy", *Science* **176**, no. 4030, 11—17.

Garber, Elisabeth. 1972. "Aspects of the Introduction of Probability into Physics", *Centaurus* **17**, 11—39.

Gibbs, J. Willard. 1902. *Elementary Principles in Statistical Mechanics.* Reprint New York: Dover 1960.

Gillispie, Charles C. 1963. "Intellectual Factors in the Background of Analysis by Probabilities", in Crombie, A. C. (ed.), *Scientific Change*, London: Heinemann, pp. 431—453.

Gold, T. and D. L. Schumacher (eds.). 1967. *The Nature of Time.* Ithaca/NY: Cornell University Press.

Grünbaum, Adolf. 1973. *Philosophical Problems of Space and Time.* 2. ed., Dordrecht: Reidel.

Heisenberg, Werner. 1967a. "Das Naturgesetz und die Struktur der Materie". Quoted from: Heisenberg, *Schritte über Grenzen*, München: Piper, 1971, pp. 223—242.

Heisenberg, Werner. 1967b. *Einführung in die einheitliche Feldtheorie der Elementarteilchen.* Stuttgart: Hirzel.

Hempel, Carl Gustav. 1969. "Reduction: Linguistic and Ontological Facets", In S. Morgenbesser *et al.*, *Philosophy, Science, and Method — Essays in Honor of Ernest Nagel*, New York: Macmillan, pp. 179—199.

Kant, Immanuel. 1785. *Prolegomena.*

Kant, Immanuel. 1786. *Metaphysische Anfangsgründe der Naturwissenschaft.*

Kant, Immanuel. 1781/87. *Kritik der reinen Vernunft.*

Kemeny, J. G. and P. Oppenheim. 1956. "On Reduction", *Philosophical Studies* **7**, 6—19.

Klein, Martin J. 1970. "Maxwell, His demon, and the Second Law of Thermodynamics", *American Scientist* **58**, 84—97.

Klein, Martin J. 1973. "The Development of Boltzmann's Statistical Ideas", in Cohen, E. and W. Thirring (eds.), *The Boltzmann Equation, Acta Physica Austriaca, Supplement 10*, pp. 53—106.

Krüger, Lorenz. 1980. "Intertheoretic Relations as a Tool for the Rational Reconstruction of Scientific Development", *Studies in History and Philosophy of Science* **11**, 89—101.

Krüger, Lorenz. 1981. "Reduction as a Problem — Some Remarks on the History of Statistical Mechanics from a Philosophical Point of View", in Hintikka, J. *et al.* (eds.), *Probabilistic Thinking, Thermodynamics, and the Interaction of the History and Philosophy of Science.* Dordrecht: Reidel, pp. 147—174.

Krüger, Lorenz, G. Gigerenzer, and M. S. Morgan (eds.). 1987. *The Probabilistic Revolution, vol. 2: Ideas in the Sciences.* Cambridge/Mass.: MIT Press.

Maull, Nancy. 1977. "Unifying Science without Reduction", *Studies in History and Philosophy of Science* **8**, 143—162.

Maxwell, James Clerk. 1860. "Illustrations of the Dynamical Theory of Gases", in Maxwell 1890. vol. 1, pp. 377—409.

Maxwell, James Clerk. 1866. "On the Dynamical Theory of Gases", in Maxwell 1890, vol. 2, pp. 26—77.

Maxwell, James Clerk. 1878. "On Boltzmann's Theorem on the Average Distribution of Energy in a System of Material Points", in Maxwell 1890, vol. 2, pp. 713—741.

Maxwell, James Clerk. 1890. *The Scientific Papers*, ed. W. D. Niven. Cambridge University Press. Reprinted: New York: Dover 1965.

Moulines, Carlos Ulises. 1975. "A Logical Reconstruction of Simple Equilibrium Thermodynamics", *Erkenntnis* **9**, 101—130.

Nagel, Ernest. 1961. *The Structure of Science.* New York: Harcourt, Brace and World.

Nickles, Thomas. 1973. "Two Concepts of Intertheoretic Reduction", *Journal of Philosophy* **70**, 181—201.

Oppenheim, Paul and Hilary Putnam. 1958. "Unity of Science as a Working Hypothesis", in Feigl, H., M. Scriven, and G. Maxwell (eds.), *Minnesota Studies in the Philosophy of Science.* Vol. II, pp. 3—36.

Plato, Jan von. 1987. "Probabilistic Physics the Classical Way", in Krüger/Gigerenzer/Morgan, 1987, pp. 379—407.

Prigogine, Ilya. 1962. *Non-Equilibrium Statistical Mechanics.* New York: Interscience Publishers.

Reichenbach, Hans. 1956. *The Direction of Time.* Berkeley: University of California Press.

Schaffner, Kenneth. 1969. "The Watson-Crick-Model and Reductionism", *British Journal for the Philosphy of Science* **20**, 325—348.

Schaffner, Kenneth. 1974. "The Peripherality of Reductionism in the Development of Molecular Biology", *Journal of the History of Biology* **7**, 111—139.

Schaffner, Kenneth. 1977. "Reduction, Reductionism, Values, and Progress in the

Biomedical Sciences", in Colodny, R. S. (ed.), *Logic, Laws, and Life – Some Philosophical Complications*, Pittsburgh University Press, pp. 143–171.

Schneider, Ivo. 1974/75. "Rudolf Clausius' Beitrag zur Einführung wahrscheinlichkeitstheoretischer Methoden in die Physik der Gase nach 1856", *Archive for History of Exact Sciences* **14**, 237–261.

Sinai, Y. G. 1973. "Ergodic Theory", in Cohen, E. and W. Thirring (eds.), *The Boltzmann Equation, Acta Physica Austriaca, Supplement 10*, pp. 575–604.

Sneed, Joseph D. 1971. *The Logical Structure of Mathematical Physics*. Dordrecht: Reidel.

Tisza, Laszlo. 1966. *Generalized Thermodynamics*, Cambridge/Mass.: MIT Press.

Tisza, Laszlo. 1977. "The Foundation of Statistical Mechanics", In Suppe, F. and P. D. Asquith (eds.), *PSA 1976*, vol. 2, East Lansing/Michigan, pp. 585–608.

Tolman, Richard C. 1938. *The Principles of Statistical Mechanics*. Oxford: University Press.

Zermelo, Ernst. 1896. "Über einen Satz der Dynamik und die mechanische Wärmetheorie", *Annalen der Physik* **57**, pp. 485 ff. Reprinted in: Brush 1965–66, vol. II.

VADIM SADOVSKY

MODELS OF SCIENTIFIC KNOWLEDGE

In studying scientific knowledge, like any other object, wide use is made of simulation techniques. From this point of view the entire history of the methodology of science in the 20th century can be viewed as a successive replacement of some models of scientific knowledge with others, more perfect as a rule. At the same time, it is generally accepted that the history of science should be treated as a succession and sometimes coexistence of different philosophical conceptions of scientific knowledge. Accordingly, a question arises as to the interconnection between philosophical conceptions of scientific knowledge and the specific models used in the study of knowledge, or, to put it differently, a question of the interrelationship between models of scientific knowledge and their philosophical interpretations. The errors in the assessment of this interrelationship which unfortunately have abounded in the 20th century, led to false conclusions when, the rightful criticism of philosophical conceptions of scientific knowledge was accompanied by certain models of science studies were wrongly rejected.

The present paper considers the relationship between philosophical conceptions and models of scientific knowledge, demonstrates how one and the same model can lend itself to different philosophical interpretations, and attempts to trace the development of the models of scientific knowledge on the one hand, and philosophical conceptions, on the other.

DEDUCTIVE MODELS OF SCIENTIFIC KNOWLEDGE

Historically, one of the early methods of the analysis of scientific knowledge made use of Aristotelian formal logic with a view to getting a deeper insight into the deductive structure of scientific knowledge. Proceeding from the modern ideas on this score, we shall give a general characterization of the deductive models of scientific knowledge.

A deductive theory is a language (deductive language) organized in a special way. Syntactically, a deductive language is merely a set of graphical signs (disregarding their extralinguistic meaning). Let us call it

J. R. Brown and J. Mittelstrass (eds.), An Intimate Relation, 391—404.

a deductive system. A deductive theory is an interpreted deductive system.

A deductive system generally consists of three parts: (1) a primitive basis (i.e. the totality of primitive signs); (2) logical means used in the given language; (3) the body of the system obtained from (1) through the application of (2). The logical means used in deductive theories, by definition, are the means of deductive inference.

Let *S* be some deductive language (an object language). It is described in the metalanguage MS.

S is a non-empty set of interconnected elements. The stages of the *S* design, described in MS in more detail as compared to the aforementioned general procedure of deductive system construction, are as follows:

(1) Primitive signs of *S* (*S* alphabet) are assigned.
(2) Rules of sentence formation in *S* are described.
(3) Primitive propositions of *S* are given.
(4) Logical means (rules of inference) permitted in *S* are fixed.
(5) Construction of *S* from (3) on the basis of (4) is carried out.

If *S* is considered from the syntactic angle, i.e. only in terms of relations between its signs and sentences, ignoring their extralinguistic meaning, then the syntactic concept of the deductive system is formulated in a syntactic metalanguage.

If, however, *S* is considered from the semantic angle, requiring semantic rules determining the meaning of the signs and sentences, as well as the conditions of their truthfulness, then the semantic concept of deductive system is described in the semantic metalanguage. This system, unlike the syntactic system, is not merely a set of signs and rules of operation thereof (a kind of "game") but a certain knowledge describing some or other area of reality, hence it would be reasonable to refer to it as a deductive theory.

The general procedure of the deductive structure of scientific knowledge has a variety of forms. The latter determine the axiomatic and constructive (genetic) systems.

Axiomatic deductive systems. The distinguishing characteristics of axiomatic constructs are:

(a) a number of primitive propositions (axioms) are adopted without proof; their terms are not explicitly defined;

(b) permitted rules of inference are fixed;

(c) the axiomatic system is a successive derivation of propositions from (a) on the basis of (b).

The deductive character of the axiomatic system stems from the fact that all the rules of inference, allowed within their frameworks, are deductive rules. The basic stages of a deductive system design, in constructing an axiomatic system, take the following form:

(1) Primitive signs of S (S alphabet) are enumerated.

(2) Rules of formation of sentences in S are introduced recursively (sometimes they are supplemented with rules of definition introducing new signs instead of complex combinations of primitive signs).

(3) A certain class of primitive propositions (axioms) is selected out of the propositions in S, which are assumed without proof.

(4) Rules of inference permitted in S are indicated. They are obliged to be deductive.

(5) The whole axiomatic system is constructed by a successive derivation from (3) on the basis of (4).

Constructive (genetic) systems represent the second type of deductive systems distinguished through a specific deductive construction of scientific knowledge.

While in axiomatic systems certain terms are introduced on the basis of other terms undefined within this system, and the proposition being proved is ultimately reducible to axioms, the building of constructive (genetic) systems seeks to eliminate the propositions assumed without proof, and to introduce all signs and propositions only through real or at least possible construction thereof with the available means.

In building the constructive systems, an important role is played by the fundamental inductive definitions which, in a general case, have the following form (for constructive system K_1):

(1) 1 is an object of K_1;

(2) if a is an object of K_1, then $a1$ is also an object of K_1;

(3) there are no other objects in K_1 except for those defined according to (1)—(2).

By using various fundamental inductive definitions, one can build different systems of constructively given objects. In order to pass to statements thereof, it is necessary to introduce a new class of inductive definitions — non-fundamental inductive definitions of this type (definition of the equality relation for constructive system K_1):

(1) $1 = 1$;

(2) if $a = b$, then $a1 = b1$;

(3) equality in K_1 is valid only due to (1)—(2).

Proceeding from the aforesaid, enumerate the basic stages in the constructive building of deductive system:

(1) Objects of K are constructively designed as a set of fundamental inductive definitions.

(2) The system's predicates describing the objects of K are built with non-fundamental inductive definitions.

(3) The techniques of inference in constructive system is based on the principle of mathematical induction which is the principle of deductive inference.

(4) The constructive system is built from (1) and (2) by the use of (3).

Along with the axiomatic and genetic deductive systems there are some intermediate types of deductive theories, one being the so-called deductive definition of set-theoretic predicates. Let us illustrate this type of deductive system with an axiomatic definition of the concept of group. The set-theoretic predicate "X is a group" can be defined in the following way:

X is a group if and only if there exist such D and O that

(1) $X = \langle D, \circ \rangle$;

(2) D is a non-empty set;

(3) $\circ$ is a function whose domain is $D \times D$ and whose range is a subset of D;

(4) for all $a, b, c \subset D$, $a \circ (b \circ c) = (a \circ b) \circ c$;

(5) for all $a, b \subset D$ there is an e D such that $a = b \circ e$;

(6) for all $a, b \subset D$ there is an e D such that $a = e \circ b$.

The objects satisfying this predicate are models of the theory of groups. The objects satisfying the first three conditions of the above deductive definition of the predicate "X is a group" form a class of possible models of the theory. This method of deductive theory construction is widely used by J. Sneed in his logical reconstruction of the Kuhn's theory of scientific revolutions. (See Sneed 1979).

The above description of both the general structure of the deductive models of scientific knowledge and different types thereof convincingly show that such models are not philosophical in character. They do not touch upon the principal philosophical problems of the relationship between being and consciousness, the material and the ideal, etc. Proceeding from the conception of methodological knowledge levels, developed in the Soviet literature, we may attribute such models to the level of general scientific methodology which does not purport to resolve the general philosophical problems (See Lektorsky and Shvyrev 1971, Blauberg and Yudin 1973, Sadovsky 1974).

At the same time, these models are not yet theories of scientific knowledge. At best, they just describe its structure. To become a theory they must be philosophically interpreted with a view to determining the relation between such formal models and scientific knowledge as a really existing object of philosophical study.

Let us look at the two diametrically opposite philosophical interpretations of scientific knowledge deductive models. The first one that has evolved within the frameworks of neopositivism or logical empiricism, was developed by Bertrand Rusell, Ludwig Wittgenstein, Hans Reichenbach, and especially by Rudolf Carnap. The distinctive features of the logical positivist philosophical interpretation of deductive models of scientific knowledge are as follows:

(1) The deductive system, first and foremost the axiomatic theory, is assumed to be the most perfect, supreme form of scientific knowledge organization. As applied to empirical knowledge, the deductive system transforms into a hypothetic-deductive system, however, fully sharing the aforementioned specifics of deductive theories.

(2) The evolution of scientific knowledge involves transition from the imperfect, substantive theories to perfect, deductive or hypothetic-deductive ones.

(3) The ultimate goal of logic and methodology of science is to build a "unified science" analogous to deductive theories.

(4) The logic and methodology of science fully accomplish their objectives by identifying the structures of scientific theories while completely abstracting from consideration of the genesis of scientific knowledge, in other words, the logic and methodology of science are concerned only with the "context of justification" rather than the "context of discovery".

A variety of partial logical empiricist principles such as a sharp dichotomy of the empirical and theoretical, the conception of verificationism, etc. are associated with the basic features of the logical neopositivist interpretation of deductive models of scientific knowledge.

As is well known, the neopositivist conception gained wide acceptance in Western philosophical literature during the 1930s—1950s. Much of it is long gone today, for all of its main principles have proved to be untenable as the conception developed.

The idea of a deductive system as the ideal of scientific knowledge applies not to scientific knowledge as it exists objectively, but to a certain image of knowledge based primarily on mathematics and theoretical physics (see Yulina 1982). The transition to the more adequate ideas of science taking into account the differences between natural sciences and the humanities, the organic interconnection between social, natural, and technical sciences means an automatic rejection of the basic features of the neopositivist conception formulated in 1 and 2.

Point 3 in the above list does not stand up to logical criticism: Gödel's theorem on the incompleteness of formalized systems (1931) convincingly refuted the idea of a "unified science" as a deductive system. Finally, the renunciation by the neopositivist of the analysis of the "context of discovery" prevented them from gaining an adequate idea of scientific knowledge. Marxists have always stressed that such an abstraction is unfeasible. In the second half of the 20th century Western philosophers have adopted this view (see Kuhn 1970).

The fallacy of the logical positivist interpretation of deductive models of scientific knowledge does not mean the fallacy or deficiency of the models as such. The latter are not responsible for the collapse of neopositivism. Logical positivism as a philosophical conception, is leaving the philosophical scene. Meanwhile, the deductive models of scientific knowledge used in it (and not only in it) continue to exist, requiring a more adequate philosophical interpretation, though. This challenge is met by the Marxist conception of scientific knowledge.

Leaving alone the details, note only the basic features of the conception relevant for the present discussion:

(1) it is generally recognized that deductive systems are an important aspect of scientific knowledge, but they are by no means treated as exclusive, let alone unique.

(2) Deductive systems are closely associated with the inductive methods of scientific knowledge acquisition (the principle of unity of deduction and induction).

(3) The task of the logic and methodology of science is to identify the scientific knowledge structure and investigate its genesis (the principle of unity of the logical and the historical).

Within these frameworks, the deductive models of scientific knowledge are perfectly justifiable, what is more, they can be considerably developed and improved.

The analysis suggests an important conclusion. Though philosophical conceptions of scientific knowledge may differ substantially and even be diametrically opposite, as is the case with Marxism and neopositivism, they may use similar models such as, for example, deductive models of scientific knowledge. This implies a relative independence of models from philosophical conceptions, within which they are used and developed, and different forms of functioning of models of scientific knowledge and their philosophical interpretations. The following examples will illustrate this conclusion.

VERIFICATION AND FALSIFICATION AS MODELS OF SCIENTIFIC THEORY APPRAISAL

The increasing complexity of structure of modern scientific knowledge stipulated by a wide ranging application of formalization, simulation, and computerization techniques places a demand for a system of methods to appraise and validate the scientific theories. Accordingly, a heavy emphasis during the past 40—50 years was placed on the verification and falsification techniques.

As is well known, the verification method was intensively developed by logical positivists and used in formulating a philosophical conception of verificationism. Similarly, proceeding from falsification techniques, Karl Popper has been developing his philosophical conception of falsificationism for fifty years now. The huge literature on logical positivism and Popperianism often confuses the verification and falsification methods and the conceptions of verificationism and falsificationism. It would be reasonable, therefore, to determine the meaning of the respective models of scientific knowledge.

The logical relations of verifiability and falsifiability have a strictly definite meaning. Proposition A is verifiable by a set of propositions $O_1, O_2, \ldots, O_n$, if and only if A is logically implied by a set of propositions $O_1, O_2, \ldots, O_n$, which are compatible, i.e. do not contradict one another. Proposition A is falsifiable by a set of propositions $O_1, O_2, \ldots, O_n$, if and only if proposition $\sim A$ (negation of A) is logically implied by a set of propositions $O_1, O_2, \ldots, O_n$, which are mutually compatible. The logical relation of verifiability and falsifiability are elementary formal logical relations, and are invariant with respect to any philosophical interpretation thereof which may considerably differ from one another.

Now define the philosophical conceptions of verificationism and falsificationism. As is known, a distinguishing characteristic of the logical positivist conception of the logic of science is that it has developed criteria for the cognitive significance of scientific sentences based on the verifiability principle. According to neopositivism, proposition A is meaningful if and only if it is either analytic or is logically implied by a finite set of simple observation reports compatible with one another. In full correspondence with the initial neopositivist premises, the other criterion of cognitive significance is based on the principle of falsifiability: proposition A is meaningful if and only if its negation $\sim A$ is either analytic or logically implied by a finite set of simple observation reports compatible with one another. The criterion of falsifiability has certain advantages over the criterion of verifiability: under the latter, no general (universal) sentence can be considered meaningful since none of them follows from the final set of simple observation reports, while under the criterion of falsifiability, considered meaningful are those universal sentences whose negation logically follows from the finite set of observation reports. At the same time, neither the verifiability nor the falsifiability criterion, as formulated here, can serve as a means for an adequate methodological analysis of scientific knowledge. Their basic flaws are well known.

Thus, if A is a meaningful sentence in terms of the falsifiability criterion, and B is a deliberately meaningless sentence then, according to the above criterion of falsifiability, the conjunction of A and B —$A \& B$ is meaningful. This is due to the fact that the negation of conjunction $\sim(A \& B)$ is equivalent to disjunction $\sim A \vee \sim B$ which, like sentence $\sim A$, is a consequence of one and the same consistent finite class of simple observation reports. Thus, the criterion of falsifiability makes it impossible to distinguish the meaningful parts of complex scientific sentences from the deliberately meaningless ones. Furthermore, this criterion does not treat as meaningful sentences of the following logical form "some A are not B" because such sentences are negated by universal statements "All A are B" which cannot be derived from a finite set of observation reports. Finally, like the criterion of verifiability, the falsifiability criterion proper does not meet its own requirements: it is neither an analytic nor a synthetic statement (which is obvious), as it does not logically follow from any finite consistent set of observation reports (see Ackermann 1976).

The said logical shortcomings of the methods or criteria of verification and falsification make their universal philosophical interpretations inadequate within the frameworks of verificationism and falsificationism. Indeed, all attempts on the part of logical positivists to use verificationism with a view to solving the problem of cognitive significance of scientific sentences, including various modifications of this criterion, have failed (Shvyrev 1966, Suppe 1977). In other words, verificationism, as a philosophical theory, has proven untenable in the course of its development.

The same holds true of falsificationism which Popper interprets not as a means of solving the problem of meaning in a language, but as a method of demarcation between science and non-science (see Popper 1980). The principal methodological drawback of falsificationism is that it treats a specific methodological procedure, i.e. falsification, as a universal principle capable of solving all the problems in the methodology of science. For the sake of universality, Popper is ready to ignore the real facts from the history of science rather than try, as he should, to define the limits of this principle's validity. Before long, even the faithful Popperians such as J. Agassi and I. Lakatos had renounced the ideas of Popperian falsificationism, at least its initial version.

An equally significant flaw of Popper's falsificationism is its intrinsic inconsistency. As was convincingly shown in the literature, notably in the works by I. Lakatos, in advancing the falsification strategy as a means of demarcation between science and non-science, Popper assumed a dogmatic and uncritical stance toward the demarcation criterion he had proposed himself. Nowhere in his works does Popper say what is the potential falsifier of his own criterion of demarcation. Therefore, following his own logic, Popper's criterion of demarcation should be referred to non-science. In other to avoid this disastrous implication, Popper had to turn to appraisals of some or other theories generally accepted by the scientific community, and declare that he was ready to renounce his criterion if it contradicts such appraisals. Not to mention the fact that this strategy provides wide opportunities for the penetration of subjectivism into the methodology of science, it does not save Popper's critical falsificationism from inner inconsistency. In fact, the universal conviction of scientists in the scientific character of Newtonian dynamics comes into conflict with the fact that no one, for example, has ever described the possible observations which would refute the theory. Likewise, Popper should have treated as non-scientific not only psychoanalysis (one of Popper's pet examples) but many theories of true science, both from the standpoint of scientific community and Popper himself.

The fallacy of verificationism and falsificationism by no means implies that we should refuse to apply them to the study of scientific knowledge. As regards their logical content these methods, being a part of the proof, confirmation, and refutation techniques well known as early as antiquity, come to be important specific procedures of scientific theory appraisal within the frameworks of Marxist conception of logic and methodology of science. The fundamental questions here are compatibility of these methods and the real limits of their application.

Hence, in contrast to logical positivism and Popperianism, each focusing on one of the above methods while ignoring the other, the dialectic and materialist conception recognizes the unity of the verification and falsification techniques which means, in particular, that any scientific theory should be appraised by applying both methods, and that every individual appraisal resulting, say, from verification of a theory must be complemented by an appraisal of the same theory by attempting to refute or falsify it. As for the range of application of the considered methodological procedures, the principal question for the dialectic and materialist conception, apart from their logical limitations, is the recognition of practice as the universal criterion of scientific knowledge truthfulness. It is practice, in the final count, that justifies, validates, and sets limits to applicability of verification and falsification techniques.

THE REDUCTION MODEL IN THE STUDY OF SCIENTIFIC KNOWLEDGE

The methods of reduction occupy an important place in the arsenal of scientific research tools. The analysis of any object starts with its disintegration into component parts, facets, aspects, etc., thereby reducing it to those parts, facets, aspects, etc. In order to understand some or other phenomenon or process, it is necessary to reduce it, or at least some of its aspects, to phenomena or processes that are already understood. The standard deductive nomological explanation consists in reducing the phenomenon explained to a certain law with the corresponding boundary conditions. At the same time, it is common knowledge that reductionism, as a philosophical and methodological conception is unanimously opposed by the proponents of diverse philosophical theories. Recent studies into the reductionism problem have demonstrated both the complexity of the arising questions and the fact that many difficulties associated with the problem solution stem from the vagueness of the terms "reduction" and "reductionism" (see Bazhenov

1980, Blumenfeld 1974, Engelgardt 1970). Hence, the task is to clarify at least some aspects of the reduction model which can be done by using the concepts and the principles of the systems approach and the general systems theory.

It is well known that Ludwig von Bertalanffy, who advanced one of the first programmes for building a general systems theory in the late 1940s—early 1950s, was an ardent anti-reductionist. He viewed the world as a set of inhomogeneous spheres of reality not reducible to one another, and tried to determine the unity of scientific knowledge through the conception of perspectivism, determining the isomorphism of laws operating in various spheres of reality (see Bertalanffy 1971). A closer look at Bertalanffy's views reveals an obvious reductionist element: isomorphism of laws implies their reduction to a unified structural basis, therefore perspectivism is an expression of at least one possible meaning of the reductionist programme. As for the perception of the world as a set of spheres of reality not reducible to one another, this concept is no more than a "symbol of faith" if unsupported by an explanation of the genesis of the higher forms of reality from lower ones.

The significance and role of reduction techniques must be considered within the system of scientific research methods. The attempts of an isolated study of these methods are theoretically unjustifiable and are behind the current difficulties of reductionism. Any reduction assumes an opposite process, i.e. derivation, and only their unity constitutes a relatively independent method of scientific investigation: a method of reduction-derivation. There are the following arguments in favor of the aforesaid. Each of the two component parts of the reduction-derivation method needs the opposite part to determine the correctness and truthfulness of the conducted reduction or derivation. Only the opposite procedure can guarantee that the suggested method of reduction (derivation) is objective. The successful reduction and derivation implies a possibility of eliminating the reduced, in the first case, and the content from which a new form (reality, law, knowledge, etc.) is derived in the second case, respectively. The fact that these forms — new and old — continue to exist is further evidence in favor of the unity of reduction-derivation techniques.

The recognition of the above unity means that the problem usually referred to as that of reductionism is but a part or facet of a more complex problem of reductionism-derivationism. Any reductionist programme, successful though it may be, cannot be considered complete unless it demonstrates how the reduced content can be obtained from the reduct (i.e. the result of reduction). Similarly, any derivationist programme implies a demonstration of not only the successful derivation of a new form, but also the methods of its reduction to the respective reduct. Naturally, in the general case, only the initial and final stages of the reduction and derivation processes coincide (more precisely, the final stage of reduction coincides with the initial stage of derivation and, respectively, the final stage of derivation coincides with the initial stage of reduction) while their mechanisms may vary considerably.

The reduction and derivation procedures are elements of human activity, therefore the problem of reductionism-derivationism is of epistemological (methodological) character. It has also an ontological content, involving identification of the objective process of generation of the higher forms of reality by its lower forms, and the presence in the former of the operating mechanisms of lower forms of reality.

With regard to the aforesaid, it is possible to identify the following major aspects of the reductionism-derivationism problem, each containing its own solution to the problem.

(A) Genetic reductionism-derivationism: find out whether the lower forms of reality can lead to higher forms thereof; accordingly, whether the latter can be reduced to the former. In recognizing the natural-historical development of the Universe, without the intervention of supernatural factors, we are bound to give a positive answer to the genetic problem of reductionism-derivationism. It is this aspect of the considered problem that is properly philosophical, and its solution (on the basis of premises adopted by the rational human) is, in principle, final.

(B) Ontological reductionism-derivationism: find out whether a certain sphere of reality is reducible to some other sphere; accordingly, if the latter is derivable from the former which requires the construction of a single language for describing these spheres of reality. Ontology is an objectivized content of human images of the world. Therefore, the problem of ontological reductionism-derivationism is solved positively in the extent to which ontology coincides with the actual genesis of the Universe. For each sphere of reality, however, this problem requires a special solution which may well be negative (the considered spheres of reality belong to different branches of the genetic tree, there are links, in between them, disregarded in this case).

(C) Epistemological and methodological reductionism-derivationism: find out whether the reduction-derivation programme is feasible for the given scientific theories and methods of scientific research. As is the case with the ontological aspect, epistemological and methodological reductionism-derivationism allows both positive and negative solutions for the given systems of scientific knowledge. At the same time, in each specific case, epistemological and methodological reductionism represents a research programme which can only be appraised in the process and after completion.

The above examination of the reduction and derivation techniques and the problem of reductionism-derivationism is further a evidence in favor of the thesis concerning the relative independence of the models of scientific knowledge and their philosophical interpretations. The classical version of the reductionist programme based, in the final count, on the principles of the mechanistic world outlook, and the dialectic and materialist approach to the solution of the problem of reductionism-derivationism, suggested in this paper, fundamentally differ from the philosophical point of view though they follow a common reduction procedure. It should be stressed, however, that the dialectic and materialist solution of the problem places a demand for a whole system of specific reduction methods, and the research in this direction is a primary task of the logic and methodology of science.

We are safe in concluding that behind the relative independence of models of scientific knowledge from philosophical conceptions, within which they are created and used, is the considerably different nature of comparable theoretical constructions: an essentially philosophical knowledge in the case of philosophical conceptions of science and special scientific knowledge, often logical and mathematical, in the case of models of scientific knowledge. Such models are not as a rule derivable from philosophical conceptions of scientific knowledge, and the sphere of their application does not coincide with that of the respective philosophical conceptions. Indeed, although scientific realism and instrumentalism are almost opposites in philosophical terms, the adherents of both philosophical trends use essentially identical logical and methodological models (hypothetic and deductive models of scientific theory, a deductive model of explanation, the methods of inductive appraisal of hypotheses, etc.).

The formulated theses on relative independence should be interpreted precisely as it is stated: models, on the whole, are independent of the corresponding philosophical ideas, but the degree of independence of each specific model may vary considerably. Sometimes, a model may be an instrument for realizing some or other philosophical positions, therefore each model requires a special analysis.

At the same time, the above thesis implies that the two major components of the methodology of science follow their own ways. The philosophical conceptions are in a permanent state of conflict, succeeding one another in the course of historical development. The key question, as applied to models, is the definition of the scope and conditions of their application: then the model may coexist with less effective models which can be replaced by more perfect, sophisticated models. This means, in particular, that models can migrate from one philosophical conception to the other. This was often the case as the methodology of science developed.

BIBLIOGRAPHY

Ackermann, R. J. (1976). *The Philosophy of Karl Popper*, Amherst.

Bazhenov, L. B. (1980). "The Principle of Reductionism and Problems of Interrelationship Between Physics and Biology", Biology and Contemporary Scientific Cognition, Moscow; (in Russian).

Bertalanffy, L. von (1971). *General System Theory*, Foundations, Development, Applications, London.

Blauberg, I. V. and Yudin, E. G. (1973). *The Emergence and Essence of the Systems Approach*, Moscow (in Russian).

Blumenfeld, L. A. (1974). *Problems of Biological Physics*, Moscow (in Russian).

Engelgardt, V. A. (1970). *Integratism: the Path from the Simple to the Complex in the Cognition of the Phenomena of Life*, Moscow (in Russian).

Kuhn, Th. (1970). *The Structure of Scientific Revolutions*, Chicago.

Lakatos, I. (1970). "History of Science and Its Rational Reconstructions", in R. S. Buck and R. S. Cohen, (eds.), Boston Studies in the Philosophy of Science, *Memory of Rudolf Carnap*, Vol 8. Reidel: Dordrecht.

Lektorsky, V. A. and Shvyrev, V. S. (1971). "Current Philosophical and Methodological Problems of the Systems Approach", *Voprosy filosofii*, No. 1, Moscow (in Russian).

Popper, K. R. (1980). *The Logic of Scientific Discovery*, ch. I, IV, Hutchinson.

Sadovsky, V. N. (1974). *Foundations of the General Systems Theory of Logical and Methodological Analysis*, Moscow (in Russian).

Shvyrev, V. S. (1966). *Neopositivism and the Problems of Empirical Foundation of Science*, Moscow (in Russian).

Sneed, J. D. (1979). *The Logical Structure of Mathematical Physics*, 2nd ed., Dordrecht-Boston.

Suppe, F. (1977). "The Search for Philosophic Understanding for Scientific Theories", in F. Suppe (ed.), *The Structure of Scientific Theories*, Urbana.

Vulina, N. S. (1982). "Images of Science and Pluralism of Metaphysical Theories". *Voprosy filosofii*, No. 3 (in Russian).

MARTIN CARRIER

CIRCLES WITHOUT CIRCULARITY

Testing Theories by Theory-laden Observations

My concern in this paper is to contribute to the clarification of the possible impact that the use of theory-laden data might have on the testability of theories. There is one faction in modern philosophy of science that does not recognize any difficulty in this matter. Testing is a kind of bootstrap operation which can proceed without firm evidential ground. Another faction holds that theory-ladenness imposes severe restrictions on testability. Robert Butts aligns himself more with the second party. Reminding us of Duhem's insight that physical hypotheses are not testable in isolation but only in the context of various other physical laws, he concludes that any piece of a theoretical framework can be saved at will in the face of whatever experimental outcome. Because, in addition, the meaning of observation terms is determined by their theoretical interrelations there is no theory-independent evidential standard against which a theory could be conclusively checked. Accordingly, he contends:

> Conventionalism, though forced upon us by Duhem's thesis, leaves us with the stark realization that theory determines observational meaning, that no independent source of the meaning of observation terms is available, and that testing as 'consulting the facts' plays a relatively less distinguished role in science than we would have wished (Butts 1976, p. 48).

There is one sense (and this is presumably the intended one) in which this statement is doubtlessly correct. Appraising the theoretical consequences of a certain empirical result is not a matter of straight-forward deduction. Other factors come into play, especially methodological criteria (or pragmatic conditions as Robert calls them (cf. *ibid.* p. 55)). If it is to be decided which part of a theoretical pattern should be modified, sacrificed or replaced in the face of unfavourable and disconfirmatory data one has to take into account methodological requirements such as emprically progressive problem-shifts, problem-solving effectiveness, explanatory power or the like.

There is, however, another possible interpretation of Robert's con-

J. R. Brown and J. Mittelstrass (eds.), An Intimate Relation, 405—428.

tention quoted above. It could be read as saying that the mere occurrence of derogatory evidence is excluded (or at least made improbable) by the theory-ladenness of observation terms. If a theory determines its own empirical basis it is hard to see how, if ever, there could arise any conflict between theory and evidence. This is the problem I want to address in this article. More specifically, I do not consider the methodological questions of how to assess different theoretical options for dealing with an anomaly or how to distribute blame among the parts of a theoretical network if an experimental failure is encountered. Instead, what I want to shed light upon is how such things as experimental failures could come about at all. How can a theory contradict theory-laden observations? I wish to show that theory-ladenness may indeed but does not necessarily involve problems in checking the empirical adequacy of theories. That is, I agree with one of the factions mentioned above in maintaining that it is in fact possible to test theories by theory-laden observations and with the other one in claiming that there may arise peculiar problems which cannot be simply dismissed or waved aside. In the first section I will introduce some conceptual distinctions between various types of theory-ladenness. In the following paragraph I want to disucss a notion of theoreticity developed by Joseph Sneed. Sneed contends that this special form of theory-ladenness leads to important test problems. I will argue that this tenet is unfounded. In the third section another variant of theory-ladenness (namely, strong Einstein-Feigl-completeness) is considered and its bearing on testability is explored. I will try to establish that there are indeed some obstacles for successful testing in this case which cannot be overcome by a generally applicable method. Nevertheless, for special cases a procedure is available which provides a basis for sound experimental scrutiny.

I. THEORY-LADENNESS: SOME PRELIMINARY DISTINCTIONS

The notion of theory-ladenness encompasses two different aspects. In the *perceptual* sense it is meant to express the fact that the *observations* or experiences of a person are influenced by the set of theoretical assumptions this person holds or by his background knowledge. In the weaker *contextual* sense theory-ladenness refers to the dependence of *observation sentences* upon the pertinent theoretical context. This means that the observation basis of theories is couched in concepts whose conditions of application are formulated in terms of general theories. The following considerations are restricted to theory-ladenness in this contextual sense. I do not address to a psychological theory of sensory perception but to a study of logical relations between statements. To give an example of this feature, the position of planets or fixed stars form part of the observation basis of astronomical theory. But these positions can only be reliably determined by recourse to an optical theory which accounts for distorting influences (like atmospheric refraction) and provides a means for correcting them. Analogously, the flux of solar neutrinos constitutes an important part of the evidential underpinning of astrophysical models of the sun. But the flux intensity is only measurable by resorting to many theoretical assumptions, mainly from the domain of particle physics. Thus, to view a certain empirical quantity as indicating the presence of certain states specified within a theory relies on the use of other theories. I call a concept figuring within some theoretical framework a *theoretical term* if the association with *all* its empirical indicators rest on certain theoretical assumptions. In other words, all measuring processes linked to a theoretical quantity make use of general hypotheses. Atmospheric pressure, for instance, is a theoretical term because all pertinent measuring devices rely on some (usually mechanical) supposition. Position, on the other hand, is non-theoretical because under some circumstances positions can be perceived directly, i.e., without theoretical mediation.

Testing a hypothesis means observing or bringing about experimentally an instance of this hypothesis. A necessary prerequisite of serious testing is that the agreement between theoretically predicted and empirically detected outcome is not guaranteed from the outset. This condition (that avowedly has a Popperian ring) states that there must exist some possible empirical results that contradict the hypothesis to be tested. A test has to be able to bring a hypothesis into trouble. It must not be excluded that things could have happened otherwise. It is obvious that serious tests are not ruled out generally if all available data are theory-laden. Consider a hypothesis H within a theory T_1. All the quantities occurring in H can only be determined empirically by making use of some *other theories* T_2, T_3 etc. It is clear, then, that testing T_1 relies on T_2, T_3, etc., to be sure, but testing T_2, T_3 etc. does not necessarily in turn presuppose T_1. In the examples presented above, optical theory is confirmable without taking recourse to astronomy, and the theorems of particle physics can be checked without referring to astrophysics. I call *observation theories* those theories on which the measurement of theoretical quantities rests in a given context of theory testing. That theory, on the other hand, that explains the data supplied by the observation theories is henceforth called *explanatory theory*. In all cases mentioned so far, observation theories and explanatory theory (1) are *distinct* from each other and there is (2) a *unidirectional* dependence of the testability of the explanatory theory upon the availability of observation theories. If these two coditions are satisfied I speak of *simple theory-ladenness*. It is characteristic for this feature that the test of observation theories may rely on further theories but not on the explanatory theory itself. Simple theory-ladenness does obviously not encumber serious tests of an explanatory theory.

There are, however, other situations in which the test of theories is not exclusively founded on theories different from the one to be tested. It may rather occur that a theory figures as its own observation theory. In cases like this the link between the the theoretical terms of a theory and their empirical indicators is brought about by this theory itself. This idea was first suggested as a methodological postulate by Albert Einstein in 1949. Einstein complained that the special theory of relativity

> introduces two kinds of physical things, i.e., (1) measuring rods and clocks, (2) all other things, e.g., the electro-magnetic field, the material point etc. This in a certain sense is inconsistent; strictly speaking measuring rods and clocks would have to be represented as solutions of the basic equations [. . .], not, as it were, as theoretically self-sufficient entities (Einstein 1949, p. 59).

In the same vein, Herbert Feigl argued in 1950 for the methodological superiority of theories containing theoretical terms by claiming that only a theory of this kind allows for a treatment of its observation basis by conceptual means belonging to this theory itself. Self-sufficiency in this sense constituted for Feigl a veritable Copernican turn.

> It consists in the derivation, with corrections coming from the theoretical scheme, of the peculiarities of the very basis of confirmation [. . .] As long as a science has not attained a very high level of explanation the process of indication may not be deducible from theoretical premises [. . .] [T]his corresponds to the theory of acids and bases in chemistry at a time when the process involved in the well-known litmus paper test was not itself logically derivable [. . .] The behavior of such thermometric substances as

alcohol or mercury was not theoretically deducible until the kinetic (molecular) theory of heat put these indicator processes on a par with countless other thermodynamic processes as interpreted on the micro-level [. . .] The Copernican turn then consists in relating the observer to the observed, the indicator to the indicated, — not epistemically, — but so to speak cosmologically. What epistemically must be looked at as the confirmation bases of the hypothetical construction, will in the fullfledged theory be given a place within the cosmos of which the theory treats (Feigl 1950, pp. 40—41; partially emphasized in the original).

That is to say, measuring procedures a theory employs should be part of this very theory's domain of application. A theory should be able to explain why certain empirical processes are reliable indicators for certain theoretical states specified within its framework. If this methodological requirement is accomplished explanatory and observation theories coincide. The former essentially contributes to bridge the abyss between theory and observation. I call this peculiarity *Einstein-Feigl-completeness* (henceforth abbreviated EFC).

It is important to realize what exactly are the distinctive features of EFC. It is not sufficient that a theory merely supplies conditions of adequacy for something to count as an indicator process. This would simply amount to postulating that there exist measuring instruments which meet the requirements stated by the theory. Demands of that kind are, of course, fulfilled by any theory whatever. If an instrument yields decreasing values when applied to a body during heating one would not be apt to interpret this procedure as measuring temperature. Einstein-Feigl-complete theories must moreover be capable of analyzing theoretically the pertinent indicator processes, i.e., more specifically, to estimate and to account for possible distortions and inaccuracies and to provide means for their correction. The theory must be used for the *explanation* of measurement procedures and not just as a basis for judging the adequacy of certain empirical indicators.

EFC, as hitherto presented, does not exclude that there are *other* theories than the one in question which *also* allow us to deal with the peculiarities of the pertinent measuring devices. As Feigl mentioned above, statistical mechanics can cope with the intricacies of thermic expansion of gases and liquids and hence derive the properties of thermometers based on these phenomena. Statistical mechanics is, therefore, Einstein-Feigl-complete in this respect. There are, however, other processes which measure the same quantity but whose functioning transcends the scope of statistical mechanics. Measuring temperature, for instance, by taking recourse to the thermic variability of electrical resistance or by determining the maximum value of a black-body radiation spectrum invokes observation theories other than thermodynamics. If such an alternative treatment of the observation basis of a theory is feasible (i.e., a treatment that is independent of the explanatory theory in question) I will speak of *weak* EFC. In a weakly Einstein-Feigl-complete theory some but not all conceptual concatenations between its theoretical terms and their observational counterparts are established by means of this theory itself. It is clear that weak EFC does not give rise to test limitations. Because the observation basis is multiplically analyzable in this case and some of the theories involved are logically independent of the explanatory theory to be tested there may indeed occur conflicts between these various accounts. There remains, of course, the resort of concept separation. In case of conflict one is always free to wave aside the disconfirmatory evidence by associating different theoretical notions with the conflicting experimental procedures. If the values obtained by measuring temperature one time by the expansion of gases and another time via electrical resistance do not agree the option stands to introduce expansion temperature and resistance temperature as different concepts. But this only repeats the lesson, learned long ago from Duhem, that it is always possible to avoid falsification of any given piece of a theoretical system. It does not detract from the possibility that contradictions between theory and evidence may arise.

It is quite another thing, however, if the empirical indicators of theoretical terms are *exclusively* analyzable by using that theory to which they belong. This phenomenon shall be called *strong* EFC. In a strongly Einstein-Feigl-complete theory *all* links between theory and observation are established on the basis of this theory. Weak EFC means that the observation basis of a theory *can* be analyzed by its own means. Strong EFC entails that it *must* be so analyzed. This is to include cases in which the explanatory theory is necessary but not sufficient to deal with its evidential basis, that is, cases in which, in addition to the explanatory theory, some other theory has to be applied to obtain a comprehensive account of the pertinent observations or experiments. It is suggesting that in cases of strong EFC one may indeed encounter serious test problems. In the following sections I wish to focus attention on this alarming suspicion.

II. SNEED-THEORETICITY: THE PROBLEM OF THEORETICAL TERMS

In 1971, Sneed proposed a logical reconstruction of the nature of theoretical terms which bears a conspicuous resemblence to the concept of strong EFC as defined above. Sneed claimed that the occurrence of theoretical terms in his sense does indeed impose serious test restrictions upon the pertinent theory if the structure of this theory is conceived in the usual way (i.e., as a set of sentences). I do not want do address Sneed's main concern, that is, his plea for a novel account of theory structure. What I wish to discuss, instead, is which kind of difficulties Sneed-theoreticity may engender for theory testing if theory structure is conceived in the usual fashion. I will first outline the concept of Sneed-theoreticity and its purported implications on testing and afterwards discuss the adequacy of the claims entertained. The results of this discussion will prove fruitful for the closer inspection of strong EFC.

Sneed calls a term theoretical with respect to a given theory if all methods used to apply this term to an empirical situation rely on this theory. As Sneed puts it:

> Roughly and intuitively a concept is theoretical in a theory if and only if *all* means of determining the truth-value of statements involving the concept presuppose the truth of the laws of the theory in question (Sneed 1979, p. XVIII; emphasis in the original).

That is to say, if we conceive of a theoretical term as a function and if we understand, accordingly, the measurement of a theoretical quantity as the evaluation of the corresponding function then the evaluation of a Sneed-theoretical function is founded exclusively upon that theory in which it is introduced or defined. This implies the following peculiarity. For evaluating a theoretical function in a given instance we have to take recourse to the pertinent theory. But if it is asked how to justify this evaluation we are forced to refer again to this same theory. Measuring theoretical function values in one application of the corresponding theory presupposes, therefore, that some other application of the same theory is successful. This procedure has in all appearance the air of an infinite regress or a vicious circle and this excludes serious testing. These difficulties involved in empirically checking theories containing theoretical terms are called by Sneed 'the problem of theoretical terms' (cf. *ibid.*, pp. 31–38).

The claim inherent in Sneed's approach has been made more explicit by Wolfgang Stegmüller. He contends that testing a statement containing Sneed-theoretical quantities requires to measure values of these quantities. But all procedures suited for this end presuppose the validity of the very theory to be tested. The circle lurking here indicates that the agreement between theory and observation cannot be checked.[1]

In order to investigate the adequacy of these claims it seems appropriate to study some examples. Sneed considers mass and force to be theoretical relative to classical particle mechanics which he sees characterized by Newton's second law, i.e., the equation of motion ($F = ma$). This entails that all procedures for measuring masses and forces should make use of this second law. To substantiate his contention with respect to mass Sneed offers the following example. If we analyze mechanically the path of a projectile we determine the mass of this body by referring, for instance, to Atwood's machine. This device consists of a freely rotating wheel around which an inextensible string is twisted. At both ends of the string masses are fixed so that one mass is falling down when the other one is rising. The ratio of both masses can be calculated from their observed acceleration. But this calculation uses essentially the second law. A successful application of mechanics to the projectile cases presupposes, therefore, its successful application to Atwood's machine (cf. Sneed 1979, pp. 32, 115–116).

Let us look more closely at this example. It is, firstly, not quite clear what Sneed exactly has in mind here. Into usual projectile problems, i.e., problems concerning the path of particles under the action of gravity, masses do not enter. Projectile paths are mass-independent. Accordingly, no difficulties arise with respect to the determination of masses. To make sense of this case we have to slightly alter the circumstances. Consider balls accelerated by contracted and subsequently released coil springs on a smooth horizontal surface. We wish to determine the velocity the balls gain by the expansion of the springs. For this purpose we have to use the equation of motion combined with Hooke's law for springs ($F = kx$ (with k being the spring constant and x denoting the contraction of the spring)). One obtains $ma = kx$. The integration yields: $v = \sqrt{k/m}\ x$. Because spring contractions and ball velocities are non-theoretical with respect to mechanics the impression is conveyed that a non-theoretical mass measurement is possible on the basis of this equation. This impression, however, is specious. One sees more clearly what is at issue when one compares this Newtonian account with an analysis conducted on the basis of late medieval impetus physics. Impetus physics was committed to an equation of motion of roughly the form $F = mv$.[2] If one applies this equation (again in conjunction with Hooke's law) to the same problem one obtains: $v = k/m\ x$. It seems as if the distinctions between both results would provide an unambiguous basis for deciding between the two accounts and, accordingly, for singling out the right mass values. The problem, however, is that we can read the resulting contraction-velocity-relation as a *definition* of mass so that both theoretical approaches would simply define 'mass' in a different fashion. If we restrict ourselves to this one experiment there is no way to exclude this expedient. We have to refer, therefore, to a second application of the same equation of motion. And now Atwood's machine enters stage. We assume that mass is an intrinsic property of a body so that the mass ratio of any given pair of bodies is constant. Applying, then, Newton's second law to Atwood's machine we obtain the prediction that the acceleration of the two bodies is constant during falling whereas, on the impetus hypothesis, their velocity should be constant. The first prediction is born out empirically and so we may reasonably use Newton's account to assign values to the mass function. But only by referring to two different applications of both approaches is the justification of a particular evaluation possible.

A casual look at this example bestows an air of plausibility to Sneed's contention. On closer inspection, however, the analysis reveals features incompatible with Sneed's account. Where, after all, is an infinite regress or a vicious circle involved here? Apparently nowhere. The appeal to Atwood's machine brings the process of mass measurement to an end. No further application of mechanics is required. But what makes the procedure stop? In deciding between the two competing approaches we had (1) to rely on the principle that masses do not change spontaneously and (2) to measure accelerations. The invariance principle for masses is a qualification (or a Sneedian constraint) imposed by mechanical theory on plausible mass functions. And secondly, most important to note, the empirical determination of accelerations do not presuppose any law of motion. Accelerations are *non-theoretical* with respect to mechanics. Mechanical theory (containing the theoretical concept of mass) predicts relations between non-theoretical quantities (i.e., accelerations) that can be checked without relying on that theory.

It seems, therefore, that there exists indeed an option for testing theories employing Sneed-theoretical terms. These theories formulate certain requirements for suitable specifications of theoretical functions and they entail moreover lawlike correlations not containing theoretical notions. It may be objected that in the case of Atwood's machine one has no evaluation of mass independent of a presupposed equation of motion. This objection is right and constitutes the valid core of Sneed's claim of lacking testability. The determination of mass only succeeds by employing *one single* hypothesis (namely, an equation of motion). The ascription of a specific value to mass is, therefore, not directly testable. Nevertheless, one may have independent empirical reasons to prefer a certain equation of motion. That is to say, the law applied to evaluation problems may well be empirically testable (within the framework of a theoretical context) even if the specific evaluation is not. If there is no theory-independent check of *specific evaluations* of theoretical functions I will speak of *local* test problems. If there is no theory-independent way of testing the evaluation *procedures* let me speak of *global* test restrictions. The salient point of the discussion is, then, that local untestability does not imply global untestability. Even if all measurement procedures of a quantity essentially make use of one single law or hypothesis there may exist independent evidential reasons supporting this regularity and thus indirectly confirming the evaluation of this quantity. The central fault of Sneed's and Stegmüller's account is not to have realized this possibility and to identify testing with direct evaluation.

In order to prevent the objection that I rest my case on one single (maybe untypical) example let me address another one. Consider the problem of determining empirically the masses of astronomical objects. It is characteristic that all such determinations necessarily rely on the law of gravitation.[3] Let me focus attention on the measurement of the mass of the sun. Applying Newton's equation of motion and the law of gravitation to a system consisting of the sun and one planet (i.e. neglecting gravitational interactions between different planets) one obtains an improved version of Kepler's third law. If P denotes the period of a planet and a the semimajor axis of its elliptical orbit one gets the regularity: $P^2 = 4\pi/\gamma M_s\, a^3$. M_s signifies the sun's mass and γ the gravitational constant. Because the mass of the sun appears in this relation the latter could in principle be used to determine the former. Planetary periods and orbits are non-theoretical quantities, to be sure, but the straight-forward calculation of mass is thwarted by the fact that the value of the gravitational constant is unknown. In order to measure its value we have to take recourse to a second instance of the law of gravitation. Such an instance is provided by Cavendish's torsion balance. In this device masses attract one another gravitationally at known distances and the intensity of gravitational force is measured by the torsion of a fibre. For analyzing this experiment one must apply, besides the law of gravitation, a rotational version of Hooke's law which relates the torsional torque with the torsion angle of the fibre. Moreover, the quantities of the masses involved must be known but the problem of terrestrial mass measurement has been discussed in the foregoing example. Suffice it to say, therefore, that these quantities can be determined by means of Atwood's machine. Things being arranged in this fashion it is in fact possible to measure the gravitational constant and to derive subsequently the value of the mass of the sun.

Again, this case seems to buttress Sneed's allegation. The successful application of the law of gravitation to astronomical mass determinations presupposes that the same law has been applied successfully to another instance, namely, Cavendish's balance. But again, this appearance of correctness evaporates on closer scrutiny. The double employment of gravitational theory does not rule out serious testing of this theory. To realize this more clearly let me assume an alternative law of gravitation which contends that gravitational force varies inversely proportional to the distance (that is, $F \sim 1/r$ instead of $F \sim 1/r^2$). Whereas on the Newtonian account the mass of the sun should be proportional to a^3/P^2 the alternative supposition yields a proportionality to a^2/P^2. Both ratios can be determined non-theoretically. The salient point is, then, that gravitational theory conceives of the mass of the sun as an intrinsic property which, accordingly, should be constant. This offers a basis for testing the rivaling approaches. If the orbital data of different planets are compared it turns out that only the Newtonian approach leads to values for the mass of the sun which do not depend upon the planet considered. It seems, therefore, that the problems of determining terrestrial and astronomical mass values exhibit essentially the same structure. In the astronomical case all mass determinations rest upon one single hypothesis (the law of gravitation). This entails that the values obtained are not directly testable (local untestability). But the reliability of the procedures employed is testable by invoking principles specified within the theoretical framework (invariance of the sun's mass) and predicting correlations between non-theoretical quantities (orbital data of the planets). Thus, global testability is secured.

Let me put this a bit more systematically. I wish to reconstruct the logical core of Sneed-theoreticity in following way. A hypothesis containing at least one Sneed-theoretical term is exclusively testable by relying on a method which can only be analyzed (and, accordingly, whose reliability can only be supported) by invoking this very same hypothesis.[4] As to mass the two examples presented exhibit the following logical structure. In the projectile case the mass measurement was founded upon Newton's second law (N_2) along with Hooke's law (L_H). The justification of the values obtained required taking recourse to Atwood's machine whose functioning can be analyzed by applying Newton's second and third law (N_3) and a special force function for the weight of the bodies involved (W).[5] If the arrow denotes 'testable by invoking' and '$\wedge$' signifies the logical conjunction one gets a scheme of roughly the form:

$$(N_2 \wedge L_H) \rightarrow (N_2 \wedge N_3 \wedge W).$$

Analogously, in the astronomical case the mass determination rested upon Newton's second law and the law of gravitation (L_G) and for testing the values assigned it was necessary to refer to a rotational version of Hooke's law and to the law of gravitation again. Using the indicated abbreviations this means:

$$(N_2 \wedge L_G) \rightarrow (L_H \wedge L_G).$$

In both cases, therefore, the characteristic feature of Sneed-theoreticity, as specified above, is clearly present.

It is not astonishing that logical structures of this kind induce the suspicion that there might be a circle or an infinite regress inherent in them. It is suggestive to assume that the law appearing twice in the procedure is not itself tested by this procedure. The analysis pursued above reveals, however, that this impression is faulty. In both case studies it was in fact possible to back the law in question empirically by checking relations between non-theoretical quantities induced by this law. Thus, both examples lead to the same conclusion. Despite appearance to the contrary global testability is not ruled out generally by the occurrence of Sneed-theoretical terms.

In order to complete my discussion of Sneed-theoreticity let me add some remarks on the theoreticity of mass. In the light of the above examples Sneed's claim that all procedures to measure mass rely in some way on Newton's second law seems plausible. There exist, however, other methods to evaluate mass that vitiate this claim. There is first a method to measure masses dynamically by reference to collision processes. Restricting ourselves to two particle collisions and assuming the mass of the particles to be constant one gets by applying the principle of conservation of momentum:

$$\frac{m_1}{m_2} = \frac{|\Delta v_2|}{|\Delta v_1|}.$$

Measuring, therefore, the velocities gained or lost by the particles suffices to determine masses unambiguously. This procedure relies in no way on the second law (although, of course, it does rely on other mechanical principles).

Another method deserving interest is the use of the ordinary equal arm balance. Here we compare two masses attracted gravitationally by the earth. Applying the law of gravitation to this case one obtains that the ratio of gravitational forces on these masses is as the ratio of the mass values. Invoking, in addition, the law of the levers (or, couched in more technical terms, the principle of the conservation of angular momentum) one gets the result that a torque is only exerted on the arms if the two masses are unequal. Otherwise stated, the equilibrium of the balance allows us to conclude that the two masses compared are equal. Again, the second law appears nowhere.

In fact, Sneed notices that as regards the balance something is going wrong with his account. It might be objected, he sets forth, that determining masses by means of a balance does not presuppose any physical laws. However, it has to be recognized, as he retorts to this feigned objection, that the claim that mass ratios are determined accurately by a balance amounts to the following expectation. If we regard the bodies on the pans of the balance as particles we obtain a model of classical particle mechanics, i.e. a system governed by the second law (cf. Sneed 1971, p. 117). In this way the equation of motion enters this procedure.

It will become clear immediately that both objection and retort are misconceived. It is to be noted, firstly, that the balance by no means constitutes a pre-theoretical procedure for comparing masses. In fact, as described above, mechanics can cope with mass determinations by means of a balance and, accordingly, justify the employment of this device. Secondly, Sneed's reply undermines considerably his notion of theoreticity. Simply assessing the adequacy of a certain procedure by taking recourse to the second law amounts only to saying that in case of conflict between observation theory and explanatory theory we always reject the former and retain the latter. Concerning the balance this means: If it turns out that the results of mass measurements carried through with the help of a balance do not fit the equation of motion we would rather give up the law of gravitation or the principle of conservation of momentum than Newton's second law. This may or may not be so, it is, at any rate, not a problem of theory-testing but of appraising methodologically. the clash between various hypotheses. Sneed's 'problem of theoretical terms', understood in this modified sense, does not transcend Duhem's conventionality thesis. I will not further consider this trivializing interpretation but retain the notion of Sneed-theoreticity in its original sense. It follows that mass is not Sneed-theoretical relative to mechanics.

III. EINSTEIN-FEIGL-COMPLETENESS: HOW TO PROCEED IN CIRCLES

Let me now resume the discussion of strong EFC as announced in the first section. It may be objected that it took a rather long time for me to eventually treat my central topic. In fact, however, I talked about strong EFC all the time. The subject of the second section could also be subsumed under the heading: Strong EFC of mechanical theory with respect to mass measurements. If we understand 'mechanical theory' a bit more comprehensively than Sneed does and use it to denote a set of hypotheses containing, for instance, all the Newtonian axioms and moreover special force laws (like the law of gravitation) the analysis pursued in the foregoing section aimed at showing how mass measurements could be carried out in relying (almost) exclusively on mechanical theory. These examples can thus be conceived as case studies on how a (nearly) strongly Einstein-Feigl-complete theory — despite the appearance of circularity — manages to obtain unambiguous measuring values which can in principle contradict this theory.

It seems appropriate to add some remarks concerning the relation between Sneed-theoreticity and strong EFC. First of all, Sneed-theoreticity is a property of *concepts* relative to a certain theory whereas EFC is an attribute of *theories*. Furthermore, in the only example Sneed discusses in some detail, namely, classical particle mechanics, the identity of this theory is defined by one single law, i.e., the equation of motion. Strong EFC, on the other hand, does not require that observational meaning is conferred to theoretical terms by using one single hypothesis. To qualify for strong EFC it is sufficient that the theory as a whole is necessarily appealed to for analyzing its observation basis. It may employ several hypotheses to fulfil this task. As the examples discussed in Section II reveal this applies to mechanical theory with respect to mass measurements. Different measuring procedures rely on different mechanical hypotheses. We may conclude, therefore, that although mass is non-theoretical relative to mechanics, mechanics is strongly Einstein-Feigl-complete with respect to mass measurements. In fact, this conclusion will have to be qualified immediately but the main point remains unimpaired: Strongly Einstein-Feigl-complete theories do not necessarily contain Sneed-theoretical terms. On the other hand, the occurrence of Sneed-theoretical terms (in the non-trivial sense) does guarantee that the pertinent theory is strongly Einstein-Feigl-complete. If all measurement procedures associated with a theoretical term t make use of one hypothesis *H* of the pertinent theory *T*, this term t is Sneed-theoretical and *T* is, consequently, Einstein-Feigl-complete relative to the relevant parts of the observation basis. It may happen, to be sure, that in addition to *H* some other laws alien to the conceptual framework of *T* must be employed. But this additional use of other hypotheses does not detract from the strong EFC of *T* (as this notion was introduced at the end of Section I).

What about the strong EFC of classical particle mechanics in general? The foregoing discussion may have conveyed the impression that mechanics is strongly Einstein-Feigl-complete. Actually, it is not. Consider, firstly, mass measurements that rely on non-mechanical force functions. A mass spectrometer for charged particles, for instance, employs Lorentz's force law for electromagnetic forces which is obviously not derivable from mechanical principles. In addition, position measurements and length determinations require that deformations of the measuring rods applied can be corrected. But to account for the influences of thermic expansion, for example, evidently transcends the scope of mechanical theory. In the light of this consideration it seems appropriate to use EFC as a comparative rather than a categorial notion. There are degrees of EFC. A theory *T* is the more Einstein-Feigl-complete the greater is the part of its observation basis that can be analyzed by invoking *T* and the less this analysis rests on the additional application of theories other than *T*.

It should be realized that the first of the two qualifications mentioned (that is, the existence of non-mechanical force laws) do not impair the results obtained concerning the testability of strongly Einstein-Feigl-complete theories. In the discussion of mechancial mass measurements only mechanical methods were envisaged. Mass values can be determined unambiguously without relying on extra-mechanical means. In other words, the testability of mechanics would not suffer if there existed only mechanical interactions in the world.

What looks more suspicious is the second one of the previously mentioned qualifications. All mass measurements sketched so far ultimately relied on relations between masses and positions (or their time derivatives such as velocities or accelerations). That is to say, the gist of these procedures was that mechanics managed to establish relations between parts of its observation basis it can account for (i.e., mass measurements) and those parts it cannot (i.e., position measurements). This peculiarity may induce apprehensions to the effect that the testability of mechanical mass measurements is essentially founded upon the fact that mechanics cannot deal with its observation basis *in its entirety*. Testability of a theory might be assured by the fact that this theory is not sufficient for analyzing the *whole* evidential basis. For such a theory which exhibits the *maximum degree* of strong EFC the method sketched for evaluating theory-dependent quantities breaks down. For this method essentially relied on the availability of theory-*in*dependent quantities, i.e., it relied on exactly those features that fall short of maximally strong EFC. So one might suppose that a theory of the latter kind is untestable.

I will address immediately to the shortcomings of the method hitherto described but I should like to show, at first, that its domain of application is more extended than it may appear so far. The discussion of measurements in Einstein-Feigl-complete theories was hitherto restricted to cases in which these measurements relied on one single hypothesis. This restriction stemmed from my concern with Sneed-theoreticity (the latter being characterized by this reference to one hypothesis). More often, however, we encounter a *mutual* dependence of testability. A hypothesis H_1 can only be tested by relying on a hypothesis H_2 and the test of H_2 in turn requires the use of H_1. To realize more clearly the specific features of a situation of this kind let me again consider an example. In Kepler's time, most of the astronomical data suited to confirm the hypothesis that planets move in elliptical orbits (i.e., Kepler's first law of planetary motion) could not be gained by direct observation but had to be derived with the help of other astronomical assumptions. The only adequate method available to gather relevant data consisted in relying on the area rule (i.e., Kepler's second law of planetary motion). Applying the ellipse hypothesis *together with* the area rule allowed for predictions of planetary positions that could be tested empirically.[6] The essential feature of this case is that the shape of the planetary orbit could only be determined by resorting to the velocities of this planet on its orbit, *and vice versa*. Otherwise stated, associating empirical indicators with the theoretical notion of 'orbit shape' had to make use of a hypothesis (i.e., the area rule) belonging to the same theory. Reciprocally, the theoretical term 'orbital velocity' could only be linked to experience by relying on another axiom of the same theory (i.e., the ellipse hypothesis). Put this way, it becomes obvious that all the marks of strong EFC are present in this case although it exhibits a structure different from the one considered in connection with Sneed-theoreticity. These differences notwithstanding, testability is assured by exactly the same token as in the examples discussed above. In addition to relations confirmable only within the pertinent theoretical framework, the theory specifies predictions which can be tested without making use of of its conceptual apparatus. Again, local untestability does not imply global untestability.

This extension in scope makes only more urgent the worries adduced above. What happens if there exist no empirical relations which can be tested without relying on the corresponding theory? Do such cases exist at all? In fact, they do. And they have been well-known in the philosophy of science for some time — although they were subsumed under a different heading, namely, Reichenbach conventionality. I will seek to argue that (1) in such cases we may indeed run into serious test problems but that (2) under contingent (i.e., not generally realized) circumstances these difficulties can be overcome.

In order to avoid technical complexities I want to discuss that most simple example of Reichenbach conventionality, that is, the problem of establishing empirically the simultaneity of distant events in the special theory of relativity (abbreviated STR in the sequel). The problem emerging in this context is that the only possible way to compare the data supplied by distant clocks seems inevitably to make use of signals, i.e., of light rays. If the velocity of light is known one can calculate at which time a light signal received was emitted at a place of known distance. This procedure would provide a means for synchronizing distant clocks. But to measure the signal velocity in turn presupposes the availability of distant synchronized clocks and at this point the ensuing circles becomes obvious.

It might be argued that one single clock is sufficient to determine the velocity of light propagation. Consider the following experimental set up. A mirror is situated at known distance from a light source. At a time t_1 a light ray is emitted into the direction of the mirror. This ray is reflected at the mirror and its time of return to the origin t_3 is recorded. It seems plausible to assume that the time t_2 of the distant reflection can be obtained by simply taking: $t_2 = t_1 + \frac{1}{2}(t_3 - t_1)$. The problem is, however, that the method sketched presupposes that the forward and backward velocity of light are equal. Otherwise stated, this procedure only allows for measuring the *mean* velocity or *two-way*-velocity of light. The underlying supposition of the isotropy of light propagation cannot be tested by such an experiment. Hence, on the basis of the available data we may only conclude that reflection occurs at any time between emission and return. The correct equation for t_2, therefore, assumes the form: $t_2 = t_1 + \varepsilon(t_3 - t_1)$ with ε varying between 0 and 1. Within this interval the choice of ε is arbitrary, that is, each ε-value has equal status. Accordingly, distant simultaneity (characterized by a certain choice of ε) cannot be established experimentally but has to be *defined*. Distant simultaneity is a matter of *convention*. The usual choice of $\varepsilon = \frac{1}{2}$ (i.e., isotropy of light propagation) is more convenient than alternative values but none of them is more valid than any other (cf. Reichenbach 1958, pp. 123—129).

As stated above, the essential feature of this scenario consists of a *reciprocal* dependence between the determination of distant simultaneity and the measurement of the one-way-velocity of light. This peculiarity induces Reichenbach to think that an empirical evaluation of both quantities involved is precluded. One of them has to be fixed by an arbitrary definition and the other one must be adjusted in accordance with this choice. As he says:

Thus we are faced with a circular argument. To determine the simultaneity of distant events we need to know a velocity, and to measure a velocity we require knowledge of the simultaneity of distant events. The occurrence of this circularity proves that simultaneity is not a matter of knowledge, but of a coordinative definition, since the logical circle shows that a knowledge of simultaneity is impossible in principle (Reichenbach 1958, pp. 126—127).

To put things more generally Reichenbach assumes that a mutual dependence of the testability of hypotheses implies their untestability and holds that it follows from the latter that essential conventional elements are inherent in the hypotheses involved.

This case clearly exhibits the features of strong EFC. To assign a value to the ε-variable (i.e., to establish a relation of distant simultaneity) presupposes that c (i.e., the one-way-velocity of light) has been evaluated. But to justify the latter evaluation one has to rely on a specific ε-value. Both ε and c are thus theoretical quantities which can be measured exclusively by recourse to that theory of which they form part. With respect to the empirical determination of these quantities STR is strongly Einstein-Feigl-complete. Reichenbach conventionality, therefore, appears to be a special case of strong EFC.[7]

To appreciate the import of Reichenbach's contention it should be noted first that, as far as hitherto presented, it is in fact true. The mutual test dependencies emerging with respect to the signal method of synchronizing clocks do indeed preclude an unambiguous determination of signal velocities or synchrony relations. It is to be realized, on the other hand, that reciprocal test dependence is not generally sufficient to create such problems. After all, we encountered the same logical structure in the case of Kepler's laws without any test problems being entailed there. Reciprocality of testing is, therefore, at most necessary for the occurrence of test difficulties.[8] But what, then, is the additional distinctive feature responsible for the emergence of test problems? Apparently, it is the absence of any relations induced by the corresponding theory but testable without recourse to its conceptual framework. Whereas in the astronomical example one could rely on position measurements to check Kepler's laws nothing similar is available here. For this reason *any* evaluation of the theoretical variables under discussion is compatible with experience. That is to say, with respect to simultaneity and signal velocity STR exhibits the maximum degree of strong EFC. And accordingly, as suspected above, untestability arises. Here local untestability (i.e., the impossibility of a theory-independent measurement of theoretical quantities) indeed implies global untestability (i.e., the impossibility of any contradictions between (the relevant parts of) theory and evidence).[9] It has to be admitted, therefore, that strong EFC may induce serious test problems.

On the other hand, strong EFC (even in its maximum degree) does not necessarily entail such problems or, put more precisely, such problems can at least sometimes be solved. In order to make this assertion plausible let me continue the discussion of the foregoing example. Despite the difficulties expounded above it is in fact possible to establish an unambiguous simultaneity relation. But one must not use for this purpose the signal method sketched but another procedure relying on the *slow transport of clocks*. To begin with, I want to outline the basic features of this procedure and afterwards analyze its characteristic aspects and assess its bearing on the problem of testability of maximally strong Einstein-Feigl-complete theories.

A casual look at the simultaneity problem suggests that two distant clocks C_A and C_B can readily be synchronized by using a third clock C_C that is locally synchronized with C_A and subsequently transported to C_B where the latter is appropriately adjusted. As a result C_A and C_B seem to be synchronous to each other. The problem arising in the context of STR is, however, the existence of distortions due to time dilational effects. Because of these effects a transported clock loses time relative to a clock at rest. The amount of slowing depends upon the transport velocity v and the length of the path s along which it is transported. On the basis of STR one obtains for the time the transported clock loses relative to a clock with standard synchrony ($\varepsilon = \frac{1}{2}$): $\Delta t = \frac{1}{2}\, vs/c$ (*). This implies that transport synchrony generally depends upon the history of the transported clock and leads, accordingly, to inconsistent results. This unhappy peculiarity caused Reichenbach to reject transport synchrony and to focus on signal synchrony with all its (meanwhile notorious) difficulties (cf. Reichenbach 1958, p. 133).

In fact, however, transport synchrony can be upheld if some refinements are introduced. As Brian Ellis and Peter Bowman (following an earlier suggestion of Percy Bridgman) have shown, slowly transported clocks are suited to establish an adequate relation of distant simultaneity. The central difficulty at issue is that the transport velocity of the clock cannot be measured reliably without invoking a notion of distant simultaneity. But this problem can be overcome if one employs 'self-measured' velocities. Let me consider again two distant clocks C_A and C_B and a third clock C_C locally synchronized with C_A and afterwards transferred to C_B. The self-measured velocity of C_C is obtained by determining the distance between C_A and C_B and referring to the transport time as measured by C_C. This self-measured velocity does *not* in general agree with the one obtained by using the standardly synchronized clocks C_A and C_B. This disagreement characterizes the relativistic influence of transport velocity on the results of clock readings. The essential point is, however, that it can be shown by making use of the conceptual framework of STR that if the self-measured velocity v_s converges to zero the standard velocity v also converges to zero. STR implies $v = v_s\,(1 + v_s^2/c^2)^{-1/2}$ and this equation entails that from $v_s \rightarrow 0$ it follows that $v \rightarrow 0$. That is to say, the extrapolation to infinitely small self-measured transport velocities leads to the same results as clock transport with infinitely small standard velocities. But for the latter STR predicts that dilational influences are negligible as is apparent from equation (*) (cf. Ellis/Bowman 1967, p. 130).

In other words, STR implies that an infinitely slowly transported clock leads to standard signal synchrony ($\varepsilon = \frac{1}{2}$). In practice, distant synchrony can be established by multiplically transporting clocks with different self-measured velocities and by extrapolating the results obtained to vanishing transport velocity. The agreement of both procedures can be checked empirically. This shows that STR specifies predictions for that part of its domain of application relevant here and these predictions may prove incorrect in the light of experimental scrutiny. In other words, by applying STR to its own observation basis an unambiguous and testable simultaneity relation can be obtained.

It is apparent that the logical structures of signal synchrony and transport synchrony essentially coincide. In both cases we encounter a situation in which the measurement of a velocity (i.e., transport velocity or signal velocity) and the determination of distant simultaneity are mutually dependent upon each other. In addition, no relevant relations testable without recourse to the pertinent theory are available. The discussion of the transport method thus confirms the results obtained as regards the signal method: STR exhibits the maximum degree of strong EFC as to the theoretical quantities considered here. But despite their basic conceptual similarity both methods are distinct in one decisive respect: One of them succeeds in establishing empirically a relation of distant simultaneity and the other one does not. What is the reason for this remarkable difference? On the basis of the transport method it is possible to construct (by making use of STR) a procedure of *successive approximation* that yields a *convergent* series of experimental values. Such a consistent approximation process cannot be specified with respect to signal synchrony and this is why the latter method fails. If such an approximation procedure can be developed it is in fact possible to attain a direct empirical evaluation of a theoretical quantity in a maximally strong Einstein-Feigl-complete theory. This sounds, admittedly, rather miraculous because such cases are characterized by a striking appearance of circularity. Nevertheless, as I sought to show, it is sometimes possible to break open the circle. Procedures of successive approximation transform, as it were, the circle into an iteration. The unambiguous empirical outcome resulting from such an approach is obtained as the limiting value of this iteration process. It should be noted, on the other hand, that the existence of a convergent approximation procedure is not guaranteed from the outset. No such method can be specified for signal synchrony. The existence of a suitable process is thus a contingent feature and the feasibility of measurements has, accordingly, to be examined separately for each case presented.

What is to be concluded from all that? It should be realized, firstly, that theory-ladenness of observations does not in general preclude serious tests of theories. Secondly, this is true to a much greater extent than it may have appeared at the outset. Even in cases which inevitably convey the impression of an ensuing circularity of testing closer inspection reveals that unambiguous measuring values are attainable that can in principle contradict the pertinent theory. Thirdly, however, in cases of maximally strong EFC it may indeed happen that serious tests are no longer possible. In such cases the appeal to experience fails. In the face of a failure of this kind the assessment of the pertinent (parts of the) theory would have to be founded *exclusively* upon methodological criteria. Here Robert Butts's contention, as quoted in the introduction, that consulting the facts plays a less dintinguished role than we would have wished, is true in a far more dramatic sense than he might have intended. And, after all, what accrues from the consideration of all these peculiarities and intricacies of theory-testing is the general insight into the difficulties involved in systematically explicating what scientists have constantly done since the time of Galileo: Testing theories by theory-laden observations.

University of Konstanz

NOTES

[1] Stegmüller 1979, pp. 17–18, 21. To prevent misunderstandings it should be

reemphasized that, according to Sneed and Stegmüller, these test problems are confined to the so-called statement view of theories.

[2] This is, admittedly, a distortion of the historical situation which I beg to be granted for the sake of argument. Actually, impetus physics knew nothing of Newtonian mass but introduced in its stead a geometrical factor (depending on the form of the body acted upon) and the resistance of the surrounding medium.

[3] Gadenne 1985, pp. 20—21 also uses astronomical mass measurements as an example of Sneed-theoreticity but his discussion differs considerably from mine.

[4] This rendering implies a slight change of Sneed's original conception that refers to theories in general (which is rather unprecise) and identifies in the case of particle mechanics this theory with Newton's second law (which is rather restricted). It should be noted that in the astronomical example Sneed-theoreticity appears primarily with respect to the law of gravitation (and not to the second law). The torsion balance does not invoke the equation of motion although, admittedly, justifying the mass values assigned requires appeal to something like Atwood's machine (that does indeed invoke Newton's second law). It seems more accurate, at any rate, to relativize theoreticity to certain specific hypotheses.

[5] I omitted reference to these further laws in the above description of Atwood's machine. They are adduced here for reasons of completeness.

[6] Cf. Wilson 1969, pp. 95—99. Glymour 1979, p. 234 also points to this fact and uses it as an example of his 'bootstrap-procedure'. This procedure consists of deriving instances of one hypothesis by using other hypotheses of the same theory (cf. Glymour 1983). Glymours notion of bootstrapping evidently bears some resemblance to my concept of testing Einstein-Feigl-complete theories but, to my mind, Glymour never stated with sufficient clarity how we do perform the feat of successfully pulling ourselves up by our own bootstraps.

[7] By contrast, Stegmüller considers Reichenbach conventionality to be an instance of Sneed-theoreticity (Stegmüller 1979, p. 20). In the light of my reconstruction this is a misconstrual.

[8] May be Reichenbach would have endorsed this qualification for he mentions (although somewhat obscure) besides reciprocality a second condition for there being conventional aspects involved, namely, arbitrariness of the definition (cf. Reichenbach 1958, p. 127). Accordingly, conventionality would not be implied by reciprocality.

[9] This is not to say, of course, that STR is untestable in its entirety.

BIBLIOGRAPHY

Butts, R. (1976). "The Hypothetico-Deductive Method of Scientific Theories: A Sympathetic Disclaimer", in W. R. Shea (ed.), *Basic Issues in the Philosophy of Science*, New York: Neale Watson, pp. 36—57.

Einstein, A. (1949). "Autobiographical Notes", in P. A. Schilpp (ed.), *Albert Einstein: Philosopher — Scientist*, La Salle: Open Court, pp. 1—94.

Ellis, B. and Bowman, P. (1967). "Conventionality in Distant Simultaneity", *Philosophy of Science* **34**, 116—136.

Gadenne, V. (1985). "Theoretische Begriffe und die Prüfbarkeit von Theorien", *Zeitschrift für allgemeine Wissenschaftstheorie* **16**, 19—24.

Glymour, C. (1979). "Reichenbach's Entanglements", in W. Salmon (ed.), *Hans Reichenbach: Logical Empiricist*, Dordrecht/Boston/London: Reidel, pp. 221—237.
Glymour, C. (1983). "Relevant Evidence", in P. Achinstein (ed.), *The Concept of Evidence*, Oxford: Oxford University Press, pp. 124—144.
Reichenbach, H. (1958). *The Philosophy of Space and Time*, New York: Dover.
Sneed, J. D. (1979). *The Logical Structure of Mathematical Physics*, 2nd ed., Dordrecht/Boston/London: Reidel.
Stegmüller, W. (1979). *The Structuralist View of Theories. A Possible Analogue of the Bourbaki Programme in Physical Science*, Berlin/Heidelberg/New York: Springer.
Wilson, C. A. (1969/70). "From Kepler's Laws, So-called, to Universal Gravitation: Empirical Factors", *Archive for History of Exact Sciences* **6**, pp. 92—170.

WILLIAM DEMOPOULOS*

ON APPLYING LEARNABILITY THEORY TO THE RATIONALISM-EMPIRICISM CONTROVERSY

INTRODUCTION

The seventeenth century dispute between "rationalists" and "empiricists" regarding the innateness of ideas and principles has received a good deal of attention from both linguists and philosphers of mind during the past two decades. Traditionally, rationalists have claimed that certain human cognitive capacities such as the capacitiy for language, for arithmetic, and for geometric reasoning, should be accorded a special status within the full repertoire of human cognitive skills. This was expressed in a combination of psychological and epistemological hypotheses about the causal and methodological bases of these components of human knowledge. Thus, while some rationalist arguments rested straight-forwardly on empirical claims, the more persuasive arguments were intended to establish what would today be regarded as foundational claims regarding the very *possibility* of acquiring these skills on the basis of exposure to the relevant experiences. Empiricist philosophers were by and large concerned to subsume these skills under a theory of general psychological mechanisms which would operate in the same way in each of the several domains of language, arithmetic, and geometry.

With the development of mathematical linguistics and Chomsky's theory of transformational grammar, there has arisen a new perspective from which traditional disputes regarding innatism might be reformulated and definitively resolved. Formal learning theory is an outcome of this work which was inspired by Chomsky's seminal contributions. The field was established by E. Mark Gold's, "Language identification in the limit". The framework within which formal learning theory considers the problem of language acquisition was there articulated by Gold, and has come to be known as the "Gold paradigm". Since Gold's early investigations, this paradigm has been elaborated and extended in several major directions. First there are developments within the field of formal learning theory proper. In addition to work on language identification, this work includes extensions of Gold's paradigm to a

J. R. Brown and J. Mittelstrass (eds.), An Intimate Relation, 429—440.

wide variety of learning situations, including the inferrability of functional relations from finite data samples (a kind of mathematical theory of inductive inference). More recently, an attempt has been made to extend these ideas to the identification of countable structures for a first order language.[1] Secondly there have been attempts, most notably by Wexler and Culicover[2] to apply the Gold paradigm to the learning of transformational grammars. We may broadly characterize these developments as (respectively) *formal* and *linguistics-based* approaches to the elucidation of the problem of first language acquisition.[3]

Many claims have been made regarding the theoretical importance of the Gold framework and much of the current work in both the formal and linguistics-based traditions has (to my mind, justifiably) been advanced as a vindication of the basic soundness of this framework. However little has been done to assess what measure of clarification learnability theory might cast on the traditional philosophical dispute between rationalist and empiricist accounts of language acquisition, although much work has been done on its psychological and linguistic importance. This chapter discusses the interpretation of the celebrated Gold unlearnability result together with a related characterization theorem of Wexler and Culicover. The conclusion at which we will eventually arrive is that some of the more ambitious claims made on behalf of the philosophical interest of the work are not adequately supported, and are unlikely to follow as easily as early investigators seem to have hoped. At the same time we try to formulate a framework for evaluating the bearing of formal learning theory on such larger questions as those surrounding the rationalism-empiricism controversy. Along the way we comment on the interrelationships between the work of Wexler and Culicover and that of Osherson, Stob and Weinstein.

1. THE GOLD PARADIGM

Gold's paradigm isolates four components of a learning situation. There are: (1) The set of objects which constitute the hypothesis space for the learner (assumed to be given a finitary representation, thus languages are represented by their grammar-names); (2) the data format and learning environment; (3) the criterion of success: the specification of what is to count as successfully learning the languages in the hypothesis space; and (4) the nature of the learner, here idealized as a function from data strings to elements of the hypothesis space. In addition to isolating the principal parameters in a theory of language acquisition, Gold introduced into the characterization of success criteria the important idealization known as "identification in the limit". Roughly speaking, the idealization amounts to thinking of learning as a process which takes place over infinite time: We think of time as a discrete sequence, $t_0, t_1, \ldots,$ of instants. At each instant a sentence of the language to be learnt is presented and the learner is asked to conjecture a name (grammar) of one of the possibly infinitely many languages the sentence is drawn from. At each instant a conjecture is advanced, so that the learner "guesses" infinitely often. Eventually *all* sentences of the target language are presented. Then the learner is said to "intensionally identify the target language in the limit" if, after some t_n, he correctly conjectures a grammar for the target language and does not move off his (correct) conjecture. (Before t_n any amount of vacillation is permissible.) The notion of identification in the limit lends itself to a variety of success criteria. For example, the learner may be required to stabilize on a single grammar for the target language (as above), or we may require only stabilization on a class of weakly equivalent grammars (sometimes called *extensional* identification), or we may only require stabilization on a class of grammars g_i such that $L(g_i)$ is only "finitely different" from the target language.[4]

2. APPLICATIONS TO THE RATIONALIST-EMPIRICIST DEBATE

Gold's unlearnability theorem establishes the surprising result that a (not necessarily computable) learning function cannot acquire (in the sense of *extensionally* identify in the limit) any class of languages which includes the finite languages together with one infinite language, provided the data is "text" data; *i.e.* so long as the data consists only of sample sentences of the target language.[5] Since this success criterion and this form of data presentation are generally regarded as plausible idealizations of the human learning situation,[6] we cannot suppose *both* that the natural languages are learnable *and* that any recursively enumerable language is a *possible* natural language; moreover, this holds even if we allow for the possibility that the learning function is not a computable function from initial segments of data sequences to languages.[7]

At first sight, there seems to be a very direct application of Gold's theorem to the issues separating rationalist and empiricist theories of language acquisition. To see this, suppose that the way to insure learnability is by restricting the hypothesis space to a set of conjectures generating a family of languages on which convergence can be shown to occur. Although convergence is relative to the appropriate success criterion and data format, let us suppose these have been independently agreed upon. As we shall see in Section 3, this way of proceeding is not forced by the unlearnability theorem; it is nevertheless a natural tack to take. In fact the argument we are about to give seems to have been first suggested by Gold in his original paper (Gold, 1967, Section 4). Then the argument against empiricism is that, within empiricist methodology, it is not possible to motivate constraints which are specific to the domain of objects learned, and thus the empiricist cannot in this way insure that acquisition will occur. But since the success criterion and data format are not in question, the empiricist position is untenable, insofar as it implies that any recursively enumerable language is learnable. Moreover, questions regarding the memory limitations of the learner, his computational capacity, and so forth, are quite irrelevant, since the Gold result requires only that the learner be representable as a functional relation from data sequences to languages. If there are restrictions on the class of learnable languages, for the empiricist these must come from the learner's peripheral processing mechanisms. But since these merely segment incoming acoustical signals according to the vocabulary of the target language, such restrictions fail to constrain the class of languages in the relevant way. (At most this restricts the learner to the class of all recursively enumerable languages over a given "alphabet"; but this does nothing to alleviate the problem.) The solution to the difficulty favored by many contemporary rationalists is to assume that the range of the learning function is constrained by the learner's knowledge of what constitutes a possible lanugage. And of course it is presumed that these constraints yield a learnable class. Since the learner considers only candidate languages from this class, this solves the problem of showing how first language acquisition can occur. But this implementation of the strategy of constraining the class of possible conjectures involves heavy rationalist commitments to innate, domain specific restrictions, and as such, is not open to the empiricist.

The argument just sketched constitutes a kind of "poverty of the stimulus" argument from formal learning theoretic consideration for supposing that the learner is innately endowed with principles which limit its available grammatical conjectures — what Chomsky calls principles of universal grammar. In this form the argument has been developed in detail and with considerable force by Matthews (1984); it is also suggested by Wexler and Culicover, (1980). In order to evaluate the argument, let us suppose the empiricist adopts the view that the right way to insure learnability is by restricting the hypothesis space to grammars generating a learnable class. Then it is reasonable to ask under what conditions a restriction on the class of languages is an acceptable restriction for an empiricist. The assumption of the poverty of the stimulus argument is that for the empiricist restrictions are acceptable or not depending on their origin; restrictions which arise because of the action of peripheral processing are both acceptable and motivated by an empiricist theory of language acquisition. Clearly if this is indeed the extent of empiricist restrictions on the hypothesis space, the view fails to delimit a *learnable* class.

In fact the restriction to recursive languages (assuming this can be motivated by empiricist principles) is insufficient to yield a learnable class *even when the data includes negative information.* Even if at each instant the learner receives a string together with a grammaticality judgment so that both the language and its complement are enumerated (what Gold calls "informant presentation") only the class of primitive recursive languages is learnable. The class of all recursive languages turns out to be unlearnable even from informant data. (This is Gold's Theorem IV.) Thus a very naive empiricism — one which leaves the learning function virtually unconstrained, save for the action of peripheral processors — is seriously compromised by Gold's work, *whatever* the form of data presentation.

While at first sight very striking, Gold's results cannot, I think, support the claims that have been made[9] for their relevance to the traditional rationalist-empiricist debate. The key difficulty is that Gold's theorems give little indication of *how much* by way of innate constraints learnability requires. Thus for all the theorem on unlearnability from text data shows, it would suffice if the learning function considered only infinite cardinality languages as candidate natural languages. For all the theorem tells us, identification in the limit on the basis of text might occur with only this minimal concession to rationalism. Moreover, the strategy, "Conjecture only infinite cardinality languages", seems not to be significantly different from the strategies: "Conjecture only recursively enumerable languages", and "Conjecture only recursive languages". For example, all three constraints are syntax-free in their specification of the family of possible languages. Nevertheless it must be granted that the restriction to infinite cardinality languages can hardly come from the peripheral processing mechanisms. So it seems safe to assume that *some* modification of naive empiricism is required. But there remains a wide gap between naive empiricism and the very strong forms of rationalism which proponents of this argument are inclined to favor.

A similar cautionary remark can be made concerning the interpretation of Wexler and Culicover's characterization theorem for learnable classes of languages. This theorem states that a class **K** of languages is *set learnable* if and only if **K** does not contain an infinite subset $\mathbf{K}' = \{L_0, L_1, L_2, \ldots\}$ such that

(*i*) $L_0 = \bigcup_{i>0} L_i$

(*ii*) *For all finite subsets F of* L_0, *there are infinitely many* $L_i \in \mathbf{K}$ *such that* $F \subset L_i$.

Wexler and Culicover's notion of *set learnability* requires that there exist a function f (not necessarily computable) from finite subsets of languages of **K** onto **K** such that for each language L in **K**, f stabilizes on L in the sense that for some finite subset D of L, $f(D) = L$ and $f(D') = L$ for all $D' \supset D$. (On the assumption that the order of the data is irrelevant, Wexler and Culicover's notion reduces to extensional identification in the limit from positive data, in the sense of Osherson and Weinstein (1982a).)

It is immediately clear that this theorem generalizes the Gold result. For let $L_0 = \{x_1, x_2, \ldots\}$ be an infinite language. Define the sequence $L_0, L_1, \ldots$ as follows: for each $i > 0$ put $L_i = L_0/\{x_i\}$. The $\{L_0, L_1, \ldots\}$ is a class of *infinite* languages which, by the theorem, is not set learnable. (As Wexler and Culicover point out, there are also set learnable classes of infinite languages.)

But although the Wexler-Culicover characterization theorem does not require that **K** contain all finite languages over a given alphabet, it is only when **K** does contain finite languages that the assumptions of the theorem are at all plausible. To see this, notice that on the hypothesis that there are finite natural languages, the existence of a countable family of languages in **K** is easily motivated: we need only take all finite subsets of some $L_0 \in \mathbf{K}$. More importantly, **K** will plausibly contain a universal language under these conditions, since there is an obvious naturalness to the supposition that if **K** contains each member of a chain of finite languages as a possible conjecture, it should include their union as a possible conjecture. But this condition becomes quite implausible if we exclude finite languages. For then, to form an unlearnable class, we must suppose the existence of a language which is universal even in a class of possibly disjoint languages.

Similarly, if the languages $L_1, L_2, \ldots$ form a chain of finite languages then of course the condition requiring infinitely many overlapping languages on any finite subset of L_0 is satisfied. But if the languages are not nested, it is difficult to see how to motivate the condition in this context. Clearly, to accommodate the phenomenon of *mistaken conjectures* we must assume that for every finite subset F of L_0 there are languages L_i, L_j $i \neq j$ which overlap on F. But we are not forced to suppose that there are infinitely many such L's available to the learner. (That there are infinitely many *logically* possible L's is, of course, quite irrelevant.)

To sum up: while the Gold unlearnability theorem depends quite strictly on the existence of finite natural languages, the Wexler-Culicover theorem applies to classes which do not include finite languages. However the plausibility with which the conditions of the theorem are fulfilled is difficult to motivate in such cases. Thus if we allow finite cardinality languages to be possible natural languages, then given a chain of finite cardinality languages it seems arbitrary to exclude any but a finite number of the members of the chain. However if we exclude the finite cardinality languages altogether, we must motivate the existence of infinitely many languages overlapping on every finite subset of some one language of the class. And it is hard to see why the class of conjectures should have such a rich membership. The obvious explanation of non-empty overlaps — *viz.* the existence of false but partially correct conjectures — is not strong enough to guarantee this condition. And condition (i) (the existence of a universal language in some countable subset) is no better off without the inclusion of finite cardinality languages.

3. RATIONALIST AND EMPIRICIST LEARNING STRATEGIES

In Section 2 we considered the possibility of constraining the range of the learning function in order to insure learnability, and we discussed an argument against the feasibility of an empiricist implementation of this solution to the problem of language acquisition. A somewhat different approach to the application of learnability theory to the rationalist and empiricist debate over first language acquisition is suggested by the elaboration of Gold's work currently undertaken by Osherson, Stob and Weinstein.

Let us first consider the problem of constraining the class of learnable languages. *A priori* we can imagine two[10] very different ways of delimiting the class of learnable languages:

(1) We may impose syntactic constraints, i.e. we may delimit the class of learnable languages by imposing restrictions on the form or function of the grammatical rules generating the languages. The important point regarding this manner of delimiting the class is that it is formulated within the language of a particular linguistic theory. One then argues apropos of the learnability problem that the sufficiency of such restrictions justifies the attribution of linguistic knowledge to the learner. But however one comes down on the issue of whether or not the linguistic constraints on the hypothesis space represent *knowledge* of the learner, and thus represent a plausible rationlist solution to the problem of language acquisition, it is an interesting and important question whether, at this level of abstraction, the characterization of the space of learnable languages must be given in the vocabulary of syntax theory.

(2) Alternatively, we may attempt to restrict the class of languages by imposing a set of constraints on the learning function (what Osherson, Stob and Weinstein call a *learning strategy*), where such constraints are to be expressed in the language of formal learning theory. (For example, "Conjecture only infinite cardinality languages" is such a learning strategy.) The program of formal learning theory is to characterize within the vocabulary of this theory — i.e. *without* appeal to syntax-theoretic properties — a class of learnable languages which includes the known natural languages. (Of course learning strategies, like syntactic constraints, may not restrict the class of languages.)

It is clear first of all, that these are two quite different programmes for characterizing the class of learnable languages which we call the natural languages. We would like to know the largest class of languages, learnable by beings psychologically similar to us on the basis of the casual exposure to data such as we have. And it seems we may approach this question from either point of view; *a priori*, there is no reason why two such different characterizations of the natural languages must coincide.

In Section 2 we considered an objection to empiricism based on the empiricist's inability to motivate domain-specific constraints on the space of possible lanugages. An alternative approach, suggested by Osherson, Stob and Weinstein, is to consider learning functions under various constraints. Such a set of constraints is said to comprise a "strategy". One then investigates the class of languages identified by such a "learning strategy" under the appropriate success criterion and data format. In principle, one would like to give a learning theoretic characterization of the natural languages; so the constraints which make up a learning strategy should themselves be expressed within the language of formal learning theory. Granted that constraints of some kind are necessary, the philosophical issues separating rationalist and empiricist theories of language acquisition turn on the specific character of the constraints which make up a learning strategy capable of acquiring a class of languages which includes the known spoken languages. The constraints may lack a determinate "rationalist" or "empiricist" flavor; in this case, the dispute will have been transcended rather than resolved. In any case, the aim of the "learning theoretic point of view" is to express sufficient constraints, and thus define the natural languages, in a vocabulary that is not tied to a specific syntactic theory. At this early stage in the development of the discipline, it is an open question whether or not this can be accomplished.

4. THE FINITENESS HYPOTHESIS

In his (1982) Wexler showed how the condition of "degree-2 learnability" forces the hypothesis space (of finite sets of transformations over a universal context free base) to be finite. (We will sketch the argument in a moment.) The concept of degree-2 learnability requires that if the target grammar **A** differs from the conjectured grammar **C** in the sense that **A** $(b) \neq$ **C** (b) for some base phrase marker b, then there is a base phrase marker b' containing at most two sentential imbeddings, such that **A** $(b') \neq$ **C** (b'). Thus degree-2 learnability constrains the hypothesis space by requiring identification of a transformational component from a restricted set of base structure-surface string pairs. Since, on Wexler and Culicover's model, all recursion in the base is through S, there are only finitely many base structures of degree-2. It then follows that there are only finitely many degree-2-inequivalent transformationally generated languages (over the universal base).

In (Osherson, Stob and Weinstein (1983)) the finiteness of the class of natural languages is derived from purely learning theoretic considerations and thus is not tied to the transformational form of the grammars which generate the possible languages. It is interesting to note that while Wexler's derivation arises from limitations on the "computational load" which the learning mechanism is capable of carrying, Osherson and Weinstein's derivation is not related to the computational capacity of the learning function, *i.e.* it is not related to data complexity or memory storage capacity. Rather, finiteness is derived from properties of the action of the control structure of the device realizing the learning function; basically, this means that finiteness is derived from the fact that the learner moves along its complexity ordering of grammars in steps bounded in advance by a fixed integer.

As Wexler has emphasized, while finiteness guarantees the existence of a learning function, it does not tell us how to characterize *feasible* learning functions, *i.e.*, it does not tell us how to characterize learning functions which solve the acquisition problem efficiently and within time constraints of realistic dimensions. But the empirical theory of first language acquisition is primarily concerned with precisely such strategies. Nevertheless, the finiteness hypothesis does seem to bear on the rationalist-empiricist debate. For suppose that as a result of considerations like those advanced by Wexler there are only finitely many available languages. Then the problem of language acquisition is not a "logical" problem in so far as the *existence* of a learning function is demonstrable; hence there is no *a priori* difficulty with the supposition that language is learned. The finiteness hypothesis is, in effect, one way of transforming the debate into an empirical investigation of *feasible* strategies. That is, constraints on the computational capacity of the learning mechanism *insure* a solution to the logical problem of whether language acquisition is in principle possible. At the same time it is not at all clear that syntax-based constraints may not be required to insure that the learning strategy hypothesized is a feasible one.

To summarize: In this discussion I have tried to outline an approach to the controversy over first language acquisition within the theoretical framework of formal learning theory. While the work of Gold shows quite drammatically that only constrained strategies will succeed in learning interesting classes of languages, there seems to be a spectrum of possible views concerning the exact nature of these constraints. Whether non-extremal points on this spectrum are happily classified as "rationalist" or "empiricist" remains to be seen. In addition to the "logical" problem of first language acquisition, there are real empirical issues involved in specifying feasible strategies so as to take account of the actual conditions under which first languages are acquired. Claims such as Chomsky's regarding the existence of innate, syntax-specific constraints may well be vindicated at this level of theorizing.

University of Western Ontario

NOTES

* I wish to thank R. J. Matthews, E. P. Stabler Jr. and S. Weinstein for comments on earlier drafts. Support of research by the Social Sciences and Humanities Research Council of Canada is gratefully acknowledged.

[1] Beginning with Blum and Blum (1975). For an overview of current work see Case (1984). Glymour (1985) contains suggestions for extending this framework to theory identification.

[2] See especially Wexler and Culicover (1980).

[3] We shall return to the difference between these two approaches in 3.

[4] See Osherson, Stob and Weinstein (1982b) for an investigation of these possibilities.

[5] The concept of text information requires that in the limit, as $t \to \infty$, every sentence of the language be presented to the learner.

[6] See (Wexler and Culicover 1980) chs 1 and 2 for an overview of the relevant empirical evidence, and Newport *et al.* (1977) for a classic study of the character of the learning environment.

[7] The requirement that the learning function be computable is a real restriction: There are classes of languages which are "function learnable" but not "computable function learnable". This is shown in Osherson and Weinstein (1982a).

[8] In Section 3 we will investigate an alternative approach.

[9] See especially Matthews (1984).

[10] In this discussion we are ignoring parsability constraints such as those discussed in Marcus (1979).

REFERENCES

Blum, L. and Blum, M. (1975). "Toward a Mathematical Theory of Inductive Inference", *Information and Control* **28**, 125—155.

Case, J. (1984). "Learning Machines", in W. Demopoulos and A. Marras (eds.), *Language Learning and Concept Acquisition*, Norwood N. J.: Ablex Publishing Co.

Glymour, C. (1985). "Inductive Inference in the Limit", in W. K. Essler *et al.* (eds), *Epistemology, Methodology and Philosophy of Science: Essays in Honor of Carl G.*

Hempel on the Occasion of His 80th Birthday January 8th, 1985, Dordrecht: D. Reidel Publ. Co.

Gold, E. M. (1967). "Language Identification in the Limit", *Information and Control* **10**, 447–474.

Marcus, M. (1979). *A Theory of Syntactic Recognition for Natural Language*, Cambridge, Massachusetts: M.I.T. Press.

Matthews, R. J. (1984). "The Plausibility of Rationalism", *Journal of Philosophy* **81**, 492–515.

E. Gleitman, H. and Gleitman, L. R. (1977). "Mother, I'd Rather do it Myself: Some Effects and Non-effects of Maternal Speech Style", in *Talking to Children*, C. E. Snow and C. A. Ferguson, (eds.), Cambridge, England: Cambridge University Press.

Osherson, D. and Weinstein, S. (1982a). "A Note on Formal Learning Theory", *Cognition* **11**, 77–88.

Osherson, D. and Weinstein, S. (1982b). "Criteria of Learning", *Information and Control* **53**.

Osherson, D., Stob, M., and Weinstein, S. (1982c). "Learning Strategies", *Information and Control* **53**, 32–51.

Osherson, D., Stob, M., and Weinstein S. (1984). "Learning Theory and Natural Language", *Cognition* **17**, 1–28.

Wexler, K. (1982). "On Extensional Learnability", *Cognition* **11**, 89–96.

Wexler, K. and Culicover, P. (1980). *Formal Principles of Language Acquisition*, Cambridge, Massachusetts: M.I.T. Press.

HUBERT SCHLEICHERT

THE RELATIONSHIP BETWEEN CONSCIOUSNESS AND LANGUAGE

1. A NOTE ON METHOD

What is the meaning of the terms "consciousness" and "to be conscious of something"? What does it mean, when we say that humans are conscious, but animals perhaps are not? The answer suggested in this paper is simple, but in order to substantiate it, some preliminary deliberations cannot be avoided. There is, of course, some vagueness in the current use of the term "consciousness" and there are practically no definitions or explanations of that term to be found in the present "philosophy of mind", yet the case is not a hopeless one. In order to analyze what is or could be meant by "consciousness" one first has to find a set of statements concerning or containing this term, which are commonly accepted. (Should this set be empty, any discussion on consciousness would be unintelligible). I call this set the "canonic phraseology" of the terms "consciousness" and "to be conscious of". This phraseology can be derived especially from philosophical texts of the 17th and 18th centuries, which may seem antiquated in the age of mass produced papers on "the phenomenon of the mind in the computer age", but there is a good reason for looking back.

In fact, our reflections on consciousness are conditioned by a number of expressions that are accepted by *all* philosophical schools and also determine our ordinary talk about consciousness. These sentences can be traced back at least as far as the philosophers of the 17th and 18th centuries. It is with these expressions, i.e. with this so called "canonical phraseology" of the word "consciousness", that one has to begin, if one wants to understand what is actually being talked about when reference is made to "the consciousness". In this context neither neurology nor computer science can be of any help.

There exists a certain type of book that might be called "synapsis- and chips literature". It deals at length with the working of nerve-cells, the structure of the brain, or with the way in which computers function. At the end of such books one finds chapters on perception, thought, consciousness, and similar phenomena. From his preceding arguments,

J. R. Brown and J. Mittelstrass (eds.), An Intimate Relation, 441—458.

(the “synapsis-and chips” part) the author more or less unconvincingly “deduces” his personal position on the mind-body problem. La Mettrie’s famous “L’homme machine” (1748) belongs to this category of books as well just like the work of Popper and Eccles (1977) or e.g. that of Delbruk (1986). Although the quality of those books is excellent for the most part, one is never told what consciousness is or could be.

The question of what *objects* are designated by the canonical phraseology is anything but trivial. The term “consciousness” cannot be learned in the same manner as, for example, the words “dog”, “joy”, “sleep” — there is nothing to point at when one uses the word “consciousness”. One learns this word by adapting conventions of speech, i.e. a phraseology, and not primarily through giving names to clearly definable situations or experiences. One does not learn the term “consciousness” and later on a series of sentences about it, but one acquires the use of those sentences right from the beginning. And here we are forced to embark on philosophical analysis.

Should one not study things themselves rather than ways of talking about them? In our case, the problem is such that it is not clear what thing is being dealt with, although everybody seems to be in command of a rich terminology and a fund of apparently necessary and valid percepts that enable him to speak about “mentality”, “consciousness” etc. Thus, this fund needs to be examined, if one wants to understand what might actually be meant by the word “consciousness”. This is an unusual procedure, but then the problem is an unusual one, too. First of all, we will establish what philosophers typically say about consciousness, and only then will we ask what “consciousness” might refer to.

2. CONSCIOUSNESS, KNOWLEDGE AND COMMUNICABILITY

Very often one comes across the claim that if a person is conscious of a thing, he also knows about this thing. Or vice versa, if a person does not know about a thing, he is not conscious of it. The nature of the relation between “being conscious of this thing” and “knowing about this thing” remains unclear. In the weakest case, the first might be thought to imply the second; in the strongest, one might consider the two as being equivalent.

J. Locke may serve as an example. In his “Essay concerning human understanding” Locke criticises Descartes’ thesis that the mind always thinks (mens semper cogitat). From a modern standpoint, this criticism only makes sense if “thinking” means the same as “thinking consciously”, and this is what Locke had in mind. In this sense he polemically asks what it is supposed to mean when thoughts which I myself know nothing of and have no perception of are ascribed to me.

"Can a man think, and not be conscious of it? . . . it is altogether as intelligible to say that a body is extended without parts, as that anything thinks without being conscious of it or perceiving that it does . . . . Thinking consists in being conscious that one thinks" Locke says and adds: "Consciousness is the perception of what passes on in a man's own mind. Can another man perceive that I am conscious of anything, when I perceive it not myself? . . ." (Locke 1689, b. 2, c. 1, §19).

If, however, one defines "thinking" as e.g. problem solving, then Locke is wrong — part of the process of solving problems takes place *unconsciously*, as is well known. If, on the other hand, "thinking" means conscious thinking, then Locke is simply stating the obvious. But Locke's formulations still contain more than the mere tautology that the thinker is conscious of conscious thought. Locke also uses the verb "to perceive" and connects it with the ability/inability to answer questions:

"Wake a man out of a sound sleep, and ask him what he was that moment thinking of. If he himself be conscious of nothing he then thought on, he must be a notable diviner of thoughts that can assure him that he was thinking." (*ibid.*)

Obviously Locke is making use of a general principle: If a man is unable to answer questions on a certain subject, then he is not conscious of this subject. He may have been conscious of it some time ago, but he has forgotten about it now. In any case he is not conscious of the subject *now* and he is unable to make statements about it. In everyday life this situation is much more easily expressed by the sentence "he does not know now", or in certain situations by "it does not occur to him" or — as we shall discuss in due course — by "he has not noticed it".

Thus it is not surprising that in the German edition of Locke's "Essay", the sentence ". . .that anything thinks without being *conscious* of it . . ." is translated as: ". . .that a being thinks, without *knowing* and without noticing it." The German edition also speaks of the sleeper who has been roused from his dreams: "he *knows* nothing of what he has been thinking", whereas the original edition says: "If he himself be conscious of nothing he then thought" (Kirchmann 1872, p. 113). The German translation is not literally correct, but presumably Locke would not have had anything against it.

We may assume that here the same situation is expressed in two different idioms. Using philosophical language, one says: "He is not conscious of it", whereas the same fact in ordinary language is expressed by "He does not know" or "He does not notice it".

However, the verb "to know" is also used in a more potential sense — namely as the content of our memory, even when one is "not thinking about this content" at that very moment. Expressed in a kind of mixed jargon of philosophical and ordinary language, one could say that what someone is thinking about right now, is what he consciously knows right now.

In our everyday language we have at our disposal a whole range of expressions for the field of sensory perception in order to formulate what philosophical jargon expressed through the terminology of consciousness. "I didn't notice the red light, I didn't see it, it didn't strike me . . . " says a driver, when asked why he did not stop at the red light. Using philosophical jargon, one would say that the driver was not conscious of the red light, that he was not aware of it etc.

Thus we have many different means of expressing one and the same basic situation: to be conscious of a thing, to notice/see/hear/perceive the thing, to think about the thing etc.

What do all these expressions have in common?

Here Locke's method can serve as an indication: "Wake a man out of a sound sleep and *ask* him . . .". Only when the man is able to give an answer, when he is able to tell us what he was thinking just now, is the terminology of consciousness applied.

If, on the contrary, upon careful inquiry an honest driver says nothing about a red light, although he has just passed it, all variations of the terminology of consciousness will be used negatively: He didn't perceive the red light, he didn't notice it, it didn't strike him, he didn't become conscious of it, etc.

What is it that all these phrases have in common? It is the fact that their use depends on the ability or inability to answer questions, or to formulate messages. In short: whatever someone is conscious of, whatever someone knows, whatever someone thinks, he is able to *talk* about it.

Christian Wolff may serve as another witness of this "thesis", which is simply one part of the canonical phraseology. Wolff, incidentally, was virtually the trend-setter for philosophical (and popular philosophical) expressions in German. A considerable amount of German philosophical terminology is derived from him, and his long-winded systematic text-books crystallize the philosophical phraseology of a whole epoch.

Let us begin with a special case, so called self-awareness; our interest lies not in self-awareness as such, but in what Wolff generally ascertained when discussing it: "... When we do not think about the operations that take place in the soul, and thus distinguish between ourselves and the things we think about; then we are not conscious of ourselves, and if someone should ask us whether we are conscious of ourselves at this moment, we would give him no other answer than that we had not been thinking about ourselves." (Wolff 1751, §730)

The ability to answer questions is thus for Wolff closely connected with the notion of consciousness, even though he does not say anything about the nature of the connection.

Wolff also relates consciousness to the ability to differentiate; he talks of it as "getting lighter inside us" when we are able to differentiate between one thing and another. This refers to perception as well as to thought. From this "getting lighter" it follows in both cases that one can make a verbal statement about it. (*ibid.* §205)

"We find", Wolff writes, "that we are conscious of things, when we can differentiate between them ... when we do not notice the difference between things that are present we are not conscious of what comes to our senses. For instance, when someone is reading a book, he is not conscious of hearing something ... And in this case we say, when we want to indicate the cause of our not being conscious: we were not paying attention to it." (*ibid.* §729)

This passage provides us in the first instance with a particular rule of Wolff's: whatever one distinguishes, one is conscious of. The following passage yields an even more interesting rule: whatever one distinguishes, one can also communicate. Assuming that we are thinking about a problem, we try — in Wolff's terminology — to mark differences. And then "it happens that we determine the difference we are thinking about, and thus if need be we are able to tell other people about it. And so our thoughts are distinct. For instance when I think about a triangle and a rectangle; I am able to mark the difference between the triangle and the rectangle, and if someone asks me, ... I am able to describe the difference." (*ibid.* §206)

Occasionally Wolff uses the term "knowledge", for instance when he writes: "Some thoughts are such that we actually know what we think, and we can differentiate them from others. Then we say they are clear.

For instance, I see buildings, people and other things right now. I am conscious indeed of what I see . . ." (*ibid.* § 198). From all this, a general rule can be extrapolated, that is in no way surprising and links up with known canonical phraseology: What can be discerned, one knows about, one is conscious of and one can also communicate, i.e. articulate. Though this rule as such may certainly be accepted straight away, its logical status is unclear. Is it an empirical statement or is it something else? I will leave this question open for now.

3. INTENTIONALITY

The intentionality of all mental operations also belongs to canonical phraseology. Whoever thinks, thinks of something; whoever has consciousness is conscious of something. All thinking, questioning, hoping has some content, some object that is referred to. There is no such thing as void thinking, void consciousness, consciousness of nothing. If someone "does not think of anything", without having lost consciousness nor being asleep, one would call him "absent-minded".

The "locus classicus" of this phraseology is Franz Brentano's "Psychologie vom empirischen Standpunkt". In his chapter on "The Distinction between mental and physical phenomena" Brentano writes: "Every mental phenomenon is characterized by what the scholastics of the Middle Ages called the intentional (and also mental) inexistence of an object (Gegenstand), and what we could call, although in not entirely unambiguous terms, the reference to a content, a direction upon an object (by which we are not to understand a reality in this case), or an immanent objectivity. (Brentano 1874, v. 1, p. 124)

One frequently connects the thesis of intentionality with the thesis that intentionality is characteristic of mental phenomena, and thus can never be attributed to physical phenomena i.e. can never be explained through physical phenomena. But this is obviously no longer part of canonical phraseology. It is a strong and also controversial hypothesis, and need not trouble us at this point. However, surely nobody will deny that each conscious thought is always made up of some content; it is this content which one is conscious of.

4. CONSCIOUSNESS AND REFLECTION

The use of the expression "knowing that . . ." leads to another canonical phrase relating to "consciousness". Chr. Wolff's contemporary Reimarus, for instance, writes: "As soon as we get to know of a noticed thing in our imagination that we have compared with and distinguish from other things, we are conscious of it, i.e. we know that we are imagining something and we know what we imagine. (Reimarus 1756, pt. I, 1, §60) Here we come across a double meaning of "knowing": "Knowing what we are imagining" and "knowing that we are imagining something".

Let us now focus on the term "reflection". "Thinking", according to Leibniz, "is that which is conscious of its activity or has reflexivity". (Leibniz 1) He thus closely connects consciousness with reflexion. Reflexion is, then, thinking of or about one's own thinking. As he says at one point: "The operation of the mind which seems most astounding to me is when I think that I am thinking and while I am thinking realize that I am thinking about my thoughts, and shortly afterwards wonder at his threefold reflexion." (Leibniz 1923, A VI 3, p. 516)[1]

It is doubtful whether anybody, apart from Leibniz, ever had the experience of a threefold reflexion and one might wonder what Leibniz went through when he experienced it. But even if Leibniz may not claim to have had an experience like this in reality, he may nevertheless be invoking a real principle: There is nothing that could not become an object of thinking, i.e. a content of consciousness. Thus thinking can also turn into an object of thinking, consciousness can turn into a content of consciousness. Putting it in traditional terms, one might say that man (in opposition to animals) can have not only ideas, but also ideas about ideas, ideas about ideas about ideas, etc.

This ability to have ideas about ideas is often considered as the basic characteristic of consciousness. Spinoza takes this widespread concept to its logical conclusion when he states: "Whoever knows something, knows eo ipso that he knows that, and at the same time knows that he knows what he knows etc. ad infinitum". (Spinoza 1677, b. 2, prop. 21, schol.)[2] An infinite regress is generally supposed to be an indicator that the argument is not sound. The unlimited addition of layers of reflexion should thus be looked upon as a mere possibility, not as something real. The possibility of reflexion (and its occasional realization) however, marks a popular "criterion" for the existence of consciousness.

This possibility is often said to distinguish man from beast and machine. One sentence from canonical phraseology goes like this: "The animal sees something, but only man sees something and knows at the same time that he is seeing something." "Seeing and simultaneously knowing that one sees" is also expressed by the phrase: "being conscious of what one sees". However, that man also knows that he knows that he is seeing something is not generally accepted as part of the phraseology.

In fact, one quickly run into difficulties when developing the aforementioned phraseology systematically. Thus, Leibniz already mentioned "that it is not possible for us always to reflect on all our thoughts: otherwise our minds would be reflecting on each new reflection endlessly, without ever being able to start a new thought. While I am conscious, for instance, of some present perception, I would have to think that I am thinking about it, and also think that I am thinking about thinking about it, and so on ad infinitum. But it appears that I have to stop reflecting on all of these reflections and finally have one idea that goes past without being thought about; otherwise one would come to a permanent standstill with the same things." (Leibniz 1665, b. 2, c. 2, §19)

The passage is a reply to Locke's question as to how someone can think without being conscious of it. Leibniz implicitly equates consciousness with reflection, i.e. thinking that one is thinking. Along with the majority of philosophers, Leibniz presupposes that "thinking that one is thinking" is possible as well as the unlimited addition of layers of reflection, but that such a case will not happen very often since thinking would get stuck. And so it must happen that one thinks of something without thinking that one is thinking. This is what Leibniz (unlike the majority of philosophers) calls a mode of thinking which one is not conscious of. This way of expression is problematic because the top level of every reflection would have to be non-conscious. Each reflection — even the cleverest — ultimately comprises only a number of layers along the lines of: "I think that I think . . .". Locke's way of putting it is much simpler: if someone says "here is a dog", then we say that he is conscious of the dog — and we do not require further reflections from him.

The canonical concepts of reflection are misleading in some way. On the one hand, it seems reasonable to place consciousness in relation to reflection, or at least the possibility of reflection; on the other, such reflections are certainly not present in most of our experiences or thoughts. It does not seem proper to speak of "non-conscious" experiences or thoughts just because one is not reflecting on them.

5. SEARCHING FOR THE LOST OBJECT OF THE CANONICAL PHRASEOLOGY OF CONSCIOUSNESS

We have been selecting elements from widespread everyday as well as philosophical phraseology. They suffice to raise the question of what is really being talked about? This search for an object may seem peculiar, since the object is explicitly stated all the time: consciousness. Consciousness, however, is anything but an ordinary object. What everybody "knows" about it, or accepts, are these same canonical expressions we have been discussing — and nothing else. Is there an object to be found at all that corresponds to all these expressions? Simply speaking of "consciousness" does not provide a satisfactory answer to our question.

There is, in fact, such an object: language. All canonical phrases concerning consciousness apply to language and to language only, they express characteristics of language, and this is what we are about to demonstrate in detail now.

The objection might be raised that this is self-evident, since language is known to be the "expression of consciousness".

I will ignore this objection for now and only show that language really has the desired qualities, no matter what theory or metaphysics may be set behind or above language.

6. INTENTIONALITY OF LANGUAGE

If someone speaks, he usually says something; what he says has a certain content and means something. Apart from rare border-line cases, one can say that there are no sentences devoid of meaning — just as one can say that there is no such thing as empty consciousness. Intentionality (bearing a content) is a characteristic of sign systems, and language is just such a system.

The problem of what actually constitutes the *content* of a sentence is totally analogous to the earlier question of what constitutes the content of a thought or of consciousness. An attempt to find an object called "content", leads in both cases to tangled metaphysics. But in both cases it is possible to explain the consequences of the fact that the content of the sentence (or of a thought) is comprehensible to me. I can handle this sentence, get messages across with it, react to it as others expect me to do, etc.

There are, however, nonsensical remarks, meaningless words and sentences that make no sense, whereas we do not speak of thoughts having no content! There is a popular, slightly ironic concept underlying this objection according to which thoughts always make sense and are good, whereas words can be senseless and bad. A frequently quoted *bon mot* says language serves to conceal our thoughts. Some people also say it is due to language that our innocent way of thinking gets confused or bewitched, because nonsense can only be *said* and *not thought*. According to A. Stöhr, "one thus does not say: don't think nonsense, but: don't talk nonsense! This does not mean that these or other terms cannot be combined but that for a moment the speaker changes into a thoughtless talking-machine." (Stöhr 1910, p. 131)

Viewed from the perspective of the problem of consciousness, this is a very misleading way of putting it. When I say something totally meaningless, am I then unconscious of what I say? Of course I am not, and the content of this consciousness is simply the meaningless passage I am just reciting.

What is different from meaningful talk is the fact that I cannot do anything with a meaningless expression. I can neither draw consequences from it, nor relate it to other sentences or patterns of behaviour. I cannot induce someone to behave in a certain way either. The expression will therefore soon vanish from my consciousness, and I will stop saying it. But as long as I *talk* nonsense, I also *think* nonsense, and that is the content of my consciousness.

In passing, I should like to point out that there is a widespread technique of meditation consisting of permanently reciting a formula that makes no sense as such (a mantra). Is the speaker without consciousness? Not at all, but the content of consciousness does not give rise to any other content, because it is nonsense. This leads to an unusual state of consciousness, but not to a non-conscious state. When asked, the person reciting is well able to tell us what he has just been thinking of — which is the mantra and nothing else.

7. REFLEXIVITY AND META-LANGUAGE

According to canonical phraseology, conscious operations always entail the possibility of an endless succession of reflection levels. These are, for example:

0. I see a dog.

1. I know that I see a dog.
2. I know that I know that I see a dog.
   etc.

Now, without using this terminology (knowing, being conscious of), it is possible to construct an analogous succession of levels:

0′. I see a dog.
1′. I say that I see a dog.
2′. I say that I say that I see a dog.
    etc.

Here again it is theoretically possible to continue ad infinitum with this succession. In both cases the cause for this potential infinity of successive levels is the fact that one can "think of *everything*" or "talk about *everything*". Thus by thinking or talking, one creates a new object which can be thought or talked about.

The linguistic version clears up problems left open by the mentalistic version. The fact that one can talk ad infinitum about language by using it, is not essential to language. It is a consequence of the principle that anything and everything can be talked about. Language also functions when it is not used to speak about language but only about the world. Analogously, thinking that always thinks about the world, never about thinking, lacks nothing — not even consciousness!

There is another aspect that is also worth noting: Where does this succession of levels 0, 1, 2, . . . originate? Do we get to them because of particular experiences of knowledge, do they result from particularly profound states of consciousness? As this is obviously absurd it would appear that the sequence 0, 1, 2, . . . , as well as the less problematic sequence 0′, 1′, 2′, . . . , is only learned through language.

No mysterious faculty is required in order to learn the sequence 0, 1, 2, . . . One simply learns it through learning how to speak. Through the learning of language, one finds out that one may always say "I know that I see a dog" if one may say "I see a dog" — this is part of the rules governing the use of the verb "to know".

Reflexion starts to become mysterious when a linguistic phenomenon is "translated" into mental phraseology. The mysterious reflexivity of consciousness, the possibility of mounting to ever higher levels of reflexion like an acrobat mounting to the top of the circus tent, means from a linguistic viewpoint (which is the really accessible stage) nothing but the possibility of constructing a meta-language for every language.

Similarly, we also learn that one may always say: "It is true that I see a dog", "It is true that it is true that I see a dog", etc. as soon as we see a dog. If one wanted to transform this into mental phraseology, for instance, by introducing a certain "ability to judge truthfulness", one would be delving into unfathomable methaphysical depths!

8. LANGUAGE AND COMMUNICABILITY

According to canonical phraseology, we are able to communicate everything we are conscious of. This ability is the very essence of language. However, a second person asking questions and listening to the answers is not essential. One can speak to oneself, and does not do so differently from when people are around. Furthermore, loudness is irrelevant, since silent "internal" speaking (when we are alone) is enough. If speaking is understood as any kind of speech, including speaking to oneself in silence, then everything that can be said about the communicability of contents of consciousness, when asked, turns into the ability to repeat aloud what one has silently been saying to oneself. And the ability to call to mind, when asked, something one knows of but has not been thinking of at this moment, turns into the ability to answer a question silently or aloud.

It sometimes happens that I cannot think of a certain word, no matter how hard I search my memory for it. What actually happens during this search we do not know, i.e. we are not conscious of it and cannot say anything about it. Suddenly I remember the word. What happens at this particular moment? All that happens is that I *say* the word!

Would it make sense to suggest that I become conscious of the word first and then I put it in words? "Becoming conscious" of the word is only the way philosophical jargon expresses a phenomenon that is called "saying out loud" or "putting in words" in another jargon. But is it really necessary to assume two different fields of phenomena, simply because two kinds of jargon exist? Entia non sunt multiplicanda . . .

9. THE IDENTITY OF CONSCIOUSNESS AND LANGUAGE

These considerations lead to the conclusion that all canonical phrases concerning "consciousness", as far as they are intelligible at all, are also applicable to inner speech. When applied to inner speech, the canonical phraseology is even more intelligible since language is a public, generally accessible phenomenon. The argument thus amounts to consciousness and inner speech being identical, i.e. what in philosophical jargon is called a process of consciousness is just the same as inner speech. Thus it is not necessary to look for another mysterious state of being which is indicated through mental expressions — this sphere of phenomena is simply language.

The thesis of identity is based on the principle of the identity of the indistinguishable. Whatever shows the same properties is indistinguishable, which in our case is that which is encompassed by the same canonical phraseology. There are two ways of arguing against such an identity thesis. One can either try to find some quality (characteristic) that distinguishes consciousness from inner speech, or alternatively one withdraws to an indisputable metaphysical position. The latter might be as follows: perhaps inner speech really shares its properties with consciousness. This is because inner speech is taken for the outer form of consciousness, and nothing else. The consciousness underlying inner speech is something completely different and defies all examination. It goes without saying that one cannot argue cogently against such a metaphysical position, as one cannot argue against metaphysics in general. Therein lies the strength as well as the weakness of all metaphysics. That is why metaphysical sentences are not disproved, but are brushed aside and forgotten.

Interestingly enough, the thesis of identity touches on the Cartesian position. For Descartes, language was the only certain criterion for another being apart from myself possessing a soul (= mind = consciousness) and not just being an automaton. Only man is in command of language; even the most stupid man is able to speak, but no animal, not even the brightest, is remotely able to speak, so Descartes maintained — and who would want to contradict him? "Haec enim loquela unicum est cogitationis in corpore latentis signum" (Descartes 1649).

But Descartes persists in his metaphysics: language is only an expression, a manifestation of thought, no more. Nevertheless it is only that which has a mind (consciousness, thought) that is able to speak. According to Descartes, *machines* will never really be able to speak (Descartes 1637, b. 5, 10—11), whereas deaf and dumb people will invent some other system of signs in order to express their thoughts. (Descartes 1646) Animals, however, do not lack vocal cords, larynx etc. . . . , but they lack consciousness.

From a logical point of view, the Cartesian position cannot be attacked unless one could construct a speaking machine — which does not really enter the reckoning at the moment. The question of the possibility of a machine having the (not just partial) ability to speak, would be a starting point from which Cartesian Dualism might be refuted.

10. APHASIACS

The philosophical jargon of consciousness is taught and learned in relation to normal people. But there are thousands of illnesses, derangements, and injuries that affect mental abilities. In particular, the manifold possible injuries to the faculty of speech (aphasia) could be brought forward as an argument against the thesis of identity. Someone suffering from aphasia may not, for instance, be able to name the objects he sees, but he can use them in a reasonable manner. The faculty of speech of someone who in no way seems to be an imbecile, is restricted to simple situations. Thus there are people who cannot speak well or cannot speak at all, but of whom one would hesitate to say that they are not conscious of things. Does this refute the thesis of identity?

The thesis of the identity of consciousness and language is not empirical (e.g. psychological) but meta-theoretical. It maintains that all our philosophical jargon of "consciousness" is regulated and channelled through our knowledge of language and that it is acquired only through talking about language. If we discount metaphysics, this implies the identity of language and consciousness.

Let us now apply the thesis of identity to our problem. The question: "Is the aphasiac conscious of the things he cannot say?" is not viewed as empirical but as a metatheoretical. The thesis of identity questions what might be *meant* by such a question. The *usual* meaning of this question has little to do with the word "consciousness" as used by canonical phraseology. Generally it means, what abilities are left in spite of the illness. It is *unusual* to ask seriously in philosophical jargon about the consciousness of an aphasiac. There seems to be some deep, dark mystery here, but the thesis of identity removes the mystery by asking what the question actually means. If the questioner is referring to canonical phraseology, where the properties of "consciousness" are pinned down, he is really asking about the aphasiac's *faculty of speech* — which, as we know, has been damaged.

In brief: no problem exists beyond the questions of how well or how badly the aphasiac is able to speak, what the disturbance is caused by, how it can be remedied, etc. The clinical psychologist has a complete grasp of the case, and there is no additional underlying mystery or unfathomable problem of consciousness.

11. THE SO CALLED "CLOSE RELATIONSHIP" BETWEEN THOUGHT AND SPEECH

I would like to point out that the identity thesis is nothing but the logical conclusion to a widespread and almost canonical view. That there is a "close relationship" between thought (consciousness) and speech has been referred to often enough. The identity thesis contains only an additional claim as to how this famous close relationship is to be considered.

Regarding this question, Heinrich von Kleist called a short essay of his rather provocatively: "On the gradual composing of thoughts during speech". He maintains "that many a great orator did not know what he was going to say at the moment when he opened his mouth", and he very vividly describes how this occurence can be seen in politics as well as in fables. (Kleist 1878, vol. 3, p. 453—459) "Knowing" here can only mean knowing consciously, and in the light of this more precise formulation Kleist's thesis clearly applies not only to great orators but also to everybody, no matter whether he speaks loudly, or softly, or just to himself. To make oneself (or perhaps one's listeners) conscious of something, i.e. exposing something, requires time and is only possible when done step by step, word by word, sentence by sentence.

Two great German thinkers who are not very congenial to one another have expressed strikingly similar views regarding this topic: F. Nietzsche and K. Marx. According to the latter: "language is as old as consciousness — language *is* real, practical consciousness that also exists for other people, and language rises initially, just like consciousness, from the necessity to communicate with other people." (Marx-Engles 1845, p. 20) Nietzsche writes: "Man like any living creature, thinks continually but does not know it; the part of thinking that we become *conscious* of is only the smallest part of it, let's say: the most superficial, worst part — since only this conscious thinking *happens in words, i.e. in signs of communication*, thus revealing the origin of consciousness. In short, the development of speech and the development of consciousness (*not* the development of reason, but only the becoming-conscious-of reason) go hand in hand." (Nietzsche 1882, No. 354)

The logical outcome of these ideas is the thesis of identity. And yet it seems that nobody has yet dared to say so.

The question of the consciousness of animals can also be answered much more clearly now. Fritz Mauthner must have met with general agreement when he conceived this question as a purely terminological one. It arises because our (human) "conscious memory purely consists of speech, our signs for remembering something are made of words; thus human consciousness is nothing but the speech everybody is capable of." (Mauthner 1923, vol. 1, p. 631) The question of animal consciousness should, according to Mauthner, be defined as follows: "Do we *want* to call the wordless memory of lower animals consciousness, and can we not detect other signs of memory in animals. For only speaking people have to equate their consciousness with speech. Other animals may and must have other signals. The self-awareness of an innkeeper is his tally." (*ibid.*)

In spite of the general agreement with this view, a slight correction cannot be helped. If canonical consciousness phraseology is to be retained (and we assume that it is) then the identity thesis is valid; and thus the statement "animals have no speech" is identical with the statement "animals have no consciousness".

It is interesting to note in this connexion the thesis held by Herder, Hamann and Humboldt regarding the "unity of speech and thought". This thesis is an attack on the popular dualistic conception of speech as a mere medium for naming ideas, whereas the ideas as such are independent of speech. On the contrary, there exists, according to the unity thesis, an inherent connexion between speech and thought. As one can read in Herder for example, "I cannot think the very first human thought, cannot make the very first rational judgement, without entering into a dialogue in my soul . . ." (Herder 1772, v. 1, p. 733)[3]

This is getting to the bottom of the identity thesis, though the keyword "unity" demands some terminological focusing. By "unity" one could, for instance, mean that "speech and thought are two sides of one and the same mental operation". (Hassler 1986) According to the identity thesis, however, there are not two sides, but two terms used for the same process.

12. THE RELATION OF IDENTITY THESIS TO OTHER THEORIES

The identity thesis refers to a statement about what people may mean when using the word "consciousness" in accordance with canonical phraseology. The identity thesis does not make any empirical statements about the origin of human speech, its prerequisites, structures, etc. This thesis remains neutral in relation to all theories about the concrete function of concrete languages. This applies, for example, to Fodor's theory about a "language of mind". (Fodor 1975)

It goes without saying that each particular language is not innate, but has to be learned, although there certainly have to be many highly complex, inborn pre-conditions for a man to be able to learn a language. But in this respect the identity thesis has nothing to say; it maintains only that "consciousness" and inner speech are the same. *This* inner speaking always happens in an ordinary, i.e. learned language. When I talk to myself, I do so in my mother tongue. I will then speak more or less carefully and extensively, i.e. I am more or less conscious of things.

Let me drop one final remark on the so called subconscious! If something is unpleasant, embarrasing or frightening, one does not want to talk about it. Sometimes one even manages not to say anything about such things when speaking internally. In this case one is not thinking about them. This is very hard, though, because "our thoughts are not in our power". But even when a thing has been totally driven out of our consciousness (our inner speech), it might still be taking effect – at least this is the psychoanalytic point of view. The theory of identity is not able to make contributions to psychoanalytic theories; its use is purely metatheoretical. According to the identity thesis, the "subconscious" is the same as the unspoken.

NOTES

[1] Operatio mentis maxime mira mihi ills videtur, cum cogito me cogitare, et inter cogitandum, hoc ipsum iam noto, quod de cogitatione mea cogitem, et paulo post mirio hanc triplicationem reflexionis . . . .

[2] simulac enim quis aliquid scit, eo ipso scit, se id scire, & simul scit, se scire, quod scit, & sic in infinitum . . . .

[3] Concerning the unity of speech and language, there is an excellent historic and systematic presentation in: Welke (1986).

BIBLIOGRAPHY

Brentano, F. (1874). *Psychologie vom empirischen Standpunkt*, Leipzig.
Delbruck, M. (1986). *Mind from Matter*, Palo Alto: Blackwell.
Descartes, R. (1637). *Discours de la Méthode.*
Descartes, R. (1646). *Letter to the Marquis of Newcastle*, Nov. 23rd, 1649 (Adam-Tannery 4, p. 575).
Descartes, R. (1649). *Letter to H. Morus*, Febr. 5th, 1649 (Adam-Tannery 5, p. 278).
Fodor, I. (1975). *The Language of Thought*, New York: Crowell.
Hassler, G. (1986). *Die These von der Sprachrelativität des Denkens in der Aufklärung und bei W. v. Humboldt*; in Welke (1986a), pp. 154—177.
Herder, J. A. (1772). "Abhandlung über den Ursprung der Sprache", in J. G. Herder, Werke in 10 Bänden, Deutscher Klassiker Verl., Frankf./M. 1985, vol 1.
Kirchmann, J. H. v. (1872). J. Locke, *Versuch über den menschlichen Verstand* (Translation of Locke's Essay), Berlin: Heimann.
Kleist, H. (1878). "Über die allmähliche Verfertigung der Gedanken beim Reden", in H. v. Kleist, Werke und Briefe in 4 Bänden, Aufbau-Verl. Berlin 1978.
Leibniz, G. W. (1). unpublished manuscript, quoted from Schneider (1985).
Leibniz, G. W. (1665). Nouveaux essais sur l'entendement par l'auteur du systeme de l'harmonie preestablie.
Leibniz, G. W. (1923). Sämtliche Schriften und Briefe, hg. v. d. Preussischen Akad. Wiss., Darmstadt 1923ff.
Locke, J. (1689). *Essay Concerning Human Understanding.*
Marx. K. and Engels, F. (1845). "Die deutsche Ideologie", in *MEGA*, vol. I 5 (written 1845, published 1932).
Mauthner, F. (1923). *Beiträge zu einer Kritik der Sprache*, 3rd edit. Leipzig 1923.
La Mettrie (1748). *L'homme machine.*
Nietzsche (1882). *Die fröhliche Wissenschaft.*
Popper, K. and Eccles, J. (1977). *The Self and Its Brain*, Springer Internat.
Reimarus, H. S. (1756). *Die Vernunftlehre, als eine Anweisung zum richtigen Gebrauche der Vernunft.*
Schneider, M. (1985). "Leibniz über Geist und Maschine", *Philos. Jahrb. der Görres-Gesellschaft* **92**, 335—352.
Spinoza, B. (1677). *Ethica ordine geometrico demonstrata.*
Stöhr, A. (1910). *Lehrbuch der Logik in psychologisierender Darstellung*, Leipzig/Wien: Deuticke.
Welke, K. (1986). "Zur philosophischen und sprachtheoretischen Begründung der Einheit von Sprache und Denken bei W. v. Humboldt", in Welke (1986a), pp. 9—67.
Welke, K. (1986a) (Edit.): *Sprache — Bewußtsein — Tätigkeit*, Berlin (DDR).
Wolff, Chr. (1751). *Vernünftige Gedanken von Gott, der Welt und der Seele des Menschen*, Frankf./M.

JOHN M. NICHOLAS

REALISM FOR SHOPKEEPERS; BEHAVIOURALIST NOTES ON CONSTRUCTIVE EMPIRICISM

1. STATISTICS FOR SHOPKEEPERS AND SCIENTIFIC CHOICE

A new phase arrives when we decide to apply these formulae and to enjoy the consequences of their properties. This phase is marked by an act of will (not reasoning) and, therefore, if it is desired to use the adjective "inductive" in relation to methods of estimation, it should be used in connection with the noun "behavior" rather than "reasoning". (Neyman, 1952, p. 210)

In his *Statistical Methods and Scientific Inference*, R. A. Fisher (1956) complained that (then) recent accounts of the rationale of tests of significance attempted to assimilate them to a quite different model from that intended by the originators. The founders of the test, actively committed to research in the natural sciences and attempting to employ the tests to improve their comprehension of experimental material, had as their goal "improved scientific understanding". The new rationale for significance tests was that of the "acceptance procedure", involving sampling of materials from industrial consignments, with a view to reducing the rate of acceptance — in a physical and commercial sense — of faulty goods, in an economically efficient way without unduly rejecting satisfactory ones.

Fisher proposed that there are fundamental differences between the contexts to which the different models apply, as, for example, in the character of the populations from which sampling is to be carried out. While he pointed up the differences, it was "not at all in disdain of an artifice of proven value, in commerce and technology", but he took the opportunity to take a swipe at the authors of the new rationale (Neyman and Wald) in noting that they would not have overlooked the relevant differences, had they "had any real familiarity with work in the natural sciences, or consciousness of those features of an observational record which permit an improved scientific understanding, such as are particularly in view in the design of experiments".

Fisher's observations belong in a long tradition of the demarcation of practical or applied reasoning from purely scientific inference and understanding. Despite Fisher's attempt to drive a wedge between pure

J. R. Brown and J. Mittelstrass (eds.), An Intimate Relation, 459—476.

and applied methodologies, an important question mark still hangs over the view that they, as forms of human deliberation, are indeed fundamentally distinct.

Decision procedures concern themselves with choices between acts, informed by the weight of statistical evidence relevant to states of nature, and guided by values associated with consequences of acts in given states, and statistical tests appear to be aimed at establishing hypotheses on the basis of experimental data. For this reason, statistical tests tempt us to view them as procedures of inference and to class them as instantiations of "a logical process of the kind we call inductive, in that [we are] attempting to draw inferences from the particular to the general; or, as we more usually say in statistics, from the sample to the population" (Fisher 1935, p. 39).

The contrast has been set out rather differently from the other side, as it were, by Neyman and Leonard Savage. Neyman, in discussing estimation by confidence intervals (and denying that we should "believe" that a true value is between the confidence limits, insisted that

> ... the word "conclude" has been wrongly used in that part of statistical literature dealing with what has been termed "inductive reasoning". Moreover, the expression "inductive reasoning" itself seems to involve a contradictory adjective. The word "reasoning" generally seems to denote the mental processes leading to knowledge. As such, it can only be deductive. Therefore the description "inductive" seems to exclude both "reasoning" and also its final step, the "conclusion". If we wish to use the word "inductive" to describe the results of statistical inquiries, then we should apply it to "behaviour" and not to "reasoning". (Neyman 1952, p. 235.)

Savage contrasted Behavioralistic and Verbalistic outlooks in statistics, and traced the latter, a methodology aimed at yielding assertions, to "an ancient tradition in epistemology that deductive and inductive inference are closely analogous processes". Savage, opposing himself to a traditional assimilation of induction to something like deductive implication, proposed that:

> whatever an assertion maybe, it is an act, and deciding what to assert is an instance of deciding how to act. (Savage 1972, p. 160)

Consequently, the Verbalistic approach would be improved if it were to be assimilated to the more general Behavioristic outlook, and represent deliberation in terms of choices of acts, given states of the world and "income" associated with consequences. Indeed one of the vices of the Verbalistic approach had been that "emphasis on assertion distracts from the indispensable concept of consequences" (Savage 1972, p. 160).

However, even the proponents of decision theoretic approaches have seemed to resist the idea that their methodologies and procedures of choice might characterize deliberation in basic or pure research, where the goal is proposed to be the disinterested pursuit of the truth. Rudolf Carnap, a proponent of decision-theoretic "inductive methodologies", nicely epitomized the view that action is one thing, thought another, in his qualified rejection of inductive acceptance rules. The qualification was that it was permissible for thinkers to employ acceptance rules when nothing of consequence hung upon the acceptance of the theory, rather than "where the utilities of the possible outcomes of decisions are relevant" (Carnap, 1978, p. 149). As Savage observed "it has been seen as an objection to the behavioralist outlook that the consequences of some assertions, particularly those of pure science, are extremely subtle and difficult to appraise" (Savage 1972, p. 161). But subtlety and difficulty are one thing: the absence of values — nothing hanging on the acceptance! — is another. We cannot deny that there are acts in question, and their consequences are not to be ignored, for all their subtlety. But like Carnap, Savage toyed with trying to purify the scientific context of value laden consequences. For example, he proposed that no scientist would publish an assertion such as "the velocity of light is between 2.99×10^{10} and 3.01×10^{10} centimeters per second". If he tried, he would not get accepted. He would render his data and experimental design, albeit in an economical way "to put his reader rapidly in position to *act* as though the reader himself had made the observations" (Savage 1972, p. 161). The only consequences (and utilities) involved would be those of the consumer — the reader. Savage's remarks seem overly sanguine. Estimates of magnitudes like the speed of light do in fact appear in handbooks and textbooks unadorned by accounts of data, methods, and sometimes without a corresponding interval of estimation. Looking more widely across a range of both qualitative and quantitative fields, we find much information rendered categorically without evidential sources and without disclaimers in proportion to uncertainty. Scanning encyclopedias, handbooks and textbooks quickly reveal marked deviations from Savage's paradigm. Savage's reply, an interesting excursion into socially constrained professional responsibilities, is that "the seriousness of the consequences of their assertions is limited because of the possibility of referring to original research publications, a possibility serious textbooks and handbooks facilitate by the inclusion of bibliographies" (Savage 1972, p. 161).

Even if scientific reports could be purified of their risk of having bad consequences, we might reasonably wonder whether that would leave the scientist in Carnap's value free situation where one might freely accept hypotheses. It seems not. If nothing were to hang on acceptance decisions, there would be no limit to what could be accepted: no costs, no constraints. It is preposterous to think, however, that scientists are free to accept what they will, and this testifies to the fact that something does hang on the acceptance decisions. The grapevine and serious academic study of the motivations of basic scientists and their "reward structure" indicate that there are powerful values at work, and, among other things, they profoundly constrain what can be asserted or communicated. It even matters where and in what form the communication appears, as Savage himself noted in contrasting original journal publication with handbook reviews.

Programmatically, Behaviouralism takes the view that the actions of the pure as well as the applied scientists are assimilable to the decision-theoretic model. Where scientific action is not immediately practical — build the cyclotron! — this is because scientific work revolves around the communication and evaluation of information and those that supply it. Scientific communities function as gigantic systems of gossip, in which payoffs are associated with the exchange of valuable information. Obviously this suggests a rather different picture of inductive inference from traditional ones, which have modelled themselves, *mutatis mutandis*, on (a certain understanding of) deductive "inference". Within this approach, the speech act, the "book act" (as Imre Lakatos called it (Levi 1976, p. 1)), the *samizdat* act, the Society's Presidential address act, and so on, become the *termini ad quem* of the canonical deliberation, each differentiated in its idiosyncratic circumstances by the choices made available and the losses and gain attached to its consequences.[1]

There is a division of labour in the Inductive Behaviouralist approach between the theory support side and the action side. Part of the burden of the traditional epistemologies is shouldered by the partial belief function which expresses uncertainty about the possible states of the world relevant to the consequences of one's acts. Part of the rest is borne by *acts* (of reporting, communicating, professing and so on). The Behaviouralist's ultimate output is an action, rather than a proposition.

2. CONSTRUCTIVE EMPIRICISM

Van Fraassen opposes the AntiRealism of his Constructive Empiricism to the core principle of a Realist view as follows:

> *Science aims to give us, in its theories, a literally true story of what the world is like; and acceptance of a scientific theory involves the belief that it is true.* This is the correct statement of scientific realism (van Fraassen 1980, p. 8).

Constructive Empiricism may be defined in terms of the same factors — the goal of scientific enquiry and what it means for a scientist to accept a scientific theory:

> *Science aims to give us theories which are empirically adequate; and acceptance of a theory involves as belief only that it is empirically adequate.* This is the statement of the anti-realist position I advocate; I shall call it *constructive empiricism* (van Fraassen 1980, p. 12).

The Constructive Empiricist asserts that theories are not to be believed in so far as they describe what is not humanly observable, but may be believed in so far as they model what is observable. We may reasonably believe the subtheories of theories which give us the structure of experiments and other observable events, even including those events so far away in space, time, or opportunity that they are never in fact observed. It is not, or should not be, any part of our scientific aspirations to believe the parts of theories which mirror only what is in principle beyond our powers of observation.

The Empiricist denies that acceptance is belief. It does have an element of belief — the belief that a theory is empirically adequate. It also has important elements of active, practical commitment undertaken by the scientist. These are (i) a commitment to maintain the theory as part of the body of science, (ii) a commitment to confront the theory with any relevant phenomena, (iii) a wager by the scientist that relevant phenomena can be assimilated to the theory, and (iv) an obligation to serve in the role of an explainer, employing the theory as a basis for answering highly context sensitive demands for explanatory information. Thus van Fraassen locates his Constructive Empiricist scientist in a network of commitments and contracts — no doubt with genuine "income" — gains or losses to enforce them, reminiscent of the kind of Behaviouralist picture outlined in the earlier section.

In addition, van Fraassen allows that the scientist who accepts a theory "immerses" himself in the language of the theory, and makes assertions as if the accepted theory were indeed taken to be true. Thus he writes:

> In a certain historical context in which a certain scientific theory has been accepted, a certain form of language is naturally adopted. We do not say, the events of spontaneous decay of radium atoms are represented by such and such elements of this model, and there is a probability function defined on those elements. We say instead: the probability of spontaneous decay of a radium atom is thus and so" (van Fraassen 1980, p. 198).[2]

Given the division of labour between belief and acts within the Inductive Behaviourist framework, the question arises whether the Empiricist's opposition between belief and acceptance is not more productively understood and realigned that way. The Behaviouralist recognizes from the beginning that believing theories is one thing and acting on them, in doing or saying things, is another.

Beyond the practical commitments subsumed under the rubric of *acceptance*, van Fraassen has given a coherence argument for a striking element of engagement or commitment in the regimentation of partial belief and the embedding of the appraisal of judgments in a framework of value laden obligation and performance. I shall now turn to review his argument and an alternative one.

3. THE WILL AND BELIEVING

> At other times, they stress economic fundamentals and thereby drive a wedge between the spot rate and the forward rate. Some people are persistently fundamentalist; others may be dedicated random walkers. But most investors switch from one way of thinking to the other according to the circumstances. (The Economist, April 9, 1988, p. 59)

van Fraassen (1984) has drawn attention to an instructive application of a Dutch Book strategy where the condition for revision of belief is a new degree of belief itself. That is to say, an agent who does not take certain steps to protect himself (which steps precisely we shall examine below) in having degrees of belief conditional upon his future beliefs falls liable to a Dutch Strategy similar to that invoked in David Lewis' Dutch Book argument for diachronic coherence, that is, for a certain interpretation of Bayes' rule for updating beliefs in the face of new evidence. van Fraassen's argument is that liability to this Dutch Strategy points to a surprising principle, the Reflection Principle, not itself part of probability theory as conventionally understood, which when properly interpreted has profound implications for our understanding of conditionalization, and even the social regimentation of belief. Further, his interpretation of this Principle provides one of the major points of contact between the kind of Inductive Behaviourism which I have described above in Section 1 and the epistemology which van Fraassen has recently sketched for Constructive Empiricism. Before considering his argument, let us sketch Lewis' form of the Dutch Book argument for conditionalization following the exposition of Paul Teller (1976).

The aim of the argument is to make clear the reasonableness of an agent who has a belief function, P, and considers revising it, to P', in the light of possible future experiences, E_i, by a rule or policy such that $p'(S) = p(S/E_i)$, for each E_i from the partition $\{E_i\}$. $p(S/E_i)$ is a current or prior conditional probability, and in the standard representation is shorthand for the quotient, $p(S\&E_i)/p(E_i)$. The probabilities are measures of belief understood as a disposition to bet, such that if the agent has probability p for event or proposition A, then he will be willing to trade (buy or sell) the bet [win \$1, if A; win 0, otherwise] for \$$p$.

Let the agent declare his betting rates on the joint event $A\&E_1$, and the event E_1 (and by assumed synchronic coherence, on $-E_1$), and also declare p' (A) in case of the possible future condition, E_1. If p' (A) does not equal $p(A/E_1)$, then the agent will be disposed to consider fair each of a combination of bets which together will lead to an inevitable loss, regardless of the outcomes of the bets. If he wins, he loses; if he loses, he loses.

The bets, the stakes of which and the buying or selling of which are dictated by the Bookie, are as follows:

$$\text{[win \$1, if } A\&E_1\text{; win 0, otherwise]} \quad \text{(I)}$$

$$\text{[win } p(A/E_1)\text{, if } -E_1\text{; win 0, otherwise]} \quad \text{(II)}$$

$$\text{[win } \{p(A/E_1) - p'(A)\}\text{, if } E_1\text{; win 0, otherwise].} \quad \text{(III)}$$

If the agent has $p'(A) < p(A/E_1)$, the Bookie sells the bets to the agent, and the agent is assumed to be willing to pay as his degrees of belief dictate, a sum, $p(A\&E_1) + p(-E_1)\, p(A/E_1) + p(E_1)\, \{p(A/E_1) - p'(A)\}$, for the combination of the bets.

If $-E_1$ comes to pass, then bets I and III are lost, and $p(A/E_1)$ is grossed from bet II. If E_1 shows, then the agent collects $\{p(A/E_1) - p'(A)\}$ from bet III, and because $p'(A)$ is now the declared rate on A, the Bookie can now buy back bet I on A for a price of $p'(A)$. Whether $-E_1$ or E_1 turns up, the net loss to the agent will be $p(E_1)\ \{p(A/E_1) - p'(A)\}$. (If $p'(A) > p(A/E_1)$, then the Bookie buys the bets, and if E_1 appears sells bet I — to the same effect: inevitable loss for the agent.)

Setting $p'(A) = p(A/E_1)$ spikes the combination, and conditionalization may therefore be thought of as a device in the planning of beliefs, for avoiding the absurd irrationality of otherwise making loss inevitable.

Consider the following, numerical illustration (anticipating van Fraassen's own example). Let A and E_1 be:

> A: Table Hands will not win at Hollywood Park tomorrow
> E_1: The going is rough at the time of the race.

Let the agent offer rates as follows: $p(A\&E_1) = 0.3$, $p(E_1) = 0.4$ (hence, $p(A/E_1) = 0.75$), and $p'(A) = 2/3$. Then the Dutch Bookie will sell the gambles

$$[\text{win } \$1, \text{if } (A\&E_1); \text{win } 0, \text{otherwise}], \text{price } 0.3 \qquad \text{(IV)}$$

$$[\text{win } 0.75, \text{if } -E_1; \text{win } 0, \text{otherwise}], \text{price } 0.45 \qquad \text{(V)}$$

$$[\text{win } 1/12, \text{if } E; \text{win } 0, \text{otherwise}], \text{price } 1/30. \qquad \text{(VI)}$$

The price of the combination is $0.75 + 1/30$. If the going is not rough $(-E_1)$, then the agent grosses 0.75, and thus nets a loss of 1/30. If the going is rough (E_1), then the Bookie buys back the bet on A for 2/3. The agent grosses $2/3 + 1/12 = 0.75$, and so, again, nets a loss of 1/30. It could have been avoided: one prudent choice would have been to set $p'(A) = 0.75$, as conditionalization dictates.

Let us, after this illustration of the paradigmatic argument of Lewis, proceed to van Fraassen's original version of the Table Hands bets. In the above version, E_1 was taken to be some information about the condition of the track, "the going is rough", but in van Fraassen's example, E_1 is taken to be the agent's having a certain degree of belief in A *immediately prior* to the race, say, $p'(\text{A}) = 2/3$.

There is an important difference between the example of the Lewis paradigm and van Fraassen's case. In the former, the system of bets operates on the possibility of a disparity between the conditional probability, $p(A/E_1)$, and the rate, $p'(A)$, intended should E_1 come about. To employ slightly different terminology, we might say that the

Lewis argument attends to the divergence between 'spot' prices and (conditional) 'future' prices for bets on A, both explicitly stated simultaneously on the day before the race. The very design of the Bookie's assignments of stakes and the direction of betting is dependent on the declaration of the policy of the agent which sets tomorrow's price. *In van Fraassen's example, consideration of any policy for $p'(A)$ is set aside; the matter is out of the hands of the agent*!

For if it should simply turn out that the agent's new degree of belief on the day of the race is other than 2/3, that is $-E_1$, then he will net a loss of 1/30 (for exactly the same arithmetical reasons as in the example of the Lewis paradigm above!). Alternatively, if $p'(A)$ does turn out to be 2/3, then he will for that very reason happily surrender the bet D to the Bookie for 2/3, and again net a loss of 1/30.

van Fraassen sets aside reservations about the alleged unintelligibility of agents having degrees of belief about their degrees of belief, and proposes that what we have learned from his example and the general liabilities to Dutch strategies which come from cases where we bet conditionally upon our future beliefs is that coherence leads us to a new requirement of rationality, the Principle of Reflection:

$$p(A/p'(A) = r) = r.$$

If this principle is indeed a rationality principle, dictated by coherence, then some unfamiliar consequence follow. Consider, for example, the instance where $r = 1.0$. Reflection dictates that $p(A/p'(A) = 1.0) = 1.0$ and derivatively that $p(-A/p'(A) = 1.0) = 0$, so that $p(-A\&[p'(A) = 1.0]) = 0$, assuming that $p(p'(A) = 1.0) \neq 0$. Reflection requires that we cannot currently give any credence to the denial of an hypothesis (say Darwin's theory) on the assumption that we come fully to believe it, or more particularly, we cannot give any credence to the possibility of our coming fully to believe some false theory! Traditional epistemologies seem to require that we should not unreasonably be able to contemplate making future errors where we are fully confident of theories which are in fact false.

van Fraassen proposes an interpretation which reconciles Reflection with an account of epistemic judgments, which bears a strong resemblance to the Inductive Behaviourist picture sketched earlier. He writes:

if I express my opinion, I invite the world to rely on my integrity and to infer from this what advice to myself and anyone else in like circumstances, concerning the carrying of umbrellas, purchase of insurance policies, entering wagers, I would presently consider the best . . .

. . . my integrity, qua judging agent, requires that, if I am presently asked to express my opinion about whether A will come true, on the supposition that I will think it likely tomorrow morning, I must stand by my own cognitive *engagement* as much as I must stand by my own expressions of commitment of any sort. . . . I can no more say that I regard A as unlikely on the supposition that tomorrow morning I shall express my high expectation of A, than I can today make the same statement on the supposition that tomorrow morning I shall promise to bring it about that A. To do so would mean that I am now less than fully committed (a) to giving due regard to the felicity conditions for this act, or (b) to standing by the committments I shall overtly enter (van Fraassen 1984, p. 255).[3]

van Fraassen eschews the idea that epistemic judgments are reports of our inner mental states. They are rather statements of intentions and commitments, having somewhat of a performative character even if they are not exactly promises.

Despite the family resemblance of the above interpretation of Reflection to Inductive Behaviourist methodologies, which emphasise acts constrained by losses and gains attached to outcomes, and associate acts of communication with a general framework of authority and responsibility (such that all utterances associated with non-trivial values for consequences have a performative character), the Principle of Reflection is not the unique base for motivating that conception. Indeed, I think that Reflection is not uniquely demanded by coherence, but is merely one of a number of solutions, some of which highlight the extent to which the Dutch Book arguments as traditionally conceived push us towards the same framework of ideas. Let us begin by examining a rather clumsy solution to the Table Hands example of van Fraassen, where the condition is the future degree of belief rather than the comparatively innocuous information about the condition of the track.

(a) *The Clumsy Solution*

Suppose that on the day of the race, the agent meets the Bookie who offers, as part of his scheming plan, to buy back the bet on A, if indeed $p'(A) = 2/3$. Suppose that the degree of belief of the agent is indeed 2/3. The Bookie is ready to pounce, but now there is a surprise: the agent demands at least 21/30 for for the gamble. "I'll have to think", says the Bookie. The Bookie may want the bet (perhaps he has information that Table Hands is in poor shape or performs poorly in the rough), and if he does pay 21/30, then the agent will be back at his *status quo* from the day before when no bets had yet been contracted. On the other hand, the price may be too steep for him, and he declines. The agent is now left with a bet against Table Hands winning, and must wait for the race. If Table Hands wins, then the agent is down by 0.75 overall, but if the horse fails, then he grosses 1 from bet IV, and nets 0.3. There is no inevitable loss, and the Dutch Book is sidestepped.

The Bookie is bound to ask, however, how the agent could demand a price for gamble IV which was higher than what was dictated by his expectation value of 2/3. The proper reply, I think, is this: the Bookie designed his combination of bets on the assumption that our (Bayesian) agent traded bets at their subjective expectation, *regardless of what combinations he was entering into*. That is to say, each bet was to be evaluated as fair *on an individual basis*, without attention to the effects that simultaneously taking on other bets would have. The Bookie (and van Fraassen) assume that the agent is what I shall call a *Bayesian Blockhead*. Real Bayesians are not Bayesian Blockheads; they do not look at bets piecemeal, and take on any combinations of bets which individually seem fair (unless they have uniformly Risk Neutral utility functions).

Consider an agent who is risk averse over money in a suitable range. He may, for example, be willing to trade a bet on a dollar stake for winning (status quo for losing) for \$0.40, but firmly resist trading a bet on \$100 for \$40. And yet the latter bet is simply the first bet made forty times. A Bayesian Blockhead, who is suitably risk averse, say, and took bets piecemeal, myopically examining each for its fairness without examining the combination, would be liable to a Dutch Book, by buying small bets expensively, and, at the same time, cheaply selling a bet equivalent to the combination of the small ones.

The Clumsy Solution simply has the agent not satisfy the conditions laid down by van Fraassen, and to that extent it does not solve van Fraassen's own difficulty. It does, it seems to me, suggest that van Fraassen's is not the right problem in any case, because if we are to learn from the prudence of Bayesian deliberation what difficulties there are for traditional epistemology, we should not set out with a view of the agent as procedurally impaired from the outset.

With the Clumsy Solution the agent has given up any concern for fixing his prices on gambles in accord with his degrees of belief. The price, 21/30, which he asks for the gamble on *A*, when his belief in *A* is 2/3 on the day of the race, is not a betting rate. For example, there is no supposition that (1—21/30) is the price of a bet on $-A$. The configuration of bets and his (perhaps foolishly and inconsiderately) having entered into contract on bets *A, B*, and *C*, on the day prior to the race constrain him to give up betting on the day of the race.

One more observation seems appropriate, and provides a suitable point of departure for looking at less vulgar solutions. A variant of the Clumsy Solution, which is suitable for weak-willed Bayesians who can't help being Blockheads when they see a lottery ticket, is the "Talk to My Lawyer" solution. Worried at whether his resolve to demand 21/30 on the next day will hold out in the heat of the moment, he leaves town, giving his lawyer strict instructions that gamble IV is not to be parted with at a price less than that amount. Effectively, the difference between the Clumsy Solution and the Lawyer Solution is the difference between policies, charmingly named by Savage, "You can cross that bridge when you come to it" and "Look before you leap" (Savage 1972, pp. 16—17). In the former, the agent decides on the day of the race, how to dispose of gamble IV; in the latter, he reviews the entire sequence of his intended actions and their consequences, and locks himself into a policy for future choices *right at the outset*. This suggests that we might look for less vulgar solutions which involve looking before leaping. There is one which fits the bill and brings us back to the strategy in Lewis' paradigmatic coherence arguments.

(b) *The Natural Solution*

At first sight, van Fraassen's example and the paradigmatic example seem to differ because the conditioning event, E_1, seems to make the choice of $p'(A)$ irrelevant in the former case. The outcome of the Dutch Strategy appears to be out of the hands of the agent. But this need not be the case, if the rate at which the agent is willing to trade bet IV on the morrow can be selected in advance, by policy. Following the strategy on which the paradigm Dutch Book arguments operate, the agent takes steps at the time of contracting the initial set of bets, to fix the rate on *A*. $p'(A)$ is fixed from the outset, by choice.

To what value should it be set? The Natural Solution is to set $p'(A)$ at whatever is the current or prior belief in *A*, $p(A)$. Let us see what follows from this. First, we should note that all uncertainty about tomorrow's rate vanishes today. We know, before entering into any bets with the Bookie, exactly what the future rate will be, and this will be reflected in betting rates at the start of betting as follows:

$$p(p'(A) = r) = 0, \qquad r \neq t = p\,(A)$$

$$p(p'(A) = t) = 1, \qquad t = p\,(A).$$

It follows from these commitments that the agent will "have no position" on bets conditional on his having degrees of belief in A which deviate from $p(A)$, since $p(A/p'(A) = r)$ will not be well defined for all r distinct from the value of $p\,(A)$. In that sense, the Natural Solution is reminiscent of a solution, rejected by van Fraassen, which denies that agents reasonably have degrees of belief about their degrees of belief. However, here there is no blanket prohibition of higher order beliefs, only the commitment to a *particular* set of such beliefs which undermine the forming of *particular* conditional probabilities where the condition is the future rate's deviating from the present rate. One conditional probability will be well defined, and it automatically satisfies the Principle of Reflection:

$$p\,(A/p'(A) = t) = t, \quad \text{where} \quad t = p(A).$$

Here the Principle of Reflection seems to invite a different interpretation from the one van Fraassen offers for his own Principle. van Fraassen's Principle is grounded by him as part of the cognitive engagement which would be entailed in the future event of his taking on a certain new disposition to bet, motivated by concern for the social fabric of appraisal of integrity and commitment of judging agents. The Principle which follows from the Natural Solution is rather simpler. It reflects the fact that the agent's mind is made up at the outset, and that there can be no question about what tomorrow's rate will in fact be. There are indeed issues of commitment, and acts of will underlying the Principle, but they are rather more straightforward. There is commitment simply because agent and Bookie are entering into a betting contract, and there is will because the rate for the day of the race is chosen voluntarily in advance, just as in the conventional or paradigmatic diachronic Dutch Book arguments.

The Natural Solution is "natural" for the following reason: it reflects an epistemological judgment that where one is contemplating updating one's probabilities (betting rates) in the light of possible new information, then we should only do so on the basis of evidence which is of material relevance to the issue. In particular, if the only condition on which the agent is considering revising the odds on a hypothesis is the revised odds themselves, then the revision should be the trivial or vacuous one of *no* revision. Other things being equal, the beliefs which we profess should be stable.

Dutch Book arguments clearly illustrate some of the fundamental ties between rationality and the statistics of shopkeepers because rational deliberation is shaped by the logic of prices. To the extent that those arguments provide part of the grounding for prudence in the sciences, they vindicate the Behaviouralist's judgment that even pure science is applied science, so to speak. The policy choice of the Lewis paradigm argument is foremost a marketing decision.

This is plain in the paradigmatic cases where the issue is the policy for regimenting the relation of currently declared future prices (contingent upon some condition) to current prices. $p'(A)$ is not operative as one's future belief; it is one's currently stated future price. That it is so seems to me to be dictated not only by the superficial appearance of the gambling apparatus, payoffs, fees, contracts and other impedimenta of the enforcement of the book or system at best, but by a seemingly paradoxical irrelevance of our true beliefs to our strategic choices.

Either we can choose our beliefs or we cannot. If we can choose our beliefs (in our hearts), either at will or by taking degree of belief pills (time release for p'; quick acting for p), then we can bring them into line with the dictates of pricing prudence. In this case the Lewis paradigms may be thought of as setting up future *rationalizations*: current circumstances and prices dictate future (conditional) prices and, with our assumed ability to do so, we take steps to ensure that our tomorrow's degrees of belief in the heart line up with tomorrow's prices in our "shop window". If we can choose beliefs, beliefs must follow prices. Sincerity — betting felicitously to our degrees of belief — comes because we have rationalized our actions by adjusting the beliefs.

Alternatively, if we cannot choose our beliefs, then there will be circumstances when we must ditch our degrees of belief as the arbiters of future betting rates. We must bet infelicitously and merely strategically, if we must bet at all. To say otherwise is, I think, tacitly to introduce a covert payoff to sincerity or felicitousness in a model in which all the payoffs are supposed to be associated with the gambles entertained.

What this puts into relief is the centrality of choice, a slight relocation of van Fraassen's voluntarism, in the operation of coherence constraints. van Fraassen represents Reflection as uniquely dictated by coherence constraints related to future beliefs. I think it is not. If future beliefs can be chosen, then setting, or committing one's self to setting one's belief on A equal to current belief/betting rate on A, will spike the Dutch book. The agent may, of course, choose not to choose, and then van Fraassen's Dutch Strategy will point to the desirability of having already in place current partial beliefs which satisfy the non-trivial Principle of Reflection. It is neither a unique solution, nor, I think, the best for most circumstances. If agents cannot change their beliefs, then the poor souls who have current beliefs which do not satisfy Reflection, and are also condemned (by some Author of Their Being) to be Bayesian Blockheads who go by expectations and one bet at a time, are committed to an eternity of lack of rationality (Grace?). That is another kind of Augustinianism distinct from the Voluntarism which van Fraassen proffers us. It is Augustine's Calvinist streak.

4. CONCLUDING REMARKS

(a) *Stability and Leaps of Faith*

van Fraassen's epistemological aim is to keep probability theory as a logic of epistemic judgment, but preserve central elements of the traditional epistemologies. One important element of the traditional epistemologies is to answer in the positive the question of whether "we can rationally come to believe a proposition that is entailed neither by those which we have believed heretofore nor by our previous opinions conjoined to the evidence before us". van Fraassen appears therefore to defend the notion of an inductive inference rule as conceived in traditional epistemologies, by which we may reasonably "proceed" from given beliefs to other beliefs which are not deducible from them or at least he desires enough slack in inductive inferences not to bar our arriving at full belief in propositions which are not deducible from our given data and other beliefs.

van Fraassen proposes that his Reflection Principle shows how there might be room for going beyond what is deductively warranted. Under the protection of the Principle, we are defended from liability to Dutch Strategies, but are not constrained to conditionalization. We may make "leaps of faith", and in so doing we may reconcile the rationality of coherence with "independence and enterprise" (van Fraassen 1984, p. 256) in scientific development. No doubt van Fraassen's Principle licences otherwise unmotivated swings in partial belief without the penalties of incoherence. But it is not immediately obvious that it gives us a picture of what traditional epistemologies thought of as "ampliative" inference where:

> general theories, such as Darwin's, Einstein's, or Bohr's, cannot be established on the basis of the evidence, but we may rationally come to believe that they are true (van Fraassen 1984, pp. 235—236)

I suspect that it does not, because the intent of the traditional models was not to see fallible (non-deductive) inferences as arbitrary jumps protected by a device such as Reflection. Rather, they intended something like (the ill-fated) inductive acceptance rules in which essential appeal is made to non-trivial evidence in support of the theory to be accepted.

There is another option, and it is one offered by the Behaviouralist. The semblance of "inductive detachment" comes from the appropriateness under many circumstances of acts of categorical assertion as outputs of a process of deliberation which does not require that some hypothesis be given full (partial) belief. The traditional "leap", seen this way, is a leap from chalk to cheese — that is, from partial *belief* to act of *assertion*, not as van Fraassen would seem to render it from partial belief to full (partial) belief.[4]

(b) *Realism for Shopkeepers*[5]

In the title of this paper, I have taken the liberty of representing the Behaviouralist as a scientific realist of sorts, despite his pragmatic stripe, his emphasis on the concrete acts (almost at the expense of belief as an inner mental condition), and its commercial emphasis on losses and gains. Traditionally such pragmatic approaches have been thought of as Antirealist.

Why I do so is because it seems to be a concession to the spirit of realist aspirations to give *any* (even partial) credence to theories about unobservables. Even if current and historical theories are improbable, there need be no reason of principle which bars low credences from climbing (with evidence, rather than artificial "leaps of faith") above evens or closer to unity than not. With its fundamental demarcation of acts from (partial) beliefs, there is no paradox to seeming to *act* as if a theory were true despite having belief in it well short of full belief. Even if a theory were a relatively long shot, in comparison with its contradictory, it might be many orders of magnitude more probable than its nearest rival contraries (Jeffrey 1975). In the right circumstances, with the appropriate loss functions in place, "immersion" in and categorical assertion of the language of the theory might be quite rational *acts*, whose appearance of firmness did not require an underlying full belief.

Like the Constructive Empiricist, the Behaviouralist insists that Acceptance is not Belief. That is because Acceptance stands for doing, and Belief for believing or being uncertain about. In the decision theoretic apparatus, their roles are quite methodologically distinct. In the Behaviouralist's division of decision-theoretic labour, there is perhaps one fundamental disagreement with the Empiricist. Belief belongs with belief, and not with acceptance. Partial belief, whether in theories or in their mere empirical adequacy is merely one of the inputs to decision making, and acts including assertions are outputs. Hence, the Behaviouralist divorces belief of any kind from what gets lumped under Empiricist's notion of acceptance, like wagers on scientific claims, commitments to research programs and so on.

The University of Western Ontario

NOTES

[1] In recent years, perhaps the most remarkable insight into scientific choice as a living, day by day balance of private aspiration and public judgment is June Goodfield's *An Imagined World* (Goodfield 1981), which epitomizes many of the academic insights of sociologists such as Merton, Hagstrom, Bourdieu, Latour, Woolgar, Barnes, Bloor, Shapin, Mulkay, Edge, Mullins, the Coles, Gaston and others. It is noteworthy that van Fraassen, the Constructive Empiricist, remarks of the "larger enterprise" of science that conflicting pressures within methodology have their resolution in practical terms "through a non-theoretical, quasi-political process" (van Fraassen 1983, p. 326).

[2] van Fraassen continues "We do not say that burning of copper at room temperature and pressure has no counterpart in any model of our physics; we say simply that it is impossible". Both the examples of such "immersion" in the language of the theory concern probability and possibility, both modal forms as van Fraassen would have it. I take it however that the literal and categorical assertion of parts of the accepted theory is not restricted to modal talk, and that stating a value for the wavelength of light under certain conditions is equally appropriate. The fact that an AntiRealist thinks it

appropriate to *speak* as if an accepted theory is true is of particular interest for an Inductive Behaviourist, who thinks that induction is the theory of rational acts of communication.

[3] van Fraassen meets certain of the demands of the traditional expectation that we should be able to contemplate our making future errors, by employing chances rather than degrees of belief for the deviation from rectitude (van Fraassen 1984, pp. 249–250).

[4] Behaviouralism also offers some interesting possibilities for understanding the how it is that some scientists can tenaciously but rationally cling to theories which the majority defect from (Priestley on phlogiston as oxygen theory becomes the paradigm?) or rashly but rationally leap to embrace theories to which the profession is not yet sympathetic (Einstein on photons during the hegemony of Maxwell's theory?). It is exactly because the kinematics or dynamics of partial belief does not exhaust scientific choice, and because scientists function in different contexts which face different choices, outcomes, and values, and so are led to different acts although sometimes their partial beliefs may be close or the same.

[5] "Statistics for Shopkeepers" has sometimes been used as an unkind characterization of statistical approaches of a decision theoretic stripe, and has been attributed to R. A. Fisher. I have adapted it here to reflect a viewpoint, embraced by Bruno de Finetti, that the engine which drives coherence is a logic of prices rather than a psycho-logic of ideal inner lives.

REFERENCES

Fisher, R. A. (1935). "The Logic of Scientific Inference", *J. Roy. Stat. Soc.* **98**, 39–54.

Goodfield, June (1981). *An Imagined World*, Harmondsworth: Penguin. Jeffrey, Richard C. (1975). "Critique of the Popper Program", *Synthese* **30**, 113.

Levi, Isaac (1976). "Acceptance Revisited", in Bogdan, Radu (ed.), *Local Induction*, Dordrecht: Reidel, pp. 1–71.

Neyman, Jerzy (1952). *Lectures and Conferences on Mathematical Statistics and Probability*, Second Edition, Washington: U.S. Department of Agriculture.

Savage, Leonard J. (1972). *The Foundations of Statistics*, Second Edition, New York: Dover.

Teller, Paul (1976). "Conditionalization, Observation, and Change of Preference", in Harper, W. A., and C. A. Hooker (eds.), *Foundations of Probability Theory, Statistical Inference, and Statistical Theories of Science*, Vol. I. Dordrecht: Reidel, pp. 205–259.

van Fraassen, Bas C. (1980). *The Scientific Image*, Oxford: Clarendon Press.

van Fraassen, Bas C. (1983). "Theory Confirmation: Tension and Conflict", in Weingartner, P. and M. Czermak (eds.), *Epistemology and Philosophy of Science: Proceedings of the Seventh International Wittgenstein Symposium*. Vienna: Holder-Pichler-Tempsky, pp. 319–329.

van Fraassen, Bas C. (1984). "Belief and the Will", *Journal of Philosophy* **81**, 235–256.

ADOLF GRÜNBAUM

WHY THEMATIC KINSHIPS BETWEEN EVENTS DO NOT ATTEST THEIR CAUSAL LINKAGE*

I. INTRODUCTION

There are strongly diverging diagnoses of the defects of psychoanalytic theory. In a paper on schizophrenia, the German philosopher and professional psychiatrist Karl Jaspers (1974, p. 91) wrote: "In Freud's work we are dealing in fact with [a] *psychology of meaning*, not *causal explanation* as Freud himself thinks". The father of psychoanalysis, we learn, fell into a "confusion of meaning connections with causal connections". After Jaspers, Paul Ricoeur has elaborated the patronizing claim that Freud basically misunderstood what he himself had wrought. As Ricoeur tells us in his book *Freud and Philosophy* (1970, p. 359), psychoanalytic theory is a hermeneutic enterprise, as opposed to a natural science: "psychoanalysis is an exegetical science dealing with the relationships of meaning between symptoms and repressed instinctual mentation".

The noun "hermeneutics" was originally introduced in the 17th century as a name for Biblical exegesis, and was then broadened to characterize textual interpretations quite generally. Ricoeur believes that he can replace the psychoanalytic sense in which *symptoms* are said to have "meaning" by the semantic sense in which *words* or texts have meaning. But in psychoanalytic theory, symptoms have socalled meaning in virtue of having unconscious causes, whereas semantic reference is a matter of designating or naming objects. Indeed, in my view, Ricoeur is just exploiting the ambiguity of the term "meaning" to shift from its causal sense in psychoanalysis to its linguistic sense. In short, he misassimilates the causal diagnosis of symptoms and dreams to the activity of textual exegesis. But compulsive handwashing, for example, is simply different from a piece of linguistic communication. And, as Freud (S.E. 1916, 15: 31) told us, "A dream does not want to say anything to anyone. It is not a vehicle of communication." Thus, Ricoeur is mistaken, I claim, in believing that interpreting symptoms is logically on a par with interpreting texts. It is true but unavailing to

J. R. Brown and J. Mittelstrass (eds.), An Intimate Relation, 477–494.

Ricoeur that both activities are interpretive in *some* respect. After all, neuropathologists too are interpreters of slides of brain cells.

Hailing the *failure* of Freud's theory to pass muster qua natural science, Ricoeur (1970, p. 358) called for a "counterattack" against those who *deplore* this failure. Finally, in this vein, Habermas advanced arguments of his own to conclude that Freud had fallen prey to a far-reaching "scientistic self-misunderstanding" (1971, chap. 10). And recently, Habermas (1984, p. 249, fn. 25) tried to make short shrift of my critique of his theses, contending that I had made "the simple decision not to consider the hermeneutic character" of psychoanalytic theory. As he sees it, my disregard of the role of socalled meaning in psychoanalysis enabled me to adopt the gambit of "assimilating Freudian theory to the standard model of unified science, only to reject it for failing to measure up to its standards".

Clearly, there is a basic divergence here as to both the source and the import of Freud's scientific shortcomings. As Jaspers, Ricoeur, and Habermas would have it, hermeneutic victory can be snatched from the jaws of scientific defeat, once we appreciate that the discernment of socalled meaning connections, as distinct from causal connections, is at the heart of the psychoanalytic enterprise.

The issues raised in this debate go far beyond psychoanalysis. In my view, their proper resolution not only spells a major general moral for the human sciences, but also has instructive counterparts in biology and physics, for example. One key lesson I shall draw will be this: Meaning connections between the mental states of a person by *themselves* never attest their causal linkage. This precept will emerge, I trust, from my analysis of just how Freud failed in his account of the ontological and epistemic relations between meaning kinships, on the one hand, and causal linkages, on the other. One corollary of his miscarriage will be that he gave *much too much explanatory weight* to meaning affinities, rather than much too little weight, as charged by Jaspers and the hermeneutic phenomenologists. By the same token, their diagnosis of his failure hardly rehabilitates psychoanalysis on new, viable foundations. Being a misdiagnosis, it merely baptizes his fallacious reasoning by labeling it "hermeneutic science". Besides, Freud often displayed striking intellectual brilliance, even as he drew fallacious causal inferences from mere meaning bonds. Yet, I submit, his critics from the camp of descriptive and hermeneutic phenomenologists have treated us to a miasma of confusions for the sake of achieving forced ideological distortions of psychoanalytic theory *and* therapy. Their attempt to force psychoanalysis onto their Procrustean bed derives, of course, from their desire to make the human sciences epistemologically independent from the sort of cognitive accountability featured by the natural sciences.

But just what *are* "meaning connections"? First, let us consider some paradigmatic illustrations from psychoanalysis, and then some illuminating examples from several other fields, some of which are quite remote from psychoanalysis. Interestingly enough, these non-psychoanalytic instances will enable us to appraise the validity of Freud's major attempts to *parlay* meaning kinships into causal linkages, epistemically and ontologically. Yet let me emphasize at the outset that I deplore the use of the term "meaning" in this context, because it is very misleading. And I use it *only* because the phenomenologists have seen fit to employ it. The word "meaning" is tricky, if only because it conjures up the notion of *semantic* meaning, which is irrelevant here, but facilitates the hermeneutic misassimilation of an epistemological and ontological issue to a linguistic one. Hence much of the time, I shall use the locution "thematic kinship", instead of the slippery term "meaning connection".

When Freud proposed to uncover the "meaning" of a symptom, he was concerned to fathom its unconscious *motives* so as to give a causal explanation of its occurrence. For brevity, we can also say that he was concerned to uncover the unconscious significance of the symptom. This kind of causal explanation is clearly different from the semantic sort of explanation of the meaning of a word that is found in a dictionary. Yet, in criticism of me, the hermeneutic psychoanalyst Anthony Storr wrote preposterously (see Grünbaum 1986, p. 260) that Freud's "expertise lay in semantics." And the psychologist Matthew Erdelyi thought that he can get rid of the etiologic, causal sense of the term "meaning" in psychoanalysis by enunciating the following utter banality (p. 234): "When one establishes the meaning of an unknown word from its context, one does not establish that the context has caused the unknown word."

II. PARADIGMATIC CASES OF MEANING CONNECTIONS FROM PSYCHOANALYSIS

(1) In 1896, Freud used the mere thematic kinship between a patient's specific experience and one of her individual symptoms to illustrate the

suitability of the given repressed experience as an *explanatory etiologic determinant* of the pertinent symptom (S.E. 1896, 3: 193—194; Grünbaum 1984, pp. 149—150). In particular, he gives the following example:

> Let us suppose that the symptom under consideration is hysterical vomiting; in that case we shall feel that we have been able to understand its causation (except for a certain [hereditary] residue) if the analysis traces the symptom back to an experience which *justifiably produced a high amount of disgust* — for instance, the sight of a decomposing dead body. (S.E. 1896, 3: 193—194)

Here Freud construes the symptom etiologically as outwardly expressing the *aversion* felt during the repressed *disgusting* experience.

The theme of aversion is likewise common to a traumatic experience and a subsequent hysterical symptom in the life of Josef Breuer's famous first patient Anna O. As reported in her case history, she had silently endured traumatic disgust on seeing a dog lapping water from a companion's glass (S.E. 1893, 2: 6—7; 1893, 3: 29—30). And later, she almost died of thirst, because of her phobic aversion for drinking water. In Jaspers' infelicitous parlance, we can say that the shared theme of aversion makes for a "meaning connection" between the original trauma and her later symptom. But I shall speak of such episodes instead as exhibiting "thematic kinship or affinity." And the main question will be what epistemic and ontological relevance, if any, these thematic kinships between mental events have to causal linkages between them.

(2) Freud's famous 1909 case history of the Rat Man provides a cardinal exemplar of his inferential reliance on a *thematic connection*. In this instance, he used the thematic kinship between a childhood episode and some of the patient's later obsessive fears to claim that the early episode had been the *crucial causal factor* in the pathogenesis of these obsessions.[1]

The obsessive fears that Freud singles out for such etiologic explanation featured two people of whom the patient was very fond: his father, and a woman he courted. During the Rat Man's army service, he had become aware of "an oriental punishment in which *rats* are allowed to bore their way into the criminal's *anus*" (S.E. 1909, 10: 166). And one of the dreadful thoughts with which he was obsessed was that just this rat punishment would victimize both the woman who attracted him, and his father, who had actually been dead for years by then. Furthermore, the patient was plagued by equally intrusive obsessive fears that his father and other loved ones would *die*, fears that likewise persisted after his father's death.

But how does Freud propose to explain those of the patient's obsessions that featured the awful rat theme? As we learn, at the age of three or four, the Rat Man had misbehaved *like a rat* by *biting* someone, presumably his nurse. Thereupon he had been soundly beaten for it by his father, and had therefore borne him an abiding unconscious hatred ever since. Freud elaborates on the dual theme of biting and punishment:

> The notion of a rat is inseparably bound up with the fact that it has sharp teeth with which it gnaws and bites [footnote omitted]. But rats cannot be sharp-toothed, greedy and dirty with impunity: they are cruelly persecuted and mercilessly put to death by man, as the patient had often observed with horror. He had often pitied the poor creatures. But he himself had been just such a nasty, dirty little wretch, who was apt to bite people when he was in a rage, and had been fearfully punished for doing so. He could truly be said to find "a living likeness of himself" in the rat (footnote omitted). (S.E. 1909. 10: 215—216)

Freud then explicitly draws an *etiologic* inference partly from the *thematic kinship* between the culpability of roguish *biting*, on the one hand, and the intolerability of a *rat's* gnawing behavior, on the other. As he reasoned, the patient's latent memory of the cruel paternal castigation for biting had engendered repressed hostility toward the father. This antagonism, in turn, had generated the *unconcious wish* — and thereby the *conscious fear* — that the father undergo the monstrous oriental punishment of anal rat penetration. The hostile *wish* that the father suffer this punishment had been morally unacceptable to the patient's consciousness. Therefore, he had repressed it, and turned it into a conscious obsessive *fear* of the father's punitive victimization.

In brief, Freud sees rat obsessions as the patient's *neurotic defense* against his own unacceptable punitive wishes toward his father. But assuming the occurrence of the punitive childhood scenario, the important issue of causation posed by this inference is *not* whether the severe paternal castigation for biting engendered hatred toward the father; instead the etiologic issue is whether that particular presumed hatred was the pathogen of the obsessive fear of the father's victimization by the oriental punishment. Therefore, when we address that issue in due course, we shall have to ask the following question: Granting the existence of a causal link between the punitive childhood experience and hatred toward the father, does it support at all the *further* etiologic hypothesis that this hatred, in turn, was the pathogen of the rat obsessions? My answer will be a clear "*No*"!

You will recall that the Rat Man was also plagued by intrusive fears that his father, and women of whom he was fond, would *die*. Without going into any of the details, let me just say that Freud attributed these death-obsessions to the patient's repressed hatred of the father for having interfered with his sex-life.[1]

(3) Inferences from thematic affinity also play a central, though logically somewhat different role in Freud's theory of the socalled "transference neurosis", a theory that is fundamental to the hypothesized dynamics of psychoanalytic *therapy*. These inferences will likewise turn out to be fallacious.

According to this part of psychoanalytic theory, the patient *transfers* onto the psychoanalyst feelings and thoughts that originally pertained to important figures in her or his earlier life. Thus, the patient is believed to behave toward her or his analyst *as though* the doctor were her or his father, mother, sister, or other significant person from the past. In this important sense, the phantasies woven around the psychoanalyst by the analysand, and quite generally the latter's conduct toward his doctor, are hypothesized to be *thematically recapitulatory* of childhood episodes. And by thus being recapitulatory, the patient's behavior during treatment can be said to exhibit a thematic *kinship* to such very early episodes. Therefore, when the analyst interprets these supposed reenactments, the ensuing interpretations are called "transference interpretations".

Astonishingly enough, Freud and his followers have inveterately drawn the following causal inference: Precisely in virtue of being thematically recapitulated in the patient-doctor interaction, the hypothesized earlier scenario in the patient's life can cogently be held to have *originally* been a pathogenic factor in the patient's affliction. For example, in his case history of the Rat Man, Freud infers that a certain emotional conflict had been the *precipitating* cause of the patient's inability to work, merely because this conflict had been thematically reenacted in a phantasy, which the Rat Man had woven around Freud.

In short, here the causal inference takes a somewhat different logical form from the one we encountered in our previous examples. There, Freud inferred the existence of a *direct* causal nexus between thematically kindred mental events from the mere fact of their thematic kinship. But in the context of his transference interpretations, the thematic reenactment is held to show that the early scenario had *originally* been pathogenic. And once this *etiologic* conclusion has been drawn, the patient's thematic reenactment in the treatment-setting is also asserted to be PATHOGENICALLY recapitulatory, rather than only thematically. Freud extols this dubious reasoning in his 1914 "History of the Psychoanalytic Movement":

> The fact of the emergence of the transference in its crudely sexual form, whether affectionate or hostile, in every treatment of a neurosis, although this is neither desired nor induced by either doctor or patient, has always seemed to me the most irrefragable proof [original German: "*unerschütterlichste Beweis*"] that the source of the driving forces of neurosis lies in sexual life. This argument has never received anything approaching the degree of attention that it merits, for if it had, investigations in this field would leave no other conclusion open. As far as I am concerned, this argument has remained the decisive one, over and above the more specific findings of analytic work. (S.E. 1914, 14: 12)

For just this sort of reason, Freud claims that investigatively, the psychoanalyst is much better off than the archaeologist, who has to rely on excavations. The analyst, he tells us, "has at his disposal . . . the repetitions of reactions dating from infancy and all that is indicated by the transference [onto the analyst] in connection with these repetitions" (S. E. 1937, 23: 259).

Freud contended that the later reenactments of childhood reactions are *pathogenically* recapitulatory. But this etiologic notion then inspired two of his further fundamental tenets: First, the *investigative* thesis that the psychoanalytic dissection of the patient's transference behavior toward the analyst can reliably identify the *original pathogens* of her or his long-term neurosis, and secondly, the cardinal *therapeutic* doctrine that the resolution of the analysand's socalled "transference neurosis" is the key to overcoming her or his perennial problems.

III. THEMATIC KINSHIPS VIS-À-VIS CAUSAL CONNECTIONS

Now we can turn to the following pivotal question: To what extent, if any, do *mere* thematic kinships bespeak causal connections? As I shall illustrate presently, thematic kinships are not only of various sorts but are also encountered in *varying degrees*, ranging from very high to very tenuous. Yet it will be crucial to appreciate the following impending moral: Even when the thematic kinship is indeed of very high degree, it does not *itself* license the inference of a causal linkage between the thematically kindred events. Thus, consider the following series of examples from various fields in their bearing on the inferrability of causal relatedness among events or states featuring *diverse sorts* of affinities or isomorphisms.

(1) My first example uses a trivially familiar fact to draw a moral that is far from trivial. Upon reading a young student's 10-page paper, the course instructor notices that not only the details of its argument but even its wording are uncannily reminiscent of an old encyclopedia article. Indeed, it turns out that the two texts agree verbatim. And despite the student's protestation of the originality of his wording, the instructor confidently charges plagiarism.

Of course, this indictment relies on the *causal inference* that the student had cribbed his paper. But what is the justification for this inference? It is pertinent to spell out the implicit background knowledge on which it relies.

Consider all hand or typewritten texts exceeding, say, 2500 words. In this class, focus on two subsets: (a) Those which their authors generated entirely by endeavoring to produce a verbatim transcription of another text, and (b) texts *not* produced entirely — or almost entirely — by *copying* from another one. As we know, people make errors of transcription. Hence, in the case of attempted transcriptions, the frequency (probability) of verbatim agreement between two or more texts is not necessarily 100%, though it is rather high. By contrast, the incidence of verbatim agreement among independently produced texts is essentially zero. Indeed, *every* case of verbatim agreement is produced by copying. Speaking elliptically, we can say that verbatim agreement is surely not a case of "*mere chance*" coincidence (correlation). There is simply no Leibnizian prëestablished verbatim harmony between two *independently* produced literary outputs. Naturally, that's why the instructor was entitled to charge plagiarism.

But the moral I am concerned to draw from this otherwise trivial example is as follows: The instructor's entitlement to draw the causal inference that the student had cribbed the paper does *not* derive from the *mere* thematic affinity inherent in the verbatim agreement of the texts; instead, *that inference relies crucially on the ADDITIONAL fact that essentially every case of verbatim agreement results from copying.*

(2) A tourist looking at an otherwise desolate beach notes that the sand reveals a string of configurations exhibiting the same shapes as the left and right shoes worn by humans. In short, the tourist observes a

geometric isomorphism — or "thematic affinity" — between the sand configurations and the shoes. And he will then infer causally that a person wearing shoes had walked on the beach and had thereby produced the sandy shapes we call "footprints".

The lesson of this example is essentially the same as that of the first one: The striking geometric kinship between the two shapes does not *itself* suffice to license the tourist's inference that the foot-*like* configurations were, in fact, caused (produced) by the impact of human feet on the beach. To draw that inference, the tourist avails himself of a crucial piece of *additional* information: Foot-like beach formations in the sand never or hardly ever result from the "mere chance" collocation of sand particles under the action of air, such as some gust of wind. Indeed, the incursion of a pedestrian *makes the difference* between the absence and presence of the foot-like beach formations.

In short, the tourist relies on the very high probability that the shape kinship was not a matter of mere chance, when drawing the *causal* inference that the sandy simulacrum of a human foot is, in fact, actually the trace or mark left by a human foot, and thus a bona fide foot*print* (cf. Grünbaum 1984, p. 63).

(3) Two significantly different dreams will now serve to show that reliance on *mere* thematic connections to draw causal inferences is a snare and a delusion. This moral will, of course, also apply to Freud's dream theory as a special case. In the first of the two dream examples, we shall have license to draw the causal inference that the manifest dream content was shaped by a salient component of the waking experience on the day before. And this warrant will be seen to derive from reasoning of just the sort I articulated à propos of the case of verbatim agreement and of the footprint.

But my point in giving this dream example will be to *contrast* it with another kind: one in which it is demonstrably fallacious to invoke a thematic connection between the waking experience of the previous day and the manifest dream content as a basis for inferring a causal linkage between them. And the latter dream illustration will be only one of several that will exhibit the serious pitfalls of causal inferences in psychoanalysis that rely on the mere thematic affinity between the mental states of a person. Yet the socalled "meaning connections" *as such* are the stock-in-trade of those champions of hermeneutics who are eager to nail psychoanalysis to their mast.

Let us turn to the first of my two dream examples. The dreamer is a woman named Agnes. The night after her first visit to Frank Lloyd Wright's famous house "Falling Water" (in Ohiopyle, Pa.), she dreams about a house *just like it,* down to many of the fine details of its interior appointments. Assume that Agnes had never heard of Falling Water until the day of her visit, let alone seen a picture or description of it. It is crucial that the very first time that Agnes's manifest dream content ever contained such a simulacrum was the night after her daytime visit to that Frank Lloyd Wright house. Without this additional fact, the strong thematic affinity between Falling Water and the dream content would not warrant the inference that Agnes's visit to Falling Water was *causally relevant* to the presence of a simulacrum of that mansion in her dream. To see why not, observe that the warrant can be gotten essentially by the use of J.S. Mill's joint method, being analogous to the evidence in the case of the footprint. In short, Agnes's visit *made a difference* to her having that dream. But now consider a related dream example with the opposite inferential moral.

(4) Assume that last night, my manifest dream content included the image of some kind of house or other. In my urban life, I routinely see and frequent houses of some sort almost daily. Thus, my impressions on the day before this dream featured visual and tactile impressions of at least one dwelling. Indeed, over the years, on the day before a dream, my waking experience practically always includes seeing some domicile or other, *regardless of whether the ensuing manifest dream content then features the image of an abode or not*! In this case, seeing a house during the day *does not make any difference* to dreaming about a house the night after.

Evidently, when a house of some kind is an element of the manifest dream, the mere generic presence of a house theme in the prior day's waking experience does not meet the key requirement for being causally relevant to the presence of a house image in the dream: Seeing a house on the day before a dream does not divide the class of the day's waking experiences on the prior day into two subclasses, such that the probabilities (frequency) of the appearance of a house in the next dream *differ as between the two subclasses.* Indeed, precisely because I see houses practically daily, year in and out, my seeing a house on the day before a dream does not even divide the class of my waking experiences on the prior day into two subclasses. A fortiori, there is no division such that the two subclasses exhibit different probabilities of the stated sort.

Thus, there is a sharp contrast with the situation in the Falling Water dream: When some house image occurs in my dream last night, it is a mistake to attribute that image causally to my having seen one or more houses yesterday. In sum, despite their thematic affinity, it is eminently reasonable to conclude that the dual presence of the generic house theme both in yesterday's daytime experience and in last night's dream was a happenstance, rather than a case of causal linkage.

To avoid a serious misunderstanding, it needs to be emphasized that this denial of causal relevance fully allows the following state of affairs: People might never have dreamt about any house, unless they had seen one *at some time or other* in their lives. But this *mere* necessary condition is not tantamount to causal relevance, as can be seen at once from an analogy: Breathing is a necessary condition for being paranoid; but breathing is hardly causally relevant to becoming paranoid, since non-paranoiacs breathe no less than paranoiacs do. Breathing does make a difference to being alive, but not to being paranoid *rather than* non-paranoid. Thus, in the context of dreams as well, a state of type *X* may be a *necessary condition* for the occurrence of some other sort of state *Y*, although *X* is not causally relevant to *Y*. On the other hand, there is evidence that when women are going through a divorce, their dreams are thematically affected by it in a causally relevant way.

It is true that, in the case of the generic house dream now at issue, the thematic affinity between the day's waking experience and the next dream is clearly much weaker than in the Falling Water example. But recall my earlier *caveat* that even in the examples featuring very strong thematic affinity, the mere presence of a very high degree of such kinship was quite insufficient to validate the causal linkage. Hence it would be a momentous error to believe that causal inferrability goes hand-in-hand with a very high *degree* of *mere* thematic kinship.

(5) As we know from American history, Thomas Jefferson and John Adams, who had been personal (though not political) friends, died within a few hours of each other on the 50th anniversary of the Declaration of Independence, written by Jefferson with the aid of a draft to which Adams had contributed. Within the reference class of, say, long-term personal friends, the temporal coincidence of events having the thematic affinities of death, careers in politics, and co-authorship is quite improbable. Nonetheless, presumably no professional historian would infer the existence of a causal linkage between the two deaths. Nor would it do to infer that there was an *indirect* causal connection between these deaths, such that each of them is a partial effect of one and the same shared cause. Thus, even if there

were both a God and a Satan, there is no basis for seeing the hand of either in the multiple correlations between the two deaths. On the other hand, a *common cause* is indeed implicated in the familiar fact that the probability of a storm occurring soon after a sudden barometric drop is appreciably greater than the probability of a storm in general: A pressure drop over a wider area is the *common cause* of both the sharp barometric drop and the storm, which are thereby linked indirectly. Thus, their common cause accounts for the positive statistical relevance of falling barometer readings to storms, as expressed by the stated comparison of the probabilities.

Instead of calling for such a common cause explanation, Jefferson's and Adams' deaths call for *separate* causal explanations. The conjunction of these explanations may then also account for the otherwise improbable coincidence of the two deaths. In a recent article, Elliott Sober (1987) gave two other interesting types of examples to show that "Correlations, *per se*, do not cry out for common cause explanation".

(6) As he points out in a case drawn from evolutionary theory (p. 466), there is a method of inference (so-called "cladistic parsimony") that operates on the following principle: When species match with respect to *ancestral* characteristics — which is one kind of thematic affinity — this similarity is *not* evidence of common descent; yet a match in regard to derived characteristics — which is another sort of thematic affinity — does qualify as evidence of a shared genealogy. Thus, "the mere existence of a [thematic] correlation between the traits of two species is not enough to justify a claim of common descent" (p. 466). In a still more recent and more extensive study, Sober (1988) has dealt with the problem of inferring common ancestry among species as a mere instance of a more general problem of just when one should postulate common causes. Sober's moral is to issue some important caveats in regard to the inferrability of a common cause. He develops this lesson by reference to a number of defects he claims to have detected in the "principle of the common cause" espoused by Hans Reichenbach and Wesley Salmon (Salmon 1984, chap. 6). But Forster (1988) has criticized Sober's own version of this principle.

(7) Sober's other example (1987, p. 465) is furnished by the positive correlation between the cost of bread in England and the Venetian sea level: Both of the quantities have increased monotonically with time. However, this *quantitative affinity* hardly betokens that it resulted from a common cause. Here again, *mere* affinity does not even warrant the

presumption of such an *indirect* causal connection, let alone the inference of a direct linkage.

(8) In the history of medical treatment, a *mere* thematic affinity has sometimes been taken to betoken or presage therapeutic efficacy, much as a sign may attest to something. Thus, in the 16th century, Paracelsus gave the following advice: To cure the liver, treat with an herb that is *shaped* like a liver (Hacking 1975, p. 42).[2] Nowadays, this inference of causal relevance strikes us as primitive rather than as quaint. But I claim that its logical defects are no worse than those of causal inferences from *mere* thematic connections, which abound nowadays in some theorizing about human behavior.

We are now ready to appraise Freud's various causal inferences from mere thematic connections. As a corollary, we can reach a verdict on the animadversions of Freud's hermeneutic critics.

IV. CRITIQUE OF FREUD'S CAUSAL CONSTRUAL OF MERE THEMATIC KINSHIPS

An instructive inferential caveat for Freud is spelled by the previously mentioned aversive affinity between Anna O.'s inability to drink water, and the traumatic disgust she had felt silently on seeing a companion's dog lapping water from a glass (S.E. 1893, 2: 6—7; 1893, 3: 29—30). For his part, Breuer himself had the wisdom *not* to adduce the aversive kinship *itself* as his evidence for inferring that the repression and affective strangulation of the traumatic disgust had been the specific pathogen of Anna O.'s hysterical inability to drink (Grünbaum 1984, pp. 177—180). Instead, he pointed to allegedly positive therapeutic results from his cathartic treatment as his evidence, rather than to the thematic kinship of aversion. But, as we know from Henri Ellenberger's research (1970, pp. 483—484; 1972), Breuer's treatment of this patient was a therapeutic fiasco, instead of a "talking cure". And, as I have argued elsewhere (1984, pp. 180—184), Breuer's therapeutic failure showed that he had *not* identified and removed the pathogens of Anna O.'s aversion to drinking water or of her other symptoms. Moreover, we know from Jung that Freud was quite aware of this debacle. Yet, as we recall, Freud told us that if we can uncover a disgusting experience, its mere aversive affinity to hysterical vomiting tells us "that we have been able to understand [the] . . . causation" of the vomiting.

Even the Anna O. episode alone should have given Freud much

pause when he felt tempted to invoke such thematic connections as a justification for etiologic inferences. Alas, as we can now see very clearly, Freud simply did not heed that particular sobering moral from the demise of Breuer's proto-psychoanalytical cathartic theory. Thus, as we saw in the case of the Rat Man, Freud appealed to the thematic kinship between the punitive biting episode and the adult rat obsessions, even as he succumbed to the explanatory blandishment of an inferred etiologic linkage between them. But, as is now clear, the thematic connection adduced by Freud does not vouch for the etiologic role of the paternal punishment in the pathogenesis of the rat obsessions. And Freud simply begs the etiologic question here by trading on thematic affinity.

An equally unfavorable epistemic judgment applies to the web of causal inferences that were drawn in Freud's theory of transference, which I have articulated. The patient's thematically recapitulatory behavior toward his doctor does not itself show that it is also *pathogenically* recapitulatory. How, for example, does the reenactment, during treatment, of a patient's early conflict show at all that the original conflict had been pathogenic in the first place? Nor does such an etiologic inference gain cogency from the fact that the setting for the recapitulation is a bizarre phantasy focused on the analyst.

Therefore, I submit, the purportedly "irrefragable proof" of which Freud spoke in our earlier quotation from him deserves more attention *not* because its appreciation "would leave no other conclusion open", as he would have it; instead, I believe, the Rat Man case and other such concrete case histories show how baffling it is that Freud deemed the etiologic transference argument cogent *at all*, let alone unshakably so.

Freud (S.E. 1909, 10: 155, 207n, 220, 249) explicitly adduced the allegedly successful outcome of the Rat Man's therapy as evidence for the correctness of the psychoanalytic etiology of obsessional neurosis (p. 220). But, as we know from Mahoney's 1986 book *Freud and the Rat Man*, Freud's purported cure did not materialize. Indeed, more generally, verbal psychotherapy is notoriously ineffective for obsessive-compulsive neurosis. On the other hand, a recent article in *The New York Times* (February 25, 1988) contains the following sobering report: A 19-year old man, who despaired over his compulsion to wash his hands hundreds of times per day, made an unsuccessful suicide attempt when he shot himself through the brain. The 22-caliber bullet destroyed only one section of the left front lobe of his brain, but left both his I.Q. and the rest of his brain intact. Almost immediately thereafter, the young man ceased to have hand-washing obsessions. And now he is a straight-A college student, whose grasp of *meaning connections* is presumably likewise unimpaired. Could it be, therefore, that unconscious or conscious *meaning connections* have *nothing to do* with the causation of obsessional neurosis? (Flor-Henry, 1983, ch. 10)

V. CONCLUSIONS

I have argued that it is always fallacious to infer a causal linkage between thematically kindred events from their *mere* thematic kinship. Yet, as illustrated by my example of Agnes's dream about the Falling Water mansion (example 3), the existence of a strong thematic connection between two mental events, or two series of such events, hardly militates against there *also* being a causal linkage between them. Indeed, as shown by the example of verbatim syntactic and semantic agreement between two student papers (example 1), the direct or indirect causal linkage between them can be perspicuously law-like. Thus, Freud should surely *not* be faulted for asserting, in principle, that some mental events can be linked *both* thematically *and* causally, though he mistakenly claimed entitlement to *infer* the latter linkage from the former alone. And when he explicitly asserted such entitlement, he was fully clear on the *distinction* between thematic connections and causal connections.

Yet, as I remarked at the outset, Karl Jaspers (1974, p. 91) chided Freud: "In Freud's work we are dealing in fact with *psychology of meaning*, not *causal explanations* as Freud himself thinks". But since causal relevance is entirely compatible with thematic or "meaning" relevance, Jaspers' objection to Freud here rests on a pseudo-antithesis of "either . . . or" (cf. Grünbaum 1984, pp. 69—83). Thus, there is no merit in Jaspers' indictment of Freud as having incurred a "confusion of meaningful connexions with causal connexions" (Jaspers 1974, p. 91). Nor in his claim that "Freud's psychoanalysis" is being vitiated by "a misunderstanding of itself" (1974, p. 80), a patronizing charge echoed later on by Ricoeur and Habermas, as we recall (cf. Grünbaum 1984, pp. 2—43; Jaspers 1973, Part 3, pp. 374—385).

As against these phenomenologists, it emerges from Freud's inferential failings that he gave *much too much explanatory weight* to thematic affinities, rather than too little, as they have charged. Indeed, *mere* meaning connections tell us *nothing* about the supposed unconscious *motives* for symptom-formation, dream-genesis and the provenance of Freudian slips. Yet such a motivational account is precisely what psychoanalytic theory claimed to offer.

To his credit, Ricoeur belatedly, albeit inadequately, recognized the causal character of psychoanalytic motivational explanations. Well after the appearance of his book *Freud and Philosophy*, he admitted that Freud's theory "forbids distinguishing between motives and causes because it calls for an *explanation* by means of causes in order to reach an *understanding* in terms of motives" (1981, p. 263). But the grudging character of this recognition soon becomes evident. As Ricoeur tells us, psychoanalytic theory "requires that the hermeneutics of self-understanding [in terms of a narrative highlighting meaning connections] take the detour of causal explanation" (p. 264). Here the grudging term "detour" is quite misleading. Imagine a philosopher of human physiology saying that when our blood circulates between our head and our feet, it takes "the *detour*" through our heart pump!

Worse, Habermas's treatment of causality in psychoanalysis is simply incoherent. When a patient overcomes his neurosis in treatment, we learn, she or he achieves something totally without parallel in the causality of nature: The patient allegedly "dissolves" or "overcomes" the very causal connection that links the pathogen etiologically to its effect. But this claim boomerangs, if only because, by parity of specious reasoning, one could say that whenever we cut a circuit by throwing an appropriate switch, causality in electric current flow is dissolved *as such* (Grünbaum 1984, Introd. p. 2). And the latter example clearly comes from the causality of nature.

I draw a two-fold moral for the human sciences from the stated failures of Freud's enterprise: (1) Let us be alert to thematic connections, but beware of their beguiling causal pitfalls; *a fortiori*, (2) narratives replete with mere hermeneutic pablum are explanatorily bankrupt; at best, they have literary value. Patronizing sermons against alleged scientistic misunderstandings of the role of meanings do nothing for the psychoanalytic enterprise or for any other theories of human nature, though they foster ideological hostility to scientific thought. After a veritable cornucopia of meaning connections in Freud's case history of the Rat Man, the etiology of the patient's obsessions remains deeply obscure to this very day. Yet encomia keep on being heaped upon that case history: Witness the (1988) *Freud* book by the Yale intellectual historian Peter Gay, the (1986) book *Freud and the Rat Man* by the Montreal professor of English Patrick Mahony, and the (1984) book *Freud and the Culture of Psychoanalysis* by Stephen Marcus, professor of English at Columbia. Indeed, Peter Gay (1988, p. 745) complains in his latest book that my scepticism is itself an obsessional symptom, which I have exhibited for a decade. Yet he presumably does not regard his own advocacy of psychoanalysis — which has been both unflagging and uncritical — as such a symptom.

University of Pittsburgh

NOTES

* The original version of this paper was delivered at the Workshop on SCIENTIFIC FAILURE, held on April 23—24, 1988 at the Center for Philosophy of Science, University of Pittsburgh. The present revised text was delivered at the Florence Center for the History & Philosophy of Science in June, 1988.

[1] For my much fuller critique of Freud's etiologic inferences in the Rat Man case, see A. Grünbaum, "The Role of the Case Study Method in the Foundations of Psychoanalysis", in L. Nagl and H. Vetter (eds.), *Die Philosophen und Freud*, a volume in the series *Wiener Reihe*. Vienna, Austria: R. Oldenbourg Verlag, 1988. This essay is reprinted in the *Canadian Journal of Philosophy* **18**: 623—658, 1988.

[2] I am indebted to Wesley Salmon for this reference.

BIBLIOGRAPHY

Ellenberger, H. F. (1970). *The Discovery of the Unconscious*. New York: Basic Books.

Ellenberger, H. F. (1972). "The Story of 'Anna O.': A Critical Review with New Data". *Journal of the History of the Behavioral Sciences* **8**: 267—279.

Flor-Henry, P. (1983). *Cerebral Basis of Psychopathology*. Boston: John Wright.

Foster, M. (1988). "Sober's Principle of Common Cause and the Problem of Comparing Incomplete Hypotheses". *Philosophy of Science* **55**: 538—559.

Gay, P. (1988). *Freud*. New York: W. W. Norton.

Grünbaum, A. (1984). *The Foundations of Psychoanalysis: A Philosophical Critique*. Berkeley, Los Angeles: University of California Press.

Grünbaum, A. (1986). "Précis of *The Foundations of Psychoanalysis*", and "Author's Response". *Behavioral and Brain Sciences* **9**: 217—284.

Habermas, J. (1971). *Knowledge and Human Interests*, translated by J. J. Shapiro. Boston: Beacon Press.

Habermas, J. (1984). "Questions and Counterquestions". *Praxis International* **4**: 229—249.

Hacking, I. (1975). *The Emergence of Probability*. New York: Cambridge University Press.

Jaspers, K. (1973). *Allgemeine Psychopathologie*. 9th ed. New York: Springer Verlag.
Jaspers, K. (1974). "Causal and 'Meaningful' Connexions Between Life History and Psychosis". In: S. R. Hirsch and M. Shepherd (eds.), *Themes and Variations in European Psychiatry*, translated by H. Hoenig. Charlottesville: University of Virginia Press, pp. 80—93.
Mahony, P. (1986). *Freud and the Rat Man*. New Haven: Yale University Press.
Marcus, S. (1984). *Freud and the Culture of Psychoanalysis*, chap. 4: "Freud and the Rat Man". London: George Allen & Unwin.
Ricoeur, P. (1970). *Freud and Philosophy*. New Haven: Yale University Press.
Ricoeur, P. (1981). *Hermeneutics and the Human Sciences*, translated by J. B. Thompson. New York: Cambridge University Press.
Salmon, W. (1984). *Scientific Explanation and the Causal Structure of the World*. Princeton: Princeton University Press.
Sober, E. (1987). "Parsimony, Likelihood, and the Principle of the Common Cause", *Philosophy of Science* **54**: 465—469.
Sober, E. (1988). "The Principle of the Common Cause", in J. Fetzer (ed.), *Probability and Causality*. Dordrecht, Boston: Reidel.

MICHAEL RUSE AND PAUL THOMPSON

NEO-DARWINISM: FORM AND CONTENT

Charles Darwin's *Origin of Species* was published in 1859. It is well known that it caused instant controversy, with Darwin's supporter Thomas Henry Huxley debating the Bishop of Oxford over our own supposedly ape origins (Lucas 1979). Less well known is the fact that many of Darwin's ideas, particularly of the occurrence of evolution *per se*, rapidly won acceptance by nearly all segments of Victorian society. Even clergymen came to think in evolutionary terms, so long as they were permitted to believe that God had miraculously breathed immortal souls into human frames, or some such thing (Ellegard 1958, Ruse 1979).

Yet, from the beginning there were serious scientific objections to Darwinism, particularly to the mechanism of selection. Two problems in particular stood out. On the one hand, the physicists calculated that the earth is far too young to have allowed a slow gradual progress of development, as the theory of evolution through selection supposes (Burchfield 1975). On the other hand, Darwin (by his own admission) was dreadfully confused about the principles of heredity, especially about the ways in which organic features are transmitted from one generation to the next and about how new features, the "raw stuff" of evolution, come into existence in the first place. Darwin was adamant that such new features or variations are "random", in the sense of not appearing to order according to organisms' needs. But of their true nature and causal origin, he could only speculate wildly and unsuccessfully (Vorzimmer 1970).

With respect to the problem posed by the earth's age, we know now that the physicists were quite wrong. A major source of the error was the omission from their calculations the heat generated by radio-active decay. When once this is considered, there is plenty of time for evolutionary processes, even by such a mechanism as natural selection. The problems of heredity have proven more significant for evolutionary studies. Although we now have a comprehensive theory of heredity, accommodating the science of heredity ("genetics") has required a complete and comprehensive transformation of the evolutionary theory

J. R. Brown and J. Mittelstrass (eds.), An Intimate Relation, 495—512.

found in the *Origin*. This "neo-Darwinism" still makes natural selection of random variation the central causal focus of organic change, but there has been considerable shifting and augmentation of Darwin's own claims.

In this essay, we shall first sketch what we believe to be the nature and structure of neo-Darwinism, showing where and how it differs from Darwin's own version of evolutionary theory. Although we are in substantial agreement on the elements of neo-Darwinism and their roles in the theory, we differ on the logical relationship between them. We do not attempt to resolve this dispute, prefering instead to identify the points of agreement and make clear the nature of the disagreement. Then, second, we shall look equally briefly at the major philosophical controversy which rages around the true nature of this revised theory.

POPULATION GENETICS

Let us begin with that of which Darwin was ignorant: heredity. In what follows, we shall treat heredity in a non-molecular way, for that was the approach when the neo-Darwinian synthesis was formed in the 1930s and 1940s; but, nothing is intended to deny or preclude a molecular backing. Indeed, today, neo-Darwinians are among the most enthusiastic advocates of using our knowledge of life's molecular processes to unlock difficult evolutionary problems. (See, for instance, Dobzhansky *et al.* 1977, Ayala 1985. Ayala and Kiger 1984 is a good introduction to genetics.)

Following a trail of discovery which dates back to the Moravian monk Gregor Mendel, who was working unrecognized on the problems of heredity even at the time of Darwin, it is now thought that the crucial unit of heredity, that which is in some sense responsible for organic features (although always in causal conjunction with the effects of the environment), is the "gene". This unit of function, today identified with the DNA molecule, is carried within the nuclei of the cells of the body on long strand-like elements, known as "chromosomes", each cell (with the exception of the sex cells) having an identical set of genes, arranged on the chromosomes in the same way. These gene sets (known as the "genotype") are activated to supply the information to produce the components of all aspects of living things, and in the differences between genes (and the ways they are ordered on the chromosomes) lie the physical differences between organisms (known as the "phenotypes").

Chromosomes (generally) occur in pairs, which means genes have mates on corresponding chromosomes (both being at the same "locus"). In (sexual) reproduction, one and only one of the members of each pair is transmitted (via the sex cells, which have only single sets of chromosomes) to the offspring (which, being started by the conjunction of a maternal and a paternal sex cell, again have a full "diploid" set of chromosomes, with parental genes at corresponding loci). Since the genes (or, rather, copies) are transmitted entire from generation to generation, and since, as it turns out, it is purely a matter of chance which one of any pair (considered on its own) will be transmitted to an offspring, we have a simple but powerful rule of transmission. This is known as "Mendel's first law" or the law of "segregation". There is also a second law (the law of "independent assortment") which states that the transmission of genes from one locus is independent of the transmission of genes from another locus. Actually, however, this second law has many exceptions, especially when dealing with loci on the same chromosome.

In evolution, especially Darwinian evolution, one is dealing not with individuals but rather with groups, most particularly with groups of interbreeding organisms ("populations") of the same kind ("species"). Can the ideas we have just been considering be extended out to the group situation? As it happens, such an extension comes quite readily (thus giving rise to the appropriately named "population genetics"). The crucial move involves the broadening of the focus of Mendel's first law from the individual to the group — a derivation first made independently in 1908 by G. H. Hardy and W. Weinberg (Provine 1971). Their "law" shows that in a large population with no impinging external factors, if the members therein are breeding quite randomly, then the ratios of the various genes within the group and their distribution between individuals will be at equilibrium — whatever the proportions, they will stay there, going neither up nor down.

One may think that the Hardy-Weinberg law sounds a bit disappointing. Basically, it would seem to say that if nothing happens, nothing happens. If, say, you start in a population with 10% of a particular kind of gene (genes which can occupy the same locus of members of a group are known as "alleles"), then after no matter how many generations, you will still have 10% of that kind. No more, no less. Moreover, if say 25% of your population had identical allele pairs at some locus, then there would continue to be the same proportion in all subsequent generations. (Actually, this result holds only after the first generation. Organisms with pairs of identical alleles are known as "homozygotes". With different alleles, as "heterozygotes".)

However, the Hardy-Weinberg law's importance lies in its very guarantee of *lack* of change. Like Newton's first law of motion (which is also an "if nothing happens then nothing happens" type of law), it provides a background of stability against which biologists can work (Ruse 1973, Sober 1984a, Thompson 1989). In particular, they can introduce forces for genetic change, knowing that the effects of these forces will not be lost in a randomly fluctuating genetic landscape. What forces are these that biologists would want to introduce? There are several, but two are of particular significance.

First, there is the change of a particular gene of one kind into a gene of another kind (that is, change from one allele to another). This phenomenon, known as "mutation", is of course precisely that which causes new variations — the "raw stuff" of evolution. Much is now known about how mutation happens, involving ultimately (obviously) a change in the DNA structure of an organism. Important to us here is the knowledge that (as Darwin insisted) the process is indeed random, in the sense that the effects of mutations are quite without respect to the needs of organisms. However, although mutation is also random in the sense that one cannot predict its happening in some specific organism at some specific time, on average mutation is regular and can thus be quantified and incorporated within population genetical theory.

The other major causal factor is selection: some organisms, because of the features produced by their genes, have more offspring than others, and thus (again on average) alleles of one kind will tend to be better represented in future generations (on an allele for allele basis) than alleles of another kind. Whereas, say, under Hardy-Weinberg equilibrium, a population may stay balanced 50:50 between two alleles, selection may be favouring one allele to the extent that (say) for every representative the other allele has in the next generation, it has two. Hence, because of selection the ratio will change to 66 2/3:33 1/3 in one generation, and correspondingly more in subsequent generations. (Note that the absolute population size may stay constant. It is the ratios which change. No one is saying that all of the more successful, the "fitter", reproduce, or that all of the less successful fail to reproduce.)

Incidentally, we see here one of the points at which today's evolutionists have moved Darwin's thinking, shall we say, sideways. For him, in the *Origin*, natural selection was something to be proven from certain wide-sweeping premises (Ruse 1971). Start with claims about arithmetical progressive limits on food and space supplies and geometrical progression possibilities for population growth, and you then get the struggle. Next, go with the ubiquitous struggle and with variation, and only then do you get selection. Neo-Darwinians tend not to offer these calculations at all — indeed you can get gene-ratio changes even if food and space constraints never come into play, so long as one group is out-reproducing the other. In population genetics, it is not so much that selection is unproven, but rather that it is introduced as a primitive within the formal theory, with the justification left for outside discussion. The basic background law is the Hardy-Weinberg law.

Tangentially, this means that neo-Darwinsim differs from Darwin's theory in another way. In the *Origin*, it is the individual organism which is the unit of selection. In neo-Darwinism, selection (and other causal processes) is considered primarily in terms of its effect on the gene: the gene is the fundamental unit of heredity. Since in reality genes do not appear naked, as it were, but rather packaged in organisms with other genes, there is always the implicit understanding that at some point the genetical theory can and will be translated up into claims about the physical features of organisms and that genic selection is often a function of organismic selection. We should note that there has been controversy between neo-Darwinians themselves about how easily and readily such translation can be done and about the actual level at which selection causally occurs. (See the discussions in Sober 1984b.)

Returning now to selection in population genetics, it obviously results in changes in gene ratios. What is perhaps surprising is that it can hold gene ratios constant, so that *because* of selection there may well be equilibrium within a population with alleles of different kinds remaining indefinitely. One of the best-known ways in which this can happen is due to so-called "balanced superior heterozygote fitness". Suppose you have two alleles, and that their heterozygote is fitter (that is, on average, has more offspring) than either homozygote. One can readily show that the heterozygote will persist indefinitely (which is just what you would expect), but also that — even if none of the homozygotes reproduce — a certain stable percentage of homozygotes will appear in each generation.

The root cause of this phenomenon may seem incredible. Alleles which occur homozygously produce none or few offspring, whereas the heterozygote flourishes? It is like two wrongs making a right. But it can undoubtedly happen — neo-Darwinians suspect it happens often. The classic case occurs in humans, where possession of one particular allele gives you a natural immunity against malaria, not possessed if you are a homozygote for the normal-type allele. Unfortunately, possession of two such genes homozygously gives you a particularly vicious genetic ailment: sickle-cell anaemia. Under usual conditions, you die by your fourth birthday. However, in certain parts of the world, malaria is such a threat there is a stable balance between the reproductive benefits for the heterozygote, and the deleterious consequences for the homozygotes (the sickling homozygote particularly) (Bodmer and Cavalli-Sforza 1976).

In the eyes of neo-Darwinians, the significance of the genetic variation brought on and maintained by selection cannot be overemphasized. It is this variation, held at the ready in populations, that gets called upon should a new need arise. Given that mutation is random with respect to need, many people cannot imagine how selection could respond to provide new adaptations should the demand come. The answer is that — apart from the fact that often the demand cannot be satisfied and organisms go extinct — there is no need to wait for an appropriate change. Variation of some kind or another is always there, ready for action.

Let us give an analogy to illustrate the point. Suppose you were asked (in a course) to write an essay on, say, dictatorship. If your only source material were the offerings of the Book of the Month Club, the course would probably be long over before you could get hold of an appropriate book. But if you have a library at your disposal, then you could surely write your paper without difficulty. If there be nothing on Hitler, then there would be something on Napoleon. And if not on him, then on someone else. Most people think that mutation works like the Book Club situation, where you are waiting desperately for the right allele/book. However, if selection is promoting variation, the better analogy is truly more like that facing a person with a card to a good library.

Do note, nevertheless, how selection still does differ from the good God of Archdeacon Paley. Both selection and Paley's God, are out to promote adaptation. But, whereas Paley's God could (if He wished) start always from scratch, making the perfect design, selection has to work with what it has at hand. Change, therefore, must be a string-and-sealing-wax phenomenon, where new adaptations are cobbled together out of whatever may be in the tool-box at the moment. If a new predator turns up, ideally the potential prey might develop added speed so it can escape. But, it might have to be satisfied with better camouflage, becoming nocturnal, or becoming poisonous. It all depends on the available variation.

THE STRUCTURE OF NEO-DARWINISM

Thus far, we have been speaking of population genetics. It contains ideas about selection and variation and so forth, so it is a good start. But, what has it to do with neo-Darwinism taken as a whole, where this is understood as the modern-day successor to Darwin's theory, which later dealt with the full spectrum of biological phenomena — biogeography, paleontology, embryology, systematics, and more? There is some controversy about this, with different commentators (biologists and philosophers) answering in different ways. Indeed, it is on this point that we, the authors, differ with each other. We shall present the two alternatives on which we differ and set the respective reasons given in support of each. Later, however, we shall admit that our arguments for our respective alternatives address the issues at two different levels. Both in spirit and in substance there may be much in the other option.

It is clear to both of us that population genetics is a part, an important part, of neo-Darwinian theory. The crucial question is whether it is an especially important part, having a privileged status over other parts, like say biogeography. Some rather think not. For instance, in a highly influential book written several years ago, the philosopher Morton Beckner (1959) wrote as follows:

> ... if we look in evolutionary theory for the pattern of theoretical explanation exemplified in that paradigm of theory formation, Newton's explanation of Galileo's and Kepler's laws, we shall be disappointed.
>
> My own view is that evolution theory consists of a family of related models; that most evolutionary explanations are based upon assumptions that, in the individual case, are not highly confirmed; but that the various models in the theory provide evidential support for their neighbors. The subsidiary hypotheses and assumptions that are made for the sake of particular explanations in one model recur again and again in other related models, with or without modification and local adaptation. (Beckner, 1959, 159—60).

We can fairly and readily illustrate this view with a diagram. (Figure 1) See how Beckner proposes a kind of net-like picture, where different

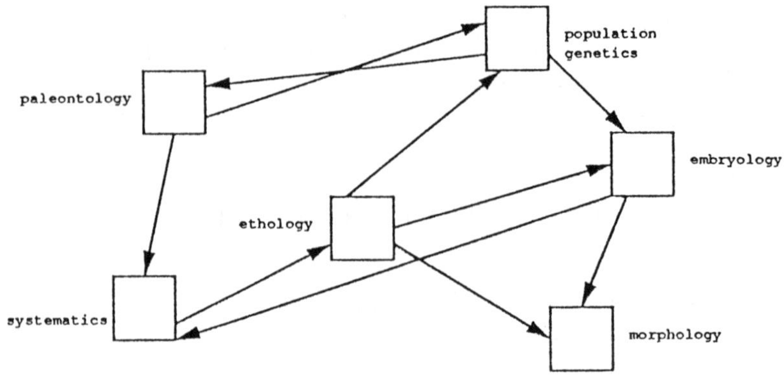

Fig. 1.

parts of the theory are joined by ropes — the ropes standing for informal links of inference of various kinds. One of us has articulated and defended a view similar to this (Thompson 1986, 1989).

The alternative picture (which was articulated fully in a book by the other one of us, written several years ago) gives population genetics a special status (Ruse 1973). It is seen as the background theory, as incorporating the mechanisms, which make sense of the various evolutionary explanations of all of the sub-disciplines of biology. It is, therefore, the connecting *core* of neo-Darwinism, as our second diagram shows (Figure 2). It is not that population genetics is better than (say) paleontology. It does have a different logical status within the theory.

How does one decide between the alternatives? We will first give the arguments for the second option. Although the arguments are reasonably compelling, in a sense they will later prove to be at one level only. We offer three reasons for giving population genetics a special place.

First, it fits with what neo-Darwinians themselves say about their theory. Of course, one should not take as gospel what anybody says about his or her work, even (especially even?) the leading creative minds. However, chief spokespeople for neo-Darwinism such as Theodosius Dobzhansky in the U.S. and Julian Huxley (grandson of Thomas Henry) in England were always adamant that the key to understanding large-scale change ("macro-evolution") lies in under-

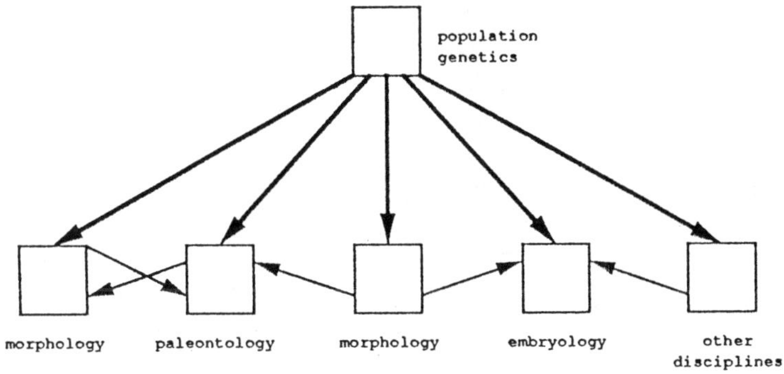

Fig. 2.

standing small-scale change ("micro-evolution") (Dobzhansky 1951; Huxley 1942). The big is no more than the small writ large. In this belief, Dobzhansky and Huxley and their fellows may have been wrong. There are those today who think they were. But it was their theory. They thought that to understand evolution, however it manifests itself, you start with the changes of gene ratios in populations between generations, primarily through such causal factors as selection and mutation. And, this latter is population genetics.

Second, this picture giving population genetics primacy makes sense of history, particularly inasmuch as one regards neo-Darwinism as a growth or development from the theory of the *Origin*. We ourselves have seen major theoretical changes from Darwin's work; but, this is at the level of mechanism. What comes next — more particularly, what came next for Darwin? The answer brooks no alternatives. Once having derived natural selection, as a mechanism explaining adaptively directed evolution, Darwin then applied his idea through the spectrum of biological studies, from behaviour, through biogeography to embryology. This, he rightly saw, as a consilience of inductions, where the mechanism throws light on the various parts of biology, which in turn confirm the truth of the central theory (Ruse 1975). That neo-Darwinians should do this also is what we would expect. Were they to go with Beckner's picture, then we have a deviation demanding explanation.

Third, such a picture of neo-Darwinism's structure makes sense of the particular explanations offered in the evolutionary sub-disciplines. Take, for instance, the classic explanation of the finches of the Galapagos, breaking into different groups from island to island. Neo-Darwinians pick up on precisely Darwin's argument, arguing that birds went from island to island, starting new colonies, and under the intense pressure of natural selection they evolved in different ways, guided always by adaptive necessity (Lack 1947, Boag and Grant 1981). In noe-Darwinism, this is supplemented by the brilliant hypothesis of the leading evolutionist, Ernst Mayr (1942), the so-called "founder principle". Mayr notes that, thanks to population genetics, our modern understanding of variation proves that there is no such thing as a typical member of a group. All organisms will be different, having their own unique combinations of genes. Hence, any small founder groups (which may be just one pregnant female) will already be tipped away from the norm, and in their new surroundings with their limited genetic variability, may be subject to strong selective pressures. These will lead to rapid evolution and the formation of new basic groups, species.

Three swallows do not a summer make, nor do three arguments prove a philosophical point. But, the reason for holding this view should now be clear.

WHAT KIND OF THEORY IS IT?

Yet, as already hinted, the discussion thus far has been at one level only. We have not really addressed the fundamental worry raised by Beckner. Whether his solution be correct, he has surely put his finger on something important. Even its greatest admirers must confess that neo-Darwinism is a little on the loose side, if not downright sloppy at times. The links between premises and conclusions do tend to be, shall we say, tenuous, often driven more by hope and expectation than by firm derivation. Indeed, in the opinion of many, it cannot be denied that neo-Darwinism does not always seem to have (as one commentator put it) quite the class of the better theories of physics and chemistry (Scriven 1959).

What should we say to this charge? The answer, of course, depends a little on what you think about the physical sciences, and in particular what you think is their true structure. The conventional answer — today often referred to as the "received view" (although, since it is generally critics who use this term, it might better be the "not-received-by-me view") — is that physical theories are hypothetico-deductive (or axiom) systems, where a few high-powered general statements act as premises or axioms for all else (Braithwaite 1953, Hempel 1966). What distinguishes a scientific system from a purely theoretical edifice, as one might find in mathematics, is that science consists in necessary generalizations — "laws" — about the empirical world. The paradigm, and this is what Beckner has in mind, is Newtonian mechanics, where from a few basic laws — those of motion and of gravitational attraction — all else, like Kepler's and Galileo's laws, can be deduced.

How does neo-Darwinism hold up against this picture? Not terribly well, we are afraid. It is not hopeless. There are parts of the theory which are reasonably formal, with conclusions following deductively from premises. This is particularly true of that area of which we have been talking most, namely population genetics. And, there are other areas of formalization, for instance to do with ecology (Roughgarden 1979). But, generally, the critics are right — much of neo-Darwinian theorizing proceeds very loosely and informally. Moreover, we tend not to get the universality and general applicability that is demanded in the hypothetico-deductive picture of theories. Such a view supposes that *all* bodies attract each other with an inverse square relationship, and that *all* planets go round the sun in ellipses, and so forth — necessarily. However, as soon as you look at the organic world, counter-examples and exceptions start to proliferate.

So, what does one do at this point, short of giving up and confessing that neo-Darwinism is pretty second-rate science? There seem to be two options. Most obvious, perhaps, is to suggest that the hypothetico-deductive picture functions as an *ideal* for neo-Darwinians. The real theory exhibits some of the appropriate features, albeit often imperfectly. But, there are reasons for this falling short, like the complexity of the material, the newness of the theory, and so forth. Hence, speaking technically, neo-Darwinism is less a theory than a "theory-sketch" — but this is not to deny the relevance of the received view, which helps you to understand the underlying logic of evolutionary studies.

What if you felt less than enthused by the received view, thinking it unduly influenced by the example of the formal sciences like logic and mathematics, and virtually determined to make a second-class science of biology, almost by definition. A number of philosophers and biologists believe an alternative analysis of theories more appropriate for biology — an analysis which neo-Darwinism exemplifies without need of tampering or special argument. This approach is generally known as the "semantic conception" of theories, and is based on a distinction, found in logic, mathematics and linguistics, between "syntax" and "semantics" (Beatty 1981, Thompson 1983, 1986, 1989). In linguistics for example, you distinguish between "syntax" which tells you of the structure of language (All sentences in English must contain a verb.) and "semantics", which tells you how the words of a language tie on to the real world. (By "red" in English we mean red things, like fires, and not green things, like grass.)

The semantic view of theories suggests that scientists run up theoretical "models" (that is, limited systems which may or may not be formal), which in themselves have no specification of an empirical application. They are just like games, which are fun to play but which in themselves tell us nothing of the real world. Then, scientists try to find empirical applications for their models, showing that the model's premises fit real life (this is where the empirical part comes in, as one tries to tie the theoretical picture on to the physical world). One major distinguishing feature between this analysis and the received view is that there is no presupposition that any particular model will apply universally — the only hope is that the model will apply to one or a range of limited situations. (In real life, obviously, life's problems and model building go hand in hand. The former do not wait helplessly on the latter, or *vice versa*.)

Semantic theorists argue that their picture captures precisely the nature of neo-Darwinism, and all must agree that there is something in what they say. Take, for instance, the phenomenon of balanced superior heterozygote fitness. As we have seen, evolutionists have produced a model showing how such balance will come about, given certain initial conditions, namely that the heterozygote be fitter than either homozygote. This result holds formally, even if there never ever be a heterozygote, let alone a super-fit one, in the world. Then the aim was to see if one could get real-life instantiations of the model, and in sickle-cell anaemia evolutionists certainly succeeded. One might add that there has been some strenuous debate between neo-Darwinians and their critics about how many more cases of balance there are (Lewontin 1974). Semantic theorists argue that this very debate shows the appropriateness of their analysis (Thompson 1983, 1989). No one disputes the model. It is the links between theory and reality that cause controversy. One of us has argued that the semantic conception also makes possible a formalization of evolutionary theory understood in a way similar to Beckner's characterization (Thompson 1986, 1989).

One final point about the semantic approach. As just presented, it has been portrayed as an escape for those who would analyze neo-Darwinism according to its own mode, without reference to the appropriate analysis for the theories of the physical sciences. However, the enthusiasts for the approach in the biological sciences usually begin or conclude their analysis by suggesting that the semantic view is the most appropriate for the physical sciences also (Beatty 1981; Thompson 1986, 1989)! There, likewise, one finds less effort put into grandiose system building and rather more effort given to the application of limited models. And, if this be so, the correct conclusion is not that biology is different from physics, but rather that for too long we have been using inappropriate tools of analysis in both physics and biology (For discussions of the semantic conception's appropriateness for physics see: Suppe 1972, 1974; Van Fraassen 1970).

A (PARTIAL) ANSWER TO THE QUESTION

So, what kind of theory is it? Is neo-Darwinism an approximation to a hypothetico-deductive system with population genetics as its core? Something which has such a system as an ideal? Or is it better analyzed through the semantic approach with the component part organized more along the lines suggested by Beckner? Is it a cluster of models, which together combine to make the overall theory? We confess that, at this point, we can give you no united answer. One of us (Ruse) inclines one way, towards the received view. The other (Thompson) inclines the other way, towards the semantic view.

On behalf of the received view, what would be said is two things. First, its adoption makes sense of history, for it was clearly what Darwin thought of theories and also it explains why the transition from Darwinism to neo-Darwinism has been marked by an increase in formalism. Second, its adoption makes sense of the present for it is clear that evolutionists do have an overall vision before them, for all that in actuality they have to work piecemeal. (The most stimulating work in evolution in the last two decades was Edward O. Wilson's *Sociobiology*: *The Modern Synthesis*, in which he tried to put the whole of animal social behaviour on a firm evolutionary bed of theory and evidence. Although his treatment of humans caused much upset, many biologists were consumed with admiration for a theory ranging from slime molds to *Homo sapiens*.)

On behalf of the semantic view, what would be said is two things. First, its adoption makes sense of history, for whatever Darwin may have said, the way that he actually worked was by trying to run up models of varying degrees of generality, and then seeing if they fit the world. And this is a practise which continued in the years after Darwin. Second, its adoption makes sense of the present, for biologists simply do not go out looking for major laws. They ferret away at particular bits of theory and reality, and tend to shun the general like the plague. Those who get too sweeping are doomed as synthesizers and popularizers. (The most controversial work in evolution in the last two decades was Edward O. Wilson's, *Sociobiology*: *The Modern Synthesis*, in which he tried to put the whole of animal social behaviour on a firm evolutionary bed of theory and evidence. Although it was his treatment of humans which caused most upset, many biologists were concerned at the sheer presumption that one could offer a theory ranging from slime molds to *Homo sapiens*.)

Could it be that what separates us at this point is a dichotomy which haunts all discussions in the philosophy of science? Always, those who talk about science must reach a fine line between *describing* what scientists actually do and *prescribing* what they should do. One wants to do more than merely give a report on scientific activity and achievement. One wants, for instance, to be able to distinguish good science from bad science. And yet, one must stay in contact with the subject. Simply to go off on one's own flight of fancy is to lose touch with one's starting point. It is true that no one has been particularly fanciful in this discussion. Nevertheless, the defence of the received view centres on an exhortation to go beyond the present reality to the as-yet-unrealized ideal. In short, it has a strong element of prescription about it. Perhaps, therefore, both sides are right in respects, and hence for a full picture of neo-Darwinism one need not (ought not) take philosophical sides.

Such an ecumenical resolution of a philosophical debate appeals to us both. We are sure that it contains a good 90% of the truth. The received view sets goals. The semantic view tells of realities. Yet, there is still another 10% of the truth. This, we are afraid, must cover tiny but persistent voices within our respective selves which tells us that the other chap is dead wrong! For us, at least, the philosophical understanding of neo-Darwinism is an unended task.

Departments of Philosophy and Zoology
University of Guelph

Department of Philosophy and Division of Humanities
University of Toronto

APPENDIX. ELEMENTARY POPULATION GENETICS

Let us represent alleles of one kind as A and alleles of another kind as a. The respective homozygotes can be represented as AA and aa, and the heterozygote as Aa. If a large population is divided between A and a in the ratio $p{:}q$ $(p + q = 1)$, then the Hardy-Weinberg law states that in the absence of disturbing factors, with random mating after the first generation (no matter what the initial distribution so long as it is the same for males and females), the distribution of genotypes will be:

$$p^2\,AA : 2pq\,Aa : q^2\,aa.$$

How do you get this result? Suppose that in your initial population, the genotypes are divided in the ratio:

$$D\,AA : H\,Aa : R\,aa.$$

This means that $D + H/2 = p$, $H/2 + R = q$.

If mating is random, we can draw up a simple matrix thus:

Random-mating frequencies

Females		Males		
		AA	*Aa*	*aa*
		D	H	R
AA	D	D^2	DH	DR
Aa	H	HD	H^2	HR
aa	R	RD	RH	R^2

Next, translate this into action, using Mendel's first law. For instance, the H^2 products of heterozygote mating ($Aa \times Aa$) will divide 1/4 AA : 1/2 Aa : 1/4 aa.

This gives us a new matrix.

Establishment of equilibrium under random mating

Type of mating	Frequency of mating	Offspring		
		AA	Aa	aa
$AA \times AA$	D^2	D^2		
$AA \times Aa$	$2DH$	DH	DH	
$Aa \times Aa$	H^2	$\frac{1}{4}H^2$	$\frac{1}{2}H^2$	$\frac{1}{4}H^2$
$AA \times aa$	$2DR$		$2DR$	
$Aa \times aa$	$2HR$		HR	HR
$aa \times aa$	R^2			R^2
Total	1.00	$(D+\frac{1}{2}H)^2$	$2(D+\frac{1}{2}H)(\frac{1}{2}H+R)$	$(\frac{1}{2}H+R)^2$
	or	p^2	$2pq$	q^2

This is the desired result, and it can be seen that the new gene ratio is $p{:}q$ as before.

Now let us go on to show balance, given superior heterozygote fitness. Selection is a relative matter, so let us speak of the "selective coefficient" of some particular genotype, meaning the reduced reproductive success that it has compared to the fittest genotype in the group. In our case, by hypothesis the heterozygote is fittest. If we stipulate its fitness as 1, then the fitness of the homozygotes is $1 - S_1$, and $1 - S_2$ where S_1, is the selection coefficient against the one homozygote, and S_2 the selection coefficient against the other. Using the knowledge that we have of genotypes where there is no selection (*i.e.* Hardy-Weinberg equilibrium), we can draw up the appropriate matrix showing the effects of selection:

Selection favoring heterozygotes

	AA	Aa	aa	Total
Initial proportion	p^2	$2pq$	q^2	1.00
Relative fitness	$1 - s_1$	1	$1 - s_2$	
After selection	$p^2(1 - s_1)$	$2pq$	$q^2(1 - s_2)$	$1 - s_1p^2 - s_2q^2$

This enables us to calculate the rate of change, Δp, of the frequency of the A genes in the population. Remembering that AA homozygotes contain two alleles and Aa contain one A allele, we get:

$$\Delta p = \frac{p^2(1 - s_1) + pq}{1 - s_1 p - s_2 q^2} - (p^2 + pq)$$

$$= \frac{p(1 - s_1 p)}{1 - s_1 p^2 - s_2 q^2} - p.$$

Since $p + q = 1$

If $p = s_2/s_1 + s_2$, then $\Delta p = 0$

which gives us a point of balanced equilibrium.

To give an example, suppose that no Aa homozygotes survive ($s_2 = 1$) and, compared to heterozygotes only 3 out of 4 AA homozygotes survive ($s_1 = 1/4$). Then if $\Delta p = 0$, $p = 4/5$, meaning that there will be a balance between A and a of 4:1. In fact, this is approximately the balance one does get in the sickle cell case.

BIBLIOGRAPHY

Ayala, F. (1985). "The Theory of Evolution: Recent Successes and Challenges", in E. McMullin (ed.), *Evolution and Creation* (Notre Dame: University of Notre Dame Press), pp. 59—90.

Ayala, F. J. and Kiger, J. (1984). *Modern Genetics*, 2nd ed. (Reading, Mass.: Addison-Wesley).

Beatty, J. (1981). "What's Wrong with the Received View of Evolutionary Theory?", in P. D. Asquith and R. N. Giere (eds.), *PSA 1980* **2**, 397—426.

Beckner, M. (1959). *The Biological Way of Thought*, (New York: Columbia University Press).

Boag, P. and P. Grant (1981). "Intense Natural Selection in a Population of Darwin's Finches (Geospizinae) in the Galapagos", *Science* **214**, 82—5.

Bodmer, W. and L. Cavalli-Sforza (1976). *Genetics, Evolution, and Man*, (San Francisco: Freeman).

Braithwaite, R. B. (1953). *Scientific Explanation*, (Cambridge: Cambridge University Press).

Burchfield, J. D. (1975). *Lord Kelvin and the Age of the Earth*, (New York: Science History Publications).

Darwin, C. (1859). *On the Origin of Species by Means of Natural Selection*, (London: Murray).

Dobzhansky, Th. (1951). *Genetics and the Origin of Species*, 3rd ed. (New York: Columbia University Press).

Dobzhansky, Th., Ayala, F., Stebbins, G., and Valentine, J. (1977). *Evolution*, (San Francisco: Freeman).

Ellegard, A. (1958). *Darwin and the General Reader*, (Göteborg: Göteborgs Universitets Arsskrift).

Hempel, C. (1966). *The Philosophy of Natural Science*, (Englewood Cliffs: Prentice-Hall).

Huxley, J. S. (1942). *Evolution: The Modern Synthesis*, (London).

Lack, D. (1947). *Darwin's Finches: An Essay on the General Biological Theory of Evolution*, (Cambridge: Cambridge University Press).

Lewontin, R. (1974). *The Genetic Basis of Evolutionary Change*, (New York: Columbia University Press).

Lucas, J. R. (1979). "Wilberforce and Huxley: A Legendary Encounter", *Historical Journal* **22**, 313—330.

Mayr, E. (1942). *Systematics and the Origin of Species*. (New York: Columbia University Press).

Provine, W. B. (1971). *The Origins of Theoretical Population Genetics*, (Chicago: Chicago University Press).

Roughgarden, J. (1979). *Theory of Population Genetics and Evolutionary Ecology: An Introduction*, (New York: Macmillan).

Ruse, M. (1971). "Natural selection in *The Origin of Species*", *Studies in History and Philosophy of Science* **1**, pp. 311—51.

Ruse, M. (1973). *The Philosophy of Biology*, (London: Hutchinson).

Ruse, M. (1975). "Darwin's Debt to Philosophy", *Studies in History and Philosophy of Science* **6**, 159—81.

Ruse, M. (1979). *The Darwinian Revolution*, (Chicago: University of Chicago Press).

Scriven, M. (1959). "Explanation and Prediction in Evolutionary Theory", *Science* **130**, 477—82.

Sober, E. (1984a). *The Nature of Selection: Evolutionary Theory in Philosophical Focus*, (Cambridge, Mass.: MIT Press).

Sober, E., ed. (1984b). *Conceptual Issues in Evolutionary Biology*, (Cambridge, Mass.: MIT Press).

Suppe, F. (1972). "What's Wrong with the Revised View on the Structure of Scientific Theories?", *Philosophy of Science*, **39**, 1—19.

Suppe, F. (1974). *The Structure of Scientific Theories*, (Urbana, Ill.: University of Illinois Press).

Thompson, P. (1983). "The Structure of Evolutionary Theory: A Semantic Approach", *Studies in History and Philosophy of Science* **14**, 215—229.

Thompson, P. (1986). "The Interaction of Theories and the Semantic Conception of Evolutionary Theory", *Philosophica* **37**, 73—86.

Thompson, P. (1989). *The Structure of Biological Theories*, (New York: State University of New York Press).

van Fraassen, B. C. (1970). "On the Extension of Beth's Semantics of Physical Theories", *Philosophy of Science* **37**, 325—339.

Vorzimmer, P. J. (1970). *Charles Darwin: The Years of Controversy*, (Philadelphia: Temple University Press).

Wilson, E. O. (1975). *Sociobiology: The New Synthesis*, (Cambridge, Mass.: Harvard University Press).

PUBLICATIONS OF ROBERT E. BUTTS

1957

1. *Husserl's Criticisms of Hume's Theory of Knowledge*, University of Michigan Microfilms.

1958

2. 'Does 'intentionality' imply 'being'? A paralogism in Sartre's ontology', (abstract) *The Journal of Philosophy* **LV**, 21.

1959

3. 'Rationalism in modern science: d'Alembert and the *esprit simpliste*', *Bucknell Review* **VIII**, 2.
4. 'Hume's scepticism', *Journal of the History of Ideas* **XX**, 3.
5. 'Husserl's critique of Hume's notion of *distinctions of reason*', *Philosophy and Phenomenological Research* **XX**, 2.

1960

6. 'Does 'intentionality' imply 'being'? A paralogism in Sartre's ontology', *Kant-Studien* **52**, 4.

1961

7. 'Hypothesis and explanation in Kant's philosophy of science', *Archiv für Geschichte der Philosophie* **43**, 2.

1962

8. 'Kant on hypotheses in the 'Doctrine of Method' and the Logik', *Archiv für Geschichte der Philosophie* **44**, 2.
9. ''Indoctrination *in*', 'indoctrination *with*', and 'indoctrination *into*'', *Bucknell Review* **X**, 4.

1963

10. (Review article) 'Immanuel Kant, Vorlesungen, Abteilung I. Vorlesungen über Enzyklopädie und Logik; Band I, Vorlesungen über Philosophische Enzyklopädie. (Deutsche Akademie der Wissenschaften zu Berlin 1961)', *Archiv für Geschichte der Philosophie* **45**, 3.

1965

11. 'Necessary truth in Whewell's theory of science', *American Philosophical Quarterly* **2**, 3.
12. 'On Walsh's reading of Whewell's view of necessity', *Philosophy of Science* **32**, 2.

J. R. Brown and J. Mittelstrass (eds.), An Intimate Relation, 513–517.

1966

13. 'Feyerabend and the pragmatic theory of observation', *Philosophy of Science* **33**, 4.

1967

14. 'Professor Marcucci on Whewell's idealism', *Philosophy of Science* **34**, 2.
15. 'The purposes of Kant's talk about purposes' (abstract), *Proceedings of the Third International Congress of Logic, Methodology* and *Philosophy of Science* (Amsterdam).

1968

16. 'Kant's schemata as semantical rules', *Proceedings of the 14th International Congress of Philosophy* **5** (Vienna).
17. *William Whewell's Theory of Scientific Method*, editor, introduction. Pittsburgh: University of Pittsburgh Press.

1969

18. 'Kant's schemata as semantical rules', in *Kant Studies Today*, ed. Lewis W. Beck. La Salle Illinois: Open Court.

1970

19. *The Methodological Heritage of Newton*, editor, introduction (with J. W. Davis). Toronto: University of Toronto Press.
20. 'Whewell on Newton's rules of philosophizing', in 19.

1971

21. 'On Buchdahl's and Palter's papers' (part of a symposium on Kant's philosophy of science), *Synthese* **23**.
22. (Review) 'History of philosophy of science', by Rom Harré, in *The Encyclopedia of Philosophy*, ed. Paul Edwards; *Philosophy of Science* **38**, 4.

1972

23. [paper in 21.] *Proceedings of the Third International Kant Congress*, ed. Lewis W. Beck. Dordrecht: D. Reidel.

1973

24. 'Whewell's logic of induction', in R. Giere and R. Westfall, eds., *Foundations of Scientific Method: the Nineteenth Century*. Bloomington, Indiana: Indiana University Press.
25. *Science, Decision and Value*, editor (with J. Leach and G. Pearce). Dordrecht: D. Reidel.
26. 'Reply to David Wilson: Was Whewell interested in true causes?' *Philosophy of Science* **40**, 1.
27. (Review) *Galileo's Intellectual Revolution*, W. Shea; Dialogue XII, 3.

1974
28. 'Philosophy of science in Canada', *Zeitschrift für Allgemeine Wissenschaftstheorie* **V**, 2.
29. 'Bibliographie, veröffentlichungen kanadischer Wissenschaftstheoretiker', (with J. Galinaitis) *Zeitschrift für Allgemeine Wissenschaftstheorie* **V**, 2.

1975
30. 'Kant's mathematization of scientific experience', *Proceedings of the Heraclitean Society* **I**, 4 (Western Michigan University).
31. 'William Whewell', *Dictionary of Scientific Biography*, ed. C. C. Gillispie, Vol. XIV (New York 1972—75).

1976
32. 'Kant and the problem of scientific methodology', *Proceedings of the 1974 Ottawa Congress on Kant*, ed. P. Laberge, F. Duchesneau and B. Morrisey. Ottawa University Press.
33. 'Filozofia nauki w Kanadzie', *Ruch Filozoficzny* **XXXIV**, 3 (Torun, Poland). [Polish translation of 28.]
34. 'The hypothetico-deductive model of scientific theories: a sympathetic disclaimer', in *Basic Issues in Philosophy of Science*, ed. W. Shea (Canadian Contemporary Philosophy Series). New York: Science History.

1977
35. 'Consilience of inductions and the problem of conceptual change in science', in R. Colodny, ed., *Logic, Laws and Life* (Vol. 6 in the University of Pittsburgh Series in the Philosophy of Science). Pittsburgh: University of Pittsburgh Press.
36. *Logic, Foundations of Mathematics and Computability Theory*, Pt. I of the Proceedings of the 5th International Congress of Logic, Methodology and Philosophy of Science; London, Canada, 1975; edited (with J. Hintikka). Dordrecht: D. Reidel.
37. *Foundational Problems in the Special Sciences* [Pt. II, entry 36.].
38. *Basic Problems in Methodology and Linguistics* [Pt. III, entry 36.].
39. *Historical and Philosophical Dimensions of Logic, Methodology and Philosophy of Science* [Pt. IV, entry 36.].

1978
40. *New Perspectives on Galileo*, edited (with J. Pitt). Dordrecht: D. Reidel.
41. 'Some tactics in Galileo's propaganda for the mathematization of scientific experience', in 40.

1979
42. 'Rescher and Kant: some common themes in philosophy of science', in E. Sosa, ed., *The Philosophy of Nicholas Rescher: Discussion and Replies*. Dordrecht: D. Reidel.

43. 'Scientific progress: the Laudan manifesto', *Philosophy of the Social Sciences* **9**.

1980

44. 'Leibniz' monads: a heritage of gnoticism and a source of rational science', *Canadian Journal of Philosophy* **X**.

1981

45. 'Comments on Zev Bechler's paper 'What have they done to Kuhn?'' *Theory Change, Ancient Axiomatics, and Galileo's Methodology, Proceedings of the 1978 Pisa Conference on the History of Philosophy of Science*, Vol. I, ed. J. Hintikka, D. Gruender and E. Agazzi; Dordrecht: D. Reidel.
46. 'Methodology and the functional identity of science and philosophy', *Probabilistic Thinking, Thermodynamics, and the Interaction of the History and Philosophy of Science*, Vol. II [as in 45.].
47. 'Rules, examples and constructions: Kant's theory of mathematics', *Synthese* **47**, 2.
48. *Kant's Critique of Pure Reason, 1781—1981* (editor): *Synthese*: Pt. I, **47**, 2; Pt. II, **47**, 3.

1983

49. 'Leibniz on empirical methodology', Abstracts, *7th International Congress of Logic, Methodology and Philosophy of Science*, Vol. 6 (Salzburg, Austria).

1984

50. *Kant and the Double Government Methodology. Supersensibility and Method in Kant's Philosophy of Science*. Dordrecht: D. Reidel.
51. 'The rôle of arational factors in interpretive history: the case of Kant and ESP', in J. R. Brown, ed., *Scientific Rationality: the Sociological Turn*. Dordrecht: D. Reidel.
52. 'John Frederick William Herschel', in *Enzyklopädie Philosophie und Wissenschaftstheorie* **II**, ed. Jürgen Mittelstrass. Mannheim/Wien/Zürich: Bibliographisches Institut.

1985

53. ''A Purely scientific temper': Victorian expressions of the ideal of an autonomous science', in Nicholas Rescher ed., *Reason and Rationality in Natural Science* (University of Pittsburgh Series in Philosophy of Science). University Press of America.
54. 'Leibniz on the side of the angels', in K. Okruhlik and J. R. Brown, eds., *The Natural Philosophy of Leibniz*. Dordrecht: D. Reidel.
55. 'Logic, epistemology, philosophy of science: English Canada'. *The Canadian Encyclopedia*. Edmonton: Hurtig, pp. 1401—1402.

1986

56. *Kant's Philosophy of Physical Science: Metaphysische Anfangsgründe der Naturwissenschaft, 1786—1986*, editor. Dordrecht: D. Reidel.
57. 'Introduction: Kant's quest for a method for metaphysics', in 56.: 1—22.

58. 'The methodological structure of Kant's metaphysics of science', in 56.: 163—199.
59. 'Kant's philosophy of science: the transition from metaphysics to science', in P. Asquith and P. Kitcher eds., PSA 1984, Vol. 2. East Lansing, Michigan: Philosophy of Science Association, pp. 685—705.

1987
60. 'Pragmatism in theories of induction in the Victorian Era: Herschel, Whewell, Mach and Mill', in H. Stachowiak, ed., *Pragmatik. Handbuch pragmatischen Denkens. Band II. Der Aufstieg pragmatischen Denkens im 19. und 20 Jahrhundert.* Hamburg: Felix Meiner Verlag, pp. 40—58.

1988
61. *Thought and Language in the Philosophy of the Enlightenment*, edited (with Thomas M. Lennon), *Synthese* **75**, 2 (1988).
62. 'The grammar of reason: Hamann's challenge to Kant', in 61, pp. 251—283.
63. 'Logic, Epistemology, Philosophy of Science: English Canada', *The Canadian Encyclopedia*, 2nd Ed. Edmonton: Hurtig. Vol. III, p. 1661.

Index of Names and Subjects